WEATHER RADAR NETWORKING

This seminar on COST project no. 73 has been organised by the Commission
of the European Communities, Directorate-General 'Science, Research and
Development' and held at the Economic and Social Committee of the European
Communities, Brussels, from 5 to 8 September 1989.

Description and brief explanation of the composite COST image on the cover

(source : Meteorological Office - Bracknell - UK)
10 April 1989 - 2.00 A.M.

Each coloured square or "pixel" represents a surface area of approximately
5km x 5km = 25 km^2.

The light and dark blue areas represents cloud top temperatures derived
from a METEOSAT satellite infra-red image.

The other colours : pink, green, yellow, red and black specify the surface
rainfall intensity measured by radar in millimetres per hour (mm/h).

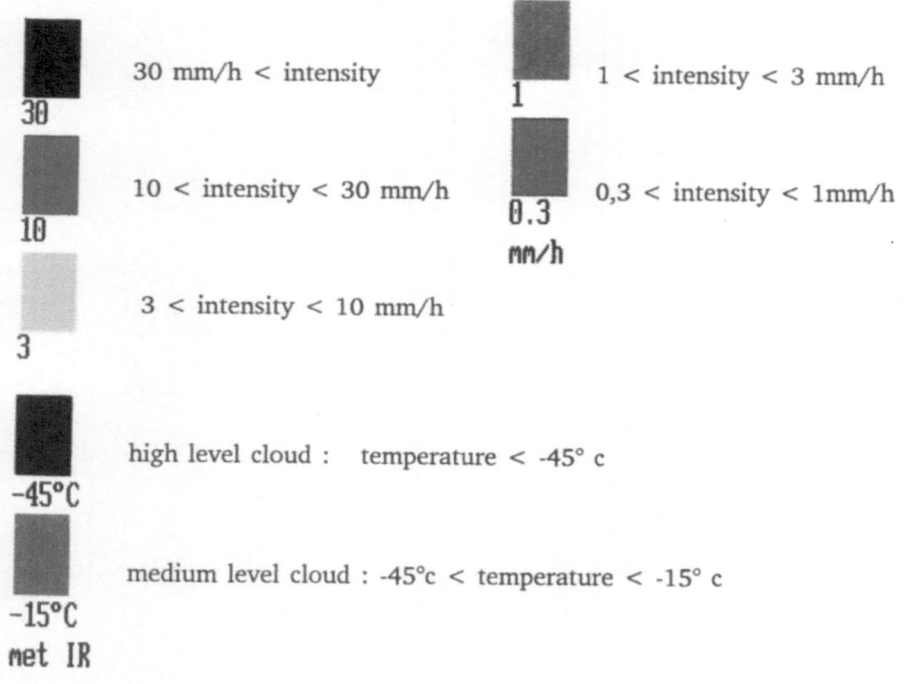

30 mm/h < intensity

30

1 < intensity < 3 mm/h

1

10 < intensity < 30 mm/h

10

0,3 < intensity < 1mm/h

0.3
mm/h

3 < intensity < 10 mm/h

3

high level cloud :   temperature < -45° c

-45°C

medium level cloud : -45°c < temperature < -15° c

-15°C
net IR

The image shows a depression over the English Channel moving North East.
The associated warm and cold fronts are producing a broad band of cloud
with patchy rain over much of Western Europe. Simultaneously, an active
eastward moving bent back occlusion is producing a narrow band of heavy
precipitation from South West England to the West coast of France.

COMMISSION OF THE EUROPEAN COMMUNITIES

Weather
Radar Networking

Seminar on COST Project 73

Edited by

C. G. COLLIER

Chairman of COST 73,
Meteorological Office, Bracknell, U.K.

and

M. CHAPUIS

Commission of the European Communities,
Directorate-General Science, Research and Development, Brussels, Belgium

KLUWER ACADEMIC PUBLISHERS
DORDRECHT / BOSTON / LONDON

ISBN-13:978-94-010-6735-5   e-ISBN-13:978-94-009-0551-1
DOI: 10.1007/978-94-009-0551-1

Publication arrangements by
Commission of the European Communities
Directorate-General Telecommunications, Information Industries and Innovation,
Scientific and Technical Communications Unit, Luxembourg

EUR 12414 EN-FR
© 1990 ECSC, EEC, EAEC, Brussels and Luxembourg
Softcover reprint of the hardcover 1st edition 1990
Document EUCO-COST 73/52/90

LEGAL NOTICE
Neither the Commission of the European Communities nor any person acting on behalf of the
Commission is responsible for the use which might be made of the following information.

Published by Kluwer Academic Publishers,
P.O. Box 17, 3300 AA Dordrecht, The Netherlands.

Kluwer Academic Publishers incorporates the publishing programmes of
D. Reidel, Martinus Nijhoff, Dr W. Junk and MTP Press.

Sold and distributed in the U.S.A. and Canada
by Kluwer Academic Publishers,
101 Philip Drive, Norwell, MA 02061, U.S.A.

In all other countries, sold and distributed
by Kluwer Academic Publishers Group,
P.O. Box 322, 3300 AH Dordrecht, The Netherlands.

Printed on acid-free paper

PREFACE

Meteorology is by nature a multidisciplinary and transnational subject and COST cooperation has proved to be a flexible and suitable framework at European level for meteorological activities such as the standardisation of observation techniques and harmonised transmission of meteorological data.

Although meteorology is not covered by a specific Community programme as such, various Community actions dealing with meteorology are now included in the EEC research programme on climatology (the "EPOCH" programme - 1989-92) concerning particularly the study of mechanisms of extreme and sudden meteorological events, in order to predict catastrophies and consequently to reduce human and material losses.

In the context of COST cooperation, which is supported by the Commission of the European Communities, the COST 73 project (1986-1991) associates 16 countries in Western Europe with the aim of setting up a weather radar network providing real-time measurements of rain, snow or hail precipitations. In this project, radar data are transmitted and combined if appropriate with satellite data - in one or more "compositing centres" of the participating countries, in order to improve weather forecasting.

Together with the COST 73 Management Committee, the Commission of the European Communities organized a seminar on this matter, in Brussels on 5-8 September 1989, at the half-way stage of the project.

This seminar was an opportunity to review what has been achieved recently in Europe, but also in other parts of the world where this research is carried out (Eastern Europe, United States, Canada, Japan as well as in some developing countries). Nearly 60 papers - included in these proceedings - were presented and discussed during the six sessions of the seminar : radar networking programmes, the role of reflectivity base techniques and of other new techniques in operational radar networks, combining radar, satellite and conventional meteorological data, meteorology, hydrology and other applications of weather radar data.

P. Fasella

P. FASELLA
Director General
"Science, Research and Development"

**Conference Dinner
Château Ste-Anne (6/09/89)
A view of the participants**

ORGANIZING COMMITTEE

Chairman : C.G. COLLIER (UK)

Members : B. BERINGUER (F)
M. CHAPUIS (CEC)
D. NEWSOME (UK)
F.P. SCHEINS (CEC)
R. SORANI (I)
A. VAN GYSEGEM (B)
H. WESSELS (NL)

HONORARY COMMITTEE

Mr. F.M. PANDOLFI,   Vice-President of the Commission of the European Communities.

Mr. H. SCHILTZ,   Belgian Minister for Science Policy.

Mr. P. CHEVALIER,   Belgian State Secretary for Science Policy.

Mr. A. MASPRONE,   President of the Economic and Social Committee of the European Communities.

Mr. G.O.P. OBASI,   Secretary-General of the World Meteorological Organization.

Mr. H. BROUHON,   Mayor of Brussels.

Mr. H. MALCORPS,   Director of the Belgian Royal Meteorological Institute.

SPONSORS

ERICSSON, (S)
PLESSEY, (UK)
SMA, (I)
DATAMAT, (I)

CONTENTS

SESSION 2 -   THE ROLE OF REFLECTIVITY-BASED TECHNIQUES IN
             OPERATIONAL RADAR NETWORKS

Chairman :   C.G. COLLIER, Meteorological Office, Bracknell,
          United Kingdom

SESSION 3 - THE ROLE OF NEW TECHNIQUES IN OPERATIONAL RADAR NETWORKS

Chairman : B. BERINGUER, Direction de la Météorologie Nationale, Trappes, France

OPENING SESSION

Opening Session (5/09/89)
Left to right :
MM. P. FASELLA, A. MASPRONE, C.G. COLLIER,
G.O.P. OBASI, H. MALCORPS and R. HOOGEWIJS
(representing Mr CHEVALIER)
Introductory statement by Prof. G.O.P. OBASI

SPEECH OF WELCOME BY THE PRESIDENT OF
THE ECONOMIC AND SOCIAL COMMITTEE, EUROPEAN COMMUNITIES,
MR ALBERTO MASPRONE,

Ladies and Gentlemen,

It gives me great pleasure to welcome you to the Economic and Social Committee of the European Communities and to this "COST" Seminar entitled "COST 73 International Seminar on Weather Radar Networking". As some of those taking part will know, this Seminar is running in parallel with the "TECIMO" Conference on Meteorological methods and instrumentation which is taking place in the Palais des Congres not far from here. The Economic and Social Committee wishes you every success to your Seminar and to the "TECIMO" Conference.

The Committee is, as some of you may know, an assembly of Socio-Economic interests set up under the Treaties which established the European Communities, to represent the interests of employers, trade-unionists, farmers, the liberal professions and all other socio-economic groupings, so that their opinions can be taken into account in the formulation of European Policy. The idea of such an advisory body, an Economic and Social assembly as a complement to the Parliamentary assembly, grew up in the early years of this century but so far it is a concept which has not found wide acceptance outside Western Europe. Our job is to prepare an opinion - a written document which can be as short as three pages or as long as thirty - which provides a consensus view of the one hundred and eighty nine members of the Committee, who represent the socio-economic interests of the 12 Members States of the European Community. Like weather forecasting, this task sounds easy until you actually begin to investigate all the variables that can influence the final result and although there are now computer programmes to assist long-range weather forecasting, in the work of our Committee we have not yet discovered a computer that would help us to plot human behaviour or give us an advance indication on how our members who represent so many varied types of economic and social interest, will view the proposals of the Commission.

Although in our Committee we have - among others - sections for agriculture and transport, I must be frank and say that in the past we have not been overmuch concerned with the topics you are to discuss over the next few days. However, the science of weather and weather forecasting has advanced so much over the thirty years during which our Committee has been in existence that it is at a point where people in industries like farming and tourism can rely with some considerable confidence on the weather forecast they can now obtain. That means more efficient farming, better tourism, and in the long run the very things that are of interest to the members of our Committee - so you can see that our interests are really very closely linked with your own.

I would also like to offer a very special welcome to all those delegates from outside the European Community, and particularly those from Eastern Europe and the Soviet Union. Their presence, together with that of delegates from the continents of North and South America, from the continent of Asia makes this a truly international conference.

Let me conclude by wishing you a very successful stay in Brussels and I hope that the Brussels weather - which has a very bad reputation - will not disappoint you too much. I now call upon Professor Paolo Fasella, Director General for the Directorate General of Science, Research and Development at the Commission of the European Communities, to open the conference.

INTRODUCTORY STATEMENT BY PROFESSOR G.O.P. OBASI, SECRETARY-GENERAL
OF THE WORLD METEOROLOGICAL ORGANIZATION

Professor P. Fasella, Director General of GD XII,
Mr. A. Masprone, President of the Economic and Social Committee,
Dr Malcorps, Permanent Representative of Belgium with W.M.O,
Dr Collier, Chairman of the COST 73 project and Organizing Committee,
Ladies and gentlemen,

I have great pleasure in addressing you at the opening of the "COST 73 International Seminar on Weather Radar Networking" on behalf of the World Meteorological Organisation and to extend to you the best wishes of the Organization for a successful seminar.

As you are aware WMO is co-sponsoring this COST 73 Seminar which has been closely co-ordinated with the WMO Technical Conference on Instruments and Methods of Observation (TECIMO-IV), which opened yesterday.

WMO has closely followed the work of the COST 73 project since it began in 1986 and I am glad to note that there are now 16 European countries participating in this project.

I understand that the tasks of the project fall into two broad categories namely, those concerned with software and hardware specification and improved radar data processing and those tasks related to the real-time exchange of weather radar information between countries.

WMO has been actively engaged in promoting the development and use of remote sensing techniques, including weather radars for many years. Through its Commission for Instruments and Methods of Observation the Organization has gathered a great deal of information on the use of radars by its Member countries throughout the world and perhaps I could mention here some of the more interesting aspects.

There are now over 100 countries operating more than 600 weather radars. The principal users are for flash flood warnings, severe weather warnings including wind, hail and tornados, light scale weather monitoring and air-terminal surveillance. Conventional radar utilisation has been increasing worldwide, especially during the past decade. 35 per cent of the total number of radars have been installed since 1980.

Progress in the development of radar systems during the past 10 years is also remarkable. Techniques such as real-time rain measurements, Doppler analysis methods (like Velocity Azimuth Displays) and the compositing of radar data are now used operationally in many countries thanks to availability of mini-computers. Many radar manufacturers offer a wide variety of possibilities in radar data processing. Almost all manufacturers offer digital displays of surface rainfall intensity and accumulation, constant altitude displays showing storm structure and evolution, and a variety of products designed to assist operational meteorologists and hydrologists in carrying out their forecasting and warning responsibilities.

At least 17 countries have started or are planning to start operating radar networks producing composite images. Remarkable advances have been made in the availability and utilization of Doppler data and further progress is expected in this field with the installation of operational systems such as the NEXRAD programme of the USA.

Full details of the information gathered by WMO are published in the WMO Instruments and Observing Methods series No.37 which is being distributed now.

As you may be aware, the organisation has adopted a new general format for data transmission - the FM 94 BUFR Code, which should be considered for international exchange of radar data. WMO could then propose for its Members a radar software library, including the radar BUFR transmission format as well as a variety of data-handling software.

There is considerable scope for improving and extending the use of weather radars and this is particularly true of the developing countries. However, before installing weather radars, it is necessary to consider two questions.

First, is radar the most appropriate and cost-effective technology to address the problem? For instance, to monitor tropical cyclones, it will usually be more appropriate to purchase good satellite imagery equipment before installing a Doppler radar.

And second, are the radar operators and users properly trained, organized and equipped to insure reliable operation, good data interpretation and adequate dissemination of the radar derived data?

If there are doubts on any of these points it is recommended that information be obtained from other users who have had experience in a similar situation and if possible, a pilot experiment should be undertaken with a single radar system in order to develop a sound basis for decision.

Weather radar operators and users represented at this seminar can contribute a great deal to our understanding and application of this powerful tool. I am sure that as the European Meteorological and Hydrological Services implement and operate these systems, share their experiences, and assist in promoting common standards, all Services throughout the world will benefit.

I am confident that your seminar which is dealing mainly with operational aspects will make a considerable contribution to progress in this important field of meteorological data acquisition. The applications of weather radar data, radar networking programmes, telecommunication aspects of radar networks and the integration of data from radars, satellites and conventional meteorological observations are all essential issues being faced by Services today. We in the WMO will be greatly interested in your progress.

I wish you all every success.

Thank you.

SESSION 1

Radar Networking Programmes

Chairman : J. Joss

Dr C.G. COLLIER
Meteorological Office (UK)
Chairman of COST 73 project
and of Seminar Organizing Committee

COST-73: THE DEVELOPMENT OF A WEATHER RADAR NETWORK IN WESTERN EUROPE

C.G. Collier

Chairman COST-73 Management Committee
Meteorological Office, U.K.

Summary

The last few years have seen a rapid development of operational weather radar networks in Western Europe funded in the main by national meteorological services. The need to exchange data with neighbouring countries was clear from the outset, and effective international cooperation was stimulated by the Commission of the European Communities (CEC) sponsored COST-72 Project. This has led to the setting up of a continuation project on weather radar networking known as COST-73. The need for a wider exchange of radar data is now recognised, and COST-73 is establishing guidelines which will underpin future developments. In this paper we outline the research aims of COST-73, and illustrate how data from many countries have already been combined in real-time for use in several application areas. Finally possible future organizational structures which might lead to the establishment of an operational exchange of data throughout Europe and associated research programmes, are noted.

1. Background

The COST programme (CO-operation in Science and Technology) is a programme for West European States which see advantage in pursuing joint research or development projects. It is organised under the aegis of the European Commission who supply the Secretariat, although it is funded from within existing national programmes.

COST Porject 72, or COST 72, as it was generally known, was concerned with the measurement of precipitation and the cost effectiveness of an integrated weather radar network in western Europe. The Memorandum of Understanding of this project was signed by 13 countries [Austria, Belgium, Denmark, Finland, France, Federal Republic of Germany, Italy, the Netherlands, Portugal, Spain, Sweden, Switzerland and UK. Greece, Republic of Ireland and Yugoslavia also participated in some of the meetings]. It was a six year project ending in December 1985 with the publication of a final report (EUR 10171 EN), and a final seminar held in Sicily, Italy (EUR 10353 EN-FR).

Work on hardware specifications, quality of radar data and network organisation was carried out, including a demonstration of the feasibility, and to some extent the utility, of exchanging and compositing data in real-time from several countries (Collier et al, 1988). However, it was recognised that further work on data quality, communications and product specification would be needed if a robust operational radar network were to be established within the existing framework of national programmes. A recommendation for a continuation programme, COST-73, explicitly on weather radar networking was accepted, and by the end of 1986 eight countries (Belgium, Federal Republic of Germany, Finland, France, Italy, Switzerland, The Netherlands, United Kingdom) had signed a Memorandum of Understanding (MOU) bringing the project into being with a secretariat supplied by the CEC. The funding arrangements were as for COST-72. These countries were joined in 1987 by Austria, Denmark, Ireland, Portugal and Sweden, during 1988 by Spain and Yugoslavia, and in 1989 by Norway, bringing the total number of countries participating to sixteen.

2.   COST-73 research programme

The MOU contained a research programme divided into five main areas (1) **radar systems**: performance characteristics of different radar techniques, display requirements, equipment standardization and investigation of new techniques; (2) **radar site and national network centre data processing**: computer requirements, meteorological calibration and data correction algorithms, software specification and compositing different data types and data from different radars; (3) **data transmission**: standardization of formats and protocols and testing different transmission media; (4) **bilateral radar data excahnges**: coordination of installations and operations across national boundaries and studies of the properties of radar data; and (5) **European network investigations**: operational requirements for European radar composite data, archiving, real-time trials, commercial exploitation and proposals for a modus operandi for a coordinated European weather radar network based upon national plans.

The Management Committee felt that this work could best be tackled within a framework of activities divided into work which could be undertaken via desk top studies using archive data and existing experience, and work which required the production and distribution of wide area radar data in real-time. A number of **off-line** and **real-time** studies were specified as follows;

Off-line activities:-

* Under certain circumstances, data from Doppler radars will be useful for forecasting purposes in Europe. This will be studied.
* A guideline hardware specification has already been accepted (COST-72) but no similar activity has been undertaken in the software field. Areas that might be investigated include software dealing with conventional surface precipitation, volumetric scanning and Doppler processing.
* Alternative algorithms to facilitate the combination of radar data with those from satellites will be investigated.
* Experiments with the existing COST image are worthwhile. At the moment satellite data are obtained from infra-red level slicing, but other options, eg a combination of IR and visible data calibrated for use with radar data might prove to be more useful.
* Radar networking software will be examined with a view to improving its overall flexibility and efficiency.

* The detection of hazardous meteorological phenomena will be
reviewed and the testing of appropriate algorithms will be
investigated.

Real-time research projects:-
* Improvements to the real-time COST image such as the
production of images over the whole of western Europe
including the Nordic countries.
* Severe weather algorithms; testing algorithms for
identifying phenomena associated with thunderstorms,
snowfall, hail and other severe weather.
* Investigation of the benefits of forecasting European
maritime weather in, for example, the North Sea, the South
West Approaches (including the Channel) and the western
Mediterranean.
* Trials of the usefulness of COST images for meteorological
numerical model initialisation.
* Tests of the usefulness of COST images for monitoring and
predicting wet deposition (acid rain and nuclear fallout).
* Definition of avalanche-prone regions in cases of heavy
snowfall.
* Investigations into the relevance, or otherwise, of COST
images in the forecasting of severe convective storms
(their lifetime and movement).

To manage this work programme three district sub-projects were
identified, and working groups were formed to allocate tasks, coordinate
the preparation of data and working paper, and present findings for the
consideration of the whole Management Committee. These working groups
are:-
WG1: Working Group on Telecommunications - chair: France
WG2: Working Group for Coordination, Compositing and Data exchange -
                              chair: The Netherlands.
WG3: Working Group on Further Development and Application of COST
images - chair: United Kingdom.
In addition the CEC have appointed an independent Project Coordinator to
aid the Management Committee in carrying out this work.

The main aim of WG1 was to consider and produce proposals for the
standardization of codes used to exchange weather radar information
between different countries. A major achievement has been the
modification of the BUFR-94 code for use with radar data. These
modifications are currently being considered by the appropriate groups in
the World Meteorological Organisation (WMO) and are the subject of a paper
in these Proceedings by B Beringuer, the Chairman of this working group.
the proposals provide a foundation for the worldwide exchange of radar
data on the WMO Global Telecommunications system (GTS), and will underpin
future international radar networking in Europe.

Given that radar data can be readily exchanged between countries, it
is necessary to understand the characteristics of these data in order to
use them in a reliable operational way. Working Group 2 has acquired
information on operational European radars and the real-time data
processing associated with them. A standard method of collecting
occultation (ground sceening of low radar beam elevations) has been
proposed (Wessels, 1989), and work is well advanced to produce an

occultation map for all west European radars currently in operation. Such
information is essential if European radar composite data are to be used
quantitatively.

The third Working Group seeks to investigate the logistics of
exchanging data internationally and the production of composite images
over a very wide area. Until recently this area covered only that part of
north west Europe shown by the small frame in Figure 1. However, products

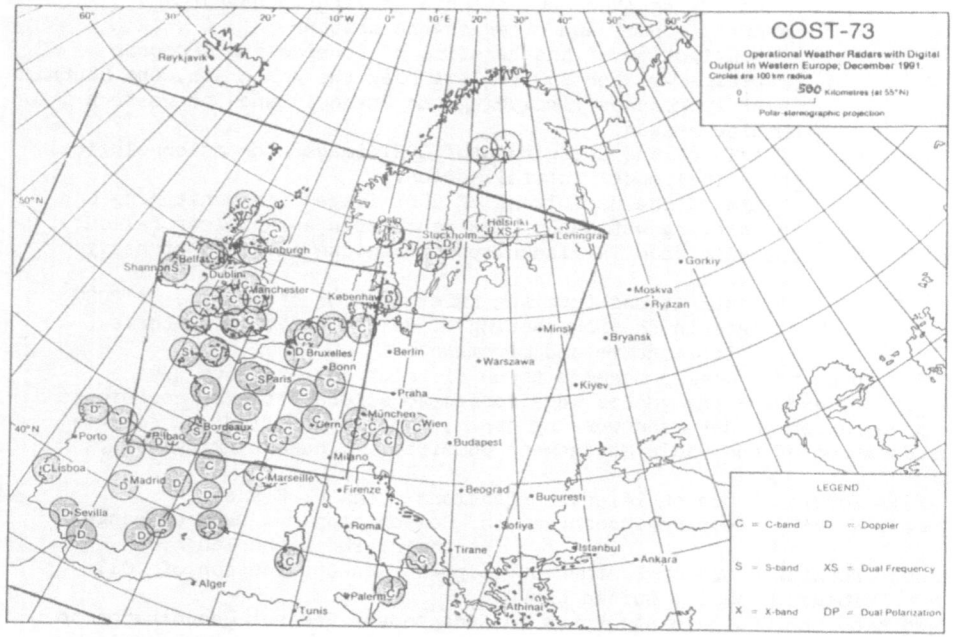

Figure 1: Radar networks in Europe by 1991 and areas for which COST
images are produced

have recently been produced for the large frame in Figure 1 namely
covering most of Europe. As in the COST-72 Project, the United Kingdom
offered to undertake the collection of data and its composition into wide
area products which are returned to other European countries. Much of the
rest of this paper is concerned with the investigations which have been,
and continue to be, undertaken with this real-time west European network
operation. The national radar network programmes underpin this work, and
these are truly operational programmes. However, the real-time
compositing for Europe is regarded, at present, as research, although
every effort is made to maintain the production of images every hour of
every day.

3. Development of European weather radar networks

Throughout the COST-72 Project there was a steady increase in the
installation of digital weather radar systems in western Europe.
Guideline hardware specifications were prepared which contributed in a

small way to this growth. By 1989 there were about 40 operational systems of which four were Doppler radars. It is expected that by 1991 there will be about 80 systems in western Europe of which some 30 are expected to be Doppler radars, Figure 1. This represents a significant increase in the deployment of Doppler radar.

Most west European countries have plans to develop radar networking systems, and some networks have already been installed and are outlined at this seminar. Recognising the importance of software systems, an outline functional software specification for radar network operations has been developed within COST-73. This is intended to provide a starting point for those countries embarking upon such work, and has been prepared in a way which does not favour any particular manufacturer.

Although many Doppler radars are being installed they will inevitably be operated within a framework of more conventional radar reflectivity observations. One challenge in the future will be to use Doppler wind and turbulence information operationally whilst still using the reflectivity data. It should be noted that Doppler radars in Europe are unlikely to be of the high power Nexrad type (Golden, 1989), but will have characteristics similar to those given in Table 1.

Table 1: Typical characterisitcs of low power Doppler radar (after EUCO-COST 73/25/88 revised 2.3.89)

antenna type:	parabolic, diameter 4.2 m
antenna gain:	43dB
antenna lobewidth:	0.9 degrees
antenna sidelobes:	$<$ -27dB
polarization:	linear horizontal
transmitter type:	coxial magnetron
peak power:	250 kw
pulse length:	0.5 µsec
PRF:	900/1200 HZ, 32 pulses at each prf
receiver dynamic range:	$>$ 85dB
MDS:	$<$ -114BM
signal processing:	32 point FFT processor
no of range cells:	120
velocity coverage:	-48 m/s to +48 m/s
velocity accuracy:	better than 0.3 m/s at S/N $>$ 10dB
	better than 0:6 m/s at S/N $>$ 0dB
spectrum width:	4 classes: 0-2, 2-4, 4-6 AND $>$6 m/s
total noise figure:	$<$ 5dB

4.   International exchange of radar data

Bilateral exchanges of radar data between neighbouring countries were developed some years ago, for example between the United Kingdom and Ireland; France and Switzerland; and France and the United Kingdom. Existing national codes, developed for particular radar installations, were used on dedicated communication links. These ad hoc arrangements were encouraged by the COST programmes, but it was recognised that the use of non-standard codes introduced high overheads in networking software systems. The Pilot Project of COST-72 (Collier et al, 1988) saw the production in real-time of composite images covering much of north-west Europe, achieved using four different code formats.

COST-73 has devoted considerable effort to proposing modifications to the WMO BUFR-94 code in order that, in the future, all international exchanges of radar data will use the same code. At present data are

received at Bracknell, UK from Ireland, France, the Netherlands,
Switzerland (via France) and soon from Belgium and Denmark. The intention
is to switch to the use of BUFR-94 as soon as is practicable.

These links are computer-to-computer, and software has been written to
receive data, transfer them from radar coordinates to polar stereographic
coordinates, composite all the data and broadcast the product so-produced
to several other countries, presently Belgium, Denmark, Finland, Ireland
and the Netherlands. Dedicated WMO GTS links are used in most cases, but
work has begun to investigate the potential of the CODE Experiment to
provide satellite communication facilities. The Olympus Satellite to be
launched in late 1989 will enable near real-time point-to-point
communications provided the appropriate ground station equipment is in
place. If the necessary equipment can be put in place it is hoped to send
Austrian radar data to the United Kingdom. Interest has also been
expressed by Italy and Yugoslavia, although such links may not become
active during COST-73.

The early COST products were produced up to two hours after the
nominal observation time. This was clearly unsatisfactory, and recently
completed work reorganising the software has enabled radar-only products
to be produced and disseminated within minutes of the observation time.
Other products to which further information has been added or their value
enhanced by various types of analysis can be made available within a
maximum of one hour of the observation time.

5. COST-73 Products

The product produced in real-time by the United Kingdom was developed
within COST-72 using the lowest elevation reflectivity data available or
the maximum reflectivity in the vertical at any grid point. Infra-red
satellite data from the European geostationary satellite Meteosat are
level sliced to show approximately medium and high level cloud areas.
Recent examples of this product are shown in Figure 2, and include data
from Ireland, France, the Netherlands, Switzerland and the United Kingdom.
It is likely that images over the larger area shown in Figure 1 will be
produced by the time of the Seminar.

Whilst the present product can be interpreted to give instantaneous or
cumulative precipitation amount, it is not satisfactory for all users.
Hence it is regarded as the first of a **product line** which will include
images making use of three-dimensional data and later Doppler data. All
products will be in polar stereographic coordinates having pixel sizes of
roughly 5 km x 5 km. The prime longitude will be 0° so that they can be
compared directly with numerical forecast information available from the
European Centre for Medium Range Weather Forecasting (ECMWF) and other
forecast centres. At present the following products are envisaged some of
which make use of independent data and value-added analysis:

* instantaneous   The present image in which areas within
  precipitation: radar coverage will be divided (by the
                         supplier of the data) into quantitative and
                         qualitative data with "cloudiness"
                         designated separately. The aim would also
                         be to separately identify precipitation type
                         using knowledge of the bright-band and
                         appropriate hail algorithms.

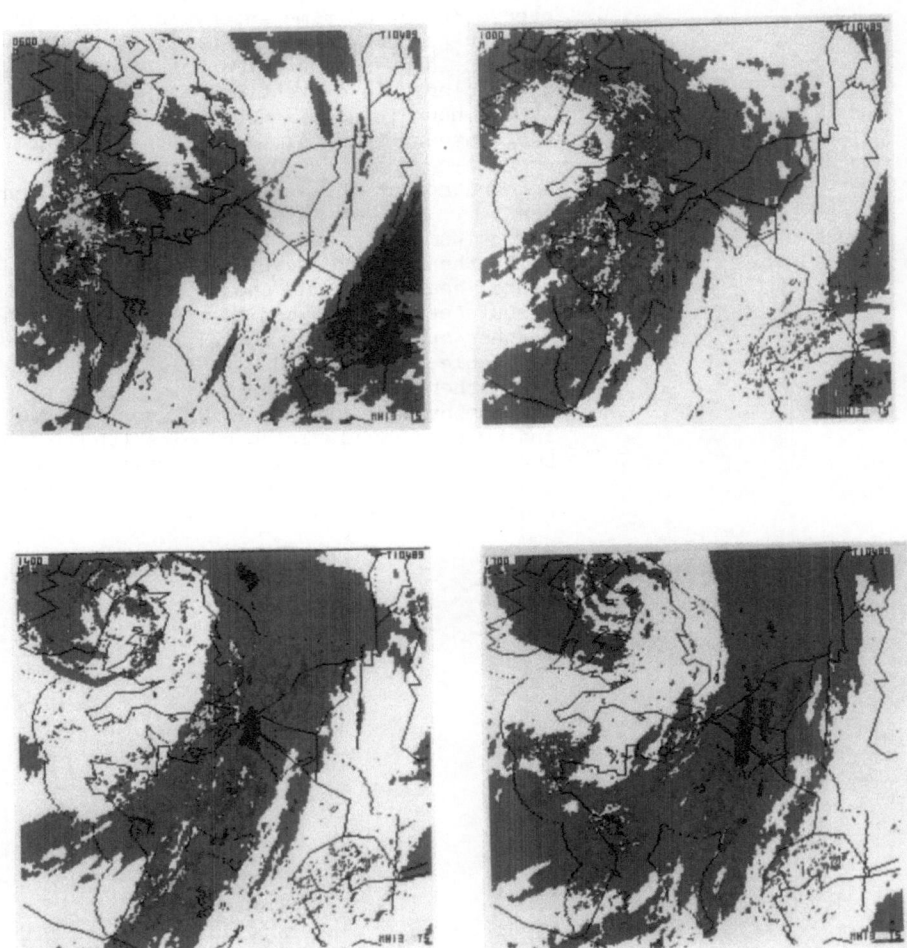

Figure 2:   Sequence of COST images on 11 April 1989. The time (GMT) is
            shown on the top lefthand corner of each picture. Blue is
            medium level cloud (-15°C to-45°C) and dark blue is high level
            cloud (< -45°C). Other colours are rainfall rates derived from
            radar data. Pink is $\leq$ 1 mmh^{-1}; green 1-3 mmh^{-1}; yellow 3-10
            mmh^{-1}; red 10-30 mmh^{-1} and black > 30 mmh^{-1}.

| * cumulative precipitation: | Cumulative precipitation would be derived from instantaneous information by summation over a number of time intervals determined by user requirements (Newsome, 1989). However large inaccuracies are likely, particularly in convective rainfall, if data at one hour intervals are used. Either a large amount interval is employed and the product regarded as qualitative, or data integrated national before compositinbg are used. |

* cumulative precipitation:

Cumulative precipitation would be derived from instantaneous information by summation over a number of time intervals determined by user requirements (Newsome, 1989). However large inaccuracies are likely, particularly in convective rainfall, if data at one hour intervals are used. Either a large amount interval is employed and the product regarded as qualitative, or data integrated national before compositinbg are used.

* hazardous or severe weather:

Algortithms to assess the lielihood of hail, thunderstorm development, strong winds, heavy rain and snow, and severe turbulence will be used to prepare a severe weather summary every hour. Where it is available, Doppler radar data will be used with other national information passed to the compositing centre. Likewise lightning information available as sferics (the radio transmissions from lightning) will also be combined with the radar data. A schematic example is shown in Figure 4.

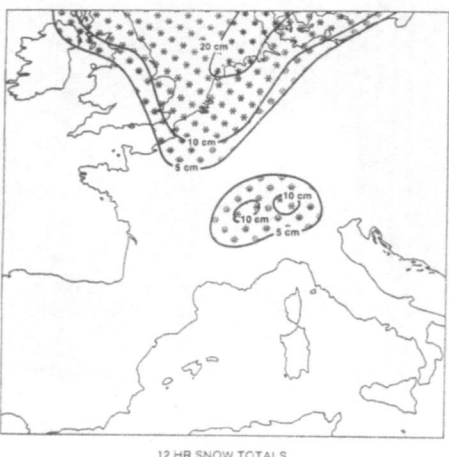

12 HR SNOW TOTALS

Figure 3: Schematic representation of the 12 hour accumulation of snow derived from European radar and satellite data.

*echo top: The heights of radar echoes will be plotted
                    wherever they are available. In recognition
                    of the difficulties in interpreting echo-top
                    images due to range dependent radar
                    sensitivity, the data may be combined with
                    satellite infra-red cloud top temperatures
                    to provide some measure of quality control.

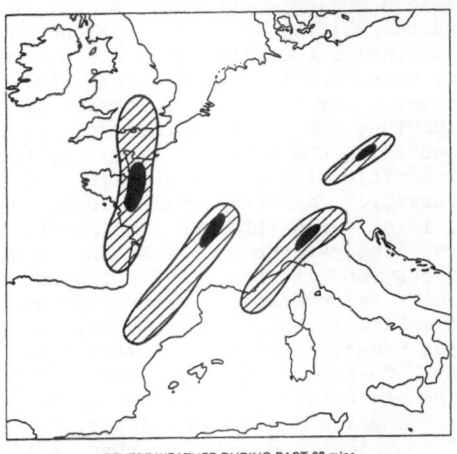

SEVERE WEATHER DURING PAST 60 mins

■■■ LIGHTNING/HAIL
▨▨ VERY HEAVY RAIN

Figure 4:   Schematic representation of severe weather summary over the
             previous hour derived from the COST images using various
             algorithms and independent data.

* echo field texture: Using constant Altitude Plan Position
                        indicator (CAPPI) data at, say, a height of
                        1.5 km, the variability of the echo texture
                        might be displayed to aid forecasters
                        identify different precipitation types and
                        their subsequent development. For example
                        the presence of convective cells embedded in
                        frontal rainfall may indicate transitions
                        between kata and ana fronts, or variations
                        in texture may indicate development or
                        decay.

* echo movement: Overall echo movement could be determined as
                        an aid to synoptic and mesoscale
                        forecasting. The use of Doppler radar data
                        in combination with reflectivity echo
                        movement might provide data leading, in
                        combination with other wind information, to
                        a more comprehensive specification of the
                        wind field.

Of these products instantaneous rainfall images are already produced and the production of a severe weather summary is being researched. Other products require more comprehensive data than are presently exchanged internationally. However the use of BUFR-94 code operationally should enable such data as are needed, to be acquired at a compositing centre.

6. Assessment of the utility of products

The real-time research projects outlined in section 2 of this paper are aimed at assessing the utility of the COST-73 products. Whilst the need for radar data from national networks is understood, and the exchange of data with neighbouring countries is clearly beneficial, the operational potential of radar products covering many countries is less clear.

Perhaps the most obvious benefit from the availability in real-time of radar composite data covering the whole of Europe is the contribution such data can make to the specification of wet deposition of pollutants particularly radio-activity. The Chernobyl accident in 1985 showed how radioactive material, if released into the atmosphere, will travel very considerable distances. The pattern of deposition is closely related to the rainfall distribution (Roesli et al, 1987, ApSimon et al, 1988, Clark and Smith, 1988) and work is underway in several countries to develop warning procedures to estimate deposition in the event of any future release. In some cases radar, satellite, conventional observations and numerical model rainfall forecasts are being combined in real-time to make the best estimate of precipitation. In the UK, COST-73 data are already being used in this way (Goddard and Conway, 1989).

The importance of numerical forecasts to changes in the initial moisture field has been stressed by Zhang and Fritsch (1986) and Golding (1987). This sensitivity has been demonstrated by Mills (1983), Bell and Hammon (1989) and Wright and Golding (1989). It seems clear that radar data over a wide area will contribute to numerical model humidity initialization and hence to forecasts for up to several days ahead. A challenge will be to investigate how best to use these data.

Assessment of the use of COST-73 radar data for forecasting over the southern North Sea has begun at the London Weather Centre in the UK, and will also soon begin in the Netherlands. Table 2 shows the type of assessment form which is being used by operational weather forecasters. In addition the contribution that COST data might make to forecasting

Table 2: Assessment form used operationally at the London Weather Centre, UK (courtesy J.M. Merson, London Weather Centre)

TIME	NO DATA	Ø=1	Ø=2	Ø=3	Ø=4	COMMENTS
0001	.					
0300	A					
0600			A			
0900			C			
1200			B			
1500				B		
1800					A	
2100					B	

A=Agrees with other data; B=Very good additional guidance; C=Misleading.
Ø=1, Fronts present; Ø=2, Unstable airmass or active front; Ø=3, Unstable airmass or active front producing precipitation; Ø=4 as for Ø=3 but producing heavy precipitation.

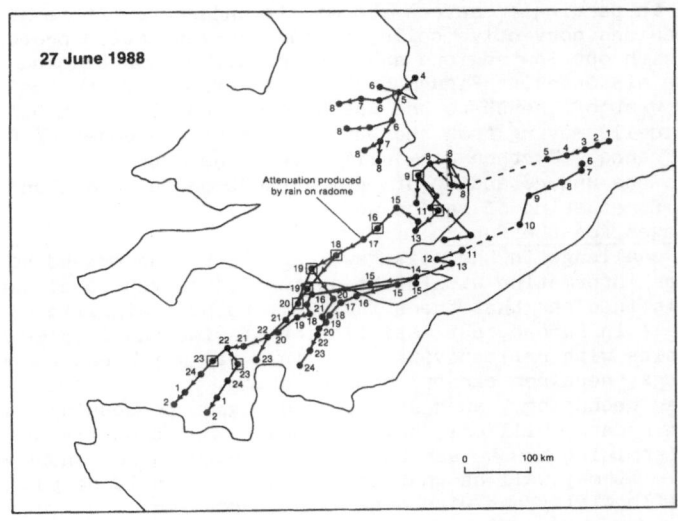

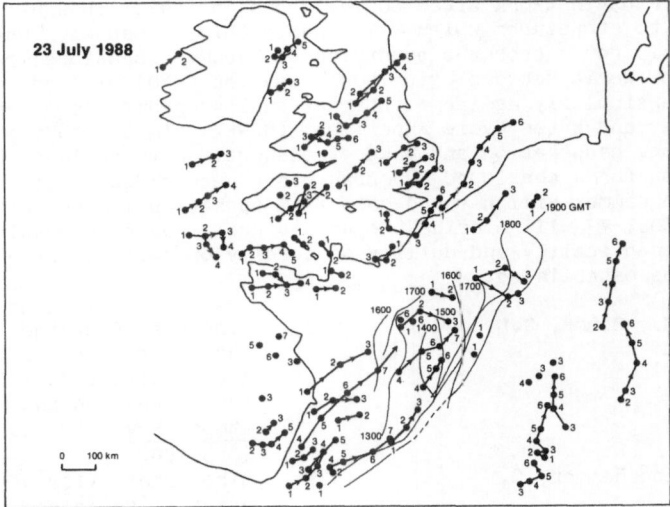

Figure 5:   Two examples of radar echo tracks over NW Europe derived from radar data from UK, the Netherlands, France and Switzerland. The direction of motion is shown by arrows and centroid positions at hourly intervals are numbered. Work is underway to investigate the interactions of cells and the effects of topography. The lower example shows the genesis of a squall line (indicated by a thin line) over France.

convection, in particular severe storms, is underway. It is already clear
that on occasions convective cells, through re-generation processes
associated with outflow regions and topography, may be tracked very
considerable distances. Figure 5 shows two examples, one case of cells
moving from west of the UK to the Dutch coast and almost to Denmark, and
one case of cells moving from the Dutch coast to the coast of Brittany in
north west France. Further work will reveal the extent that these data
can aid both the understanding of cell generation in Europe and the
operational forecasting of severe weather.

7.   <u>Challenges for the future</u>

A major challenge in the next few years is to understand how to use
Doppler radar information within the context of conventional weather radar
networks. It is clear that there is likely to be a significant growth in
Doppler radars in Europe, but assimilation of wind and Doppler spectrum
characteristics with reflectivity data, in a way which can be readily used
by forecasters, requires careful consideration.

New radar technology, such as improved signal processing and dual
polarization radars, will only have an impact operationally if effort is
put into determining how to use the data so produced for short-range
forecasting. It may well be that greater effort should be put into using
wide-area reflectivity and wind data within numerical model data
assimilation procedures.

If these challenges are to be met, then it is necessary to maintain an
operational radar network after COST-73 ends in 1991. Thought is already
being given to structures which will enable this to happen. One
possibility is for one or two countries to become operational compositing
centres. The World Meteorological Organisation (WMO) will be
involved, particularly as large international programmes such as the WMO
Global Energy and Water Cycle Experiment (GEWEX) aim to study global
precipitation, evaporation and water transport. Nevertheless, the main
justification for a continued European radar network based upon existing
national programmes will have to come from operational meteorological
services. COST-73 will provide the underpinning research results to
enable the practicality and utility of a fully **operational** European radar
network to be established.

References

ApSimon, H.M., Simms, K.L. and   1988   "The use of weather radar in
Collier, C.G.   assessing deposition of
                                              radioactivity from Chernobyl
                                              across England and Wales,"
                                              <u>Atmos. Env.</u>, 22, no 9,
                                              1895-1900.

Bell, R.S. and Hammon, O.   1989   "The sensitivity of fine-mesh
                                              rainfall forecasts to changes
                                              in the initial moisture
                                              fields," <u>Met. Mag.</u>, June.

Beringuer, B.   1989   "Review of the
                                              telecommunications work of
                                              COST-73," this volume.

Clarke, M.J. and Smith, F.B.   1988   "Wet and dry deposition of
                                              Chernobyl releases," <u>Nature</u>,
                                              332, 17 March, pp 245-249.

Collier, C.G., Fair, C.A. and   1988   "International weather-radar
Newsome, D.H.   networking in Western Europe,"

Golden, J.H.   1989   "The prospects and promise of
                                            NEXRAD: 1990s and beyond,"
                                            this volume.

Golding, B.W.   1987   "Strategies for using
                                            mesoscale data in an
                                            operational mesoscale model,"
                                            Preprint Vol., Workshop on
                                            Satellite and Radar Imagery
                                            Interpretation, Reading,
                                            England, 20-24 July, publ. by
                                            Eumetsat, pp 341-364.

Mills, G.A.   1983   "The sensitivity of a
                                            numerical prognosis to
                                            moisture detail in the initial
                                            state," Aust. Met. Mag., 31,
                                            pp 111-119.

Newsome, D.H.   1989   "Practical applications of
                                            weather radar data in Europe,"
                                            this volume.

Roesli, H-P, Joss, J. and   1987   "COST-73 and its application
Collier, C.G.   in very short range
                                            forecasting," Proc. Symp. on
                                            Mesoscale Analysis and
                                            Forecasting, Vancouver,
                                            17-19 August, European Space
                                            Agency, Special Publ. No. ESA
                                            SP-282, pp 13-18.

COST 73 Seminar :
View of the meeting-room
"Europe" (E.S.C.)

Economic and Social Committee meeting-room
Salle "Europe"
A view of the attendance

Dr J. JOSS
Swiss Meteorological Institute
Chairman of Session 1 at the Seminar

Dr J.H. GOLDEN
NOAA/National Weather Service (USA)
Key-note speaker at the Seminar (Key-note 1.2)

THE PROSPECTS AND PROMISE OF NEXRAD: 1990'S AND BEYOND

J. H. GOLDEN
National Weather Service
National Oceanic and Atmospheric Administration
United States of America

Summary

The Next Generation Weather Radar Program of the USA, NEXRAD, awarded
a contract to UNISYS Corporation in December 1987, to build and
deploy operational Doppler weather radars for the NOAA-National
Weather Service (NWS), the Federal Aviation Administration (FAA), and
the Department of Defense (DOD) Weather Services. The preproduction
prototype system is presently undergoing a final phase of operational
test and evaluation at the agency operational offices in central
Oklahoma. This operational test is scheduled for completion in
August 1989.

NEXRAD, the WSR-88D (Weather Surveillance Radar - 1988 Doppler), is a
replacement radar system for the current mix of reflectivity-only
weather radars, including the aging NWS network WSR-57's. We shall
discuss the scope for this program and the schedule for
implementation.

1. INTRODUCTION

The Next Generation Weather Radar program, NEXRAD, is a tri-agency
development, acquisition, and deployment effort which began in the late
1970's. The participating agencies are the Department of Commerce (NWS),
the Department of Defense (Air Weather Service and Naval Oceanography
Command), and the Department of Transportation (Federal Aviation
Administration). NEXRAD has its roots in the early, well-documented
successes at improved tornado detection with research Doppler radars in
Oklahoma and New England by Brown et al (1) and Donaldson (2), respectively. Research studies of the Union City, Oklahoma, tornadic storm of 24 May
1974 led to the discovery of a characteristic tornado life-cycle, its close
association with a parent mesocyclonic circulation, and the tornadic-vortex
signature (TVS) on single Doppler radar (3, 4). The earliest known Doppler
measurements across an actual tornado at relatively close range (in 1961)
were reported by Smith and Holmes (5) using a continuous-wave radar in
Kansas. The research work on Doppler measurements of tornadic storms in
Oklahoma during the mid-1970's was so impressive (6) that the Federal
agencies organized a quasi-operational experiment there in 1977-78. The
Joint Doppler Operational Project (JDOP) had two main goals: [1] to
determine real-time Doppler capabilities for improving tornado and severe
thunderstorm warnings and providing better flight safety for aircraft, and

[2] to outline specifications of a new generation radar for replacing the aging WSR-57 network radars operated by the NWS. Researchers and operational forecasters from all three agencies worked together as a team in quasi-operational experiments hosted by the NOAA National Severe Storms Laboratory (NSSL) in Norman, Oklahoma. JDOP forecasters viewed color graphics displays of three-moment data (Z, Vr, σ_v) from the 0.75° beamwidth, S-band Doppler radar at NSSL. Severe thunderstorm and tornado advisories simulating public warnings were recorded for comparison with conventional techniques (i.e., WSR-57 at the Oklahoma City Weather Service Forecast Office, OKC WSFO). Table I, taken from Burgess and his colleagues (7), shows the statistical results from the 2 years of JDOP experiments.

Table I. CRITICAL SUCCESS INDEX ANALYSIS OF SEVERE STORM DATA

Hail and Wind 1978

OKC WSFO	NSSL Doppler
POD = .47	POD = .70
FAR = .40	FAR = .16
CSI = .36	CSI = .62
LT = 13.6 min.	LT = 15.4 min

Tornado 1977-1978

OKC WSFO	NSSL Doppler
POD = .64	POD = .69
FAR = .63	FAR = .25
CSI = .30	CSI = .56
LT = 2.2 min	LT = 21.4 min

X = Forecast severe event which occurs
Y = Forecast severe event which does not occur
Z = Forecast non-severe event which occurs severe
LT = Lead time between advisory/warning issuance and event occurence (minutes)

Probability of Detection (POD) = $\dfrac{X}{X+Z}$

False Alarm Ratio (FAR) = $\dfrac{Y}{X+Y}$

Critical Success Index (CSI) = $\dfrac{X}{X+Y+Z}$

They show the improvements of using the NSSL Doppler in both tornado and severe thunderstorm warnings; specifically, the use of Doppler information gave increased lead-time as well as decreased false-alarm ratio.

2. THE ROOTS OF THE NEXRAD PROGRAM

It should be recalled that the existing United States radar network operated by the NWS consists of 44 WSR-57 tube-type radars and an additional 12 WSR-74S solid-state systems built by Enterprise Electronics. These radars operate 24 hours a day and have a nominal 2° beamwidth. In addition, there are 73 local warning radars nationwide (68 are WSR-74C, 5 cm, 1° beamwidth) which are operated only when significant weather is expected to occur. Most of the network radars are now 30 years old, and spare parts and maintenance costs became a growing concern during the mid-1970's. Sirmans and his colleagues at NSSL (8) did a study for the NWS on the feasibility of Doppler conversion of existing radars; however, they concluded that the extensive engineering changes and additional hardware and software development costs added to existing radars would not result in Doppler radars with acceptable Doppler performance, maintainability, and satisfactory life expectancy.

The U.S. Air Force Air Weather Service has been operating its AN/FSP-77 (5 cm wavelength) radars since 1964. While the NWS has the primary mission of providing weather radar surveillance for the national network, the Air Force is primarily concerned with localized warnings for severe weather threats to its air bases in the U.S. and overseas. In fact, during the SESAME '79 field experiment in Oklahoma, the NSSL Doppler operations center relayed a mesocyclone advisory to the Base Commander at Vance Air Force Base (9). A large tornado associated with the Doppler-detected mesocyclone fortunately lifted as it approached the Base's airfield from the west-northwest, but large hail fell on the base 40 minutes after the Doppler advisory was received from NSSL. Millions of dollars in potential aircraft damage were averted by the advance warning, immediately after which the Base Commander ordered most of the large aircraft moved into protective hangers. Also, during the spring 1979 severe weather season in Oklahoma, both Air Force Geophysics Laboratory's 5 cm and NSSL's 10 cm Doppler radars were operating side by side. The radar products were fully digitized, processed, and displayed for the radar meteorologists. The 1979 test was a great success and confirmed the practicality of a NEXRAD network; moreover, it became clear that the 5 cm Doppler was not the equal of the 10 cm NSSL Doppler radars when it came to accurate measurements in supercell storm outbreaks commonly found in the Central U.S. Indeed, the NEXRAD radars of choice were to be 10 cm systems (10). Finally, a Joint System Program Office (JSPO) within the NWS, staffed by representatives of, and funded by the three participating agencies was established in August 1979 to assure proper inter-agency coordination.

It is important to remember that the WSR-88D system is more than a new Doppler radar. It is being built to specifications, both hardware and software, set by the three participating agencies during the late 1970's and early 1980's. The NWS has network surveillance responsibilities across the U.S.; the DOD must issue point-severe weather warnings to protect facilities at its air bases in the U.S. and abroad; the FAA must provide en route weather advisories to aircraft pilots for flight safety. All three agencies mandated significant state-of-the-art automation in the design requirements they jointly set for NEXRAD (11) so as to operate the system with minimal human intervention, knowing that staff resources would become increasingly limited. Therefore, the WSR-88D is a replacement radar system for the current mix of reflectivity-only weather radars. It will provide new capabilities in radar remote sensing of precipitation, as well as the optically clear boundary layer; the real-time processing, communication, and display of digital radar data; and state-of-the-art meteorological radar analysis algorithms.

3. SYSTEM COMPONENTS AND PERFORMANCE

The WSR-88D is functionally composed of the Radar Data Acquisition subsystem (RDA), the Radar Product Generation and distribution subsystem (RPG), and the Principal User Processor system (PUP). These three subsystems comprise approximately 60 percent, 20 percent, and 20 percent of total unit costs, respectively.

The salient characteristics for the RDA subsystem are summarized in Table II.

Table II. RADAR DATA ACQUISITION (RDA) SUBSYSTEM SPECIFICATIONS

Frequency	2.7-3.0 GHz
Beamwidth	1 degree (maximum)
Range, reflectivity (ref.)	460 km
Range, velocity (vel.)	230 km
Transmitted power	1 MW (peak), 2 KW (average)
Pulse Length	1.5 and 4.5 microseconds
Rulse Repetition Frequency	320-1300 pulses per second
Clutter cancel (ref. & vel.)	50 db possible in both channels
Automated performance monitoring and error detection	

The RDA subsystem is composed of a steel tower, antenna pedestal, reflector, coherent S-band transmitter (klystron) and receiver, signal processor, and acquisition computer. Among the special features are an automatic, adaptive selection of a pulse repetition frequency, which minimizes obscuration of second-trip radar returns in the velocity channel, and the provision of a long-pulse (approximately 4.5 microseconds) capability for use in clear air or very-low-rate precipitation conditions. Use of the long pulse, combined with a reduction of the radar scan rate, effectively increases the sensitivity of the radar by about 10 db. The signal processing capability, which takes raw analog data and estimates the three meteorological moments, also provides clutter filtering (for both ground clutter and anomalous propagation), range unfolding, and velocity de-aliasing. (The FORTRAN-77 software needed to operate the RDA required about 600,000 lines of executable code.)

Site-adaptive clutter cancellation of at least 50 db is provided in the velocity channel and up to 50 db in the reflectivity channel. The RDA then delivers the real-time, high-resolution digital base data to the RPG. Other characteristics of these base data include: 1 degree beamwidth; better than 1 dbZ accuracy and precision for reflectivity over a 90 dB dynamic range; 1 km range resolution for reflectivity to 230 km, 2 km resolution at 460 km, and about -15 dbZ sensitivity at 30 km; ¼ km resolution to 230 km for mean radial velocity and spectrum width estimates, with accuracy and precision better than 1 meter/second; and Nyquist velocity up to approximately 32 meters/second with automatic unfolding to approximately ±64 meters/second. In addition, a substantial effort has been made to develop and incorporate a performance monitoring, fault detection, and diagnostics system capable in most cases of isolating a fault down to the component or single PC-board level. The radar antenna (a 28-ft diameter parabolic dish) can be mounted on a 100-foot tower and incorporates a 39-foot diameter rigid fiberglass radome (with an rf two-way loss of 0.6 db at 2800 MHz) as shown in Figure 1. Transmitter, receiver, and signal preprocessor are collocated with the antenna and housed in a shelter at the base of the tower. In an operational mode, the antenna rotates continuously in azimuth at a maximum speed of 5 rpm and moves over an elevation range from - 1° to 20° at a rate appropriate to selected scan strategies. The government has required that there can be up to 8 selectable and programmable scanning strategies available, but at this time only 4 have been specified. Two of the precipitation volume scan modes are illustrated in Figures 2a and 2b.

Interconnection of the RDA to the RPG is provided over a wideband communications link. The type of communication link used will depend on the terrain, the physical distance between the RDA and RPG, and a number of other factors. Technologies used will include coaxial cable, or fiber optics for shorter distances where rights-of-way present no major problems, and microwave line-of-sight for longer distances or where problems preclude the use of other technologies. The Strategic Plan for the Modernization of the NWS (12) indicates that, as much as possible, we intend to collocate future Weather Forecast Offices with the NEXRAD RDA's; however, problems with land acquisition or lease, along with other factors noted above, have resulted in approximately 44 sites having separation distances between the RDA and RPG requiring one (or even two) microwave repeater linkages.

The RPG subsystem provides the primary data processing capability in the NEXRAD system. It is the host for all the meteorological algorithms and the source of all the processed products developed by the system. The RPG supports the generation, local storage, distribution, and archiving of products. The characteristics of products produced by this subsystem are summarized in Table III, where the numbers is parenthesis indicate the number of products of that type available.

Table III. RADAR PRODUCTS GENERATION (RPG) SUBSYSTEM SPECIFICATIONS

Base products (3)	reflectivity, radial velocity, spectrum width
Derived products (19)	e.g., composite reflectivity, surface rainfall accumulation, shear
Alphanumeric products (3)	e.g., severe weather alert, radar coded message
Derived data array products (4)	e.g., reflectivity radial data, hourly radar rainfall-digital estimate
Operational position	selected units (NWS and FAA systems)
Expandability	Factor of 3 - CPU, global memory, mass storage
Storage	1½ hours minimum for all products, up to 6 hours for selected products

Since the WSR-88D is as much a signal processor as it is a Doppler radar system, the ultimate utility of the system depends on the flexibility and adaptability of the RPG. Note that the RPG consists of a Concurrent 3280 digital computer, having 6.1 million instructions per second CPU capability and expandable at the government's discretion to 3 CPU's. The

RPG software contains about one million lines of executable code. In addition to the three base products (reflectivity, radial velocity, and spectrum width) noted in Table II, a large number of derived products are produced by the RPG, with a number of range and resolution options available to the system operator. Table IV lists the meteorological NEXRAD algorithms, many already developed and tested on real Doppler data.

Table IV. NEXRAD METEOROLOGICAL ALGORITHMS

1. Storm Segment
2. Storm Centroids
3. Storm Tracking
4. Storm Position Forecast
5. Storm Structure
6. Hail
7. Mesocyclone
8. Echo Tops
9. Vertically Integrated Liquid (VIL)**
10. Severe Weather Probability**
11. Shear
12. Velocity Azimuth Display (VAD)
13. Turbulence
14. Velocity Volume Processing (VVP)
15. Precipitation Preprocessing
16. Precipitation Rate
17. Precipitation Accumulation
18. Precipitation Adjustment
19. Tornadic Vortex Signature (TVS)
20. Precipitation Products
21. Gust Front
22. Transverse Wind
23. Convergence/Divergence
24. Constant Altitude PPI: Reflectivity and Velocity (CAPPI)
25. Flash Flood Potential

[Note: Products #20-25 are not in the WSR-88D Limited Production contract, but they may be added later].

Table V lists the base and derived products required to be available from the RPG by processing of those meteorological algorithms.

Table V. NEXRAD PRODUCTS

Base Products

1. Reflectivity Maps
2. Velocity Maps
3. Spectrum Width Maps

Derived Products

4. Combined Shear
5. Combined Shear Contour
6. Echo Tops**
7. Echo Tops Contour**

8. Composite Reflectivity
9. Composite Reflectivity Contour
10. Layer Composite Reflectivity
11. Layer Turbulence
12. Hail Index*
13. Mesocyclone*
14. Tornadic Vortex Signature (TVS)
15. Storm Structure*
16. Storm Tracking Information*
17. Weak Echo Region ("stacked reflectivity plates")*
18. Velocity Azimuth Display Winds*
19. Velocity Azimuth Display Plot*
20. Combined Moment
21. Storm Relative Velocity*
22. Severe Weather Analysis Display*
23. Severe Weather Probability**
24. Vertically Integrated Liquid (VIL)**
25. 1 HR Precipitation Accumulation**
26. 3 HR Precipitation Accumulation**
27. Storm Total Precipitation Accumulation**
28. Cross Section

Alphanumeric Products

29. Severe Weather Alert Message
30. Free Text Message
31. Radar Coded Message**

Derived Data Array Products

32. Hazardous Aviation Weather Data
33. Reflectivity, Radial Data
34. Velocity, Radial Data
35. Hourly Digital Radar Rainfall Estimates

*   See reference (13) for results of winter 1985 field trials at NWS
    Boston, MA, Forecast Office, using the VAD wind profiling and other
    NEXRAD products obtained from AFGL 10 cm Doppler.

** These NEXRAD-required algorithms and derived products were originally
    developed by Techniques Development Laboratory and the Office of
    Hydrology. They have been tested in the field, especially at the
    Weather Service Forecast Offices in Oklahoma City, OK (14), and Denver,
    CO (PROFS, 15). See paper by T. Schlatter, this symposium.

It is important to emphasize that the main operational control of any
particular NEXRAD unit is accomplished at the Unit Control Position (UCP)
of the RPG. This control includes RDA operational modes, volume coverage
patterns, and monitoring the system's status. The UCP has an application
and a system console for entering commands, adaptation data or alarm/alert
thresholds for that particular site, plus a printer. This and other entry
points to the system (including the RPG Operational Position--similar to
the PUP subsystem described below) are double password-protected for
critical parameters. The adaptation data will be used to set the
adjustable parameters in the products and algorithms for geographical and

seasonal variability, derived and controlled by the NEXRAD Operational Support Facility (OSF). (See Section 5 for its functions.) It appears that optimum settings for these parameters may require the archival of RPG data at a given site for 2 years or more (16). Alarm/alert capabilities are discussed below, but they can be varied on a daily or shift basis by the local NEXRAD site official-in-charge at his UCP. The RPG is designed to support multiple user workstations, supply external systems with products, provide products to external (i.e., non-principal) users, drive all the communications, process all the meteorological algorithms, and generate all the products and meteorological alerts.

Finally, the main agency-user interface will take place at the PUP subsystem. Even though the maximum number of WSR-88D systems that can be procured by the agencies is about 195 in the U.S. and overseas, there will be more than twice as many PUP's needed. PUP's provide both graphic display and additional processing to the NEXRAD user (fixed point, 32 bit Concurrent 3212 general purpose digital computer). The PUP display entails two high-resolution color graphic monitors and one alphanumeric (operator command and data entry) terminal, shown in Figure 3a. Using either or both 19-inch color monitors (resolution: 1280 x 1024 pixels), the user can look at products; manipulate products, e.g., zoom and off center; or time-lapse one or more products. A quarter-screen display capability permits simultaneous viewing of four different products on a single screen (see Figure 3 sequence). The resolution of the products is 640 x 512 pixels.

The user-system interface is a software-controlled graphics tablet with a movable "mouse" shown in Figure 3b (3000 x 3000 cells resolution). This interface device (or, alternatively, keyboard command entry at the adjacent applications terminal) permits the operator to change display data resolution and change magnification at the same resolution to improve the analysis of small-scale features. A sequence of up to 60 user-function keys on the graphic tablet will allow the development of many local applications macros at each NEXRAD site, another added flexibility of the system. One such user-function might be a four-panel display of the base reflectivity plus low and upper-level storm-relative radial velocities, which would quickly aid the forecaster in locating mesocyclones and/or shear regions. A large selection of operator-selectable background maps (up to a maximum of four simultaneously--examples in Figure 3) will be provided to each site showing geographic and political boundaries; rivers, lakes, and streams (hydrologic watersheds); cities and towns, highways; airports and prime airways; etc. The operator can provide alphanumeric annotation for a displayed image and also has 64 special characters and symbols to speed analysis and increase the information provided in the displayed area. Other PUP features include a time-lapse capability, in which a sequence of up to 72 images for each of three selected products can be stored and continuously updated, and the local storage of all products received for a 6-hour period.

An important new capability of the NEXRAD system is the automatic area-alert feature. Using the PUP, the user can define alert areas, i.e., areas in which the detection of several predefined phenomena will auto-matically generate an audible and a visual alert to the operator. Each operational position will be able to display an operator-defined outline of up to two alert areas as an overlay and will be able to select threshold criteria for up to 10 of the alert categories. For each designated NEXRAD product, initial alert threshold value sets will be established by the OSF. When an alert condition has been met within an alert area, a text alert message and an audible and visual notification are provided to the operational position. It is important to note that each NEXRAD RPG can drive up to 21 "associated" PUP's (via dedicated lines) and up to 26 "non-associated" PUP's (via dial-up lines) at other remote agency sites. Thus, NEXRAD sites which suspect that severe weather is moving into the upstream fringe of their coverage areas can rapidly access products from adjacent NEXRAD sites closer to the echoes of interest. There are also 4 external user access ports on each RPG, but those users can receive only a fixed product set.

4. NEXRAD SCHEDULE AND COSTS

The NEXRAD program began in 1980 as a tri-agency development and acquisition effort. It has followed the Office of Management and Budget's Circular A-109 procurement procedures for large systems. Thus, during the System Definition Phase there were 3 contractors selected to develop system approaches and designs. In 1982, a Validation Phase contract was awarded to Sperry (now UNISYS) and Raytheon for each to develop independently a complete system architecture, software, and documentation, and finally to construct a pre-production NEXRAD unit during the Validation Phase. By the time NEXRAD is fielded, beginning in early 1990, the Doppler radar will have undergone four stages of testing: developmental test and evaluation (DT&E); production acceptance test and evaluation; and two major operational tests, Initial Operational Test and Evaluation (IOT&E) and Operational Test and Evaluation (OT&E). DT&E was performed during the Validation and Limited Production Phases of the NEXRAD Program. This testing included risk reduction verification, configuration item reliability predictions, configuration item performance tests, computer program configuration item tests, functional area tests, and systems tests.

Each pre-production unit was subjected to DT&E at factory sites in New England during 1985-87 by the NEXRAD JSPO. Finally, after an Initial Operational Test and Evaluation (IOT&E) conducted with participants from all 3 participating agencies during October 1986 and spring 1987, a Limited Production Phase contract was awarded to UNISYS in December 1987 to build the first 10 operational NEXRAD units. More details on the objectives of IOT&E and the OT&E are described below. Production Acceptance Test and Evaluation will be performed during the Limited Production and Full-Scale Production phases of the NEXRAD program. This testing includes configuration item and computer program configuration item tests, factory tests, factory system tests, and installation and checkout tests. Once the Full Production Phase contract is awarded to UNISYS (scheduled for fall 1989) for the remaining 165 NEXRAD systems, total system costs including development will be about 1 billion dollars. It should be emphasized that the projected life cycle for the entire system is 20 years. Approximate costs: for the 3 contractors in the System Definition Phase were $6 M; for the UNISYS and Raytheon prototype in the Validation Phase, $50 M. In addition, UNISYS' new development work, required by the government since the NEXRAD Technical Requirements document, January 1986, is estimated at $2M. During late 1989-91, the OSF NEXRAD unit will be upgraded to the Limited Production configuration, and the other ten Limited Production Phase NEXRAD systems will be installed at: Oklahoma City, Oklahoma; Melbourne, Florida; Washington, D.C. (Sterling, Virginia, near Dulles Airport); Washington, D.C. (FAA, near National Airport); Kansas City, Missouri (non-operational, for training); Frederick, Oklahoma (DOD); Chicago, Illinois (FAA); NW Florida (DOD, near Pensacola); Chanute Air Force Base, Illinois (for maintenance depot); and St. Louis, Missouri. Installations will begin early in 1990 for these first ten operational systems.

5. TESTING AND IMPLEMENTATION

During fall 1986 and spring 1987, teams of meteorologists and engineers from all three participating NEXRAD agencies conducted IOT&E. One team operated and maintained the UNISYS prototype near Hartford, Connecticut, while another team did the same with the Raytheon prototype just west of Boston, Massachusetts. We should note that these tests were quasi-operational, i.e., they were not conducted in real forecast offices or base weather stations. The purpose of these IOT&E tests was to assess the operational effectiveness and identify deficiencies impacting operational suitability of the NEXRAD. The primary focus was to assess the accuracy and usefulness of NEXRAD's user products as an aid to detecting and forecasting weather, system maintainability, supportability, software operator-machine interface, and software growth and update capability. Use of the prototype NEXRAD's was shared between the JSPO contractors and the test team at each facility. At the request of the three NEXRAD agencies, the entire IOT&E test process was planned, managed, and conducted by the Air Force Operational Test and Evaluation Center (AFOTEC), an independent testing body within the U.S. Air Force (USAF). The results of these tests were evaluated by AFOTEC in final reports (September 1987) on each of the UNISYS and Raytheon prototypes and, along with proposals for the Limited Production Phase with cost estimates (including total life-cycle cost projections), formed the basis for the Government's selection of the winning contractor, UNISYS, to build the first 10 NEXRAD systems.

A NEXRAD OSF was established during 1988 in Norman, Oklahoma, to provide centralized coordination, management, applications development, and control of technical support for the proposed 165 NEXRAD units to be deployed (17). Specifically, the OSF will provide operational systems support (technical guidance, software maintenance, and configuration management of unit components). Beginning in late 1988, UNISYS shipped the prototype NEXRAD system to the OSF, upgraded from the version tested to its factory during IOT&E.

The OT&E is also being planned, managed, and executed on behalf of the tri-agency NEXRAD program by AFOTEC (18). The OT&E tests began on March 6, 1989, and will continue through July using the OSF's first article NEXRAD in Norman, Oklahoma. This time, however, the tests are more operational in character and are being carried out with PUP displays in the NWS Forecast Office (adjacent to the OSF), the Base Weather Station at Tinker Air Force Base and the FAA Academy, both in Oklahoma City (25 miles north of the OSF). In each case, the NEXRAD displays and products are being used with other operational data sets to prepare and issue to the public (or USAF base commanders) severe weather warnings, air terminal advisories, and routine forecast products (during IOT&E, no warnings or forecast products were disseminated, although each was subsequently evaluated with verification data). The purposes of the OT&E are to: [1] evaluate the operational effectiveness and suitability of the preproduction NEXRAD for the three participating agencies; [2] review deficiencies and needed enhancements documented by the test team during IOT&E; [3] identify deficiencies and needed enhancements not previously documented; and [4] identify items to be addressed during follow-on operational test and evaluation (FOT&E). FOT&E is presently being planned by the tri-agencies to take place at one or more of their early operational NEXRAD sites in 1990 and beyond, as part of the Test and Evaluation Master Plan. The three NEXRAD agencies and NEXRAD Program Council have identified five critical operational issues: performance, availability, responsiveness, growth capability, and interoperability. There are 18 objectives for OT&E related to either operations and hardware/software aspects of the test. AFOTEC will provide test results to the tri-agency decision makers in support of the full-scale production option (165 NEXRAD's) to be exercised by early fall 1989.

6. NEXRAD SITES AND NETWORKING - GOVERNMENT AND PRIVATE USERS

It might well be asked at the outset: Why have a network of NEXRAD's at all? The rationale for having a network of NEXRAD systems, rather than implementing stand-alone, single stations, was addressed by the Advisory Committee for NEXRAD System Requirements Evaluation (16). It had been suggested that such single weather radar stations be developed with only the equipment required to meet the needs of the specific site. However, the Committee noted that the essence of the NEXRAD concept is the development and procurement of a system that will meet the minimum requirements of the NWS, FAA, Air Weather Service, and Navy. Not only did these agencies agree on a joint set of requirements, they have configured the deployment plan so that an individual site can supply the radar umbrella coverage and products needed by several user agencies in that area. In addition, there are NWS national centers in Washington, D.C.; Kansas City, Missouri; and Miami, Florida; and DOD's Global Weather Center that require real-time radar information from many regions of the country. A common set of spares, maintenance documents, procedures, and training will be utilized by the NEXRAD concept, resulting in major savings in life-cycle costs. Single unique sites would develop into software-unique sites, dependent on the knowledge of key individuals. Future maintenance and system improvements would be difficult and costly. Methods for deriving new products at the OSF and in the research community will undoubtedly require data from multiple sites in ways not presently contemplated. As noted above, products are required by national facilities over large geographic areas. In order to merge the data and produce national and regional products, data must be integrated from several radar sites to produce frequent, accurate rainfall maps and to determine the three-dimensional structure of weather phenomena. Such data integration can be accomplished in an efficient manner only if the system is part of a common network and meets uniform standards.

How was the NEXRAD installation sequence for the new NWS network prioritized? During 1985, a strategy was developed by the NWS Director, Dr. Richard Hallgren. It took into account the locations of all existing radars, especially the older, more difficult-to-maintain ones, and emphasized the severe storm climatologies and population densities across the U.S. Top priority is given to those stations that will be supporting the Modernization and Restructuring Demonstration (MARD, 12) during 1993 in the Central U.S. Figure 4 shows the expected NEXRAD radar coverage at 10,000 feet by fall 1992. The next level of priority is to install NEXRAD's at existing Weather Service Forecast Offices (WSFO), 55 in all, where we currently have meteorological staffs, completing one coverage area approximately the extent of a river basin per NWS Region, until all WSFO's have received a NEXRAD. Following WSFO installation, several NEXRAD's will be installed to meet specific hydrologic and other special needs. The final installations will be at existing Weather Service Offices (smaller than WSFO's) and new sites (e.g., Flagstaff, Arizona).

A design requirement of the NEXRAD network is to provide nearly continuous coverage over the continental U.S. at 10,000 feet above ground level for the detection of severe storms and other significant weather phenomena. Radar sites have been chosen to provide the maximum coverage possible, taking into account terrain, equipment collocations and/or accessibility, and overall cost factors. The network design allows for reflectivity estimates to 460 km and velocity and spectrum width estimates to 230 km (Figure 4).

The process of locating a NEXRAD site was essentially a three-stage process. The first stage involved an initial site assessment. During this process, a map search was done to find optimum radar sites with emphasis on government-owned facilities. Next, these sites were visited and preliminary site surveys were completed. These site surveys involved analyses for adequate radar coverage, geographic and operational suitability, roads and utilities, and environmental impact. Finally, possible radar locations were limited to one or two sites, which were visited again, and an in-depth site survey was done for the optimum location. The process to locate, perform an analysis, and acquire the land for a radar site can take up to 44 months.

There were often several environmental issues of concern when locating a radar site. Therefore, environmental issues usually included mineral, water, and recreational resources, air quality, hydrogeological processes, flora and fauna, aesthetics, cultural status of the area, radiation hazard to humans, availability of frequency allocation, and electromagnetic interference to TV, radio, and other communications equipment.

The NEXRAD RDA is optimally located relative to the required area of coverage, which has resulted in the adjustment of coverage area for several NWS sites (e.g., to cover nearby USAF air bases). The NWS attempts to locate the RDA a little downwind (east quadrant) from the population area to be covered. The reason behind this is that, even though the radar may temporarily "lose" the storm as it passes overhead (in the so-called "cone of silence," above the maximum tilt angle of the automatic volume scans), it will reacquire the storm as it moves upwind, into the area to be provided advance warnings. The NEXRAD network has been designed so that, to the maximum extent possible, an adjacent radar will be able to detect a storm that passes directly overhead to another adjacent site.

Two of the most important NEXRAD products which will form network data arrays for composites are the Precipitation Processing System (References 19, 20--see sample PUP presentation in Figure 3) and the Radar Coded Message (RCM, 21). The RCM is automatically produced in three parts.

Part A contains an intermediate graphic product and a tabular listing of alphanumerics. The graphic product contains reflectivity data for the 1/16 x 1/16 Limited Fine-Mesh Model (LFM) grid over the radar area of coverage out to 460 km. The intermediate graphic product is provided to the NWS meteorologist for possible editing by deleting data or changing data to another data level. The editing process may be envisioned as shown in Figure 3. The alphanumeric list contains a reflectivity intensity value for each grid box, the height, the position of the maximum echo top (where available), the locations of all centroids with reflectivity intensity values that exceed the minimum reflectivity threshold, and the forecast movement of each such centroid. Part B of the RCM contains a single profile of the horizontal wind information derived from the output of the VAD algorithm. Part C contains remarks in an alphanumeric format which give the position of each detected mesocyclone and uncorrelated horizontal shear region, each detected tornadic-vortex signature, the hail index, and storm top information for each centroid. These RCM's will be transmitted automatically, after manual editing (if required), over the AFOS System Z communications network to the National Meteorological Center, where a national radar summary chart will be formed. By the mid-1990's, RCM's may also be used by some offices to prepare local or regional composites on AWIPS-90 for their specific needs. RCM parts A, B, and the automated portion of Part C are generated by the NEXRAD RPG software twice each hour, in time for release at 5 and 35 minutes after the hour. Initially, as NEXRAD's are deployed in limited regional domains, composites will be produced once per hour, with horizontal resolution of about 20 km. Once the full network is deployed, we expect to produce and disseminate national summary charts, including a detailed rainfall map, twice per hour at resolutions of about 10 km.

It will be NWS policy to encourage the widest distribution, dissemination, and effective usage of WSR-88D Doppler base and derived meteorological products through media outlets such as television, radio, and private sector communications and meteorological service companies. To this end, the NWS intends to establish a NEXRAD Information Dissemination Service (NIDS), which will make 8 selected NEXRAD base and derived products available to external users, via intermediate telecommunications vendors or meteorological "service" companies. The NWS and DOD are planning to make a total of four user access ports available at each of 137 NEXRAD sites across the U.S. A Request for Proposals will be issued shortly to select, on the basis of competitive bids, the four companies allowed to access the 4 NIDS ports at each NEXRAD. The data available from the NIDS are in encoded form. It is anticipated that other service companies who subscribe to the NIDS will, after suitable conversion, make NEXRAD products available to the larger community of external users. Each of the four primary NIDS distributors must also provide the suite of "unaltered" products from all 137 NEXRAD sites to their subscribers, but they can also provide new, additional derived products as well. Under no circumstances will any company be given exclusive or proprietary rights to the NEXRAD data or products.

7. OPPORTUNITIES FOR THE FUTURE

By late 1994, the 175th (and last) scheduled NEXRAD unit will have been installed. Data integration capabilities are among the most urgent priorities of the NWS Strategic Plan for Modernization and Restructuring. Although they must await the development and deployment of the AWIPS-90 system, those capabilities will be initially tested with NEXRAD and other data sets combined on AWIPS-prototype workstations at NWS offices in Denver, Colorado, and Norman, Oklahoma, beginning in 1990. These two risk-reduction activities will expand to include the 10 earliest NWS NEXRAD installations over the Central U.S. in the 1992-93 timeframe. We are confident that the powerful, high-resolution observing capability afforded by NEXRAD in precipitation and in the clear convective boundary layer will lead to exciting new scientific discoveries of mesoscale weather disturbances and triggering mechanisms for convective storm outbreaks. There will also be applications to long-range transport, urban and topographic influences on the flow, and even bird and insect migrations. New product and algorithm development for NEXRAD will continue, with emphasis on winter storms and tropical cyclones; moreover, new, more reliable severe thunderstorm products for NEXRAD will use combined attributes, such as strong upper-level storm divergence coupled with large VIL to predict hail presence and size. Algorithms giving objective estimates of tornadogenesis potential in evolving severe thunderstorms will be developed, using diagnoses of shear-vector turning properties (wind hodographs) obtained from NEXRAD VAD's and Wind Profiler data.

More accurate regional and continental-scale precipitation maps will be derived by combining NEXRAD with rain gauge inputs, as well as with refined visible/infrared satellite estimation methods (especially those planned for the Tropical Rainfall Measuring Mission (TRMM) satellite). By the mid- to late-1990's, sophisticated meso-β scale numerical forecast models will be developed, with improved convective parameterizations and topography, capable of assimilating NEXRAD winds and precipitation data. Finally, the introduction of NEXRAD into regions with little or no prior weather radar coverage will lead to the discovery of new severe storms and clear air phenomena being reported, in a surprising array of locales and seasons!

ACKNOWLEDGEMENTS
    This paper is dedicated to the author's recently-retired NWS colleague, Paul L. Hexter, Jr., for his expertise, leadership and perserverance as NWS Radar Program Leader and NEXRAD focal point. Helpful reviews of this paper were provided by Ron McPherson, Doug Hess and Bob Saffle, NWS. Editorial assistance was given by Ms. Billie Cooper and the manuscript was ably typed by Ms. Eileen Joseph. The product figures (3C -3K) were provided by the NEXRAD OSF in Oklahoma (Messrs. Zittel, O'Bannon and Fornear). Ms. Therese Pierce, NEXRAD JSPO, provided some useful information.

REFERENCES

(1) BROWN, R.A., LEMON, L.R. and BURGESS, D.W. (1978). Tornado detection by pulsed Doppler radar. Mon. Wea. Rev., 106, 29-38.
(2) DONALDSON, R.J., Jr. (1970). Vortex signature recognition by a Doppler radar. J. Appl. Meteor., 9, 661-670.
(3) BURGESS, D.W. (1976). Single Doppler radar vortex recognition: Part I, Mesocyclone signatures. Preprints, 17th Conference on Radar Meteor. (Seattle), AMS, Boston, MA, 97-103.
(4) BROWN, R.A., and LEMON, L.R. (1976). Single Doppler radar vortex recognition: Part II - tornadic vortex signatures. Preprints, 17th Conference on Radar Meteor. (Seattle), AMS, Boston, MA, 104-109.
(5) SMITH, R.L. and HOLMES, D.W. (1961). Use of Doppler radar in meteorological observations. Mon. Wea. Rev., 89, 1-7.
(6) RAY, P.S., BROWN, R.A. and ZIEGLER, C.L. (1978). Doppler radar research at the National Severe Storms Laboratory. Weatherwise, 32.
(7) BURGESS, D.W., WILK, K.E., BONEWITZ, J.D., GLOVER, K.M., HOLMES, D.W. and HINKELMAN, J. (1979). The Joint Doppler Operational Project. Weatherwise, 33, 72-75.
(8) SIRMANS, D., BURGESS, D. and ZRNIC, D. (1976). Considerations for Doppler Conversion of NWS radars. Final Report to National Weather, NOAA, 14 pp. and Appendices (Available from: NSSL, NOAA - 1313 Halley Circle, Norman, OK 73069).
(9) ALBERTY, R.L., BURGESS, D.W., HANE, C.E. and WEAVER, J.F. (1979). SESAME '79 Operations Summary ERL Report (SESAME Documentation Series), August, 1979, 253 pp. (Available from NSSL).
(10) ALLEN, R.H., BURGESS, D.W. and DONALDSON, R.J., Jr. (1980). Severe 5 cm attenuation of the Wichita Falls storm by intervening precipitation. Preprints, 19th Conf. on Radar Meteor. (Miami Beach, FL), Amer. Meteor. Soc., Boston, MA, 87-89.

(11) NEXRAD Technical Requirements Document (1986). 160 pp. and Appendices. (Available from: NEXRAD Joint System Program Office, 8060 13th Street, Silver Spring, MD 20910).

(12) Strategic Plan for the Modernization and Associated Restructuring of the National Weather Service (March 1989). NOAA Department of Commerce, 21 pp./Available from NWS Transition Program Office, 8060 13th Street, Silver Spring, MD 20910.

(13) FORSYTH, D.E., ISTOK, M.J., O'BANNON, T.D. and GLOVER, K.M. (1985). The Boston area NEXRAD demonstration (BAND). Environmental Research Papers, No. 912 (AFGL-TR-85-0098), 59 pp. (Available from: Atmos. Sciences Div., Air Force Geophysics Lab, Hanscom AFB, MA 01731).

(14) WINSTON, H.A. and RUTHI, L.J. (1986). Evaluation of RADAP II severe storm detection algorithms Bull. Amer. Meteor. Soc., 67 (2), 145-150.

(15) RASMUSSEN, E.N., SMITH, J.K., PRATTE, J.F. and LIPSCHUTZ, R.C. (1989). Real-time precipitation accumulation estimation using the NCAR CP-2 Doppler radar. Preprints, 24th Conf. on radar Meteor. (Tallahassee, FL), Amer. Meteor., Soc., Boston, MA, 236-239. (See also 19, 20 below.)

(16) SAFFLE, R.E. (1989). Plan for the development and implementation of severe weather probability relationships at NEXRAD sites TDL Office Note 89-1. 12 pp. (Available from Techniques Development Lab, NWS, NOAA, 8060 13th Straeet, Silver Spring, MD 20910).

(17) NEXRAD Joint Program Office (1988). NEXRAD Integrated Logistics Support Plan. 100 pp. Available from NEXRAD JSPO.

(18) Air Force Operational Test and Evaluation Center (1988). NEXRAD Initial Operational Test and Evaluation, Phase II. 67 pp. and Appendices. (Available from: HQ AFOTEC/RS, Kirtland AFB, New Mexico 87117-7001).

(19) AHNERT, P.R., HUDLOW, M.D., JOHNSON, E.R., GREENE, D.R. and ROSA DIAS, M.P., (1983). Proposed "On-site" precipitation processing system for NEXRAD. Preprints, 21st Conf. on Radar Meteorology (Edmonton, Canada), Amer. Meteor. Soc., Boston, MA, 378-384.

(20) HUDLOW, M.D., GREENE, D.R., AHNERT, P.R., KRAJEWSKI, W.R., SIVARAMAKRISHNAN, T.R., JOHNSON, E.R. and DIAS, M.R. (1983). Op. cit., pp. 394-403.

(21) Federal Coordinator for Meteorological Services and Supporting Research (1989). Doppler Radar Meteorological Observations - Federal Meteorological Handbook No. 11-Part A: System Concepts, Responsibilities and Procedures. (Final draft available from NEXRAD JSPO in Fall, 1989). RCM is discussed in Chapter 6.

FIGURE LEGENDS

Figure 1   NEXRAD RDA subsystem at OSF, Norman, Oklahoma. See text for details.

Figure 2A   One of 8 selectable and programmable volume scan strategies specified for NEXRAD: Synoptic Scan - Precipitation Modes, 9 elevation scans every 6 minutes (hatched areas are beam sampling areas on range/height diagram, numbers in right/upper margins are mid-points of beam in degrees for each scan).

Figure 2B   Same as Figure 2, except a Precipitation/Severe Weather Scan: 14 scans every 5 minutes.

Figure 3A   Principal User Processor (PUP), 2 color monitors; graphics tablet (lower) and applications terminal (right).

Figure 3B   Close-up PUP of graphics tablet with mouse - the main operator - user interface for NEXRAD.

Figure 3C   Severe Weather Analysis and Display (SWAD): a four-panel NEXRAD product (synthesized from NSSL Doppler data) depicting zoom 50 x 50 km panels of reflectivity, velocity, spectrum width and storm-relative velocity. Note: Any of the 3 base-products is also available for full-screen display.

Figure 3D   NEXRAD Hydrology (Precipitation) Algorithm output: 1-hour Surface Rainfall Accumulation over Northern Oklahoma (maximum 1.45 inches).

Figure 3E   Same as 3D, except Storm Total Accumulation (maximum 15.85 inches)

Figure 3F   Vertical Cross-Section NEXRAD Product: R/Z cross-sections of reflectivity, velocity and spectrum width for PPI shown in lower-right.

Figure 3G   Weak-Echo region (WER) NEXRAD Product (vertically "stacked plates" of reflectivity, viewed in perspective.

Figure 3H   Echo Tops NEXRAD Product (threshold =18 dBZ), tops in KFT (maximum = 63,000 FT).

Figure 3I   Vertically - Integrated Liquid (VIL) NEXRAD Product (maximum = 60 kg/m^2).

Figure 3J   Velocity - Azimuth Display (VAD) wind profiles, plotted at 12 minute intervals (radiosonde winds on far left profile at 1800 CST).

Figure 3K   Radar-Coded Message (RCM) with grid/map overlay used for editing non-weather echoes. See Reference (21) for details.

Figure 4   Map of complete NEXRAD network - surveillance coverage at 10,000 ft beam - elevation, when all systems are installed across United States.

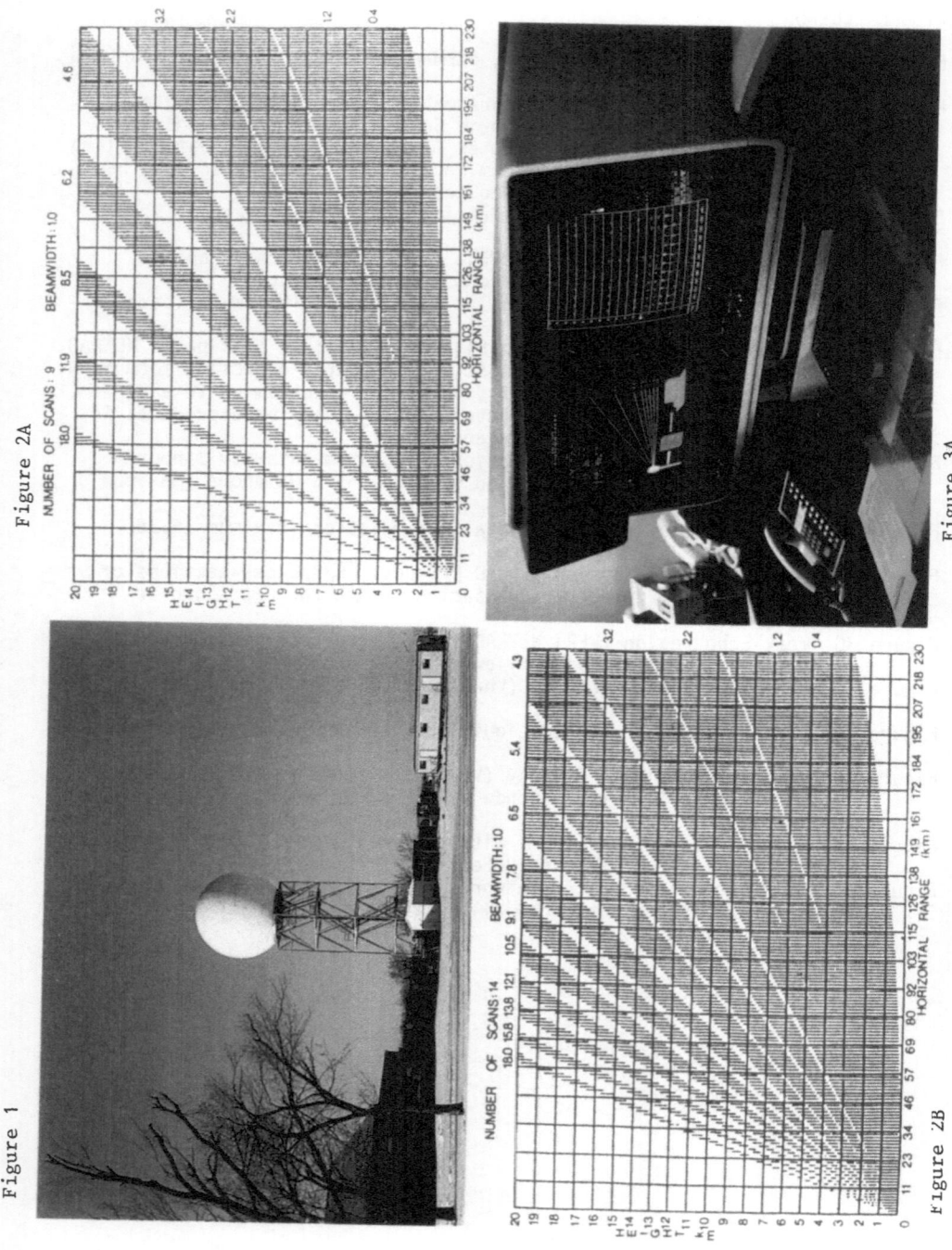

Figure 2A

Figure 3A

Figure 1

Figure 2B

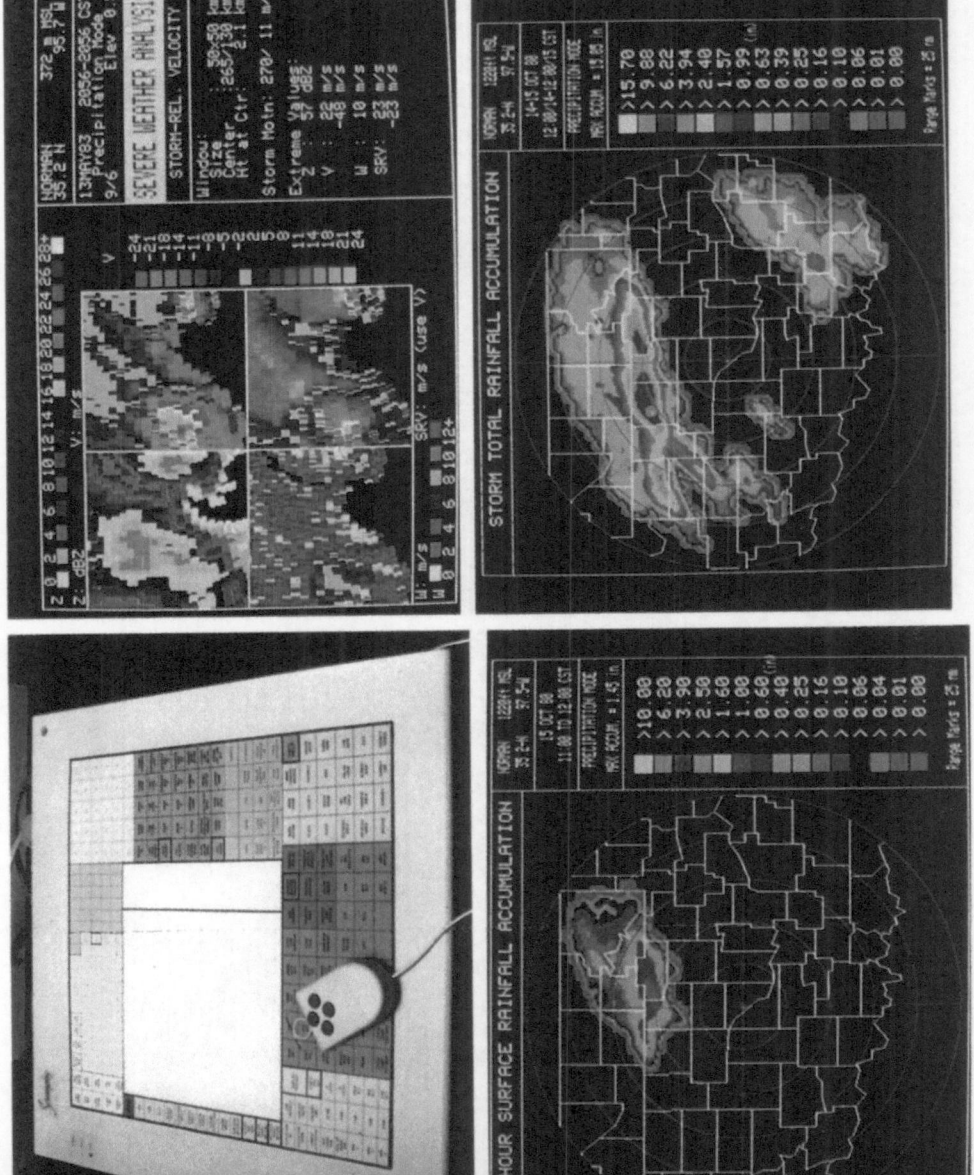

Figure 3B

Figure 3C

Figure 3D

Figure 3E

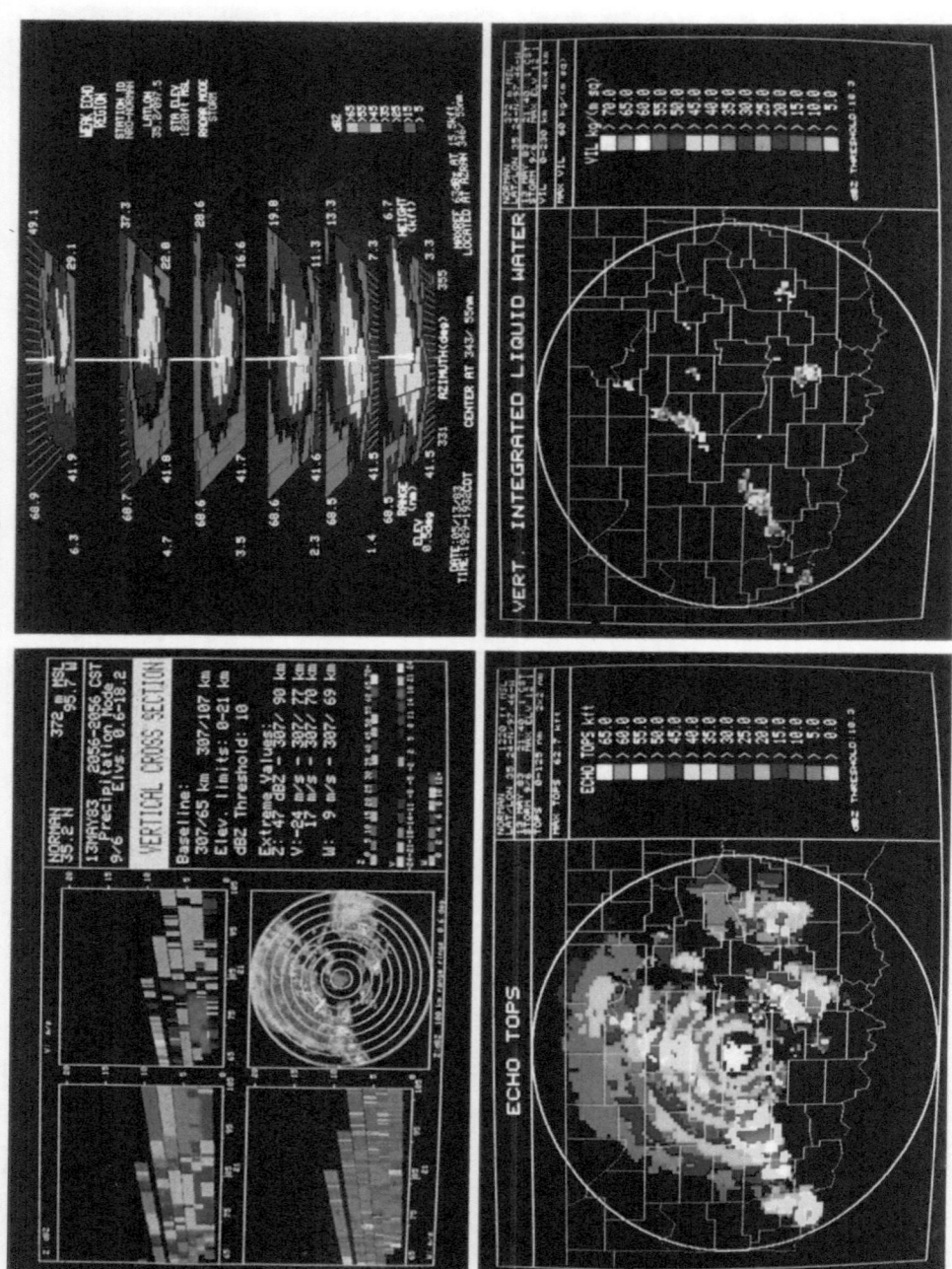

Figure 3F

Figure 3G

Figure 3H

Figure 3I

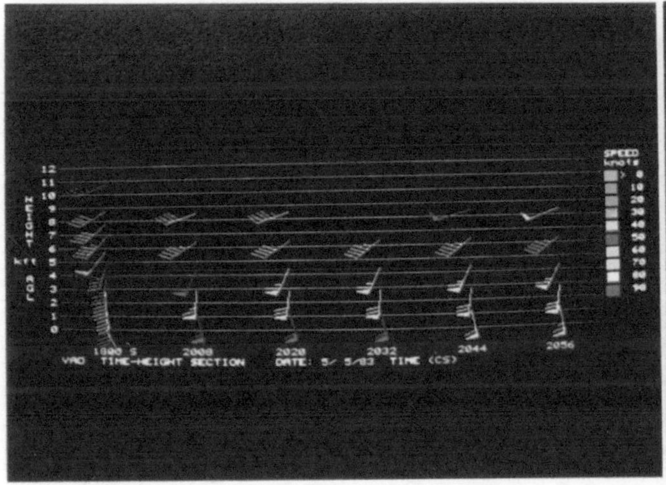

Figure 3J

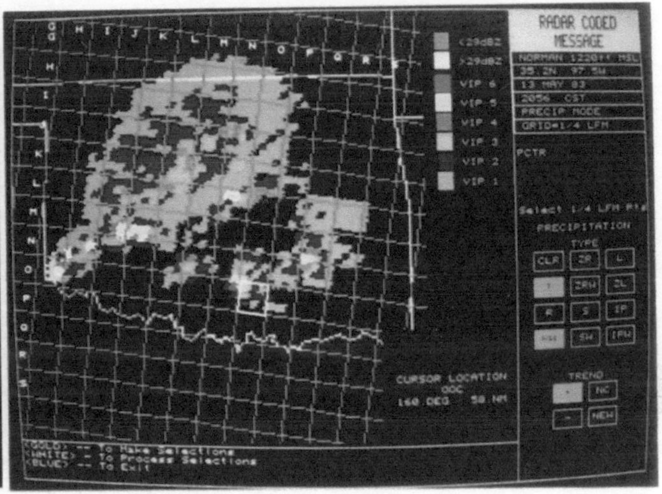

Figure 3K

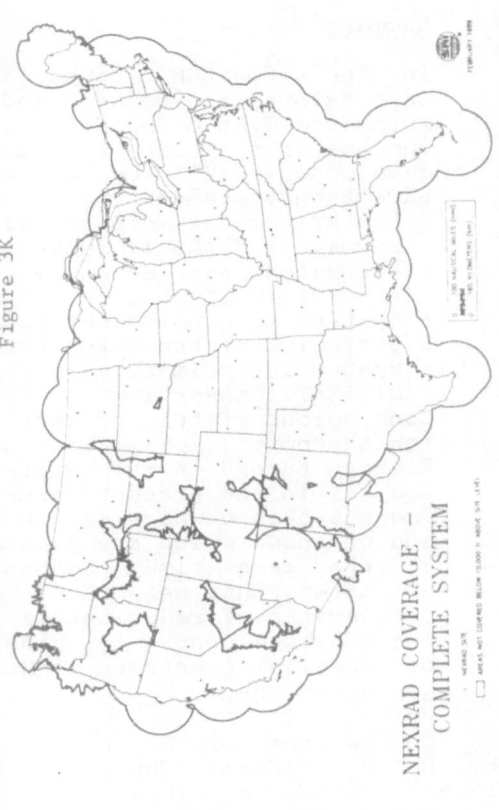

NEXRAD COVERAGE —
COMPLETE SYSTEM

Figure 4

RADAR NETWORKING IN EASTERN EUROPE

D. PODHORSKÝ

Slovak Hydrometeorological Institute

Summary

In the COMECON countries a common project of Czechoslovakia, Poland and USSR named "Comprehensive Automated System for Meteorological Air Traffic Controll and Solution of the Problems of the National Economy" was developed in 1985. Up to now the folloving subprojects have been implemented :
- the system for meteorological air traffic control on runways – KRAMS II (USSR);
- automated radar meteorological systems – ARMS (CSSR), AKSOPRI (USSR);
- central meteorological subsystem – CMS (CSSR); inclusive of the automated meteorologist work-stations;
- system for primary satellite data reception – WIPS and DIGISAT (Consortium of CSSR, GDR and Romania).

Conclusions of this project have provided the conditions for a prompt implementation of the radar network in Eastern Europe on the basis of Soviet two-wave MRL-5 radars. In the present paper a survey is given of weather radar stations which are in operation in Eastern Europe and of those which are planned to be built by 1995.

At the same time, technological prerequisites for automated radar network of COMECON countries as well as for devices of radar automation from the view point of the Regional Centre for Radar and Satellite Meteorology of Socialist Countries at Malý Javorník near Bratislava are given there.

In the sense of theoretical works and conceptual studies by E. M. Saľman (USSR) and the author of the present paper as his disciple, the principles and goals of a unified meteorological radar network in Europe were published [1,2,3] in the years 1968 – 1970. Unfortunately, the political atmosphere was not favourable for the implementation and for further advance of these concepts at that time.

Therefore the first international symposium of COMECON countries on "Applied Radar Meteorology" was held in Bratislava in 1972. At this symposium methods of building up a unified automated weather radar network in Eastern Europe were specified [4,5]. The period of automation development of Soviet radars followed, namely the types MRL-1 and MRL-2 (in the USSR under the name of Cyklon), from 1976 also in Czechoslovakia (under the name JPS) and in Poland (under the title SKORA). The situation in the advance of computer technology, the shark contrast between the trends in the world and in our country had a considerable influence upon the project's further implementation up to 1985.

At the seventh meeting of the WMO Regional Association VI (Europe) held in 1978, the Czechoslovak Hydrometeorological Service concerning the working-out of the principles of regional aspects of collection, exchange and processing of radar information in digital form. The eight WMO RA VI meeting held in 1982 adopted the report presented by the Rapporteur for this issue [6]. At the same time the Resolution N^{8}5 - "Regional Procedures for the Transmission of Digitized Meteorological Radar Data over the GTS" was adopted; this resolution ensured from the cooperation between Great Britain and Czechoslovakia, and thanks to Mr. G.C. Collier its aims were fulfilled in 1986.

The adverse situation in the field of radar automation as well as in the sphere of the implementation of cybernetics in the hydrometeorological service (HMS), particularly after the transition to MRL-5 two wave radars, was expected to change considerably as a result of COMECON project - with the participation of Czechoslovakia, Poland and the USSR - entitled "The Comprehensive Automated System for Meteorological Servicing of Aviation and for the Solution of Tasks Permitent to other Branches of the National Economy" (known under the acronym KAS METEO). This project was under way since 1985 to 1989.

With the aim of speeding up and focusing the efforts aimed at implementing an automated radar system in the socialist countries, and thanks to the exceptional organizing skills of Mr. S. Kotra from Poland, a symposium on "Digital Processing of radar Information" was held near Warsaw in 1986 The papers presented at the symposium pointed out quite a great deal of disintegration of the different countries' approaches. Seven systems - some of them in the process of development, others implemented already - were presented, out of which two in the USSR (METEOYATCHEYKA and AKSOPRI), one in Poland (SKORA), one in Czechoslovakia (ARMS), and some concepts and automation development projects at different specialist levels in Bulgaria, Hungary and Rumania. This fact can follow from both the ecomonic principle based on the assumption that brain work represents the cheapest investment and from the basis established for the assessment of the research and development cost effectiveness in our respective countries.

Within the period mentioned above, certain social and economic problems were observed in some of the COMECON countries as a result of which reductions in capital investments occured and some of the meteorological services started to operate on a (partiallly or completely) self-financing basis.

In spite of these unfavourable circumstances we are optimistic, however. I think that this opinion can also be proved by the results achieved recently in the implementation of the KAS METEO project (see enclosured KAS METEO diagram).

The subsystem for meteorological servicing of runways "KRAMS" (USSR) has been innovated from the veiwpoint of microelectronics and has been complemented with conventional accoustic radars – sodars and, particularly, with Doppler sodars. The development and manufacture is provided by the GDR, Poland and the USSR. The first comparative experimental measurement of different types, including the systems by the Sensitron company, took place in Czechoslovakia in May 1989. Special attention should be devoted to the Soviet-made lidar which is a laser system measuring the cloud base level as well as the horizontal and, especially, the oblique visibility.

As it has been mentioned before, the systems AKSOPRI (USSR) and ARMS (CSSR) have been developed for the purpose of MRL-5 two-wave radar automation. Ten of these radars systems are to be manufactured in each of the two countries in 1989. The ARMS is to secure complete automation of the meteorological radar networks in Czechoslovakia and GDR in 1990. At the same time the innovation of the Polish system SKORA is expected to take place, too. Systems with a high level of pattern recognition designed for receiving primary data from geostationary satellites and from quasi-polar orbiting satellites have been developed in Poland, Rumania and GDR.

A software system operating on the basis of primary data from geostationary satellites has been put into service in Czechoslovakia. This system calculates the top level echo values, the primary data correction for the visible channel against the level of the Sun, it also calculates the advective – convective tendency, it produces forecast based on the temperatures at the top level echo and at ground level, if provides classification of cloud types and of the associated phenomena, and it forecasts the occurence of these clouds and phenomena for the next two hours. This prognostic approach based on the physico-statistic methods, known under the name of METEOTREND, provides nowcasting both for the air traffic control and other spheres of the national economy.

In the recent years attention has been paid to the development of meteorological observers' automated work-stations as well as to the development of automated work-stations for meteorologists operating on the basis of personal computers, namely the IBM PC/XT or AT. The first outputs from these projects are available in Poland, others are under way in GDR, Bulgaria and Czechoslovakia.

The Slovak Hydrometeorological Institute has been entrusted with developing a central meteorological subsystem (CMS) which is the heart of KAS METEO. The CMS controls the collection of data from automated radars, the comprehensive processing of aerological, synoptic, satellite and radar data, it also controls the calculations for forecasts such as nowcasting and very short-range weather forecasts, and by means of the telecommunication computer it distributes information, warnings and forecasts among the users. The basic unit of this subsystem is a 32-bit VAX type computer manufactured in Czechoslovakia.

What is the current situation in the building up of a radar network in the differnt East-European countries?

1. Bulgaria – There are 12 radars in operation there, namely
-eight MRL-5 radars which are used for hail suppression services and one of them serves for scientific and research purposes;
-two MRL-2 radars are used for continuous scanning of dangerous meteorological phenomena, the output data being used for the air traffic control (Sofia and Varna);
-one MRL-2 radar is used for nowcasting and for local forecast in the Bulgarian capuital Sofia. At present all these radars are operated manually and the automation is expected after 1990 with those MRL-5 radars which are used for the weather modification purposes.

2. Hungary – The radar network consists of three MRL-5 radars (Budapest, Szentgotthard, Nyíregyháza) operated manually. Through a special warning code (WAFOR) the information is transmitted to Budapest, namely to the Weather Forecast Centre and the information is also provided for weather authorities. On the basis of bilateral agreements, the information is also transmitted to Austria and Rumania. Software has been developed for IBM PC/AT computers, which enables the reception of radar data via telex and produces animated films, stores the data and evaluates the precipitation zones for the needs of the water management. In 1988, the development of the multilevel iso-echo has been completed and its implementation and utilization has been secured by means of a new measurement metodology.

3. Poland – There are seven radars in operation there, namely:
- one MRL-2 radar in Legionowo near Warsaw, which has been automated by the SKORA system and provides information for the Okencie airport (Warsaw) as well as for other users;
- six MRL-1 radars (this type is a two-wave mobile-type radar, the wavelenghts are 0.85 and 3.2 cm). These radars are used for the air traffic control. Four of them are automated (Deblin, Elblag, Slupsk and Lodzh); and the remaining two ones (Zielona Góra and Wroclaw) have not yet been automated on the basis of the SKORA system.

4. GDR   –   The radar network in this country operates in a
                    continuous mode, the combined radar echo chart
                    is produced in one-hour intervals, the range
                    normally being 300 km, and in case of
                    dangerous   phenomena   occuring   (hail,
                    thunderstorms and showers) the intervals are 30
                    min and the range of 150 km. The radar data
                    transmission is carried out through facsimile
                    and the central work-station is situated at the
                    Berlin-Schonefeld airport. The system operates
                    manually at present, thus following the
                    simplified methodology devised by the
                    Methodological Centre for Radar Meteorology of
                    the Socialist Countries in Leningrad.
                    The MRL-5 radars are situated at the following
                    sites: Berlin-Schonefeld, Warnemünde and
                    Neuhaus (south-west of GDR, altitude 840 m).
                    Full automation based on the Czechoslovak ARMS
                    system is expected to be started at all the
                    above sites from 1990.
5. Romania –   Six manual radars operate there at present: one
                    is an MRL-5 working in Bucharest, and five of
                    them are MRL-2 radars situated at Cluj,
                    Craiova, Iasi, Tulcea and Mangalia on the coast
                    of the Black Sea. Observations are made in
                    one-hour intervals and a radar echo overlaps
                    chart, representing significant weather
                    phenomena, is produced in synoptic observation
                    times (i.e. once in 3 hours).
6. USSR   –   Weather radars in continuous operation are used
                    for the needs of the meteorological warning
                    service, for the weather forecasts, for hail
                    suppresion, for precipitation intensity
                    measurements and for science research
                    activities. At present, there are some 120
                    different radars in use there, namely the types
                    MRL-1, MRL-2, and MRL-5. Fully automated
                    operation through AKSOPRI system is provided
                    with six radars in the Moscow area and at the
                    Pulkovo airport in Leningrad.
                    Radar echo overlap chart (map) is produced at
                    synoptic observation times.
7. CSSR   – The Slovak Hydrometeorological Institute in
                    Bratislava performs the role of a Regional
                    Centre for Radar and Satellite Meteorology
                    within the framework of the socialist countries
                    of central and south-eastern Europe. The most
                    recent modification of the Cetnre's Statutes
                    was carried out by the Conference of the
                    Hydrometeorological and Meteorological
                    Services' Representatives of the Socialist
                    Coutries in Bucharest, in October 1988. The
                    Centre has been set the task – in accordance
                    with the progress in the field of automation –
                    of operatively securing the confirmation

processing, evaluation, distribution and storage of radar data from these countries and the task of receiving, processing and distributing the data from weather satellites, and, on the basis of the comprehensive processing of the individual sources of meteorological information, to develop and operate nowcasting and very short-range weather forecasts. This way good continuity has been achieved between the WMO Activity Centre for Very Short-Range Weather Forecasts with the Slovak Hydrometeorological Institute (SHMI) in Bratislava – Malý Javorník and the mission of this Regional Centre.
At present, there are two weather radars in Czechoslovakia in continuous operation:
the combined radar echo charts are produced in one-hour intervals and are then transmitted through an OLT – 22 long-way facsimile transmitter (124,6 kHz, power output 100 kW). The two radars in question are as follows: an MRL-5 radar in Bratislava – Malý Javorník (altitude 584 m above sea level)', and an MRL-2 (until July 1989; this will be substituted by an MRL-5 in September 1989) in Prague. An MRL-5 radar has been put into trial operation at Kojšovská Hoľa near Košice (altitude 1,246 m). Three more radar observatories are planned to be built by 1995, the locations being Spičák (in the Ore Mountains, near the town of Kraslice), Skalky (in the Central Moravian Drahan Highlands), and near Banská Bystrica (central Slovakia).
Automation of the Czechoslovak radar network – three such radar sites – expected from 1990.

The unified automatic radar network which is to serve the needs of hydrology and meteorology and thus – from the point of view of "the common European house" – it must rest upon common principles, calibration and measurement methods, data transformation into a unified graphic system as well as into a unified code for data transmission.A great deal of work has already been done in this respect within the WMO Regional Association VI (Europe) in the mid-1980's and therefore it will be necessary to use those basis for further development, i.e. to create mutual links between the COST-73 project and the results of the KAS METEO project.

In our view, it is necessary to first of all secure the telecommunication (computer) links between the individual radar sites and the national centre for radar data processing, with the transmission speed recommended being that of 9,600 bit/s (the minimum transmission admissible in the first stage is 2,400 bit/s).

At present, studies are under way into the amount of primary information obtained from automated radars which is necessary to be transmitted to the Processing Centre. The first theoretical and experimental results are contradictory and often contain errors due to subjective approach. The specialists try to provide the highest possible volume (amount) and frequency of the data transmitted, yet at the same time they neglect the cost-effectiveness of the technological line. In my personal view, this problem should be solved with utmost care. For example, one of the papers presented at this symposium (Podhorský, Vlčák) opens new possibilities for the provision and assessment of radar meteorological network data. It seems that the transmission of the different static and particulary dynamic features as well as the tranmission of the sets of standards to the Centre will make it possible to solve this problem more effectively.

We recommend to provide the observation synchronization with the EUMETSAT geostationary satellites within the automated radar network and to in this way secure the comprehensive assessment of both remote sensing sources. The question of coordination in the radar operations and observations is not a negligible one from the economic point of view. We can tell from our experience that in the case when METEOSAT data are available every 30 minutes, it is possible to control the switching-on of the radars in accordance with the METEOTREND results in 80 % of the cases. A new generation of EUMETSAT geostationary satellites should help no in this respect. At the same time it would be quite useful to analyze the WEFAX (SDUS) programme of these satellites. A certain part of the users would probably prefer to receive radar network data or the combination of radar and METEOSAT data, or the graphic form of nowcasting and of the very short-range weather forecast.

In Czechoslovakia a system for the transmission of radar network information and nowcasting is being developed within the KAS METEO project. The system provides the transmission of the above information by means of the long-way facsimile transmitter onto the FAX-CARD of an IBM PC/AT computer.

25 years have gone by this year since the publishing of the extensive work by David Atlas entitled "Successes in Radar Meteorology" (Advances in Geophysics, vol. 10, 1964, Academic Press). The author states in historical perspective that "...no sooner will radar meteorology became established as a scientific discipline than the radar becomes one of the standard devices used by a meteorologist, although probably not quite exactly the same as a barrometer is, but it will have to be a device whose data could then be interpreted unambiguously. As a result of that, radar meteorology will disapear as an individual discipline and it will only become a part of meteorological measurement methods.".

I think that the new point of the current application of radar meteorology in the world — in comparision with the situation in the individual countries of the World Meteorological Organization — we are only at the beginning. It is, first of all, the economic and sometimes also political problems which hamper the speeding-up of the process of automated radar implementation and the radar network built-up. For this reason it would be appropriate to provide conditions for the interconnection of the EEC project COST with the COMECON KAS METEO project. Further, it would be of interest to furnish the possibilities for cooperation also in the field of research and development. The embargo in the field of meteorological technology seems to be losing ground in the current trends towards the improvement of international relations.

The Czechoslovak Hydrometeorological Service is providing conditions for the automated network to be put into operation in 1990, namely in the area covered by COST-73 radars and by those of COMECON countries, i.e. the area between Yugoslavia, Austria, the FRG on the one hand, and the GDR and Czechoslovakia on the other.

References

(1) SALMAN, E.M. (1969) : Kompleksonoye ispolzovanye radiolokatsionnykh i sputnikovikh nablyudenii pri analize mezo- i makromashtabnykh oblatchnykh sistem, Meteorologia i gidrologia, 2, 44 - 49.
(2) SALMAN, E.M. (1968) : Printsipi postroyeniya sistemy radiolokatsionnykh nablyudenii za oblakami, oblatchnimi sistemami i opasnimi yavleniyami s pomoshtchiyu seti meteorologitcheskikh radiolokatsionnikh stancii, Trudy GGO, 231, 9 - 23.
(3) PODHORSKÝ, D. (1970) : Radar Meteorology in Czechoslovakia, Meteorologické zprávy, vol. XXII, 3 - 4, 84 - 87.
(4) ZILMANN, R. (1972) : Vorstellungen über die Planung eines optimalen Netzes von Wetterradarstationen in Bereich den sozialistischen Länder Europas, Zborník prác HMÚ, vol. 6, 125 - 134.
(5) PODHORSKÝ, D. et all. (1972) : Multilateralnii tsentr obrabotki MRL informatsii v ramkakh jedinoi seti MRL socialistitcheskikh stran ili Regionalnii tsentr radiolokatsionnoi meteorologii, Zborník prác HMÚ, vol. 6, 135 - 162.
(6) PODHORSKÝ, D. (1982) : Report of the Rapporteur on Regional Aspects of Collection Exchange and Processing of Radar Information in Digital Form, VIII - RA VI/Doc. 39, WMO.

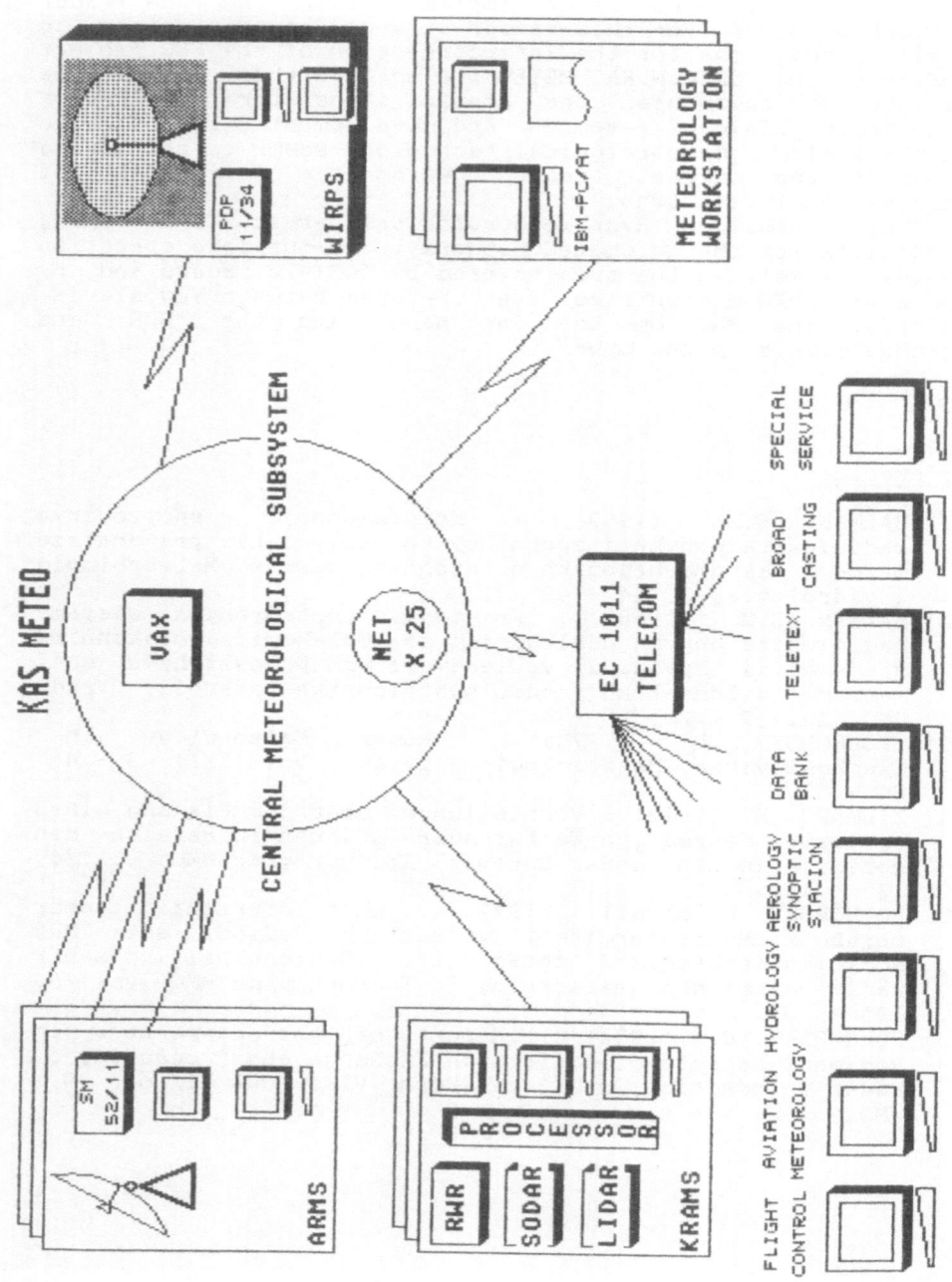

A Nordic Weather Radar Network

by

S. Overgaard, The Danish Meteorological Inst.

Summary

This paper gives a brief description of a technical specifiation for a Nordic Weather Radar Network which is going to be a fact by the end of 1991.

The specification consists of an estimate of the expected data traffic between the Nordic Countries, and the required capacity as a result of the main nodes.

The regional nets connecting the radars to the main nodes are also described.

The network has user accessible application functions in every connected node. These functions can do a variety of tasks such as extracting information of the data volumes collected by the radars, and producing composite images from images from several radars.

The described network is a very flexible one, and can be integrated into the COST 73 network.

The work is partly funded by The Nordic Council, and has been conducted by a reference group with members from each of the Nordic countries, Denmark, Finland, Norway and Sweden (Iceland is also a nordic country, but has for obvious reasons not participated in the work). The author is the chairman of the reference group.

The network will be named NORDRAD.

1. Introduction

In March 1981 a pilot scheme was undertaken as a joint effort by The Meteorological Institutes of the Nordic countries. The name of the project was 'Operational Weather Radar Data Transmission between the Nordic Countries'. The transmitted data ranged from code-telexes to telefaxes of images. This project demonstrated that this kind of data was valuable and could complement the normally distributed meteorological data transmitted via the GTS system.

This project led to a proposal for a nordic weather radar network, divided into three phases of which the first was partly funded by the Nordic Council in 1986 and the second has been founded from the same source this year.

The implementation phase is as follows:

Phase 1: Definition of requirements for the radar hardware, the software, data communication, computers, terminals in a Nordic weather Radar network. All of this was subdivided into two parts.

    1a: User requirements and a coarse technical specification.

    1b: A technical specification, to be used for a call for tender to phase 2.

Phase 2: Which is building the common part of the data network and connect existing radars to it. These task is to purchase and set-up the node computers in the network and the necessary communication and application software.

Phase 3: Establish new radar sites so that all of the Nordic countries are covered by weather radars. Also in this phase, the education of meteorologists in the use of weather radar images is taking place.

The before mentioned proposal also recommended that the implementation of the network be argumented by a research project into the operational use of weather radar.

Today phase 1a and 1b has been completed, and the resulting technical description of the system to implement will be described in the next section.

It should be stressed at this place, that until today, most of the work done have been on the data network, which is a vital part of a weather radar network.

2. The Technical description of a Nordic Weather Radar Network

The proposed system consists of several items:

- The physical and logical structure of the network.

- The netwide functions which can be performed.

- The catalogue of products.

- The administration of the net.

These items will be described in the following.

The physical structure is fairly simple, because there will be four main nodes in the network, one in Oslo, Norway, one in Norrköping, Sweden, one in Helsinki, Finland, and one in Copenhagen, Denmark. Each of the main nodes will be connected to the others, as shown in figure 1. This will reduce the load of the node computers because routing of data only takes place when one of the lines is inoperative.

The data traffic is estimated on the basis of two parameters, the updating time interval required by the users, and the data volume produced by the existing radars in the Nordic countries (see figure x). The airtrafic control has the most intensive requirements on updating time, 5 minutes between volumetric scans, whereas the more earth bounded users can do with 10 to 15 minutes between volumetric scans.

Some of the existing radars are doppler radars, and therefore they produce the double amount of data that radars without doppler do. Roughly speaking, a 240 x 240 km² area covered with resolution 2 x 2 km² in 12 CAPPI-layers produce a data volume of 170 Bytes with an average compression factor of at least 5. A doppler radar will produce the double amount, 340 kbyte.

In order to cope with the most intensive requirements and have a good safety factor for capacity in the network, 64 kbit/sec. lines are required between the main nodes.

Sweden which will have the highest number of radars in the operational network will very likely produce an output of 2.7 Mbytes of data for each volumetric scan. This amount of data can be transmitted in approximately 7 minutes over a 64 Kbit/sec. line. Data will probably be compressed prior to transmission.

The main nodes in each of the Nordic countries also serve as the centre for a national star-shaped network (shown in figure 2). All users and radar sites are connected to the national main node, either directly or through local node. A local node can be a data producing nodes (e.g. have a radar connected) and have users connected or can be a data consumption node with only users connected.

The network is designed in a wide sense, with functions ranging from communications to applications. To accomplish this in the network, a well-structured function layering, following the OSI-model has been constructed. This structure is shown in figure 3.

Between the local nodes and the main node in each country, a 9,6 Kbit/sec. line is able to handle the traffic involved in delivering a volumetric scan, but other traffic will be slowed down. Selecting a 19,2 Kbit/sec. line should be adequate.

Applications use the communication functions on a high level, both program to program messages and file transfer. Both of these types of access are available in DECNET and TCP/IP and are expected to be possible in the OSI-model. It is also expected that DECNET will be modified to follow the OSI-model. Therefore DECNET is specified as the first implemention of the lower levels of the logical structure in the network.

The functions in the network are divided into two categories. Functions which deliver data, and functions which work on data from one or more of the previously mentioned functions. The types of functions are illustrated in figure 4. Some of these functions are performed in the radar computer, and the result is always present in the node computer, others are only done on request from a user placed somewhere in the network

Some common products, produced by the network, are composite images of data from more than one radar. A list of standard composite images are given below:

- Previously defined areas, max. size 1500 x 1500 km Max. resolution 2 x 2 km².

- An image covering all of the Nordic countries. Max. resolution 2 x 2 km².

- The northern part of Europe. Max. resolution 8 x 8 km².

The images use a common map projection, and there is a possibility to include data from other sources, raingauge, satellite images etc. etc.

The network will be open to the inclusion of new application functions.

The amount of data produced by a network this size will be huge. A strategy must be defined in order to handle these data. All data from the actual day should be available from disk, where as older data should be stored on magnetic tape, optical disk, or the coming DAT tapestreamers. Storage takes place locally, and a catalogue of available stored data is available for the users.

All storage of data is done on request by a user who specifies the type of data to store and how far back in time the archive should go.

The proposed network is rather complex, and as such it must have administration capabilities. These are configuration routines even on application level, routines for monitoring the system, routines for statistical purposes, routines for controlling the radars, and routines for maintaining product catalogues.

Some of these routines are purely for administration purposes to keep the network running and are not accessible to users, but only to an operator responsible for the system. Others, however, serve the user in telling him which products are available or for which he can give a special request for a specified period in the future.

The general concept is that some products are always available in the network and on the node where they are produced. These are:

- Reflectivity image.

- Doppler image.

- Dual polarization images.

- Data from other sources.

The user can choose to transfer these images or data from other nodes to his own node and do the data processing here, for instance to produce a composite image or he can request the data processing be done on the nodes where they reside, and then transfer the result which probably contains less data than the original data set from many nodes. A user can have an outstanding request for such a calculation to be done until he terminates it.

The functions define the types of data processing that can be done on the weather radar images.

There is not specified a presentation system in the network. The main reason for this is that a variety of presentation systems already are in use in the Nordic countries, and because the market for graphic workstations is developing so quickly. It is possible, however, that X/WINDOWS will become a worldwide standard for graphical workstations that the presentation system can be standardized to X/WINDOWS.

The network will, however, specify a common geographical grid, covering all of the Nordic countries, see figure

5. This grid or map is based on a polar stereographic projection and will be adjusted to the proposed COST 73 map. All geographical references should be done in this master grid.
The map in figure 5 shows the weather radars in the Nordic countries that will be operational at the beginning of 1991. In Sweden, there will be 8 radars, in Norway 1, in Finland 4 and in Denmark 2. The fully drown circles (with a radius of 150 km) shows the existing radars today and radars which have been decided upon and will be ready at the beginning of 1991. The broken circles shows radars which are being planned to be constructed in the nineties.

The map shows a good coverage of the southern part of the Nordic countries and will be a valuable contribution to the COST 73 image of northern Europe.

3. The future

Some of the specifications are rather ambitious and will be very expensive to fulfil, especially the data link specified.

In the reference group, there has been a through discussion of the need for transmission of a full volumetric scan from one end of the network to the other. The feeling is that only processed data and single layers is likely to be transmitted.

Therefore, and because of cost from the start the main network will be a star shaped one with its centre in Norrköping, Sweden. All lines will be rated to 64 kbit/-sec. This will of course place a heavy load on the central node in Sweden but will reduce the working expenses considerable. Later, when it is required, the main network can easily be expanded, to the one shown in figure 1.

The computers that are used to control the weather radars will also be used as local nodes in the implementing phase to save cost. As a consequence of this, more powerful computers are needed, but it is estimated that for instance a MICROVAX II has the necessary computing capability.

The main node is specified to be a DEC station 3100 in all countries for the data processing and a MICROVAX II for the network connection in Norway, Finland and Denmark, and a VAX server 3600 in Sweden.

The DEC station 3100 is specified to have:

- 24 Mbyte memory.

- one 105 Mbyte disc (in Sweden 330 Mbyte).

- one tapestreamer.

- Ethernet controller.

- a DEC router 2000 connected (in Sweden two).

The MICROVAX II is specified to have:

- 5 Mbyte memory.

- two 159 Mbyte discs.

- one tapestreamer.

- A Ethernet interface.

- V.24 interface.

The VAX server 3600 in Sweden is specified to have:

- 32 Mbyte memory.

- one 622 Mbyte disc.

- one tapestreamer.

- Ethernet interface.

Additionally, there can be interface boards for local display systems.

As mentioned previous, phase 2 is funded, so the network is planed to be operationatial at the end of 1991. Also is it expected that the data from this network constitute the north eastern part of an European Weather Radar Network.

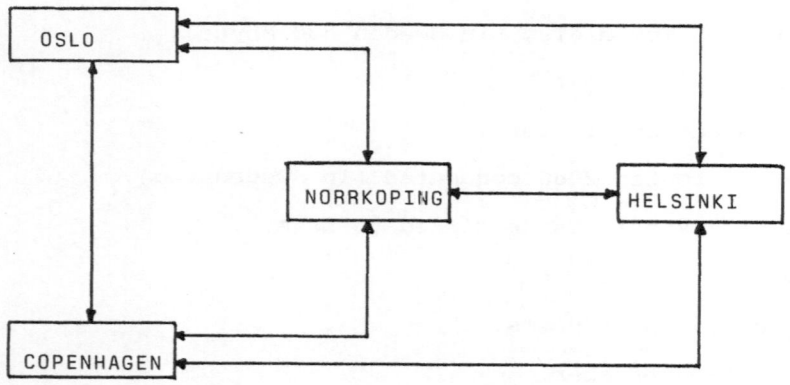

Figure 1 : The main nodes in the Nordic Weather Radar Network, placed at the National Meteorological Institutes.

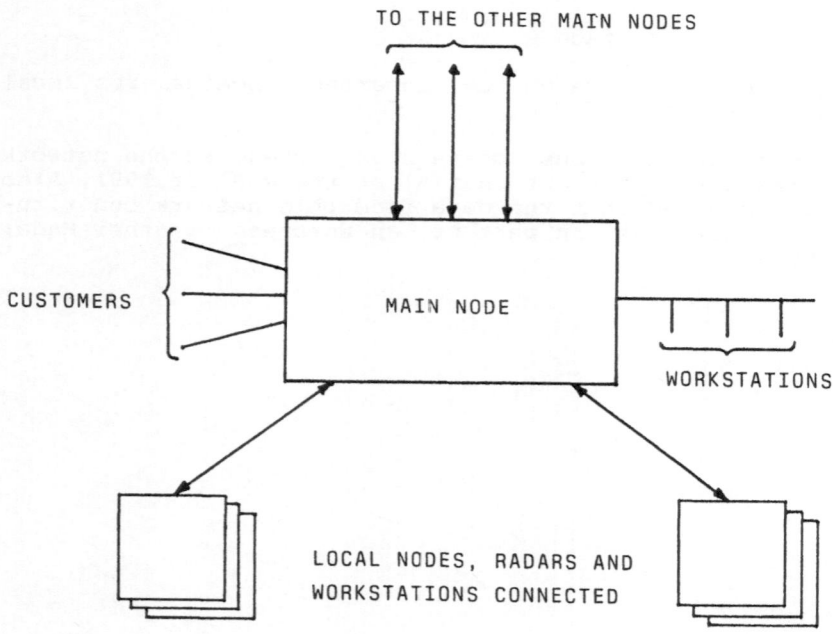

Figure 2: A main nodes with its attached regional works.

DATA SOURCES	STORAGE	PROCESSING	USERS	A
RADARS PROCESSED IM. OTHER SOURCES	ARCHIVE	COMPOSITE EXTRACTION of data CONVERSION between for- mats	DISPLAY INTERFACE	P P L I C A T I O N

PRESENTATION:
ORDERING, ADMINISTRATION AND SUPERVISION.
CODING AND COMPRESSING.

SESSION: STANDARDIZATION OF DIALOG

TRANSPORT: SECURE TRANSMISSION OF DATA, PROTOCOL

NETWORK: SELECTION OF ROUTE TO DESTINATION

LINK: SECURE TRANSMISSION OF BYTES.

PHYSICAL LINK: PTN LINES

Figure 3: The logical structure of the network.

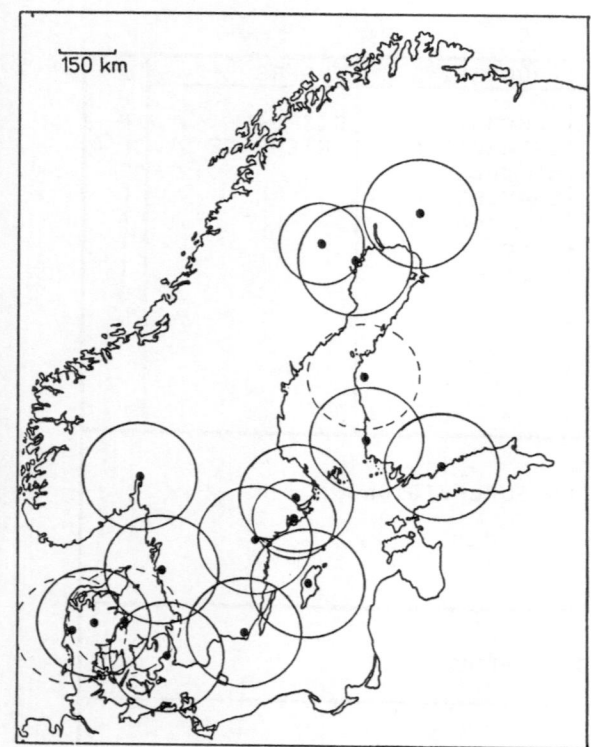

Figure 4 :
The functions schematically.
They are divided into data
producing functions and data
processing functions.

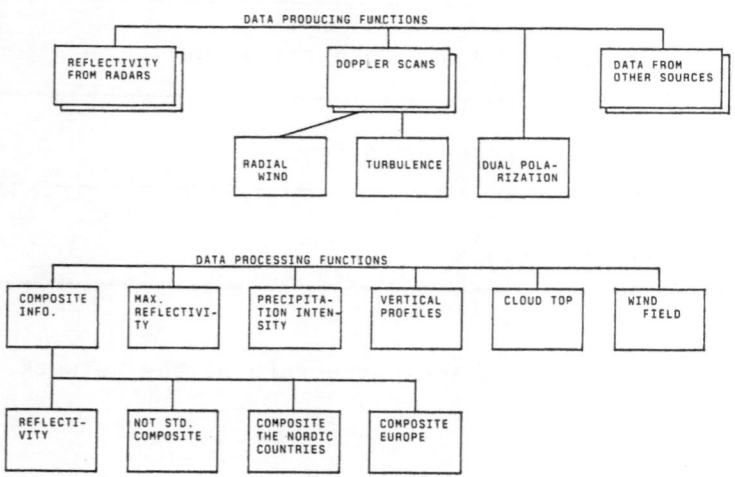

Figure 5 : Polar stereographic map showing the weather radars being operational at the end of 1991 (fully drown circles) and radars which
are planned but not yet funded. All circles are drawn with a
radius corresponding to 150 km, except for two Swedish radars run
by the Swedish Airforce, they are drown with a radius corresponding
to 100 km.

THE SÃO PAULO WEATHER RADAR NETWORK PROGRAM

R.V.CALHEIROS

Meteorological Research Institute, University of the State of São Paulo
and
Institute for Space Research, Secretary of Science and Technology

Summary

A radar network to cover the State of São Paulo, Brazil, was in consideration since the pioneer efforts in radar meteorology in the country were developed in the middle 70's with the operation of a radar at Bauru, in the central area of the State, in the context of the so denominated RADASP (Radar in São Paulo) Project. The network was planned for the second phase of the Project - RADASP II - which began in 1982. The configuration of the network anticipates the installation of three radars to cover the State, two of them are already in operation in the central (Bauru) and eastern (Ponte Nova) regions of the State while the western radar siting and acquisition is in the planning phase. Data from the radars will converge to a central computing facility at Bauru where composite rainfall maps, integration with satellite imagery and nowcasting will be executed. Research activities related to the network are being developed focusing observations at the overlapping areas and nowcasting techniques for the whole coverage region. The network is supposed to play a major role in the context of a new project for the implementation of a meteorological system recently proposed for thê State. Estimates about the potential benefits to be derived from the network from such important sectors like agriculture, water resources, building industry, water resources, and civil defense indicate reaping of high dividends. The State network program and the prospects for a nation-wide system based on that network are considered in this paper.

1. INTRODUCTION

In the early 1970s an effort was developed in the State of São Paulo to implement activities in meteorology which resulted in the instalation of a C-Band weather radar at Bauru (23,36 S; 49,03 W), a central location in the State through a project denominated RADASP (Radar in São Paulo), (1).

The radar - a low cost system with only analog processing - was aimed at both research and operational activities and permitted the observation of rainfall systems developing and sweeping the central region of the State (2).

This helped to emphasize the importance of an operational network to survey the State and adjacent areas, an idea which had been considered when the original project was conceived.

In 1978, a project was developed for the digitalization of the existing radar system, what occurred in 1980 (3).

In 1982, as a consequence of the succesfull results until then obtained, a second phase of the project, designated RADASP II (4) was planned in which the two main points were the implementation of a three radar network in the State and of a central computing facility at the Baùru radar site. As this second phase progressed, a VAX 11/780 was installed at Bauru in 1983 and the eastern radar, a S-Band system, located at the Ponte Nova Dam of the State Department of Water and Electrical Energy (23,34 S; 45,97 W) is just starting operations. Plans call for the replacement in 1989 of the C-Band Bauru radar by a S-Band equipment; the C-Band will be configured as a mobile system and stationed temporarely in a western position until a definitive S-Band radar replaces it, at which time the mobile C-Band may be transferred for a better coverage of the southern tip of the State and nearby regions for an early warning of precipitation approaching from that area.

As of now consideration is being given for a fourth S-Band radar in a fixed configuration for that southern position. Figure 1 depicts the radar positions and coverages. Very recently, a project to provide the State of São Paulo with a meteorological system has been designed of which the network is a major component; that system will comprise also a wind profiler, a 3 D acoustic sounder, a captive ballon and radiometers sited at the Bauru radar and a network of automatic surface stations covering the whole State. The system, envisaged for the mesoscale, will integrate existing radiosonde and surface stations, will make use of all meteorological information supplied by the federal government and will work in close cooperation with it.

New techniques such as Doppler and dual polarization are being evaluated for a possible future addition to the network radars.

Along the years of operation of the C-Band Bauru radar, applications were made of the observations in real time, mainly to agriculture and hydroelectric generating power but also to the building industry, transportation, civil defense and others which indicated the importance of having the State covered by a network. In the project of the before mentioned state meteorological system an estimate of the potential economic benefits to be derived from the use of forecasts was performed and is presented in this paper focusing on the radar network. The figures obtained in the estimate - above US$100 million only for agriculture, although for the whole system including the radar network - are considered highly favorable.

Research regarding specifically network aspects is beginning now and concentrates on the composite rainfall maps and the nowcasting techniques for the whole coverage area.

Also in the context of the state meteorological system attention is being given, in the research activities, in what concerns the calibration of satellite imagery with the network data. Another point of interest being considered in the research effort refers to the use of the network data in conceptual models.

This network program is being used as a basis for similar projects to be implemented in other regions of the country, in the context of the preliminarly designated National Program for Nowcasting and Very Short Range Forecasting (PRONAPRI)

2. NETWORK CONFIGURATION

The network was configured so that the main objective of surveillance of rainfall systems over the area of the State of São Paulo for civil defense purposes, as a support for economical activities and for research, could be met. The precipitation in the area was characterized through a verification of the meteorological systems giving rainfall. This was done mainly in the central area of the State were a considerable amount of observations from the Bauru radar has been accumulated since 1974. By 1976, after approximately two years of operations of that radar its capability to detect significant features of instability lines sweeping the State had been identified (2).

In one case analysed in that early work, the line had an average speed of about 85 km/h, going approximately from W to E, with ground measured peak rainfall rates of about 150 mm/h and wind gusts of 110 km/h. These values were later verified to be in the upper-limit range for lines observed in the area.

In 1980 Calheiros and Antonio (5) characterized better the prevailing directions of displacement of those lines as can be seen in Figure 2 adapted from that paper, showing that they come mainly from the south west (A, in the figure) and secondarely from the west (B, in the figure).

It was also shown that slow lines have an average speed of about 45 km/h while for the fast ones the number is approximately 65 km/h. Data from July 1974 to September 1979 were used in the analysis when a certain degree of blocking of the beam to the southeast sector existed. However, verifications made after the radar was changed to a new site minimizing that effect have confirmed the results. Figure 3 reproduces a typical line.

That verification indicated that about 50% of the lines comes from the SW and 30% from the W. Using a limited amount of data (from January, 1979 to March 1980) from a Brazilian Air force weather radar situated at São Roque (22,56 S; 47,09 W, indicated by SR in Figure 1) and surveying the eastern part of the State, Gandu (6), observed a general tendency of the echoes to move from W to E.,i.e., to the (30 - 150) degree azimuth sector, at average speeds of about 30 km/h and maximum speeds of 90 km/h.

More recently Andrade Filho (7), using digital data from the Bauru radar studied convective processes not associated with frontal systems, for two rainy seasons (October 1981 to March 1982, and October 1984 to March 1985) observing average speeds of echo motions in the 30 km/h to 50 km/h range and maxima values of about 100 km/h.

Other related studies have been performed in which similar results are found (8), (9), (10), (11).

Regarding the fronts affecting the State and their motions, the most common are cold fronts, both those well characterized and propagating from SW to NE and those of limited extension and almost undefined over land, with a S to N motion.

Also present in the area are: stationary fronts where motions can take place from NW to SE, from SW to NE, from W to E and, seldom, from N to S; warm fronts which most common propagation is from N to S and from NW to SE; and occluded fronts, in general affecting predominantly the region up to about 200 km inland, propagating from SW to NE in that region.

The vertical structure of the precipitating systems has been studied since the early stages of operation of the Bauru radar when observations of tall storms in the 320-400 km brought into consideration the question of the maximum range to be used in a network (12). Those pioneer studies which included the internal structure of storms (13) were later verified with digital data with observation of 20 dBz tops around 15 km and maximum reflectivities at about 6 km (14), (15), (16), (17).

The studies from which were derived the range corrected reflectivity rainfall rate relationships for use with the Bauru radar indicated a height of about 4 km for the CAPPIs to be used, when clutter and melting layer effects were taken into account (18).

The previous experience with the Bauru radar observations, the fact that at 240 km the beam is approximately at 3.4 km AGL (about 4 km AMSL for the Bauru radar), the required coverage of the State (an area of approximately 248000 km^2) led to the choice of a 240 km range for the radars in the configuration of the network, as shown in Figure 1.

The final position of the western radar is not yet defined and will be in or near the area shown in Figure 1; the dashed circle in that figure indicates the coverage of a fourth radar which may be added later to the network.

In a first phase, the Bauru C-Band radar will be replaced by a S-Band and, transformed into a mobile configuration, will operate temporarely at the western position.

After the planned S-Band equipment is installed in the west, the mobile C-Band will possibly be relocated to the southern position until a decision is reached on that fourth component in the network.

An early warning of the arrival of many severe weather systems at the densely populated and heavily industrialized eastern region of the State can be given by the radar "belt sector" covering the northwest - west - southwestern area adjacent to the Ponte Nova radar coverage.

The Bauru radar system is an Enterprise Electronics Corporation WR-100-5 with the following characteristics: peak power: 250 kW, pulse duration: 2.8 microsec, PRF: 259 Hz, frequency: 5,60 - 5,65 GHz, antenna beam width: (2 degree). Coupled to it there is a signal processing subsystem developed by the Alberta Research Council, Canada, with a radar-computer interface manufactured by Athabasca Research and with a software which, among other basic functions, performs the data storage and post-facto analysis. Radar computer is a PDP 11/34 which transfers the volume scan to a VAX 11/780 where products like 4,1 km AMSL CAPPIs are generated in real time, with basic cells of 4 km x 4 km, to a 157,5 km range.

The Ponte Nova radar was built by McGill University; its characteristics are: antenna beam width: 2,1 degree, frequency: 2,70 - 2,90 GHz, peak power 650 kW, pulse duration: 2 microsec, PRF: 250 Hz.

The digital subsystem performs, among other functions, data storage and CAPPI construction.

Figure 4 (a) and (b) shows the struture of both systems.

The Bauru C-Band system will be replaced by a S-Band RMT-0100 built by the Brazilian company Tecnasa Eletrônica Profissional SA. Its characteristics are: peak power: 450 kW, pulse duration: 2,0 microsec antenna beamwidth: 2,0 degree, operating frequency: 2,70 to 2,90 GHz, PRF: 250 Hz. Software products will be similar to those of the existing radars. In a first phase, the integration of the two radars now composing the network will use the products presently generated which are CAPPIs at approximatety 4 km AMSL with cells of 4 km x 4 km.

The approximate area composition of the products is shownin Figure 5.

3. RESEARCH ASPECTS AND ECONOMIC VALUES

One of the main aspects of the research effort directly related to the network concerns the optimization of the observations in the coverage area. The delimitation of the boundaries between radars will be studied in a manner similar to that considered by Browning and Collier (19).

Another research activity refers to nowcasting techniques to be used in the network coverage area. The SHARP technique developed at McGill (20) is being evaluated for the Bauru radar area. The preliminary tests evidentiated the need of many adaptations (21). This procedure will be extendend to the network as a whole.

The recent proposal of a meteorological system for the State of São Paulo (22) included estimates of the economic value of meteorological information to important productive sectors in the State. Although no distinction has been made among the different types of information which can be made available, experience with the use of the Bauru radar data in two specific areas, i.e., agriculture and water resources, and the potential benefits estimates presented in studies such as the one by Clift (23) indicate that the contribution of the radar network to the economic value is substantial.

For the building industry a 1% figure of its economic value was assumed for the meteorological information for the sector, which resulted in US$219 million yearly.

In the transformation industry the percentage used was 0,2% arriving to a yearly value of US$38.9 million.

The estimate for the hidroelectric generation of energy indicates that the benefits to be derived from an optimum use of weather forecasting could be equivalent to 750 MWh, each hour, with a corresponding value of US$49.3 million.

In the agriculture and cattle raising sectors the estimate considering only the most significant crops (24) resulted in a value of approximately US$133 million. Thus a total value of about half billion dollars is obtained not including all sectors which can benefit from the information.

In all those estimates intensive use was made of the experience available in other parts of the world as, for instance, reported by Maunder (25).

4. RADAR-SATELLITE INTEGRATION AND NEW PROJECTS

Use of the network to calibrate satellite imagery is planned, through the implementation at the Bauru site of a radar-satellite integration similar to that developed at McGill University, denominated RAINSAT.

With this, rainfall probability maps will be generated for an area of about 3 million km^2, centered approximately in the network, as shown in Figure 6.

The network is a major component of the before mentioned meteorological system proposed for the State of São Paulo. This system will include in its data collection base, in addition to the network, a wind profiler, a 3 D acoustic sounder, a captive ballon and radiometers, all installed at the Bauru radar site, and a network of 25 automatic surface stations installed in strategic positions throughout the State. This new hardware will complement the existing surface and upper air station networks.

5. FINAL COMMENTS

The São Paulo State weather radar network is the backbone of the RADASP Project, wich is a program of the Foundation of Support to the Research in the State of São Paulo (FAPESP), proposed by the Meteorological Research Institute (IPMet) of the University of the State of São Paulo (UNESP) and based there, and involving mainly the Department of Water and Electrical Energy (DAEE) - of the State, and the Institute of Astronomy and Geophysics (IAG), of the University of São Paulo (USP). Among others, the Institute for Space Research (INPE) of the Secretary of Science and Technology - Federal Government - has been collaborating with the Project.

The experience gained in the Project since the Bauru radar began operating in 1974 is being used to support the development of new radar centers which will hopefully evolve into networks in the future. This is the case with the State of Rio Grande do Sul where a radar facility is just being completed at the city of Pelotas, (31,73 S; 52,33 W), in a program based at the Federal University of Pelotas.

Basic hardware is an EEC - DWSR-88 radar with a Microvax II, and a computing facility with a Cyber 930 system. At the center of the State of Santa Catarina, in the city of Fraiburgo (26,98 S; 50,81 W) a soviet MRL-5 system is being installed for operations which include general meteorological surveillance for the center of the State. For that sake, a video digitizer and a processing system will be added to the radar.

As of now, there are plans for a radar in the northwestern region of the state of Parana and consideration is being given for a four radar network in the State of Minas Gerais. All this, together with the São Paulo network may constitute a substantial part of a future network for the center - southwest. In order to organize and coordinate efforts on a national level a project tentatively denominated National Program of Nowcasting and Very Short Range Forecasting (PRONAPRI) is being designed.

Regarding the São Paulo Network itself consideration is presently being given to the addition of capabilities such as Doppler and dual polarization to the Bauru radar, for evaluation purposes in a first phase, and studies on conceptual models are being included on the research effort.

6. ACKNOWLEDGEMENTS

The Federal organizations FINEP (Financing Agency for Studies and Projects) and CNPq (National Research Council) contribute to the RADASP Project. Thanks are due to Carlos Alberto de Agostinho Antonio for elaborating the figures and to Marlene Sueli Moya Munhoz and Regina Célia Nogueira Lima for typing.

REFERENCES

(1) CALHEIROS, R.V. (1973). Um radar de objetivos múltiplos para pesquisa Meteorológica no Estado de São Paulo - Projeto RADASP. Presented to the Foundation of Support to the Research in the State of São Paulo.

(2) CALHEIROS, R.V. (1976). Winter Instability Lines as Detected by the Bauru C-Band Radar. Proceedings, 17th Conf. on Radar Meteorology (Seattle), AMS, Boston, 390-392.

(3) IPMet (1978). Projeto de Pesquisas Meteorológicas com enfoque em Radar de Tempo. Presented to the Financing Agency for Studies and Projects.

(4) CALHEIROS, R.V. (1982). Meteorologia com Radar em São Paulo - Projeto RADASP II. Presented to the Foundation of Support to the Research in the State of São Paulo.

(5) CALHEIROS, R.V. and Antonio M. de A. (1980). The Bauru radar as an efficient warner for severe instability lines. Proceedings, 19th

Conference on Radar Meteorology (Miami Beach), AMS, Boston, 727-729.

(6) Gandu, A.W. (1984). Análise Estatística de Ecos de Radar Associados a Sistemas de Precipitaçao na Região Leste do Estado de São Paulo.

(7) ANDRADE FILHO, A.G. (1987). Determinação da Estrutura da Intensidade de Precipitação em Células Convectivas. MSc Thesis, EESC/USP, São Carlos, SP.

(8) CALHEIROS, R.V. (1978). Comparison on Prefrontal Squall Lines as Detected by Radar in Brazil and South Africa. Proceedings, III Congreso de Meteorologia (Buenos Aires), Centro Argentino de Meteorologos, Buenos Aires.

(9) LIMA, M.A. and SILVA DIAS, M.A.F. (1980). Deteção de uma linha de instabilidade prefrontal através de análise dos campos meteorológicos de superfície e do radar meteorológico de Bauru. Proceedings, 1st Congresso Brasileiro de Meteorologia (Campina Grande), Soc. Brasileira de Meteorologia, RJ.

(10) ANTONIO, M. de A. and LIMA, M.A. (1981). Acompanhamento da evolução de ecos de radar. Proceedings, 33a. Reunião Anual da SBPC (Salvador), 09 pages.

(11) SILVA DIAS, M.A.F. and LIMA, M.A. (1982). Deslocamento de Linhas de Instabilidade e sua Relação com Ventos em Altitude. Proceedings, 2nd Congresso Brasileiro de Meteorologia (Pelotas), UFPel/CNPq/CAPES/FINEP, Pelotas, vol.2, 0411-0431.

(12) CALHEIROS, R.V. (1975). The Bauru C-Band Radar Detection of Storms in the 320-400 km Range, During the August-October 1974 Period. Proceedings, 16th Conference on Radar Meteorology (Houston), AMS, Boston, 341-344.

(13) CALHEIROS, R.V. and ANTONIO, M. de A. (1979). Evolução de uma tempestade de verão conforme observada pelo radar de Bauru. Proceedings, 31st Reunião Anual da SBPC, (Fortaleza), 10 pages.

(14) CALHEIROS, R.V. and ANTONIO, M. de A. (1982). Determinação de Topos de Células de Precipitação Durante o Experimento I do Projeto RADASP II. Proceedings, 2nd Congresso Brasileiro de Meteorologia (Pelotas), vol.2, 052-058.

(15) ANTONIO, M. de A. (1986). Ecos de radar associados a eventos de chuvas intensas - Parte I. Proceedings, IV Congresso Brasileiro de Meteorologia e I Congresso Interamericano de Meteorologia (Brasília), SBMet, vol.2, 301-306.

(16) ANTONIO, M de A. (1987). Ecos de Radar Associados a Eventos de Chuvas Intensas - Parte II. Proceedings, II Congresso Interamericano de Meteorologia e V CONGREMET (Buenos Aires), CAM, 5 pages.

(17) GROSH, R.C., MASSAMBANI, O. and BRIGUENTI NETO, J. (1985). Brazilian Convective Precipitation: Evolution Study With the Bauru Radar, 6th Conf. on Hidrometeorology (Indianapolis), AMS, Boston, 5 pages.

(18) CALHEIROS, R.V. and ZAWADZKI, I. (1987). Reflectivity-Rain Rate Relationships for Radar Hydrology in Brazil. J. of Climate and Applied Meteorology, AMS (Boston), Vol.26, 118-132.

(19) BROWNING, K.A. and COLLIER, C.G. (1982). An Integrated Radar-Satellite Nowcasting System in the UK. Nowcasting, chapter 1.5, Academic Press Inc. (London), 47-61.

(20) BELLON, A. and AUSTIN, G.L. (1978). The evaluation of two years of real time operation of a short-term precipitation forecasting procedure (SHARP). J. of Applied Meteorology, vol.17, 1778-1787.

(21) LIMA, M.A.; CALHEIROS, R.V. and SACOMAN, M.A.R. (1987). Procedimento de Previsão de Precipitação a Muito Curto Prazo (SHARP): Um teste Preliminar na Area Central do Estado de São Paulo, Brazil.

Proceedings, V CONGREMET e II Congresso Interamericano de Meteorologia (Buenos Aires), CAM, 4 pages.

(22) IPMet (1989). Sistema Paulista de Meteorologia. Proposed to the Government of the State of São Paulo.

(23) CLIFT, G.A. (1985). Use of radar in meteorology. Technical Note n.181, WMO n.625.

(24) CALHEIROS, R.V. and ANTONIO, M. de A. (1989). Nowcasting for Agriculture with the São Paulo Radar Network. Elsewhere in these proceedings.

(25) MAUNDER, W.J. (1987). The uncertainly business. Methuem & Co. Ltd, New York.

(26) BELLON, A., LOVEJOY, S. and AUSTIN, G.I. (1980). Combining satellite and radar data for the short-range forecasting of precipitation. Mon. Weather Rev., vol.108, 1554-1566.

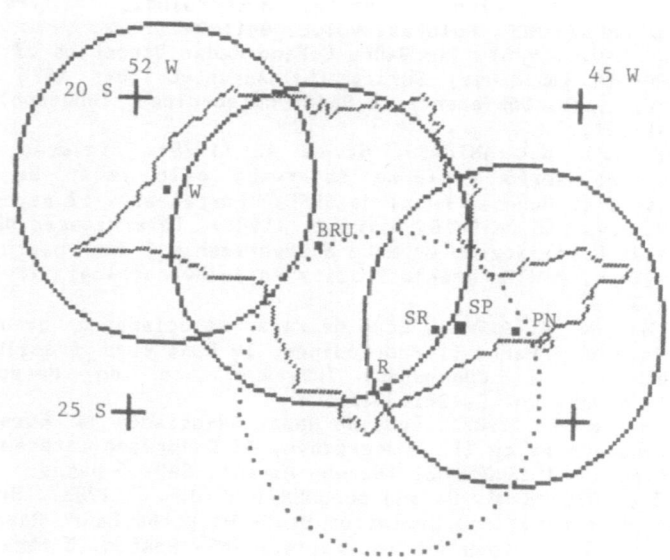

Fig. 1 - The São Paulo State weather radar network. Range of the radars is 240 Km. Dashed coverage: see text. W = West, Bru = Bauru, R = Registro, SR = São Roque, SP = São Paulo, PN = Ponte Nova.

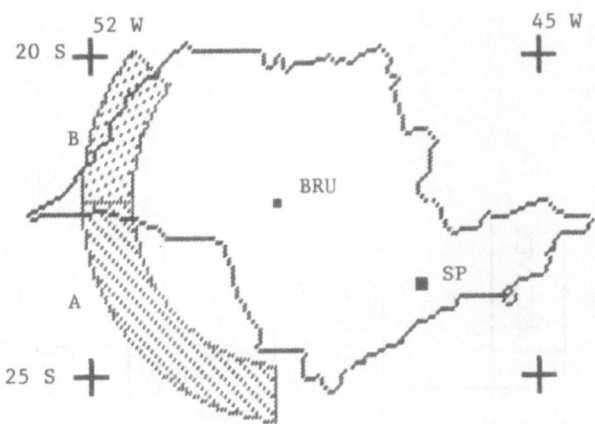

Fig. 2 - Prevailing Aproach sectors for instability lines sweeping the area of the state of São Paulo. Frequency is higher for A. Sites are same as in figure 1.

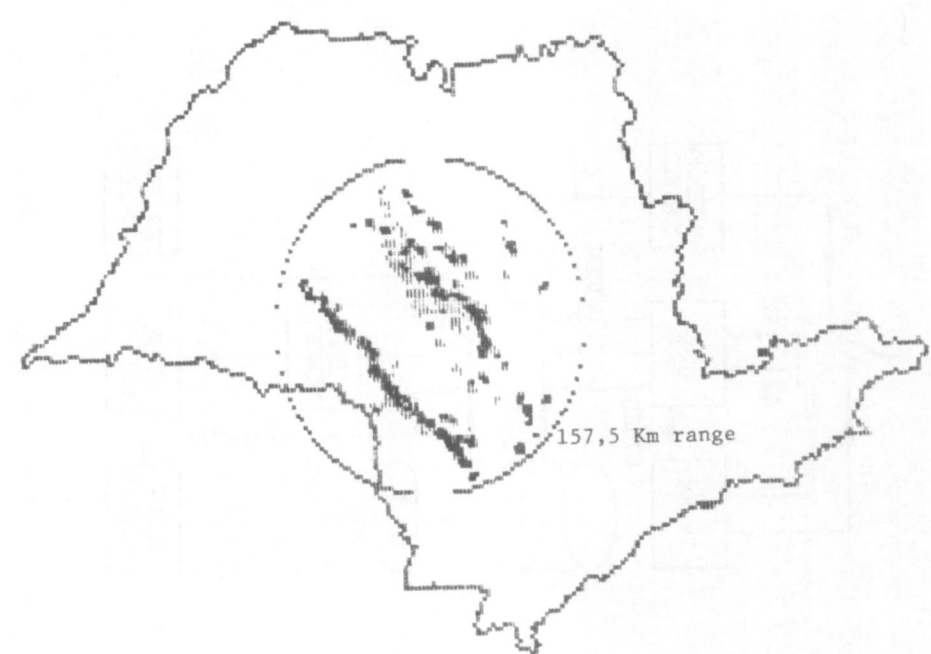

Fig. 3 - Typical instability lines over the central area of the State of São Paulo. Situation at 13:45 LT of May, 8, 1987. Range Show 157,5 Km, cell: 4 Km × 4 Km.

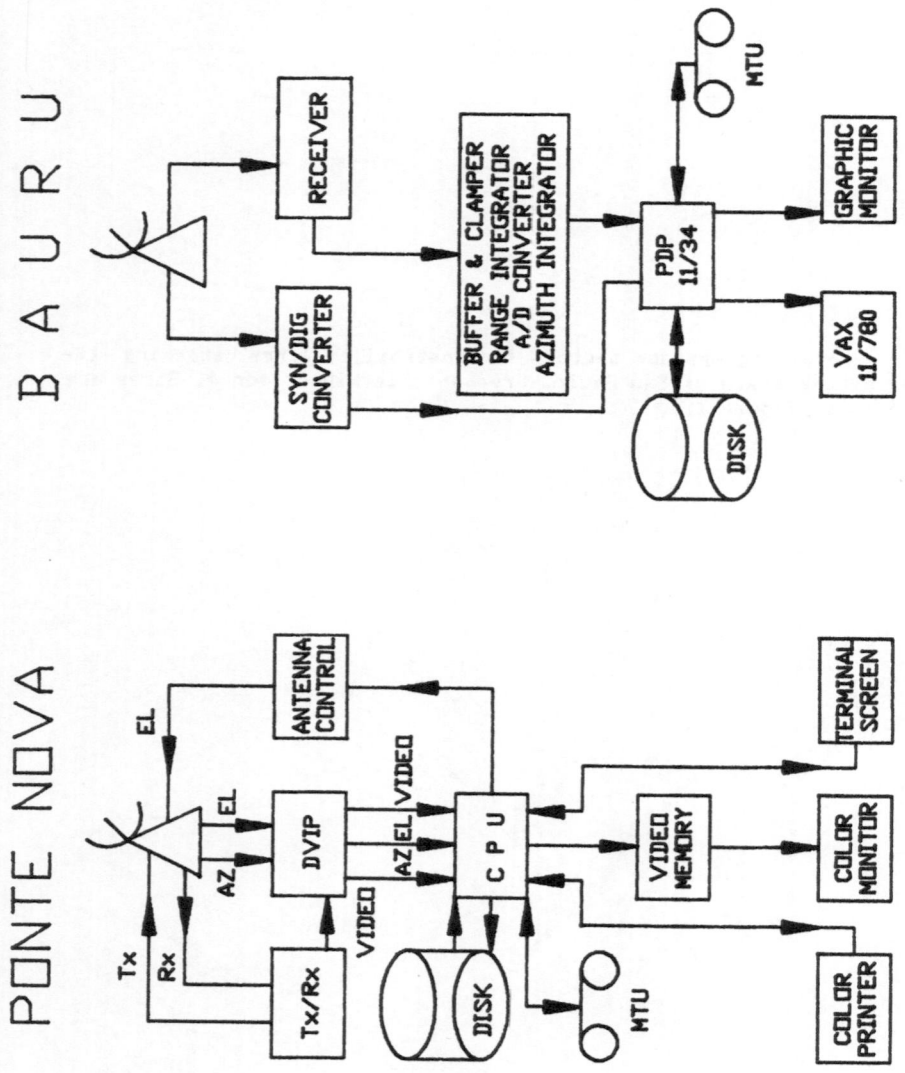

Fig. 4 – Structure of the Ponte Nova (a) and Bauru (b) radar systems.

Fig. 5 - Area Delimitation for the first phase of the composite rainfall map derived from the Bauru and Ponte Nova radars data. Coverage: 157,5 Km radius circle for the Bauru radar, 360 Km side square for the Ponte Nova radar.

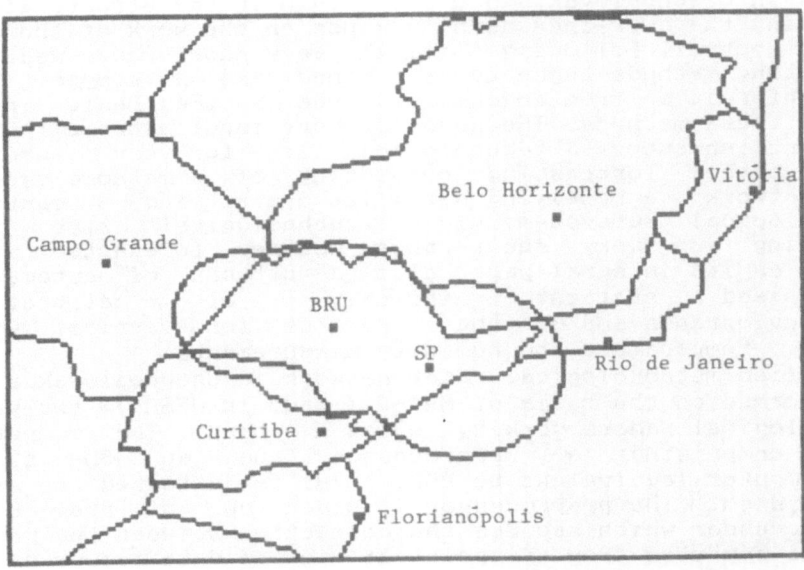

Fig. 6 - Area of rainfall probability maps to be derived from satellite imagery calibrated by the São Paulo network. Aproximate surface: 3 million Km^2. State Capitals are shown.

UNIFIED METEOROLOGICAL RADAR NETWORK OF CSSR
AND CENTRAL METEOROLOGICAL SUBSYSTEM

L. LIETAVA, J. NEMEC, D. PODHORSKY
Slovak Hydrometeorological Institute

Summary

The development of both the Automated Radar Meteorological System (ARMS) and the Central Meteorological Subsystem (CMS) has been paved the very for the establishment of a unified automated radar network in Czechoslovakia which is due to start operating in 1990. The present paper gives a brief description with the radar network, data transmission as well as the transmission to users.

The first idea of developing a meteorological radar network in Czechoslovakia has its origin in the efforts aimed at meeting the ever-increasing demands on the work of the air traffic control. Following this, the very short-range weather forecasting methods began to be defined and developed. The radar information from an observed area is the basic input data of these methods. The need for more input sources led to the establishment of a technological line for very short-range weather forecasting consisting of a meteorological radar network , a receiving satellite system and a central meteorological subsystem. This technological line for nowcasting and very short-range weather forecasts must include as its integral part also a network of automated ground-based stations, inclusive of calibration radiopluviographs and of the apparatus for vertical wind profile , temperature and humidity measurements.

Unified meteorological radar network in Czechoslovakia is being formed on the basis of MRL-5 (made in USSR) two-wave meteorological radars working in the X – band. The computer system comprising a preprocessor and an SM 52/11 mini-computer (equivalent to PDP 11/70) is attached to the radar system. The preprocessor is based on an Intel 8085 mikroprocessor which secures the connection between the radar and mini-computer from the point of view of data measured on both channels in parallel as well as of the whole radar operation control. This system, as a whole, is named ARMS (Automated Radar Meteorological System). It furnishes the possibility of measuring radar reflectivity parallelly on both channels within the radar range. On the basis of primary

data obtained from both channels, the system evaluates the information for the individual discrete squares of the radar grid (8x8, 4x4 and 2x2 km). Through the assessment of static and dynamic features the system provides the following pattern recognition: cloud types and phenomena associated with them, tendency of their development, direction and velocity of radar echo propagation as well as the determination of precipitation intensity and of precipitation total per time unit selected. The software enables to eliminate the ground-based targets and to carry out corrections to the radar beam according to the real refraction of the atmosphere. This system is the basic unit of the radar network. The layout of the radar network has been designed with the following points in mind :
- 100 % coverage of the Czechoslovak territory
- optimum location with regard to the main international airports in Czechoslovakia, in dependence on the relief, i.e. from the view-point of securing the optimum radar horizon;
- meteorological aspect, i.e. in order to cover by radar measurement (at the range of up to' 150 km) those physico-geographical areas which are associated with the occurence of dangerous weather phenomena;
- cost-effective utilization of the area in question, i.e. bearing in mind the intensity of agricultural production and for the impact on ecology of the industrial production in the area;
- mutual overlapping of the individual radars, with optimum range (150 km) being secured for the pattern recognition and nowcasting of dangerous weather phenomena .

The " Unified Radar Network " project is being developed by stages. At present two radar sites are working on the continuous operation mode; they are in Bratislava – Malý Javorník and Praha – Libuš. The radar site at Kojšovská hoľa (Eastern Slovakia) has been engaged in trial operation. During the year 1989 automated operation will be introduced at Malý Javorník (May) and Kojšovská hoľa (September). ARMS will be installed in Praha – Libuš after the change of the radar type there. In 1993 an automated meteorological radar site at Spičák (Western Bohemia) will be set up. Establishment of radar sites at Skalka (Southern Moravia) and Křížna (Central Slovakia) is under preparation.

An important function of the system, besides the radar operation itself, is the distribution of the data measured to the centre and to the nearest airport. Within the framework of the East-European project entitled KAS METEO (Comprehensive Automated System for Meteorological Servicing of Aviation) the X.25HDCL communication protocole has been adopted which is secured for both the ARMS and the Centre by means of a NET module. Transmission velocities are set with regards to remote radar sites and conventional telecommunication lines at 2400 bit/second (minimum). Communication modules serve for perfect transmission from the radar sites into the centre and for the mutual communication between the radar sites if necessary. Each radar site is directly connected both to the centre and to the airport within its range.

The Meteorological Satellite Reception System

Weather Image Receiving and Processing System (WIRPS) is installed at Malý Javorník and it has been in continuous operation since 1984. Digital meteorological satellite data from NOAA and METEOSAT 4 are received and processed there. The system as a whole is controlled by a PDP 11/34 computer. The Central Meteorological Subsystem Provides :
- gathering of radiolocation information from separate sites of the radar network;
- input of the satellite information processed ;
- combination of radar, satellite information and prognosis computation;
- link-up to the EC 1011 telecommunication computer;
- distribution of the processed radar, satellite and prognostic information.

The technological basis of this system consists of an SM 52/12 computer (equivalent to VAX 11/780) and an SM 52/11 distribution — communication minicomputer (equivalent to PDP 11/70). The system also includes units for rapid visualization and surveying of the information processed (SM 54/30, BGT and IBM PC), communication modules and modems.

With regard to the need for data processing and their real-time composition, the project has solved the high-speed interconnection between the computer subsystems by means of DMA modules with the minimum speed of 500 kByte/second and by using dual-port-type disk units.

The information distribution from the technological line takes place at several levels :
- through a radio-relay line (video signal) the information is directly distributed to the various work-stations : airport — the aviation meteorologist, forecast centre — the synoptic specialist, and the Czechoslovak television.

All of these work stations are separated both from the central meteorological subsystem and from each other by the distance of about 10 km. In spite of that, they are provided continuously and synchronously with radar or satellite information :
- by means of hardcopy and facsimile equipments this information is transmitted directly to users through telephone lines, or by means of an OLT 22 long-wave facsimile transmitter (power output 100 kW)
- by means of digital transmission channels to microprocessor image equipment of a BGT type or to screens of personal computers

– by means of a " digital facsimile " which utilizes the analogue transmission channel of the OLT 22 facsimile transmitter for the transmission of the digital information, with a possibility of access to an IBM XT/AT type of personal computer.

References

(1) PODHORSKÝ, D. (1985) : Weather Phenomenological Studies and their Aplication in Very Short—Range Weather Forecast, Proceedings of the Seminar "Remote Sensing Applications in Hydrology and Water Resources", UNESCO, WMO, FAO, Bratislava – Kočovce, 133 – 145.
(2) NEMEC, J., PODHORSKÝ, D. (1987) : Automatizirovannaya radiolokatsionnaya meteorologichgeskaya sistema i yeyo funktsia v gidrometeorologicheskoy sluzhbe, Materialy MCRM "Radiolokatsionnaya meteorologia", Leningrad, 3 – 6.
(3) PODHORSKY, D., NEMEC, J., STEFANKA, Ö. (1988) : Automated Radar Meteorological System – ARMS, WMO, TECO '88 Leipzig, GDR. 115 – 122.

UNE REVUE DU PROGRAMME ARAMIS

J.L. CHEZE

Direction de la Météorologie - S.E.T.I.M.

RESUME

Le réseau de radars météorologiques français, appelé ARAMIS, est
opérationnel depuis 1984. L'imagerie radar est maintenant largement diffusée
par les terminaux METEOTEL, à l'intérieur et à l'extérieur du service
météorologique français. Néanmoins, de nombreuses améliorations sont
programmées sur les quelques prochaines années ; elles concerneront
notamment le nombre de radars, la mise en place de nouveaux calculateurs,
appelés CASTOR, de systèmes éliminateurs d'échos de sol et de radômes, et
l'évolution du schéma de concentration des données.

SUMMARY

The French weather radar network, named ARAMIS, is operational since
1984. Radar imagery is now widely displayed on METEOTEL systems, inside and
outside the French Met. Office. Nevertheless, many improvements are planned
on these few next years, concerning the number of radars, new radar
computers, named CASTOR, ground clutter cancelers and the telecommunication
scheme.

1 - INTRODUCTION

En 1984, le projet ARAMIS (Application du RAdar à la Météorologie
Infra-Synoptique) de la Direction de la Météorologie a connu une étape
importante lors de la mise en service opérationnelle du réseau de radars
panoramiques du même nom.

L'objectif de ce projet est de permettre une large diffusion des données
radar en direction des stations météorologiques et de manière générale
vers tous les utilisateurs concernés par la détection et la signalisation en
temps réel des phénomènes dangereux liés à l'activité convective,
l'évaluation quasi-instantanée de la répartition spatiale des
précipitations, de l'estimation rapide des quantités d'eau reçues au sol à
différentes échelles de temps et d'espace.

Ce réseau a été constitué à partir de radars météorologiques déjà en
place dont l'exploitation était faite localement avec, pour objectif
principal, la protection de l'activité aéronautique. Il a donc été
nécessaire, dans une première phase, d'associer au radar un système capable
de créer une image radar numérisée et de la transmettre. Pour des raisons
économiques, le schéma de concentration et de diffusion des données a été
conçu à partir de liaisons spécialisées existantes en partage avec
d'autres applications de transmission de données météorologiques.

Cet article veut faire un bilan rapide de la mise en place et des premières années de fonctionnement du réseau ARAMIS et indiquer les principales actions engagées ou envisagées pour améliorer la couverture, la qualité des données et la fiabilité de leur diffusion.

2 - L'ETAT PRESENT DU RESEAU

En 1988, le réseau ARAMIS comprenait dix radars :

- 7 radars THOMSON en bande C de type RODIN installés à Nancy, Trappes, Nantes, Bourges, Lyon, Marignane et Toulouse,
        - 2 radars OMERA en bande S de type MELODI à Brest et Bordeaux,
        - 1 radar Plessey en bande C à Grèzes.
Ce dernier radar appartient au service hydrologique de Périgueux.

En 1989, un radar RODIN sera installé à Abbeville ainsi que 2 radars MELODI à Nîmes et Calern en remplacement de celui de Marignane et deux nouveaux radars seront achetés pour être mis en place à Arcis sur Aube dans l'est de la France et Alençon dans l'ouest, portant à 14 le nombre de radars mis en réseau (figure 1a).

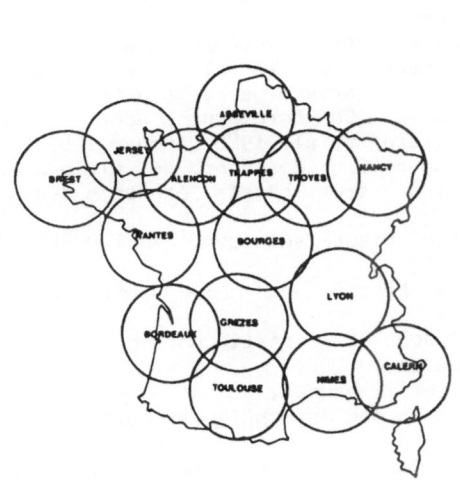

fig. 1a

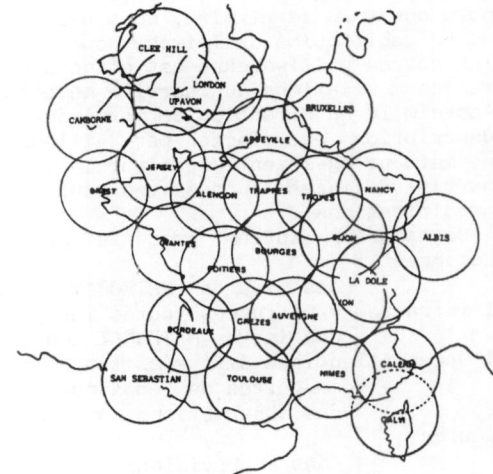

fig. 1b

	MELODI	RODIN
Manufacturer	OMERA	THOMSON
Wavelength (cm)	10.7	5.3
PRF (Hz)	250	330
Pulse width (µs)	2	2
Power (kW)	700	250
Antenna diameter (m)	4	3
Beamwidth (degrees)	1.8	1.3
Minimum detectable signal (dBm)	-106	-112
Digitization system	SAPHYR	MT750
Maximum range (km)	256	200

Table I.
Melodi and Rodin radar specifications

Les principales caractéristiques des radars RODIN et MELODI sont indiquées dans le tableau I. Les radars fournissent uniquement des données de réflectivité. Les radars de type RODIN ont pu être équipés d'un système de numérisation MT750 réalisé par leur constructeur; pour les autres types de radar, le système SAPHYR a dû être développé par le service météorologique.

Dans tous les cas, une image de 256 * 256 pixels de réflectivités codées sur 16 niveaux est transmise toutes les 15 minutes vers le système de concentration de données. Les images peuvent être fournies à une cadence plus élevée, par exemple pour des applications hydrologiques. Les données d'un radar particulier peuvent aussi être transmises directement vers l'utilisateur depuis le site radar.

Un calculateur situé au service central d'exploitation à Paris concentre les données qui transitent par le réseau de liaisons téléphoniques spécialisées et réalise toutes les 15 minutes une image composite qui intègre également des données en provenance de réseaux de radars étrangers. Des échanges de données radar sous forme d'images composites font l'objet d'accords avec les services météorologiques de Suisse et du Royaume Uni. Les images du radar de Jersey, très utiles pour la couverture du nord-ouest du territoire, sont également reçues par liaison directe.

La rediffusion de l'image composite vers les stations météorologiques et les autres utilisateurs est effectuée via le système COMETE par les voies de télécommunications du service météorologique en partage avec la diffusion fac-simile en direction de terminaux METEOTEL. Ces terminaux dont une description est faite par ailleurs (Gaillard et al.,1986,Pircher,1987) permettent également d'estimer la vitesse d'advection des échos de pluie à partir d'images successives pour la prévision à très courte échéance des précipitations.

De nombreux autres produits peuvent être visualisés sur le METEOTEL, notamment :

-des images METEOSAT sur la France dans le visible et l'infra-rouge diffusées toutes les 30 minutes,

-des images METEOSAT couvrant l'Europe et la partie nord-ouest de l'Océan Atlantique diffusées toutes les 3 heures,

-des cartes du temps prévu toutes les 12 heures,

-des observations du réseau de stations synoptiques toutes les heures,

-des prévisions des champs des paramètres atmosphériques effectuées par le modèle numérique à maille fine Péridot toutes les 24 heures,

-des cartes de localisations d'impacts de la foudre toutes les 15 minutes.

Un archivage systématique des images radar est effectué après contrôles dans le service de climatologie.

Dans la configuration actuelle du réseau, la disponibilité moyenne des images est de l'ordre de 90% sur une année comme l'ont indiqué M.Gilet et J.P. Musiedlak (Leipzig-1988). Les statistiques élaborées pour l'année 1988 font apparaître un taux moyen de disponibilité supérieur à 91 %. Le taux d'images indisponibles à la suite de problèmes de transmission avoisine 4%. Les autres images manquantes sont dues aux arrêts du radar pour maintenance ou pour cause de vents forts.

3 - LES EVOLUTIONS FUTURES DU RESEAU

3.1 - Amélioration de la couverture

La couverture que fourniront les 14 radars du réseau en 1990 devra être complétée au cours des années suivantes (figure 1b). En effet, la portée utile des radars pour la détection des précipitations de faible intensité n'excède guère une centaine de kilomètres en particulier pendant la période hivernale.

Plusieurs sites ont déjà été définis pour améliorer la couverture dans régions de Poitiers, Dijon, en Auvergne et en Corse et seront équipés en priorité.

Des échanges de données radar sont en cours de mise en place avec la Belgique (radar de Zaventem) et l'Espagne et sont envisagés avec l'Italie et la République Fédérale Allemande.

3.2 - Amélioration de la qualité des images

Tout d'abord, le service météorologique a fait développer par la société THOMSON un système éliminateur d'échos de sol pour les radars RODIN. Basé sur un principe d'analyse de la variance du signal rétrodiffusé, cet équipement traite le signal vidéo logarithmique en sortie de la baie d'émission- réception dans laquelle il est implanté. Le taux de réjection des échos de sol est de l'ordre de 30 dB. La figure 2 illustre l'efficacité du système sur des échos de propagation exceptionnelle pour le radar de Trappes. Ces systèmes sont en cours de mise en place sur tous les radars RODIN et vont être adaptés aux radars de type MELODI.

Figure 2.

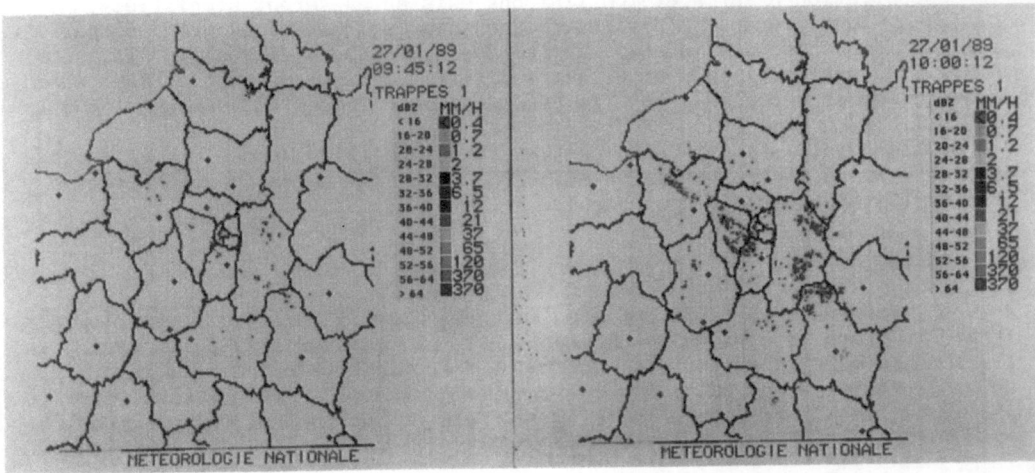

Un nouveau calculateur radar, développé par le SETIM, service technique de la Direction de la Météorologie contribuera largement par un meilleur traitement du signal à l'amélioration de la qualité des données. Ce calculateur baptisé CASTOR fait par ailleurs l'objet d'une description détaillée par J.L.Maridet dans ce séminaire. L'utilisation d'une intégration en distance par la médiane et l'identification des pixels isolés contribueront à l'élimination des échos parasites. L'intégration en azimut sera effectuée après numérisation et conversion du signal rétrodiffusé sur 12 bits dans une échelle linéaire d'intensités de précipitations et par un filtrage récursif adapté en fonction de la distance. Les données en coordonnées polaires seront installées sur une grille cartésienne correspondant à une image de 512*512 pixels de 1 kilomètre de côté en conservant la résolution de 12 bits. De cette image, seront dérivés des produits adaptés aux besoins des différents usagers. Ce nouveau calculateur permettra de prendre en compte un fonctionnement du radar en multisite, des corrections de masque partiel et pilotera des équipements de mesure (générateur hyperfréquence, milliwattmètre), assurant un contrôle automatique de la calibration de la chaine d'émission-réception. Les calculateurs CASTOR seront progressivement mis en place à partir de fin 1989 sur tous les radars du réseau ARAMIS.

Ces améliorations permettent d'envisager la mise en place de procédures de calcul automatique de cumuls de précipitations sur différents pas de temps au niveau des calculateurs radar.

3.3 - Amélioration de la fiabilité du réseau

La fiabilité du réseau concerne bien évidemment à la fois le radar et son calculateur, et le système de concentration.

La mise en place des calculateurs CASTOR contribuera à la maintenabilité des systèmes radar grâce à une homogénéisation des équipements et par les possibilités offertes de télé-transmission d'informations de maintenance.

Un programme de mise en place de râdomes permettra d'éviter les arrêts par vent fort et d'épargner les pièces mécaniques du radar.

Par ailleurs, le schéma général des télécommunications utilisées pour la concentration des images va être corrigé de façon à obtenir une meilleure fiabilité dans la transmission des données. Les nouveaux radars seront reliés directement par liaison spécialisée avec le calculateur chargé de la concentration à Paris et les liaisons existantes seront simplifiées.

A plus long terme, l'utilisation de nouveaux services proposés par le service français des postes et télécommunications (Transpac, Transcom, R.N.I.S.) est envisagée ainsi que l'utilisation du protocole X25 et du nouveau format européen pour la transmission d'images radar retenu par le C.O.S.T. 73.

Parallèlement, la diffusion par satellite des images radar et des autres produits vers les terminaux METEOTEL va être progressivement étendue à toutes les stations météorologiques.

4 - CONCLUSION

De nombreuses actions ont été engagées par le service météorologique français pour accroître les performances du réseau ARAMIS en termes de couverture géographique et de qualité et disponibilité des images radar.

La fiabilisation du réseau qui concerne à la fois les systèmes radar et la transmission des données devrait se traduire par un taux moyen annuel de disponibilité des images de l'ordre de 97%.

De nouveaux calculateurs radar permettront, outre l'amélioration de la qualité des images et une aide efficace à la maintenance, la génération d'images adaptées aux besoins de différents types d'utilisateurs et notamment à ceux des hydrologues.

REFERENCES

ANDRIEU, H., 1986. Interprétation des mesures du radar Rodin de Trappes pour la connaissance en réel temps des précipitations en Seine Saint Denis et Val de Marne. Thèse présentée pour l'obtention du diplôme du Docteur Ingénieur, Ecole Nationale des Ponts et Chaussées, Paris.

DAVID, P., J.P. MUSIEDLAK et P. BISSONNIER, 1987. Utilisation du radar Rodin en pluviomètrie : résultats des mesures de 1982. Note technique n° 12, Direction de la Météorologie Nationale, Boulogne.

GAILLARD, C., V. PIRCHER et R. PAILLISSE, 1986. The French ARAMIS project : use of microprocessor based systems for real time broadcasting of weather radar and satellite pictures. Proc. 2nd Int. Conf. on Interactive Information on Processing Systems for Meteorology, Oceanography and Hydrology, AMS, Miami, Fla., 10-13.

GILET, M, H. SAUVAGEOT et V. TESTUD, 1984. Weather radar programs in France. Proc. 22d Conf. on Radar Meteorological, AMS, Zürich, Switzerland, A, 15-20.

GILET, 1984. The French weather radar network-a status report. Proc. Nowcasting II Symposium, Norrkopping, Sweden, 3-7 sept 1984, ESA SP 208, 417-422.

GILET, 1985. Le réseau français de radars météorologiques. Proc. of the third WMO tech conf. on Instruments and Methods of Observation, TECIMO III, Ottawa, Canada, 8-12 Jul. 1985, 175-180.

GILET, M. et J.P.MUSIEDLAK, 1988. Assistance to the maintenance of meteorological radars. Proc. of the TECO, LEIPZIG, 16-20 May 1988,p 367-373.

HAUSER, D., P. AMAYENC and M.CHONG, 1984. Precipitation efficiency within a tropical squall line observced during the COPT 1981 experiment. Preprints, 22d Conf. on radar Meteorological AMS, Zürich, Switzerland, 134-139.

PIRCHER, V. 1987. Combined use in operational forecasting of animated imagery and NWP fields. Preprints, EUMETSAT workshop on satellite and radar imagery interpretation, Reading, UK, 20-24 July 1987, 365-383.

ROUX, F., 1985. The retrieval of thermodynamic fields from multiple Doppler radar data using the equation of motion and the thermodynamic equation. Mon. Wea. Rev., 113, 2142-2157.

THE WEATHER RADAR NETWORK OF THE DEUTSCHER WETTERDIENST
PROGRESS REPORT

J. RIEDL
Deutscher Wetterdienst
Meteorologisches Observatorium
Hohenpeißenberg

Zusammenfassung

Vom Wetterradar-Verbundnetz des Deutschen Wetterdienstes sind die er-
sten beiden Anlagen in Betrieb; die dritte Anlage soll im Herbst 1989
in Betrieb gehen. Das gesamte Netz wird 12 computergesteuerte, für
Dopplerbetrieb vorbereitete C-Band-Radaranlagen enthalten und soll
etwa 1995 installiert sein. Es werden sowohl qualitative Bildprodukte
als auch quantitative Datensätze generiert und verbreitet. Automa-
tisch werden alle 15 Minuten ein Bild der dreidimensionalen Echover-
teilung erzeugt und Hinweise auf starke konvektive Entwicklungen und
mögliche Hagelbildung abgeleitet. Aus Abtastungen im 5-Minuten-Raster
werden Datensätze der 1stündigen bis 24stündigen Niederschlagshöhen
im 100-km-Umkreis gewonnen. Als Sonderprodukte können vom Nutzer u.
a. abgerufen werden: CAPPI- und Schnittdarstellungen, Echotopdarstel-
lung, Falschfarbendarstellung der Niederschlagsverteilung aus den
quantitativen Datensätzen. Die nächsten Projektschritte werden die
Herstellung eines Kompositbildes und Prüfung der operationellen Ein-
setzbarkeit des Dopplerprinzips sein.

Summary

The first two systems of the radar network of the Deutscher Wetterdienst (DWD) are already operational; the third radar system shall be
added in autumn 1989. The network is planned to be completed in 1995
and will consist of 12 computer-controlled C-band weather radar systems with provision for Doppler operation. Qualitative image products
as well as quantitative data files are generated and distributed. A
quasi-three-dimensional echo distribution image is produced every
15 minutes and indications for severe convective developments and
possible hail formation are derived. Data files of the hourly and
daily precipitation amounts in the 100 km range are evaluated from
five-minute scans. Besides this users can request special products:
CAPPI and cross-section, echotop maps, false colour image of the precipitation distribution from the quantitative data files. Generation
of a composite image and investigation of the operational utility of
Doppler technology will be the next steps of the project.

1. Description of the project

As a result of the Deutscher Wetterdienst's decision to replace its weather radars by a radar network a project for computer-supported acquisition and evaluation of radar data was set up in 1985. A project team was set up from different branches of the Deutscher Wetterdienst for this purpose. Its task was to draw up technical specifications for the radar and the data processing system, select equipment, carry out acceptance tests on the first radars and put them into operation, and introduce users to the system. A Project Management Group comprising representatives from all the relevant departments within the Meteorological Office coordinates the project.

The objective is to improve and enlarge use of radar precipitation data by computer processing. Colour images of the instantaneous distribution of the precipitation echo, its development and movement will provide an important basis for operational use of radar data within the service. As the network is extended, the area covered can be enlarged by combining images from several locations. Finally, it is planned to produce composite images from the echo distribution over the Federal Republic which can be exchanged with neighbouring countries and incorporated into a European radar composite.

This is expected to prove extremely useful to the aeronautical meteorological service and regional forecasting centres, as it will provide colour images which can be immediately interpreted. Also operational area precipitiation data for hydrological purposes will be supplied for the first time.

The whole Federal Republic should be covered by the time the network is completed (1995) when there will be 12 radar locations. At present radars are operating in Munich and Frankfurt. Another is scheduled to be installed in the Hamburg area this autumn.

2. Equipment technology

The equipment consists of a radar, computer and display system. The radar system is a DWSR-88C built by EEC with provision for Doppler facility upgrading; it is designed for sub-unit process control by microprocessors. A digital video processor (DVIP) is used for echo processing; this will be replaced by a Doppler video processor when the Doppler facility is retrofitted. The radar has a real-time display for maintenance and function testing. A preprocessor controls the radar and general data flow.

The radar computer is a MicroVax II from DEC. It handles the whole process from data collection to product transmission. It prescribes important operating parameters for the radar and generates the images and data required by users from the processed echo data. A colour graphics console attached to the computer can be used for maintenance, operation and image display. A printer records system and error messages. Precipitations data, and on request image data, are stored on magnetic tape. User terminals and display units are located at the nearest department of the Meteorological Office (e.g. aeronautical weather office or regional weather centre). They are connected via an in-house modem or a telephone line (Datex-P) depending on how far away they are. A number of outside users (e.g. military weather office, water authorities) can be supplied with certain products over the Datex-P network. This interface can also be used for remote diagnosis for maintenance and error correction.

Finally, the images will be outputted on the display units of the Deutscher Wetterdienst's communication system (AFW), now being installed. Radar products will also be transmitted via this system to the headquarter and other offices. A PC will be used for image display as an interim measure until the AFW system comes into operation. If necessary, this can be connected via a dedicated line to another PC in an other meteorological office.

Images and quantitative data are transmitted to the AFW system or PC, or within the AFW system, in a uniform format using a compression algorithm. This significantly reduces the data transmission volume.

The system is operated by menu selection from the terminal. The screen is divided up into so-called windows to provide continuous operating, warning and error data. There are different levels of access to the radar computer for operational, maintenance and system-management purposes.

3. Data acquisition

Data are collected by volume scan at 15-minute intervals. The radar atenna starts scaning from the highest (around 30°) of the maximum of 20 elevation angles sweeping from top to bottom. The elevation angles are selected according to the relief of the local terrain. The data (dBZ) produced by the preprocessor in the form of polar coordinates are entered in a 400 km x 400 km x 12 km cartesian date cube based on a polar stereographic projection (60°N, 10°E) and buffer stored in this form. This cube provides the basis for most of the image products. The spatial elements with ground clutter, recorded in a separate scan, have also been previously marked in this data cube.

An additional scan is carried out at five-minute intervals to acquire precipitation data. This antenna sweep is made as close to the ground as possible. The elevation angle can be adjusted to the local relief in 45° sectors. These echo data remain in the polar coordinate system and are stored in a 360° x 100 km data disc. Precipitation intensity values (mm/5 min) are calculated later on the basis of a Z/R relation to be entered by the user.

On completion of the scan cycle the radar operates in the standby condition until the next 15-minute interval.

4. Description of products

Two sorts of products are generated: qualitative images and quantitative data. The most important products are generated and transmitted as a matter of routine; special products have to be requested by users. Images meet the meteorological office's (analysis and forecasting) requirements for immediate interpretability and display in colour technique. A number of the images for each product will be stored in the radar computer's ring memories for consultation to enable trends to be observed by reference to past data. Other facilities such as zoom and loop display are also available on the display systems, where the necessary geographical overlays can also be superimposed.

The quantitative data (hourly precipitation, interim and daily amounts) will be stored on magnetic tape but can be immediately called up on the radar computer too. They can also be displayed as colour images for rapid consultation.

4.1 Qualitative products

PL-image: this is automatically generated and provides information on the quasi-three-dimesional echo distribution in six intensity classes. Resolution is then reduced to 2 km x 2 km for horizontal projection. The highest clutter-free echo value in the lowest height level (Z = 1 km) is taken from the four 1 km x 1 km subelements and assigned to the relevant dBZ class. If no clutter-free field is found in this height level the search procedure will be continued in the height level above and so on. "Top- and side-view" show the projection of the strongest echos at each height from S to N resp. W to E. The height range and development of strong echos can be identified from these projections (e.g. showers, thunder storms). They can be located in the horizontal plane by back projection.

DW-meteorological warnings: an automatic search is made of the echo data stored in the data cube using the relevant algorithms to determine strong convective developments and possible hail formation. For each x,y - coordinate (= column in data cube) the echo values in two particular heights levels are checked to see whether they have been exceeded (40 dBZ at 8 km and 15 dBZ at 12 km at present). Neighbouring columns are combined and the ten coordinates with the highest echo values are shown in the "warning window" of the operating menu at the terminal.

The method of evaluating whether there is a risk of hail is based on the results of experimental work in Switzerland ("Grossversuch IV"). The "convective" columns found are checked to see whether a particular echo value (45 dBZ) has been exceeded in the next two height levels above the freezing level. If it has been exceeded there is considered to be a risk of hail in this cell and a warning is flashed on the terminal. The freezing level must be entered and updated by the observer.

The following special products can be generated by the radar computer on request from the operating console.
(1) PS-image: a cross-section presentation in the horizontal and vertical planes through any desired point of x, y and z in the data cube; size and resolution are the same as for PL;
(2) PX-image: a plan view (similar to the PL image but only 100 km radius and without projections) derived from the five-minute-scan data-disc which can be used for rapid consultation in critical situations;
(3) PE-image: a false-colour image of the echo height distribution in the data cube;
(4) PY-image: a false-colour image of the accumulated precipitation amount taken from the interim data file (100 km radius).

4.2 Quantitative products
(1) DH: dBZ values from five-minute scans are converted into precipitation intensity (mm/5 min) using the Z/R relation entered and summed up to hourly totals;
(2) DS: a running total is worked out from the hourly totals (DH);
(3) DD: the daily total is worked out from 0630 to 0630 UTC.
Area elements (1° x 1 km) with ground clutter are marked in these data sets. A separate scan is carried out to determine ground clutter.

Radar results can be adjusted by comparing them with raingauges at the ground. The telemetering rain gauges are not yet available, the data of which can be fed into the radar computer via the AFW system. Preliminary studies have shown that five to eight ground-truth gauges would be needed over a 180° sector (100 km radius) and these would have to be distributed as uniformly as possible, within an 85 km distance. Experience shows that the best results are achieved by adjusting with the mean logarithmic quotients (rain gauge (RS)/radar (RA) values) of the hourly data. This will improve 60% to 70% of hourly radar data. Additional improvements can be achieved by a further weighted adjustment at the end of the precipitation event (quotient of the total rain gauge measurements of all stations and total radar data).

Special products:

(1) DP-product: up to ten values from selected locations from the hourly data file to be used for comparison with weather reports;

(2) CEL-report: this provides the echo values within the data cube to be visualized in all height levels on the console by shifting the point of origin.

The characteristics of the products (automatic or separate generation, automatic (or no) archiving on disc or magnetic tape) can be determined from the radar computer's manager menu.

5. Progress report and future work

The first two radars have been operating satisfactorily on a continuous basis for 12 resp. 18 months. The software has now reached the necessary level of stability after a few minor adjustments. The simple hail algorithm did not work in its original form, producing too many false alarms due to isolated echos (aircraft, anomalous propagation). It will be therefore now used only for echos already identified as convective cells. It has become clear that greater capacity by a second disc system at the radar computer is necessary for future expansion. Account has already been taken of a second disc system in ordering subsequent radar systems.

On handover to the Operations Branch it turned out that users need considerable further training in the operation of computer equipment, product interpretation and equipment maintenance. The training effort required must not be under-estimated.

The next stage in the project will be to connect the system to the AFW network to make use of the better display facility and work on the composite image. As the preliminary work has already been carried out, it is possible even at this stage to composite images from Munich and Frankfurt into one image on a PC. The alorithms used must be developed and improved. Once the Hamburg radar comes into service it should be possible to produce an initial - albeit fragmentary - composite image for the Federal Republic of Germany.

One major task ahead is to test the operational utility of the Doppler principle. A new radar with a Doppler facility has been operating at the Hohenpeissenberg meteorological weather observatory since the end of 1988. It can produce additional products such as radial wind distribution, wind shear, vertical wind profile using VAD, and wind shear warnings. The quality of these products and the effectiveness of additional Doppler filters for ground clutter suppression have yet to be tested.

OPERATIONAL EXPERIENCES WITH THE FINNISH WEATHER RADAR NETWORK

R.H. KING
Finnish Meteorological Institute
Helsinki, Finland.

Summary

The Finnish Meteorological Institute (FMI), which is the government agency responsible for providing meteorological services in Finland, has been operating digital radar systems for the last four years. At the present time three radars operate continouously, making three-dimensional data-collection scans every half hour, and automatically processing the data to a radar-centred xyz format. Two of the radars are unattended and remotely-controlled. The primary users of the data are duty meteorologists working at three regional meteorological offices. The complete data sets are relayed to the central office of the FMI in Helsinki, where selected products are made for inclusion in the meteorological work-station menu, and for conversion to a polar-stereographic projection, from which a network composite is made. Weather radar images are used in various applications, ranging from the primary use in weather analysis and forecast at the FMI Central and regional forecast offices and their branch offices to use in TV weather presentations. Other uses include road and agricultural weather services employing videotex-type display systems. Future developments include a greater emphasis on precipitation-orientated services, such as the provision of optimised real-time intensity and cumulative total values, as well as short-term forecasts. For these applications a more frequent data-collection schedule will be essential, involving improved computing facilities and some rearrangement of priorities. Improvements in general forecasting for the 0 - 6 hour period are expected to follow when data from the Nordic Network of weather radars becomes available. The usefulness of radar data from the wider COST area is expected to be seen in a more timely pinpointing of deviations from computer model forecasts, and in the sphere of aviation meteorology, where near real-time radar data should be of immediate interest for flight route forecasting and briefing.

1 INTRODUCTION

By the beginning of the 1980's, the advantages to be obtained from digitized, as opposed to manually-derived, weather radar data was becoming very evident from the examples provided by the successful operation of the radar networks in the UK and Switzerland. The Finnish Meteorological Institute (FMI) thus decided to initiate a project for the digitalisation of the existing radar at Helsinki-Vantaa airport, which led, in 1983, to the acceptance of a successful tender, and in February, 1985, to the acceptance of the computerized radar control, data-handling and display system into operational use. The software and hardware were developed also taking into account the requirements of the Swedish Meteorological and Hydrological Institute, which at that time was also acquiring a digitized radar system. Since its introduction over four years ago, the radar data processing unit at Helsinki-Vantaa airport has been operating satisfactorily, and two more similar systems have been acquired. During this period, much operational experience has accrued, allowing an assessment to be made of the usefulness of the network. It is thought that an account of experiences with the Finnish weather radar network may be of wider interest.

2 THE NETWORK

The FMI weather radar network is shown in Fig.1, and consists of three computerized radar systems at Rovaniemi (on the Arctic Circle), Masku (near Turku, in south-western Finland), and Helsinki (at Helsinki-Vantaa airport). A fourth, non-digitized radar is operated locally at Kuopio. Plans exist to replace this with a digitized system, and to add another modern radar system in the Ostrobothnia area. These two radars are shown as "planned" in Fig.1. As can be readily seen, even addition of these two planned stations does not provide a very satisfactory coverage over many parts of the country. For completeness, it should be mentioned that the Meteorological Department of Helsinki University operates a Doppler weather radar in the city of Helsinki.

3 THE RADARS

The FMI digitized radars are all of Soviet construction, of type MRL-5. Compared with radars in use in most other (West) European countries, a surprising feature may be the use of a 3 cm wavelength. Although attenuation by moderate and heavy rainfall is sometimes significant in the warmer seasons, this is compensated in the cold season, when the narrow beamwidth (0.5°) and high antenna gain provide better performance in snow echo conditions. The dual-wavelength (3 + 10 cm) capacity of the Helsinki radar has not been utilized operationally for e.g. hail detection, as only one DVIP unit is used. The Masku radar has twin tranceivers, one being employed as a back-up unit in case of breakdown, while the Rovaniemi radar has only one transceiver. The nominal maximum range of the radars is 300 km. The radar antennae are enclosed in radomes, and operate without problems in the extremes of temperature found in the Finnish climate. Both the latter stations are unmanned, all operations, including radar switching, being normally carried out through remote computer-to-computer communications. All radars are equipped with a preprocessor of Swedish design and construction, which acts as a signal digitizer and averager (DVIP) and also as a two-way interface between the computer and the radar. Data handling is described in more detail in Section 4.

4 RADAR DATA HANDLING AND PRIMARY DISPLAY

The EWIS (Ericsson Weather Information System) software package differs somewhat from radar to radar. This is mainly due to the difference in the primary displays employed at the three "controller" sites, but also reflects the development that has taken place over the years in computers (from the VAX 11/730 to the MicroVAX II), and the change to a split system, in which the processing tasks are divided between the "radar data processor" computer at the radar site, and the "operator data processor" computer at the primary display ("controller") site. The latter configuration, used at Masku and Rovaniemi, has increased the speed of the data collection cycle (update rate) by reducing the computing load at the radar site, and will be described in the

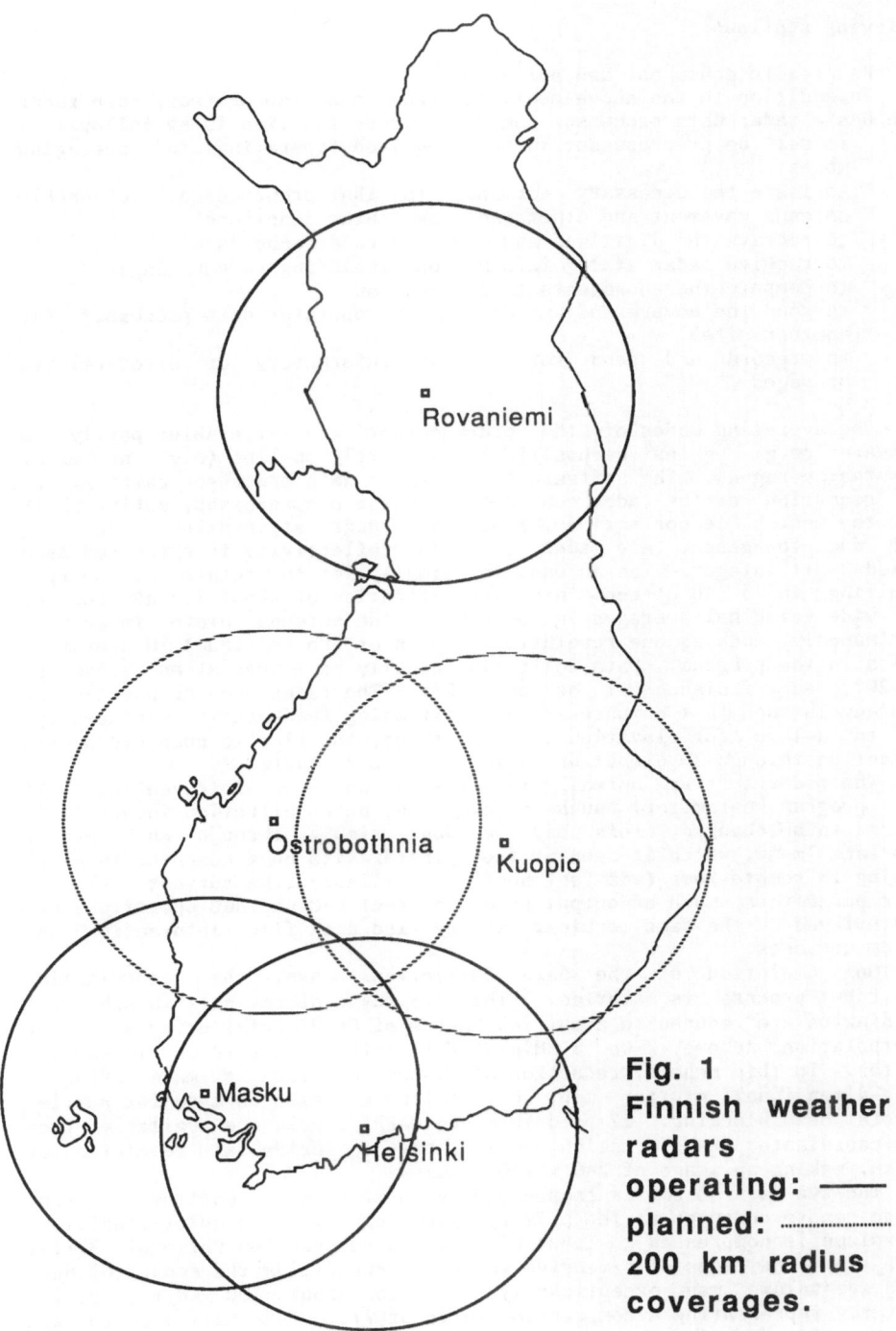

Fig. 1
Finnish weather radars operating: ──────
planned: ··················
200 km radius coverages.

following sections.

4.1 Radar-site equipment and software

In addition to the above-mentioned radar and preprocessor, each radar site has a radar data processor computer, whose function is as follows:

- to set the preprocessor in the required operating and averaging modes
- to issue the necessary commands to the preprocessor to enable antenna movement and other radar switching functions
- to receive the digitized and averaged radar echo data
- to receive radar status information, including antenna angles
- to compute the coordinate transformation
- to send the compressed xyz data to the operator data processor (at another site)
- to record and send for display informatory or error-related messages

The averaging modes of the preprocessor are alterable partly in firmware (e.g. gates' disposition), and partly on-line (e.g. number of azimuths averaged). The software in the radar data processor carries out the reception of the radar echo data from the preprocessor, during which time the most basic corrections, i.e. for range attenuation, including that due to gases, are made. The radar reflectivity is expressed as a scaled 8-bit integer. The antenna is normally set to rotate at 3 rpm, resulting in a SD of echo intensity estimation of about 1.2 dBm for the 0.5°-wide azimuthal averages in 300 gates. The antenna rotates in azimuth continuously, making one revolution at each of the (maximum) 20 elevation angles in the program. Data collection usually commences at an elevation of 20°, and finishes at or near 0°. The radar angular positioning accuracy is such that computer-radar positioning feedback is unnecessary. Due to delays for elevation angle settling, the time to complete a data collection through 20 elevation angles is over 8 minutes.

The radar software normally runs without operator intervention. If some program parameter, such as e.g. the data collection interval, is desired to be changed, this may be done simply through an operator interface menu, which is seen at the operator site on a computer terminal working in remote mode (via "set host"). Similarly, the current state of radar parameters, such as output power, correct AFC and AGC operation, may be displayed on the same terminal, as may also data file information and system messages.

On completion of the data collection phase, the coordination transform process is started. The 4.3 MByte of raw data in spherical coordinates are reduced to about 740 kBytes of CAPPI data, using a 4-point interpolation scheme. On a MicroVAX II this is carried out in about 4 minutes. In this scheme, reduction of ground clutter is made using a three-dimensional clutter map. Corrections may also be made for partial or total beam blocking. 12 predetermined CAPPI levels are available from the coordinate transformation, which has a horizontal resolution of 2.5 km, making an image of 240 x 240 pixels.

The xyz data volume is compressed by an efficient algorithm and sent to the remote computer at the primary operating site. In quiet conditions the volume is compressed to about 20 kByte (a compression ratio of 37:1), while in conditions of extensive and high echo fields the amount of data in an xyz volume may exceptionally rise to about 100 kByte, still, however, representing a compression ratio of 7:1. Information on the xyz file sent is transmitted to the expansion task in the remote computer to trigger it into action. Any informatory or warning messages are also transmitted onward to this computer.

The prepreocessor and radar data processor computer are protected from short-term (approx. 30 min.) interruptions in the power supply, but the consumption of the radar is such that the radar cannot be so protected, economically. Thus, when a disturbance occurs in the power supply, the radar automatically shuts down, and has to be restarted through the computer system on restoration of national grid power. For power cuts of longer duration than 30 minutes (rather rare), the computer automatically reboots on restoration of power. The program package has then to be restarted by an operator. Radar malfunctions result in an automatic shut-down of the part of the radar affected.

4.2 Primary display site equipment and software

The primary display site at which the operator data processor (remote) computer is situated is in every case a regional forecast office. In the case of the Masku radar, the primary display site is at the Western Finland Regional Office, about 125 km away, while the distance from the Rovaniemi radar to the Northern Finland Regional Office is only 6 km. In both cases the 9600 bd synchronous communications links are provided by the telephone companies in the areas concerned. Computer-to-computer dialog takes place within a DECNET environment.

Having received the xyz data volume, and the message relating to it, the display site software expands the compressed data, and stores it on disk. From this data extra "pictures" are prepared, as follows:
- echo top, from highest CAPPI level with echo in that pixel
- maximum value (dBZ) of echo in any CAPPI level for that pixel
- height of this maximum value (i.e. CAPPI level in which occurring)
- maximum values in NS and EW cross-sections (like Swiss products, see Joss (1))
- preselected x-y cross-section
- Pseudo-CAPPI, i.e. echoes observed at lowest elevation angle

Additionally, any CAPPI level is available for display. A selection of previous images and xyz-volumes is stored on disk to provide series for display. Removal of older data occurs automatically. Data may be also moved onto tape, if desired, for later study or demonstration purposes. The image display systems vary from station to station, and thus the display facilities also vary. All have, however, a time series loop facility, together with the possibility of viewing any of the above-mentioned products singly, either in entirety, or zoomed. The reflectivity images may be scaled to precipitation intensity using selectable (preset) Z/R relationships. Cross-sections may be selected at will, also at any angle in the later versions. A mouse-controlled cursor allows the selection of cross-sections, pixel value read-out and image-to-image determination of echo velocity. Easily-understood menus allow selection of parameters and display of picture directories and messages.

5 CENTRALIZED DATA USAGE

The primary radar data displays at the regional forecast offices give valuable information for mesoscale analysis near the station and for local forecasting in the very short term. All three stations having the primary displays are aeronautical forecasting offices, which need such information for e.g. the making of TREND forecasts and for the provision of on-line briefing of pilots. However, the full usefulness of radar data can best be obtained when that from neighbouring radars is composited. This is especially true when the radars are close enough to "pass on" the echoes to the next station without any gaps occurring. In Finland, only the two southern radars fulfill this condition, being situated about 160 km apart. For winter conditions in Finland this must be considered as about the maximum for satisfactory networking, if awkward gaps in the echo fields are to be avoided. As can be seen from Fig. 1, the positions of the projected radars in Central Finland do not fulfil this requirement, either relative to each other, or to the other existing radars. During the summer, when frontal echoes can be observed out to 200 - 250 km and cumuliform echoes to 300 km, the situation is clearly better. According to Browning et al. (2), objective forecasts derived from radar network data can provide useful information (evidently in frontal situations) even up to 6 hours ahead. Due to the longer period of the year with winter-type echoes from shallow snow-bearing systems, results from the Finnish radar network are expected to be more modest. Improvements should accrue from the addition of data from neighbouring countries, such as Sweden and the USSR.

5.1 Data collection from radars

The collection of data by all three radars is synchronized: the antenna scan commences at H - 10 and H + 20. As the collection takes about 8 - 9 mins, the last scan, which is nearest the surface, occurs at about H - 2 and H + 28. After coordinate transformation, all xyz data volumes are immediately forwarded, in compressed form, to the central computer cluster in the FMI main office in Helsinki, where expansion of the data files takes place. At 9600 bd a compressed xyz data volume is normally transferred in 20 - 30 secs, while the whole process of compressing, transmitting and expanding the file normally takes about a minute. This time is considerably increased if a file containing an appreciable amount of noise is passed for transmission (see Section 6.1). Even from the slowest computers, the xyz data is available for central processing at about H + 12 and H + 42 (from the quickest, five minutes earlier).

5.2 Archiving

The compressed xyz data volumes are stored on disk at the central computer facility, and archived weekly. The archiving has been carried on experimentally since summer 1988 in order to gain experience in the handling of such data. If the trial is considered successful, it is probable that archiving will be continued on a routine basis. In the highly-compressed form the archived data uses little room: about 16 - 18 TK50 tape casettes a year for the three radars.

5.3 Further processing

Immediately upon expansion of the xyz data volumes, derived products of the same type as available at the primary display stations are produced, as required, in the central computer facility. These single radar products are (at present) then slightly reformatted for use in the meteorological work-stations within FMI, which are situated at the regional weather offices and their branch offices. Distribution to these work-stations of meteorological data, including radar data, is by computer-to-computer links and automatically-operating software.

During certain periods of the day, true composites on a polar stereographic grid are prepared for use in a TV weather presentation, in which Meteosat images are combined with radar data. The Meteosat image resolution for the areas around Finland is quite coarse compared with its optimum value, and additionally the grid chosen for the TV image is coarser than that of the original radar data. For these reasons, the radar data is sampled with a nine-point running filter, giving very much improved echo "cohesion" compared with a raw nearest-neighbour selection algorithm.

Work is at present under way to provide a composite radar image (possibly melded with satellite data) for other applications, such as road and agricultural weather. These applications are designed for display on videotex-type terminals or PC's. They also will include forecast precipitation determined from objective echo velocity determinations. The difficulties involved in using objective algorithms to determine precipitation for road weather applications in winter are discussed in another paper in this seminar (3).

6 OVERALL NETWORK DATA AVAILABILITY
6.1 General picture
The availability of the radar data files at the central computer facility depends on the serviceability of a number of links in the chain, from the production of the data at the radar through to the final writing onto the central computer disk. Experience has shown the following to be the most important:
- serviceability of the radar
- serviceability of the preprocessor and radar data computer
- availability at the radar station of electric power from the grid
- serviceability of the radar station air-conditioning
- serviceability of communications (modems, lines)
- correct functioning of applications (and system) software
- serviceability of the operator data processor computer
- availabilty of sufficient disk space in archiving computer
- correct and prompt action by operators to react to malfunctions

The availability of the volume data at the central computer, when expressed as the percentage of all files which could have been created in the time period, was found to be 86%, 89% and 92% for the three radars, when measured over the whole of a 19-week period from October 1988 to February 1989. These are similar to figures given by Gilet and Musiedlak (4), for French radars. The main causes of the reduced availability are generally partly organisational, partly technical, but, in contrast to French experiences, very few losses are due to malfunctioning of communications. Work is continuing to pinpoint the most important factors, and reduce their impact on data availability.

The physical availability of the xyz files does not, of course, guarantee their usability. Even if the serviceability of the radar is understood to include correct and recent calibration, there is still the question of quality control, which in the narrow network context mainly means the correct adjustment of the threshold to remove radar receiver noise. In the Finnish radar systems, this threshold is adjusted before the data collection starts, using an operator interface menu. If the threshold is too low, unnecessary noise passes with data into the processing chain (including range correction in the raw data), causing corrupted images, increased transmission times and excessive disk usage in the central computer. With too high a threshold, sensitivity to weak signals (which at long ranges may represent significant reflectivity values) is diminished.

6.2 System monitoring responsibility

As can be seen from the above list, the responsibility for keeping the network running covers a number of widely-assorted factors. At the present time, the Finnish network responsibility devolves on the operators at the primary display stations (the regional forecast offices). The overall responsibility there is carried by the duty meteorologist, since he is the main user of the primary data, and thus most likely to be the first to detect a malfunction or a case of corrupted data (noise threshold too low). On observing a malfunction, he sets in motion a fault-finding operation, at first locally, and if this is unsuccessful, then seeking aid from the central facilities of FMI. He has trained assistents and technical expertise locally available, which he can call upon when needed. Responsibility for keeping sufficient disk space available in the central computer, and for monitoring the state of communications throughout the computer network, will rest with the operators in the FMI main office.

6.3 Quality control

Apart from the provision of basic quality control in the sense of the maintenance of a suitable noise threshold, as discussed above, there remains the wider problem of monitoring the radar network data for invalid data, with the necessity of removing artefacts and e.g. anomalous echoes, before the data set can be employed for basic applications. The approach at FMI has so far been that the necessary resources for an on-line monitoring system using human interpretation and intervention will not become available in the foreseeable future, and that therefore the emphasis must be on computerized methods making use, wherever necessary, of other meteorological data sources. This would seem to be a possible application for AI methods in the future.

On a more mundane level, the maintenance of a consistently high quality of any instrumental data must rest upon efficient servicing schedules and techniques for the instrument. In the case of weather radar, the question of overall serviceability is a complex one, and involves much more than merely upkeep of a good electronic calibration in the narrow sense. Work is continuing at FMI to define suitable servicing schedules, in which radar upkeep is only one of the demands on limited technical servicing resources.

7 DEVELOPMENTS
7.1 System improvements

One of the disadvantages of the interpolation scheme used is the time taken to complete the computations on the earlier computers (e.g. VAX 11/730). With the advent of faster computers, such as the MicroVAX 2, this time has been considerably reduced, allowing the possibility of more frequent data scans, which in turn allow better cumulative rainfall estimates and better cell-tracking in conditions of cumuliform echoes. It is planned to replace the older and slower computers in the near future to further enhance the working capacity of the radar systems.

All the display systems at present in use at the primary display stations have certain undesirable characteristics, and in addition they are all different. The older displays, at least, may be upgraded to be in line with current and projected FMI work-station practice and philosophy, aiming for a consistent representation and use of all observational meteorological data using standards such as X Window.

7.2 Hydrological applications

Historically, unlike the meteorological authorities in some other countries, the FMI has not used weather radars in the first instance for hydrological purposes. The radars were originally sited at airports in order to provide support for the aeronautical weather services stationed there. Nowadays there is interest in using the digital data produced by the radar network for hydrological applications, especially at critical times of the year (such as the thawing period in spring) when flooding can be a problem in certain areas. Work is under way to examine the usefulness in practice of weather radar for precipitation estimation in the climatic and hydrological conditions of Finland, when data is used to determine rainfall accumulations for specific river basins lying in the radar coverage.

7.3 Other applications

Many customers of the weather service are interested in short-term forecasts of the occurrence of precipitation for a certain locality. Concerning snow one may mention e.g. road maintenance authorities, railways, building contractors and the general public. In summer one may add ports and agriculture. As mentioned in 5.3, implementation of an objective echo velocity-measurement algorithm is ongoing, and successful completion of this will benefit the abovementioned areas of application.

Radar data can also be used to indicate the presence of, or imminent occurrence of certain types of severe weather, such as hail or thunderstorm. A test of some known algorithms, using three-dimensional radar data in conjunction with lightning-locator data, is due to be started in the current year.

7.4 Data from other networks

The importance of radar network data in contrast to those from single radars has already been mentioned, and is also discussed in (3). In order to gain the full benefit from radar data, not only in mesoscale, but also in larger-scale analysis and forecasting, data must be obtainable from a large area covered as uniformly with radars as possible. Although the usefulness of high resolution satellite data covering large areas of the globe has been accepted almost without argument, an acceptance of the need for a coverage of weather radar data complementing the satellite data has been much harder to gain, probably because of prejudices formed as a result of experiences in earlier years with single-station radars. Finland has been actively associated with the European weather radar projects COST 72 and COST 73 since their inception, and is also involved with the Nordic weather radar network, of which the nucleus should be operational in about two years. To complete the radar data field surrounding Finland, observations are also required from the Soviet Union. FMI has an ongoing cooperative project involving weather radar with the USSR State Committee for Hydrometeorology and Control of the Environment, as a result of which it is hoped that bilateral exchange of radar data will take place in the not-too-distant future.

7.5 Training

One of the most forceful impressions which has emerged from the experience of FMI with the digitized radar systems, and in fact with the use of weather radar data in general, is that of the necessity of providing adequate training for all staff involved with the radars, both those utilizing the data, and those servicing the equipment. In an environment with tight schedules, such as the duty meteorologist has in a regional weather office, it is easy to forget how to use some of the more rarely-used functions in a system, and thus to fail to obtain the full benefit from the available facilities. Also, there is a general lack of knowledge on the best methods of using all the data products according to the weather situations occurring. It is believed that much more effort will have to be put into solving this aspect of the problem, which is of crucial importance: for what use is a technically sophisticated system if the user cannot derive the full benefit from it?

8 CONCLUSIONS

The FMI has implemented a digital weather radar network, which it intends to develop in the future to improve the coverage in Finland and integrate into the networks of neighbouring countries. The usefulness of radar networking is expected to be particularly felt through improved analysis at all weather scales, and the use of data in objective nowcasting methods for several different applications. Specialized products may also be used for more accurate briefing for aircrew flying over domestic and European routes, and to provide e.g. humidity field data for use in fine-mesh limited area forecasting models. Experiences especially indicate the need for continued improvements in such areas as quality control, timely rectification of malfunctions, and training.

REFERENCES

(1) JOSS, J., 1981. Digital Radar Information in the Swiss Meteorological Institute. In: Preprints, 20th.Conference on Radar Meteorology, Boston, Mass.. American Meteorological Society.

(2) BROWNING, K.A., C.G. COLLIER, P.R. LARKE, P. MENMUIR, G.A. MONK and R.G. OWENS, 1980. On the Forecasting of Frontal Rain using a Weather Radar Network. Research Report No.22. Meteorological Office Radar Research Laboratory.

(3) NYSTEN. E. and R.H. KING, 1989. The use of weather radar data in road weather services: present and future needs. COST 73 International Seminar on Weather Radar Networking, Brussels.

(4) GILET, M. and J.-P. MUSIEDLAK, 1988. Assistance to the maintenance of meteorological radars. In: Instruments and Observing Methods. Report No. 33. WMO/TD No.222.

ATTEMPTS IN HUNGARY FOR AUTOMATION OF WEATHER RADAR OBSERVATIONS

F.Dombai, A.Kapovits   and   L.Illés
Central Institute for Weather   Upper-Tisza Valley District
Forecasting, Hungary   Authority of the Environment
1675 Budapest, POB.32   and Water Management
                                            4400 Nyiregyháza POB.14.

SUMMARY

In Hungary a weather radar network, equipped with MRL-5 dual-wavelength weather radars has been operated since 1983, based on standardized observing procedures. The manually generated weather radar data are digitized and processed in a common grid system with 20 km resolution on a map of polar-stereographic projection and transmitted by teletype in matrix from to the users.

It has become evident, that great potential of the system remains unutilized if observations are made manually.

An experimental automated system was developed in 1986 with the primarily aim of collecting experiences and providing demonstration. It makes possible real-time onservations in the horizontal and the vertical (PPI, RHI) and at a selectable level of constant altitude (CAPPI) with a resolution of 1x1 or 2x2 km out to the range of 100 km or 200 km.

Experiments with this system has revealed some defficiences. As a consequence of all these a new project started in 1988 for the development of an automated system meeting the sophisticated requirements of radar measurement of precipitation in NE Hungary, in cooperation between the Central Institute for Weather Forecasting and the Upper-Tisza Valley District Authority of the Environment and Water Management.

The new automated weather radar observing and processing system is capable to operate on two wavelengths simultaneosly, using two data processing channels, or alternatively on one of the two channels, while providing multilevel CAPPI-s in real-time and to make various corrections required for correct radar rainfall measurements.

INTRODUCTION

Installation in Hungary of a multipurpose weather radar network with the overall intention of meeting the mutual requirements of meteorology, hydrology and aviation was completed in 1983. This network consists of three weather radar stations situated at Szentgotthárd, Budapest-Ferihegy international airport and Nyiregyháza. The effective range of this weather radar network covers the whole territory of our country and even it extends over the catchment areas of the main rivers outside of the country (Fig.1). Regular weather radar observations are made hourly. The measured radar data are digitized manually, transmitted to the users via telex and there composited into a radar chart by microcomputer. The experiences gained with the use of the output of this network has revealed that data reliability and resolution, both in space and time, do not meet the users' requirements at high level.

Anticipating the limits of the manual system, the total automation of radar observations has been intended in the planning phase of the network, however, that has never been reached, because of funding shortage. After all, keeping our original intention in mind, an experimental system of low cost for the digital processing of radar data was developed in 1986. This system has very spectacularly shown all the advantages of digitization and automated processing of radar data that resulted in finding the necessary financial support to further developments. These have made possible to go on in this direction, relying on the gained experiences. A project for automation of radar observations was started in 1988 planned to be finished and operational by the middle of 1990. This Weather Radar Project of Upper-Tisza River Region makes possible real-time three-dimensional data collection simultaneously on both channels of MRL-5 radars operating in the network.

Plans for establishing high-speed data links between the radar station and the users, and the computational environment necessary for receiving digitized radar data for further processing are extended to complete the project

Block diagrams, data processing flows and environmental conditions are presented and described in this paper, for both the experimental system and the system under development in the framework of the Weather Radar Project of Upper-Tisza River Region.

EXPERIMENTAL SYSTEM OF LOW COST FOR AUTOMATION

Savings of costs of investment in implementation of the national weather radar network have been allowed to develop an experimental system to automate the basic functions of digital processing of weather radar data (Fig.2).

The main functions of this system that have been aimed at are as follows:
- averaging after digitization the radar video signals radially and from pulse to pulse, with programmable averaging number
- real-time data collection for RHI, PPI and CAPPI presentation at selected level
- data conversion from polar into Cartesian-coordinate system and display of digital radar images on colour monitor
- automated control of the manual functions of operating modes of MRL-5 radar
- provision for post-processing the digitized data.

A special device controlled by a micro-computer on 8 bits was developed, considering the small money available. In this device data processing is done by two processors built around micro-processors of Z80A types. The basic elements of the system are as follows.

Radar processor with functions and capabilities of
- digitization in 8 bits/1 MHz,
- filtering with 1-16 samples,
- resolution of 100 samples radially, out to 100 or 200 km, and 0,5 tangentially around the whole circle,
- coordinate conversion from polar into Cartesian in matrix form of 200x200 pixels by the aid of stack memory with precalculated addresses,
- switching the radar operational modes,
- control of the antenna elevation angle.

Display processor with the function of
- colour graphics display of data fields provided by the radar processor together with annotation in 16 colours in 384x256 pixels,

- manipulation of radar images (loading, saving, zooming, setting of colour tables etc.).

Micro-computer with the features as follows
- system control,
- interaction between the computer and the user
- archiving,
- transmission of data on low speed telecommunication line,
- timely execution of batch functions.
     Although this system has never been used operationally, very useful experiences have been gained with it on digital processing of radar data from MRL-5.

WEATHER RADAR PROJECT OF UPPER-TISZA RIVER REGION
     The experimental system completed in 1986 that capable for displaying the digitized radar data in real-time has clearly demonstrated the adventages of digital processing over any other manually operated system. (However, it is not suited for complex procedures to provide reliable data it has not been intended to.) In 1988 the Upper-Tisza Valley District Authority of the Environment and Water Management decided to solve the automation of the observations on the weather radar station near Nyiregyháza in the framework of a project closely related to water management within the region of its interest. To achieve this, installation of a telecommunication link of medium speed between the weather radar station and the Computer Centre of the District Authority and properly designed environment for the reception and inclusion of radar data into the water management system are under development. The project aims to put into practice reliable areal precipitation measurement by radar over the catchment areas of the Upper-Tisza river outside the country. This project well fits in the developments of a telemetering measuring system of the District Authority and provides suitable time adventage in flood conditions on the Tisza-river.
     The project is planned to be accomplished in two phases. In the first phase, digitization of weather radar, i.e. automation of weather radar observations, taking into account the requirements of areal precipitation measurements and considerable flexibility, should be realized. In the second phase of the project measuring methods and techniques should be tested and selected for putting into operation. The radar rainfall measurements are planned to be operational in 1991. Data of the automated weather radar observations should reach the LAN of micro-computers at the District Authority and should be post-processed there accordingly. The LAN provides an easy access to the processed products for all working places of the District Authority (Fig.3).
The complete system planned to be developed in this project consists of the following basic components:
- automated weather radar
- data transmission and telemetering system
- workstations at the District Authority with the capability of reception of radar data and image processing.

Automated weather radar
     Digital pre-processing of the radar data shall be made in situ at the weather radar station. Two independent channels are available for the radar video data. Complete volume scans in cycles of 5 minutes are foreseen on both channels. When planning the architecture of the system, processors of general use with good performance/cost characteristic were looked for.

With respect to the data processing functions the automated weather radar system is composed of the elements of
- intelligent interface between the radar and computer
- radar video processors
- host computer for the system control.
The composition of the planned system is seen in Fig.4.

Intelligent interface between the radar and computer
The first specimen of this interface has been manufactured and after successful testing the new version of this interface being capable to operate on both channels is planned to be developed by the end of this year.
Functions of this interface are
- control of the radar status and operating modes (transmitter-power, receiver noise level, pulse-length, attenuators etc.)
- antenna control (readout and setting of elevation and azimut, control of RHI, PPI and CAPPI scans, programmed antenna motions)
- signal processing on both channels (A/D conversion on 8 bits/10 MHz, LUT for linearization of the receiver characteristics, providing anti-log conversion with 12 bits, moving average of 1-8 samples in radial direction and from pulse tu pulse, output LUT, setting up the digital video-bursts of maximum 512 samples by pulses on both channels)
- data connections (interfacing to the host computer for purposes of control and status data transmissions, interfacing to the video processor to transmit the digital video bursts).

Radar video processor
On both channels separate radar video processors shall be applied. Processing of the digital radar video data is entrusted on a micro-processor of Motorola 68020 with the capability of fast data processing rate, lots of inner registers and flexible memory management. In our system the Motorola 68020 micro-processor is placed on a co-processor board that can be plugged in the host computer. The board has its own bus system with required memory and interface circuits. The fast data transmission to the host computer is guaranteed by dual-port realization of a part of its memory.
Functions that should be realized are as follows:
- calculating reflectivity factor values with respect to the radar status data
- making corrections for atmospheric and precipitation attenuations
- producing PPI, RHI, CAPPI, VIL and ETP data fields
- providing precipitation fields based on in pixel averages
- providing MIN-MAX fields
- composition of polar raw data fields for back-up purposes
- setting data fields for transmission to the host computer

Host computer
A mini-computer of general purposes is intended to control the automated system and to process the data from radar video processors. Considering the operational rate required for data processing the computers built on a processor of INTEL 80386 on 32 bits, which is of personal computer category, were found appropriate to be used for host. These computers have good performances for fast data processing as their operational rate is about 1-3 MIPS.
In order to have flexibility in data processing a UNIX type

multitasking operating system is going to be used on the host computer. The
planned functions of the host computer are the followings:
- control of radar-computer interface (transmision-reception of
  parameters, preparation of radar operational modes, supervision
  of radar control procedures, etc.)
- control of radar video processors (setting modes, reception of
  digital radar data, etc.)
- radar data processing (producing precipitation amounts, data
  compression of generation output data fields for displaying,
  archiving and transmitting)
- graphical presentations (displaying, loading, savings of images,
  composition of mosaics, annotation, hard-copy)
- data transmission and reception on telecommunication lines (radar
  data transmission, telemetered data reception)
- processing of general purposes (archiving, producing statistics
  etc.)
- options (for the time being, the digital processing of APT and
  WEFAX transmission of meteorological satellites is anticipated
  in this respect).

It is important to be noted that our system is not planned to operate
completely unmanned. It is quite natural, however, that the system should
be capable to execute automatically all the procedures necessary to
precipitation measurement by radar, including radar data transmission, too.

Data transmission and telemetering system

It is a prerequisite for post-processing the data from the automated
system for hydrological and other meteorological purposes to transmit them
in time to the places of work where they are needed. This requires
telecommunication links with speeds adequate to the data volume and the
time available for real-time processing. To meet this requirement two
telecommunication links of 9600 bps will be established between the radar
station and the Directorate Authority. Significant investments have been
commenced recently on the area of interest of the District Authority in
order to develop a modern flood-prevention information system. In the
framework of these investments their telemetering measuring system has been
developed and modernized and all the places of work have been equipped with
personal computers and an up-to-date telecommunication system is under
development, too. These will make possible real-time calibration of the
radar measured precipitation data by the data from the telemetering
raingauge network. By the end of 1988 seven telemetering raingauges were in
operation providing precipitation data at 5 minutes intervals and further 5
stations are expected to be installed by the end of this year.

Transmission of data from the automated weather radar and the
telemetering system and from other sources, too, is going to be solved by
establishing a microwave data link in common effort with the Hungarian PTT.
One of the relay stations of this microwave link will be situated in the
weather radar station that provides excellent conditions for transmission
of digitized radar data. The overall transmission capacity of this
microwave data link vill be 2,1 Mbits/s. Its accomplishment is expected in
the first quarter of 1990.

Workstation for radar data

Digital weather radar information for hydrological purposes can be
utilized at different levels. In this project an interactive utilization
has been supposed to, which makes possible testing and accomplishemen of
different methods and techniques. A workstation being capable for the

reception and display of weather radar data and for image processing is going to be developed and join the LAN at the District Authority. In addition, the workstation is going to execute other image processings too, related to environment protection tasks of the District Authority (multispectral images from space and aircraft, etc.).
The basic tasks of this workstation, related to radar data processing, are as follows:
- Automated functions
  It receives then evaluates the radar precipitation data over the catchment and subcatchment areas together with the telemetered data. It provides both raw and processed radar data to be disposed for the LAN at the District Authority and it takes care of archiving the data and administration of procedures executed and finally from time to time transmits the telemetered precipitation data to the radar station.
- Interactive functions
  It makes possible the graphics presentation of digital radar data and processed products, including the archived data, too. Radar observations can be initiated by the operator of the workstation by the use of its interactive capability. In interactive mode two way picture transmission is possible between the workstation and the host computer at the radar station.
  This workstation with graphics display will have other regular tasks related to the digital processing of the analogue APT and WEFAX images from meteorological satellites. This is because of the reliability of radar precipitation data can be enhanced by comparison with satellite data or the latters make possible to estimate precipitation amount over the areas beyond the radar coverage. Polar orbiting meteorological satellites also provide acceptable estimation on the snow cover which is of great importance in flood situation in early spring. The workstation is based on a personal computer, equipped with graphics adaptor of multi-image plane and enhanced image processing software package.

ALGORITHM SELECTION FOR PRECIPITATION MEASUREMENT

MRL-5 weather radars operating on X and S bands are used in our national weather radar network. This renders possible to apply of two basically different methods for precipitation measurement. One of them is based on the reflectivity factor measurement on single wavelength and the other is founded on attenuation measurements on dual-wavelength. In both cases various corrections should be applied to get reliable radar precipitation data. As the automated system will have two independent channels for data processing, it is assumed, we shall have possibilities for experimentations with different algorithms of precipitation measurement in the same weather situations. At the same time an objective method for the recognition of precipitation type is intended to be accomplished by the aid of volume scan on dual-wavelength.

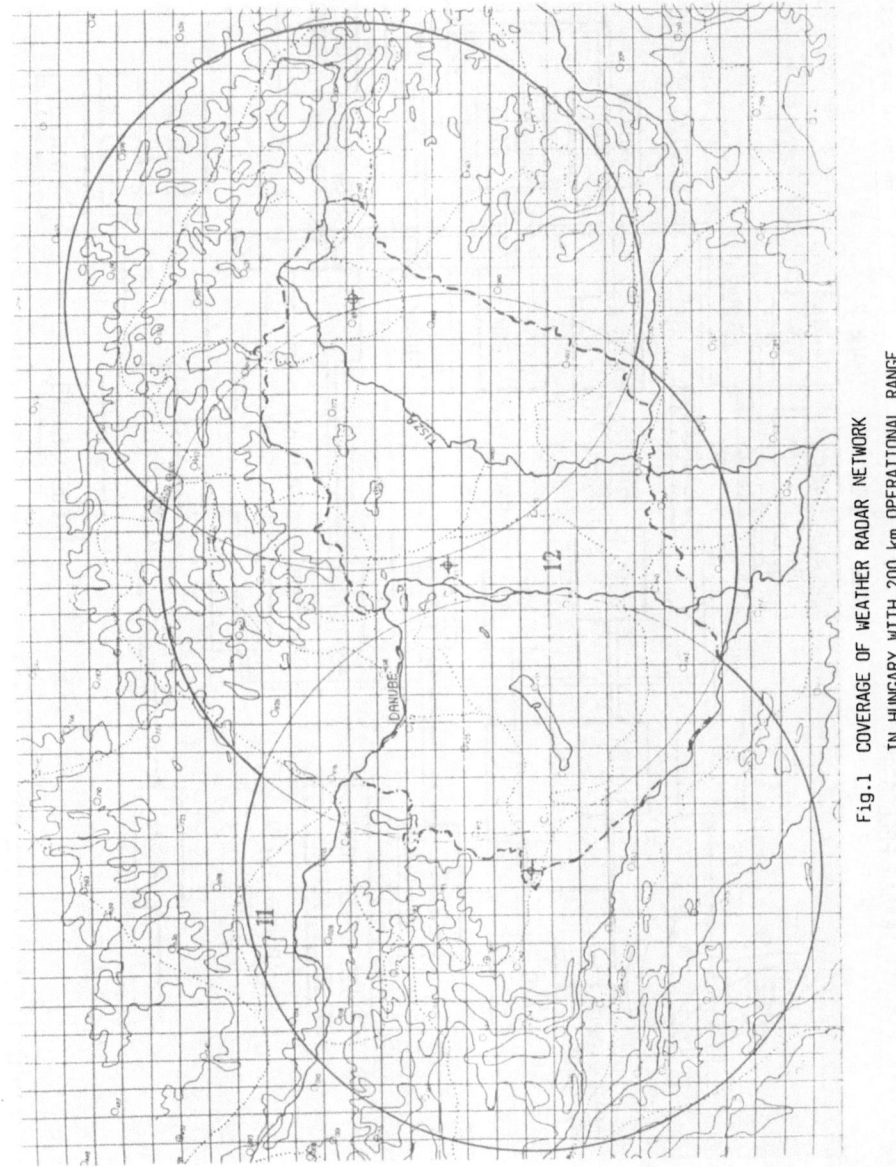

Fig.1 COVERAGE OF WEATHER RADAR NETWORK
IN HUNGARY WITH 200 km OPERATIONAL RANGF

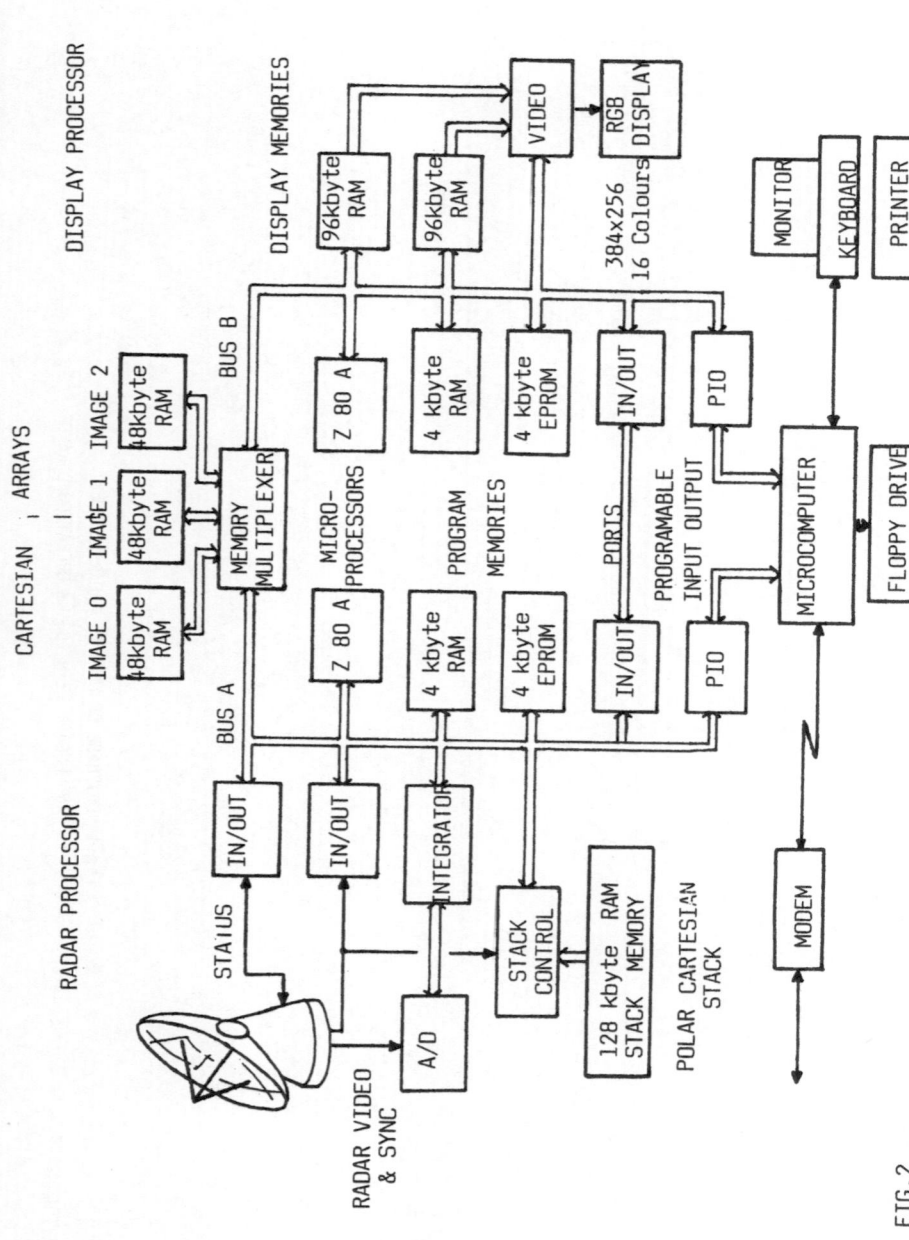

FIG.2
EXPERIMENTAL SYSTEM OF LOW COST
FOR WEATHER RADAR AUTOMATION

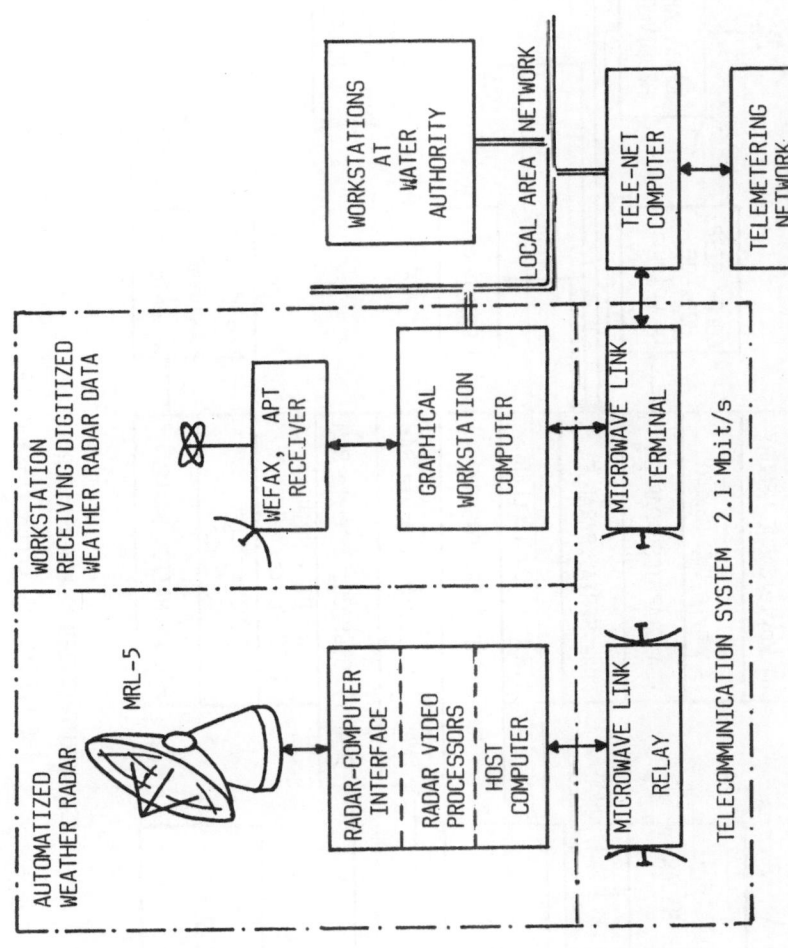

FIG.3

WEATHER RADAR PROJECT OF UPPER TISZA RIVER

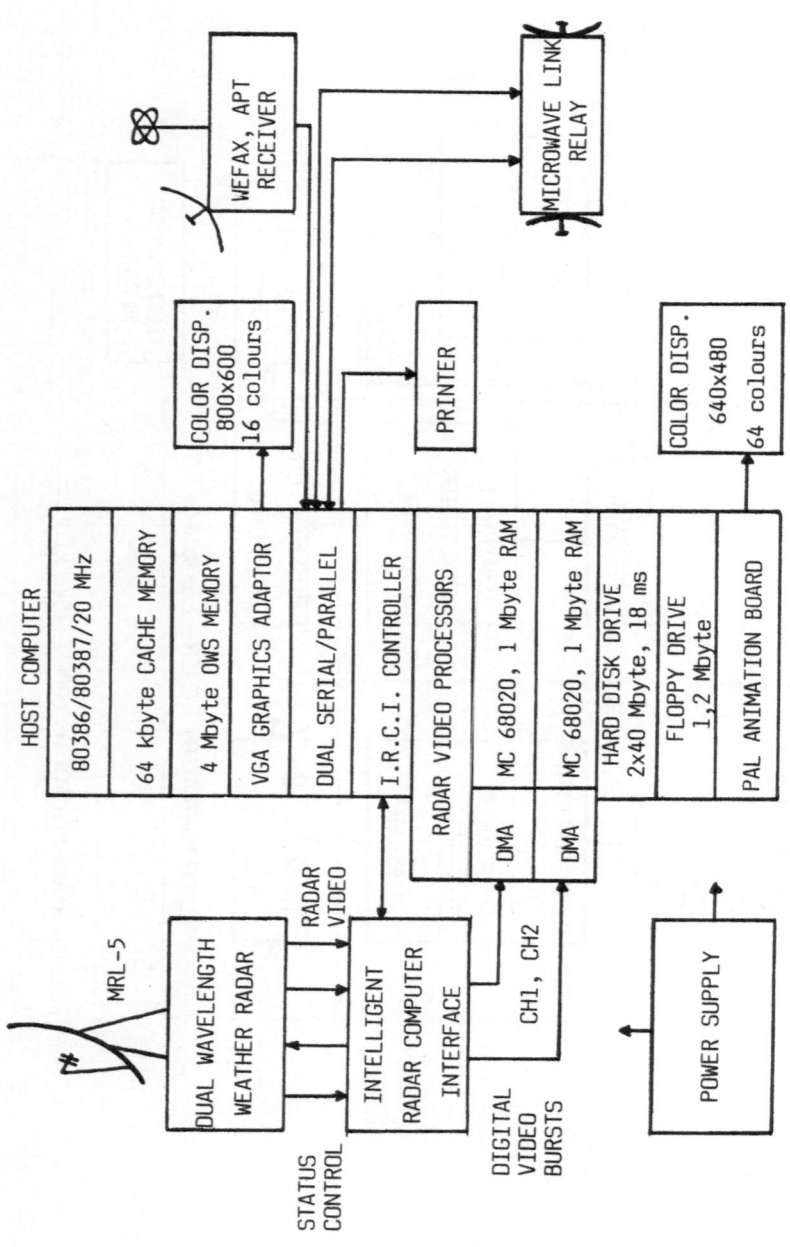

FIG.4 AUTOMATIZED WEATHER RADAR IN
THE PROJECT OF UPPER TISZA RIVER

OVERVIEW OF NATIONAL AND REGIONAL RADAR METEOROLOGICAL ACTIVITIES IN ITALY

R. Sorani and E. Dietrich
Italian Meteorological Service

Summary
The different plans of exploitation of meteorological radar systems in Italy, as well as future development of networking are reported in this paper. The use of radar data in support to economical sectors at regional and national levels is also outlined. A picture comes out in which radar meteorology is going to assume an increasing importance for assistance to aviation, crop protection and planning, water resources management and general weather forecasting. Firm proposal for inclusion of different radars into COST 73 network is a guideline to present and future actions.

1. INTRODUCTION

Many workshops, seminars, conferences and discussions have emphasized in recent years the potential benefits of a wide use of meteorological radars.

Following both the political influence and the technical expertise of COST 72 Action, requirements of modern radar systems have been identified, specially suitable for an efficient support to the different economic activities in Europe.

In Italy a few sectors which can benefit by radar data have been identified:
- flight assistance to civil and military aviation;
- meteorological support to agriculture;
- nowcasting and early warning system;
- planning of hydrological resources;
- support to weather modification experiments.

The above listed fields of application reflect the prominent interest towards radar meteorology in Italy and take into account the increasing importance of meteorology in planning social and economic development at regional and national levels.

The complexity of orography in our territory causes a number of problems in using radars. Two major mountain systems, the Alps in northern part of Italy and the Appennines along the N-S line, affect weather conditions and make particularly difficult to forecast and to track mesoscale and local scale perturbations.

It is well known that frontal and instability lines approaching Italy go into deep modification of their structure producing local cyclones, enhanced precipitations and thunderstorms of limited extension.

Several airports in Italy experience severe wind shear conditions due to orographic effects and hail causes major damages to our agriculture .

Sea storms affect coasts with high frequency in winter time determining hazards to navigation and even loss of properties and human lives.

The lack of water in several southern Italian regions is also well known; a few zones suffer from droughts and, in those areas, a careful management of water resources is mandatory.

Trying to understand, forecast and mitigate negative effects of fast variable weather conditions, several Agencies have installed their own meteorological radars and have plans of exploitation and development.

Fig. 1 summarizes the situation of radar sites in Italy, as it will appear in two years time. Operating Agencies are also indicated.

2. METEOROLOGICAL RADARS AND FLIGHT ASSISTANCE

Flight assistance in Italy is jointly operated by the National Meteorological Service, which has also a special responsibility for military flights, and the Azienda Autonoma per l'Assistenza al Volo ed al Traffico Aereo Generale (AAAVTAG).

Both Agencies have same programmes in the exploitation of meteorological radars; some new equipments will be soon installed.

Terminal areas forecasting techniques and methods, nowcasting procedure and warnings of severe and/or hazardous phenomena are the main tasks carried out by the National Meteorological Service.

Due to the presence of relieves many airports in Italy experience peculiar conditions of severe wind shear. Up to now prediction techniques are mainly based upon surface wind observations in different sites in the vicinity of runways. However limits of this approach are evident, neither the extended use of SODARs have proved to be satisfactory.

Doppler facilities in the new radar system will help greatly in solving the problem even if, at present, consolidated algorithms do not seem to be available.

Large size thunderstorms in the Po valley, as well as lee cyclogenesis over the gulf of Genova are very known meteorological phenomena affecting Italy and surrounding seas.

Storms in the Po valley originate suddenly, caused by huge heating and high concentration of humidity, and develop quickly up to 10 - 12 km height.

Cyclones originating in the gulf of Genova, lee to western Alps, often develop as quasi stationary depressions with very bad weather conditions persisting throughout 4 ÷ 5 days and affecting several regions of our territory.

Floods and damages to coastal infrastructures are relevant and inland disasters may even cause people to be injured and killed.

We know by experience that hail is often associated with storms and that gust fronts and squall lines cause temporary cutting of flights activity and large damages to crops.

Synoptic and mesoscale prediction techniques do help greatly in forecasting events; however it is only by means of sophisticated digital Doppler radar that the most dangerous areas of turbulence and heavy rain can be detected and tracked.

The possibility exists that dual polarization techniques can be utilized to detect hail in cumulus clouds.

Combined use of new radars information and satellite data will improve the capability of short term forecasting of mesoscale and local phenomena.

With reference to Fig. 1, it can easily seen that fast moving cold fronts affect Sardegna Island and quickly move to the Thyrrenian coasts.

No synoptic information is available on the sea path but geostationary satellite images; accurate digital radar data are essential for continuous monitoring of severe weather phenomena approaching our most important airports located along the western coast of Italy (Sites n° 2,12,8,11 in Fig.1).

Actions related to an international tender are on the way at the moment of publication of this paper. Therefore no technical specifications can be reported below.

All radars will be fully on line with COST requirements.

3. THE RADAR SYSTEM IN VENETO REGION

The Veneto Region has achieved an integrated system of monitoring of weather conditions which includes a raingauges network, a meteorological radar, some satellite receiving facilities and a computing Center (site n°4 in Fig.1).

The meteorological radar UBS 10304/3 has been installed since 1988, and tests have been completed; the radar now is working in an operational way, mainly for quantitative measurements of rainfall rate. Calibration in real time is carried out by intercomparison with tele-raingauges.

The radar is located on the top of "Monte Grande" and data are computed at the operational Center "CISM", 2 km apart.

A few software packages have been implemented, mainly dedicated to:
- rainfall measurements (including totals in time)
- rainfall forecasts (in terms of probability per levels of intensity)
- storm analysis
- wind field analysis

Rain measurements are performed using 1x1 km resolution up to 120 km, and 2x2 km resolution up to 240 km, and different values of A and B constants are used in the Marshall-Palmer relationship.

Forecasts of amount of precipitation are computed by cross-correlation techniques; "rain-patterns" movement (speed and direction) is evaluated by sequence of images.

Product outputs (precipitation intensity every 10 minutes and cumulated forecasted precipitation in 3 hours) feed numerical hydrological models and a water balance model for agricultural uses.

The studies on thunderstorms and showers is another important field of application: the radar processes volumetric digital data up to 12 km in height and CB clouds are analyzed using vertical projections along a direction. Top levels of maxima echoes are available too.

Within the frame of ECC project "Application of weather radar for the alleviation of climatic hazards", two main research-themes have been implemented:
- comparison between radar estimates of precipitation and ground measurements with raingauges;
- detection of hail by radar.

In Appendix A the main characteristics of the radar system are listed.

4. THE RADAR SYSTEM IN EMILIA-ROMAGNA REGION

The Regional Meteorological Service in Emilia-Romagna has recently installed a digital radar system based on GPM 500 C multiparameter sensor.

This radar will be used mainly for agriculture planning, crop protection and hydrological resources management in the Region.

The system provides the following products:
- Precipitation intensity at ground surface
- Precipitation at various altitudes
- Total precipitation at ground level (1h, ...24h)
- Area integral of precipitation for particular areas and times
- Vertical integral of liquid water
- Maxima of rainfall rates (on three cartesian planes)
- Echo tops on a plane
- Average radial wind at various altitudes

-     Spectral amplitude of wind at various altitudes
-     Transverse and radial gradient of wind
-     Maps with vectors of horizontal components of wind at various altitudes
-     Combination of precipitation intensity, radial wind and spectrum amplitude
-     reflectivity
-     differential reflectivity

Vertical sections are also available.

A radio relay system makes it possible to rapidly transmit to the Bologna Operations Room "Products" and "raw data" for further processing.

In this way the Regional Meteorological Service has radar information available which can be used directly or in conjunction with other data in the preparation of nowcasting outputs.

Other parameters will be further investigated to fulfill regional meteorological needs, including local specialized short term forecasting:

- vertical and horizontal clouds structure (dimensions, gradient of reflectivity)
- precipitation (type and intensity)
- turbulence (inside clouds and in clear air)
- wind (velocity, shear, exceptional phenomena)

In Appendix B the main characteristics of the radar system are reported.

Detailed information concerning the use of this radar are given in another session of the Seminar (Nanni, Salsi, Proceedings of COST 73 Seminar - 1989).

5. METEOROLOGICAL RADAR FOR REDUCTION OF HYDROLOGICAL RISKS: ARNO PROJECT

Keeping in mind the terrific flood that in 1966 caused enormous damages to the city of Florence, and several other destructive events, a wide program of hydrological assessment of many river basins was established and is being implemented.

A special project is concerning Arno river and measurements of precipitation by radar are regularly fed into a hydrological model for the monitoring of the Arno flow and flood forecasting (radar site nº14).

To this purpose the radar POLAR 55C of the National Research Council (CNR) will be placed by the end of 1989 in Monte Maggio (Siena).

The main characteristics of this radar are in Appendix C.

By the end of 1990 another radar system with same characteristics of site nº3 is expected to be installed in site nº14 (replacing the existing), integrated in experimental way in an advanced system for continuous monitoring of hydrologic hazards (SICIG - Sistema Integrato per il Controllo Idro Geologico).

Combining precipitation measurement by rain gauges with radar measurements and comparing those data with dual polarization measurements the theoretical possibility exists of having rainfall data over a large area with accuracy of ±25 %

Horizontal reflectivity Z_H accuracy will be ± 1.0 dB and Z_{DR} accuracy ±0.2 dB.

Algorithm used comes out from the basic relation

$$R = 0.6\pi \int_0^\infty N(D)v(D)dD \qquad (1)$$

where v(D) is the ground velocity of drops and N(D) is the drops distribution.

Assuming a distribution such as

$$N(D_i) = N_0 \exp\left(-3.67\frac{D_i}{D_0}\right) \qquad (2)$$

- 114 -

where:
D_m = drop equivalent diameter
D_o = average drop diameter
N_o = number of drop per volume unit
it can be demonstrated that Z_{DR} is approximately related to D_o by

$$Z_{DR} = 15.6\,D_0 - 0.63 \qquad\qquad (3)$$

D_o can thus be extracted by (3) and N_o by measuring Z_H.
Relation (1) can be expressed as a function of Z_H and Z_{DR}

$$R = \frac{1}{119}\ \frac{Z_H}{(Z_{DR}+0.63)^{2.45}} \qquad\qquad (4)$$

A method for decreasing down to 25% the error due to the differential attenuation of signals in polarization diversity is at present under evaluation.

6. SUPPORT TO WEATHER MODIFICATION EXPERIMENTS

6.1 The Italo-Yugoslavian Hail Project

Italy and the Yugoslavia, following the World Meteorological Organization (WMO) Convention and within the Osimo Agreements, signed on November 10th 1975, in the interest of mutual economic and technical collaboration and in the spirit of limiting hail damages, have decided to set up a Common Antihail Defense System, located in the bordering areas, in Friuli-Venezia Giulia and in Slovenia.

Such Defence System, based on the physical process of "embryos competition", will be implemented by seeding the potential hail clouds with suitable rocket-carried products in the right zone of cumulus-nimbus. The above under the temporal-spatial control of two computerized meteorological radars, located one in Italy zone (Friuli-Venezia Giulia Region) and the other in Yugoslavian territory (Republic of Slovenia).

Considering the many-fold aspects of such defence system, and in agreement with WMO recommendations, it was also decided to carry out some research activity in order to enhance the knowledge of hail-cloud physics. To this purpose a radar sensor has been committed, taking into account the results of the researches carried out in the field of radar-meteorology during the last years, particularly considering American experience acquired by NCAR, NOAA, NSSL.

In particular, the system will be able to provide both horizontal reflectivity (Z_H) measurements and the differential reflectivity (Z_{DR}), in conjunction with Doppler measurements. Further possibility to modify the system, with a minimum of structural variations, in order to enable the CDR (Circular Depolarization Ratio) or LDR (Linear Depolarization Ratio) was also requested.

As far as the data processing are concerned, here including the management of the whole Defence System, the computing system will be organized around a "central database" used by one or more "Intelligent Systems" with the aim, in the future, to become an Artificial Intelligence System.

Furthermore, considering the rapid evolution of hail clouds (complete evolution in a mean time of 30 minutes) and the consequent need to undertake defence actions in a very short time from hail detection, the radar system will operate only in real time, simultaneously acquiring Z, Z_{DR}, V e σV parameters.

To this aim, apart from standard volume scan, sectorial and alternate scans will be used, together with special techniques to reduce the time of acquisition.

At present a system based on the multiparametric GPM 500 C radar sensor is under delivery.

In Appendix D the System main characteristics are listed.

6.2 The precipitation enhancement project

A no profit Association named TECNAGRO is also very active in Italy as far as weather modification techniques are concerned.

Using mobile C band digital radars TECNAGRO, following a programme designed by prof. Gagins, and under the scientific consultancy of dr. Nania, has being performing campaigns of rain stimulation in two target areas in South of Italy (Puglie Region).

Operational campaigns have been preceded by feasibility studies and by limited experiments with in flight and ground measurements of clouds parameters.

Echo top levels, maxima of reflectivity and RHI structure are also investigated before cloud insemination with AgI agent is decided.

Aircraft are guided towards targets by the simultaneous monitoring of ground control and meteorological radar, and seeding is done upwind to selected cloud patterns.

Preliminary results of enhanced precipitation are encouraging and extension of techniques are planned to include other regions in south-Italy and Sardegna island.

Recently the European Community has shown considerable interest for the initiative and a "ad hoc" working Group has been established.

7. SUMMING UP

Radars numbered with 1, 2, 5, 7, 8, 9, 10, 11, 12, 15 in Tab.1 are regularly operated by the National Meteorological Service (A.M. - 8 radars) and by the AAAVTAG (Agency for Flight Assistance and Civil Aviation - 2 radars). All these radars, analogous at present, are going to be digitalized or replaced by multiparameters digital systems.

Tab.1

N°	SITE	BAND	AGENCY	Z	Z_{DR}	V, σV
1	Milano	C	AAAVTAG	Y	Y	Y
2	Pisa	C	Met.Serv.	Y	N	N
3	Capofiume	C	Emilia-Romagna Region	Y	Y	Y
4	Teolo	C	Veneto Region	Y	Y	Y
5	Istrana	C	Met.Serv.	Y	N	N
6	Cervignano	C	Italo-Yugo Project	Y	Y	Y
7	Cagliari	C	Met.Serv.	Y	Y	Y
8	Napoli	C	AAAVTAG	Y	N	N
9	Brindisi	C	Met.Serv.	Y	Y	Y
10	Catania	C	Met.Serv.	Y	N	N
11	Trapani	C	Met.Serv.	Y	N	N
12	Roma	C	AAAVTAG	Y	Y	Y
13	Bari	C	Puglie Region	Y	Y	Y
14	Monte Maggio	C	Arno Project	Y	Y	Y
15	Alghero	C	Met.Serv.	Y	N	N

Radar n°4 is operating and radar n°3 is already installed.
Radar n°6 is going to be delivered and will be operating within the frame of the Italo-Yugoslavian Hail Project.

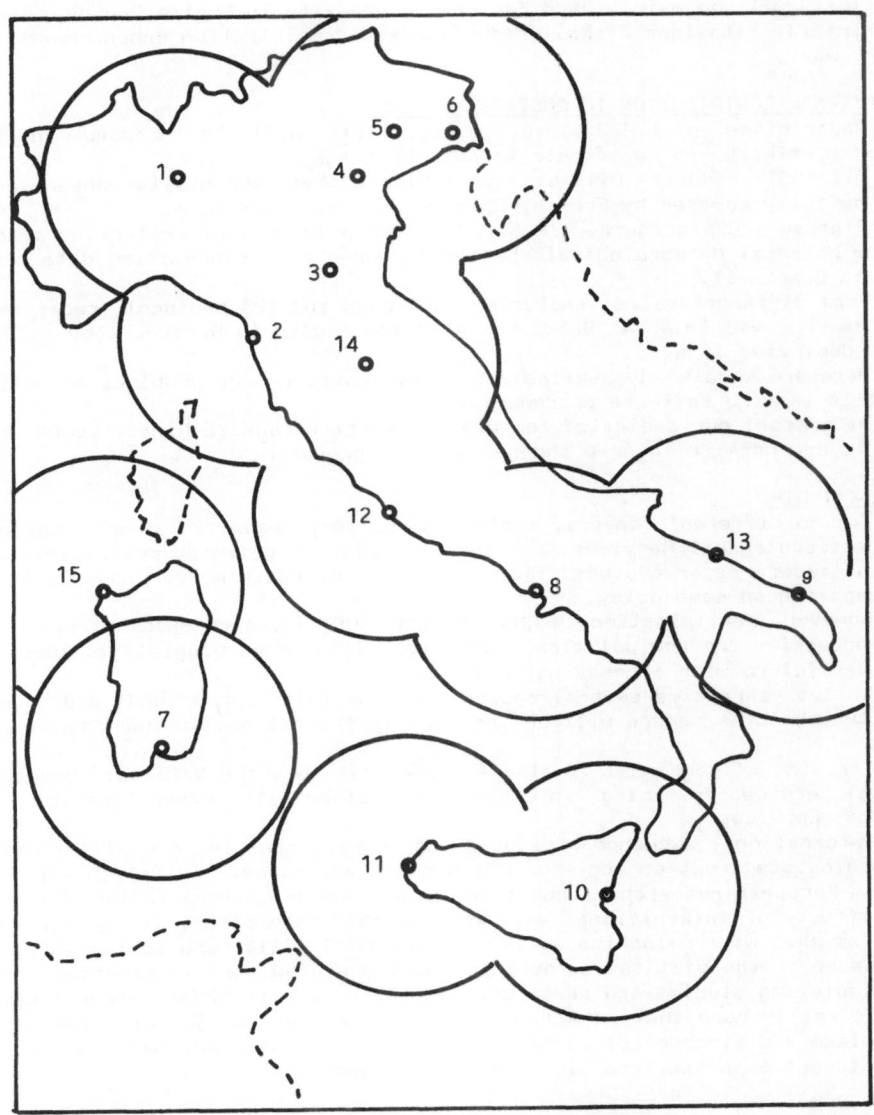

Fig. 1

Digital meteorological radars in Italy by the end of 1991.
Site denomination and operating Agencies in Table I

Radar n°13 will be mainly used for crop protection in Puglie Region, where experimental campaigns of hail reduction and precipitation enhancement are on the way.

8. NATIONAL CONTRIBUTION TO COST-73 NETWORK

Radar sites n° 1,2,3,4,5,6,7,8,9,10,11,12 will be included in the European network and contribute to data exchange.

All COST products will be available so that the Mediterranean area will be fully covered by French, Spanish and Italian radars.

Planned TLC procedures are based on the principle of collecting images at the National Meteorological Center in Rome and transmitting data from Rome to Bracknell.

From different sites, and using BUFR code and X25 protocol, radar data will be received in a VAX 8250 computer and routed to Bracknell by a 9600 bauds dedicated line.

Hardware is already available in Rome, however same problems do still exist in setting software procedures.

At present our dedicated computer does not recognize binary codes and experts are working to make the necessary improvements.

9. CONCLUSION

Due to different reasons, including the very recent birth of regional agro-meteorological Services, and the necessity of defining priorities and uses of modern radar systems, the development of radar meteorology in Italy has experienced some delay.

However, participation in COST 72 and COST 73, a renewed interest by our industries and new political decision allow meteorologists to look at the near future with increasing confidence.

In two years time several radars will be fully operational and every year two or three radars will be included in the national network up to 15 sites.

Training of personnel is also a major task that the National Meteorological Service is facing in close cooperation with other Agencies and Regional Services.

International exchange of ideas and experiences is a reality which meteorologists trust on and from which major advantages can be achieved.

As European researchers and technicians we do believe in the beneficial effects of international work and, in this connection, it is with some concerns that we are looking at 1991 when COST 73 will come to its end.

Never in the past the European Community played such an important role in stimulating studies and researches in the domain of radar meteorology.

We really hope that we will be able to continue our job with the same enthusiasm and sincere spirit of cooperation which already made our countries to get important and significant progresses.

Main characteristics of radar-system in Veneto Region

Denomination: Radar della Regione Veneto - Centro Sperimentale per l'Idrologia e la Meteorologia
Site: Località Monte Grande (TEOLO)
Radar type: UBS 10304/3
Manufacturer: ERICSSON
Operational frequency: C band
Coherent transmitter-receiver
Polarization diversity

Present configuration:
- Multiparameter Radar Sensor (MPRS)
- Radar Data Processor (RDP)
- Radar Data Storage System (RDSS) on magnetic tape

Main features:

Antenna:
4.2 meters solid parabolic disk reflector (fibre glass and epoxy)
Gain: 43 dB
Beam width: 0.86°
Sides lobes: -30 db
Radome: 6.7 meters diameter (fiber glass designed) <0.5 dB loss

Pedestal:
Continuous scan in azimuth and from -2° to +90° in elevation

Polarization switch time: <10 µs

Transmitter:
type: long life coaxial magnetron
peak power: 270 kW
pulse duration: 0.5, 2.0 µs
P.R.F.: 250, 656-875, 900-1200 Hz

Receiver:
type: supereterodyne (image reject.)
noise: 5 db
Dynamic: >85 dB

Doppler measurements:
Input video: linear "I" and "Q"
Algorithm: Fast Fourier Transform
Unambiguous velocity: ±48 m/s ±1 m/s
Clutter suppression 32 dB

Operative range: 120 km (Doppler, Z_{DR}), 240 km (Z)

RDP:
VAX 8200 computer
Cartesian cells: 1x1x1 km (Doppler), 2x2x1 km (Z)
Colors display - All COST products available

Main operational use:
- METEOROLOGY (analyses and forecasts)
- HYDROLOGY (rain measurements and forecasts)
- AGRO-METEOROLOGY (rain forecasts and thunderstorms studies)

Input video: linear "I" and "Q"
Average speed "V" using Pulse Pair algorithm.
Speed variance (σV) as a second moment of the Pulse Pair.
Unambiguous velocity: ±49 m/s ± 0.5 m/s

Radar consolle:

- Presentation of radar video, raw and unprocessed, at the signal processor output, on high-resolution (1024 pixels on color raster scan monitor) in order to display PPI and RHI maps for real time operation and maintenance purposes.
- Full sensor control by external computer.

The output data from the signal processor (RSP) go to a multiprocessor computer system (RDP) (fast parallel link and Ethernet central bus) with the following characteristics:

- radar control with optimization of operational parameters
- raingauges network interfacing
- real time raw data preprocessing (data correction, polar to xyz conversion, interpolation)
- real time COST-72 products availability (for local display and radar network or secondary users transmission)
- immediate response time on products display request
- display on high resolution graphic workstation (up to 1280x1024 pixels) with cartographic maps overlay
- single image or sequence animation (4 images per second) optionally 15 images/s, 3D animation
- database management with optimized retrieval criteria and large on-line storage capabilities
- optical removable cartridge system management for off-line historical archiving
- data and images disseminations to secondary users

Main characteristics of radar-system in Emilia-Romagna Region

Denomination: ERSA Emilia-Romagna Servizio Meteorologico Regionale
Site: S.Pietro Capofiume (BOLOGNA)
System supplier: jointly by SMA, SELENIA, DATAMAT Companies
Radar type: GPM 500 C
Operational frequency: C band
Coherent transmitter-receiver
Polarization diversity

Present configuration:
- Multi Parameter Radar Sensor (MPRS)
- Radar Signal Processor (RSP)
- Radar Data Storage System (RDSS) on optical medium
- Radar Data Processor (RDP)

Main features:
Antenna:
- 5 meters offset double reflector
- Gain: >45 dB
- Beam-width: 0.9°
- Sidelobes: <-30 db
- Cross-polarization: ≤-27 dB
- Radome: no

Pedestal:
High precision, high dynamic response (sectorial scan), good resistance to wind torque without protective radome
Polarization switch time: <5 µs
Transmitter:
- Type: coherent chain
- Power amplifier: klystron
- Peak power: 700 kW
- Pulse-width: 0.5, 1.5, 3.0 µs
- P.R.F: 1200, 600, 300 Hz
- Up-grade provision for frequency agility, phase codes, etc.
Receiver:
Type: supereterodyne (image reject.) with LNA
Noise figure: <5 db
Dynamic:
Intensity: Log >80 dB, linearity ±1 dB
Doppler: Lin >30 dB extended to 90 dB by I.A.G.C.
IAGC: Instantaneous pulse to pulse, range bin to range bin,
Automatic Gain Control (processor controlled)
R.S.P. : 1024 range cells, 32 bit floating point processing
Measure of Z:
Algorithm: linear averages up to 1024 pulses
Accuracy: ±1 dBz rms
Clutter suppression: up to 20 dB (±1 dBz)
Measure of Z_{DR}:
Algorithm: difference between linear averages H and V
Accuracy: ±0.2 dB
Doppler measurements:

Input video: linear "I" and "Q"
Average speed "V" using Pulse Pair algorithm.
Speed variance (σV) as a second moment of the Pulse Pair.
Unambiguous velocity: ±49 m/s ± 0.5 m/s

Radar consolle:
- Presentation of radar video, raw and unprocessed, at the signal processor output, on high-resolution (1024 pixels on color raster scan monitor) in order to display PPI and RHI maps for real time operation and maintenance purposes.
- Full sensor control by external computer.

The output data from the signal processor (RSP) go to a multiprocessor computer system (RDP) (fast parallel link and Ethernet central bus) with the following characteristics:
- radar control with optimization of operational parameters
- raingauges network interfacing
- real time raw data preprocessing (data correction, polar to xyz conversion, interpolation)
- real time COST-72 products availability (for local display and radar network or secondary users transmission)
- immediate response time on products display request
- display on high resolution graphic workstation (up to 1280x1024 pixels) with cartographic maps overlay
- single image or sequence animation (4 images per second) optionally 15 images/s, 3D animation
- database management with optimized retrieval criteria and large on-line storage capabilities
- optical removable cartridge system management for off-line historical archiving
- data and images disseminations to secondary users

Main characteristics of C.N.R. Polar 55C radar

Antenna:
- 4.57 meters offset
- Polarization: horizontal or vertical
- Gain: 45.5 dB
- Azimuth beam-width: 0.92°
- Elevation beam-width: 1.02°
- Sidelobe level: <-32 db
- Cross-polarization (ICR): ≤-27 dB
- Radome: no

Transmitter:
- Type: coherent chain
- Power Amplifier: klystron VCK 7762
- Frequency: 5395÷5595 MHz (C band)
- Peak power: 500 kW
- Pulse-width: 0.5, 1.5, 3.0 µs
- P.R.F: 1200, 500, 250 Hz

Receiver:
- Number of channels: 2
- Response: linear, logarithmic
- Dynamic range: 35 dB+ IAGC (Lin), 80 dB (Log)
- Noise figure: 5.5 dB
- IF center frequency: 60 MHz
- IF band-width: 2. 0.7, 0.5 MHz

Main characteristics of Italo-Yugoslavian Hail Project radar-system

Denomination: Italo-Yugoslavian Hail Project
Site: S.Pietro Cervignano (UDINE)
System supplier: jointly by SMA and SELENIA Companies
Radar type: GPM 500 C
Operational frequency: C band
Coherent transmitter-receiver
Polarization diversity

Present configuration:
- Multi Parameter Radar Sensor (MPRS)
- Radar Signal Processor (RSP)
- Radar Operation Control (ROC)
- System Operation Supervisor (ROS)
- Radar Data Storage System (RDSS) on optical medium

The main features are:
Antenna:
- 5 meters offset double reflector
- Gain: >45 dB
- Beam-width: <1°
- Sidelobes: <-30 db
- Cross-polarization: $\leq$-27 dB
- Radome: no

Pedestal:
High precision, high dynamic response (sectorial scan), good resistance to wind torque without protective radome
Polarization switch time: <5 µs
Transmitter:
- Type: coherent chain
- Power amplifier: klystron
- Peak power: 700 kW
- Pulse-width: 0.5, 1.5, 3.0 µs
- P.R.F: 1200, 600, 300 Hz
- Up-grade provision for frequency agility, phase codes, etc.
Receiver:
Type: supereterodyne (image reject.) with LNA
Noise figure: <5 db
Dynamic:
    Intensity: Log >80 dB, linearity ±1 dB
    Doppler: Lin >30 dB extended to 90 dB by I.A.G.C.
    IAGC: Instantaneous pulse to pulse, range bin to range bin,
        Automatic Gain Control (processor controlled)
R.S.P. : 1024 range cells, 32 bit floating point processing
Measure of Z:
    Algorithm: linear averages up to 1024 pulses
    Accuracy: ±1 dBz rms
    Clutter suppression: up to 20 dB (±1 dBz)
Measure of Z_{DR}:
    Algorithm: difference between linear averages H and V
    Accuracy: ±0.2 dB
Doppler measurements:

Input video: linear "I" and "Q"
Average speed "V" using Pulse Pair algorithm
Speed variance (σV) as a second moment of the Pulse Pair
Unambiguous velocity: ±49 m/s ± 0.5 m/s

Radar consolle:
- Presentation of radar video, raw and unprocessed, at the signal processor output, on high-resolution (1024 pixels on color raster scan monitor) in order to display PPI and RHI maps for real time operation and maintenance purposes.
- Full sensor control by external computer.

System Operation Supervisor (ROS):
Radar Operation Full Optimization and data flow control by interactive menus or by intervention of an external computer system

OVERVIEW OF RADAR NETWORKING BY MOC, JAPAN
AND ITS DATA DISSEMINATION SYSTEM (FRICS)

Dr. Yoshino, F. P.W.R.I., MOC.
Ichimiya, K. Economic Affairs Bureau, MOC.
Kanbayashi, Y. Dept. of Planning, FRICS
Dr. Yamaguchi, T. Dept. of Research, FRICS
Shirakawa, N. Dept. of Planning, FRICS

Summary

The present report briefly introduces the radar networking of the
Ministry of Construction (MOC) in Japan, the radar system which
provides radar rain gauge data within MOC and FRICS which supplies
radar rain gauge data to prefectures and local municipalities.

INTRODUCTION

Ministry of Construction, Japanese Govt. (MOC) is responsible for
river administration. In order to get information about heavy rainfall
areas, MOC has 16 radar rain gauges in operation (planned are 22).
But these data were distributed only to local construction offices of
MOC until a few years ago.
So, prefectures and local municipalities could not get such data
easily and promptly, which are responsible for flood defence.
Foundation of River and Basin Integrated Communications (FRICS) was
established in October 1985 as the body for processing and desseminating
hydrological data (such as radar rain gauge data and telemetered rainfall
and water level data by on-line) and river information (including flood
warning) obtained from MOC etc. under the back up of prefectures and
private organizations.
Authors introduce the radar network, characteristics as a rain gauge,
data collection and processing, and data dissemination system to local
construction offices of MOC (Radar system and terminal) and local
municipalities (FRICS system and terminal) each.

1. Radar Network

Flood forecasts and dam operations are extremely difficult because of
the characteristics of the rivers in Japan. Therefore, river
administrators (MOC and prefectures) must obtain not only meteorological
information from meteorological offices, but also quick and reliable
information about the amount of rainfall in basins and the water levels of
rivers from their own information network. They have developed and
improved various telemeter systems and radar rain gauge information
systems. Radar rain gauges have been developed independently by MOC since
1966. The first radar rain gauge system was completed at Mt. Akagi (Gunma
Pref.) in 1975. Among the 22 systems being planned, 16 are already in
operation and 2 are being constructed (see Fig. 1). The radar rain gauge
is able to obtain such quantitative information which is hard to observe

with meteorological radars (wave length 10 cm). It obtains reflected
signals from a rainfall area at 5-minute intervals, processes them by a
computer and displays them in color according to purposes. Therefore, it
never fails to catch local downpours which cannot be caought by a ground
rain gauge. In other words, it enables to watch current rainfall areas on
a display.

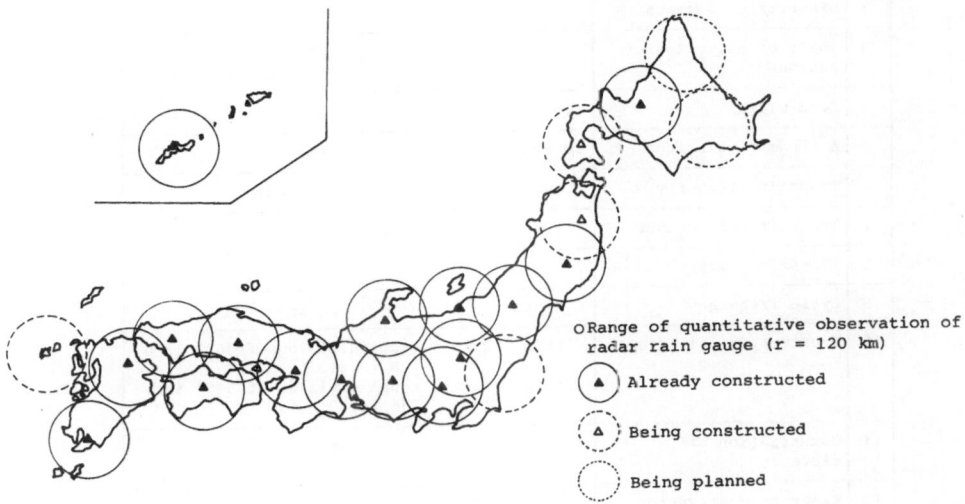

Figure 1. Locations of MOC's Radar Rain Gauges

2. Recent Radar Rain Gauges

Table 1 shows the specifications of the oldest Akagisan radar and the
newest Owasan radar systems. A radar rain gauge was installed on a high
mountain around in 1975 mainly for observing the mountain zone of a large
river. Recently, it is installed on a relatively low mountain for
improving the precision and for catching rains on a sea in addition to a
land.

Similarly, the diameter of the antenna was changed from 3 m to 4 m and
the beam width was decreased for raising the accuracy. The angle of
elevation was lowered from $0.8° - 1°$ to $0° - 0.5°$ for obtaining low
altitude data. The ground clutter (G.C.) which is increased by it is
removed by MTI (Moving Target Indicator). The shielded area which is
increased also by it is to be corrected with a correction factor.

In order to obtain equal altitude data, a radar which is capable of
simultaneous observations of 3 angles of elevation (simple Cappi) and a
Cappi are also in operation.

Since MOC attaches importance to radar data, they calibrate data with
ground rain gauges only at a few radars. However, they are conducting a
research on on-line calibration because the fluctuations of B and ß
$(Z = BR^ß)$ are found even during a rainfall. In addition, MOC owns one dual
Polarization Doppler radar (mobil type) for a research on the accuracy
improvements of radar rain gauges.

Table 1. Comparison of Performances of Akagisan Radar and Owasan Radar Systems

	Item	Akagisan radar rain gauge	Yamatosan radar rain gauge
1	Altitude	1962 m	632 m
2	Diameter of antenna	3 m	4 m
3	angle of elevation of antenna	1°	0.8°, −2° − 44.9° (remote control)
4	Beam width	1.6°	1.2°
5	Antenna revolutions	10 rpm	5 rpm
6	Transmit frequency	5,300 MHz	5,270 MHz
7	Transmit peak output	250 kw	
8	Transmit pulse	2 µs	
9	Cycle frequency	450 pps	260 pps
10	Quantization of azimuth	Within 120 km: Equal division of 360° into 128 (2.81225°) Within 120 km: Equal division of 360° into 256 (1.40625°)	
11	Quantization of distance	3 km	
12	Range of observation	Quantitative: 120 km in radius Quantitative: Appr. 200 km in radius	
13	Data updating interval	5-minute intervals	
14	Clutter removal	log subtraction	log subtraction and MTI

3. Collection and Processing of Data

Figure 2 shows the basic system and configuration of a radar rain gauge. Functionally, it consists of a radar unit, a data transmission unit, a data processing unit and an operation monitor & control unit.

(1) Radar unit

It is a data collecting subsystem which emits radio waves, receives and amplifies reflected waves from rain drops and converts them into digital quantity for data processing. It operates continuously for 24 hours.

(2) Data transmission unit

It is a data transmission subsystem which obtains mean data of each mesh from the above digital data according to the mesh structure shown in Figure 3 Mesh Configuration on Polar Cordinates (omitted). It raises the data quality by obtaining mean data of a fixed time and sends them to the data processing unit. It also sends control signals for monitoring and controlling operations, time data and other necessary information.

The mesh size is determined by the density which is required for grasping rainfall areas as well as the data transmission and processing time. In principle, the polar coordinates around a radar site are used for mesh division, as shown in Figure 3 (omitted). Within the area of 120 km in radious, a mesh is formed by divinding the azimuth (360°) equally into 128 and cutting them with concentric circles of 3 km intervals in the distance direction from the center. Within the area of 120 km – 198 km in radious, it is divided equally into 256 in the azimuth direction and by 3 km in the distance direction.

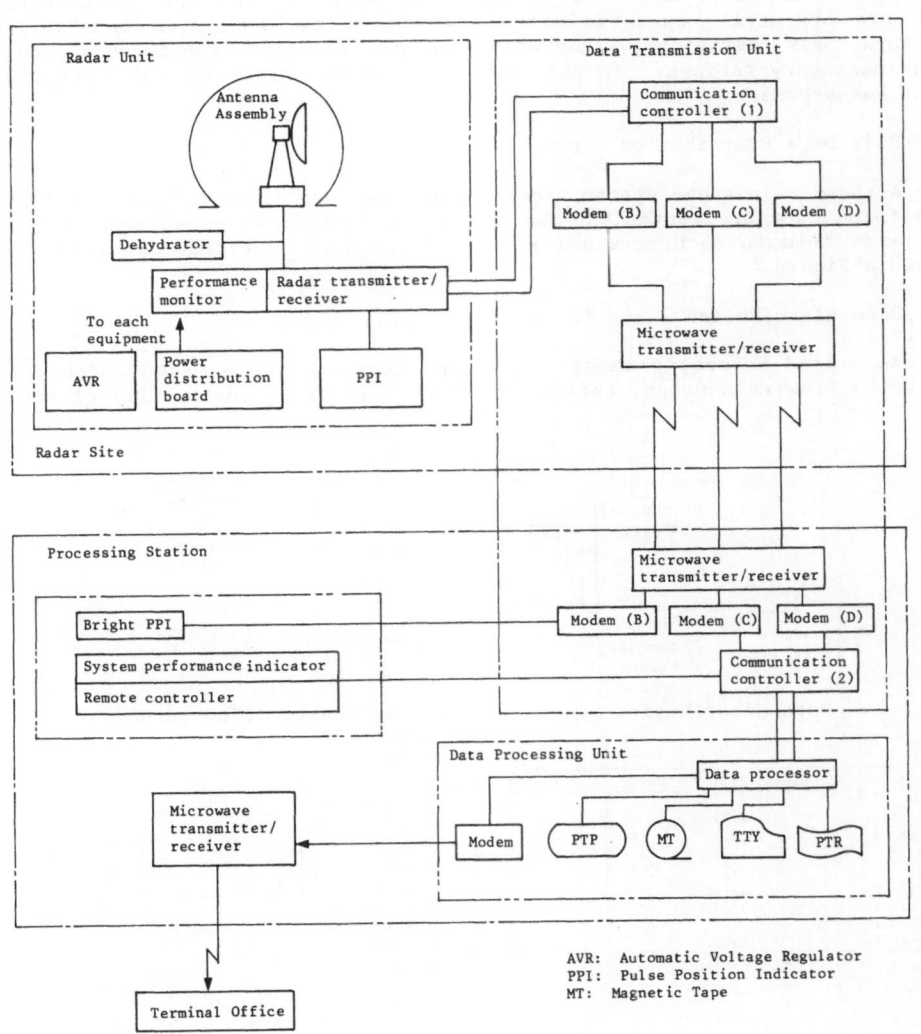

AVR: Automatic Voltage Regulator
PPI: Pulse Position Indicator
MT: Magnetic Tape

Figure 2. Radar Rain Gauge System

(3) Data processing unit

It is a data processing subsystem which converts transmitted data into the amount of rainfall by radar equations to calculate the rainfall intensity data of all the meshes. It also processes data into qualitative display data of several stages and quantitative display data to facilitate the uses.

(4) Operation monitor and control unit

The radar unit, the data transmission unit and the data processing unit are ordinarily operated without a man. This subsystem constantly monitors their operations, sends alarms at the detection of a trouble, and, takes necessary actions. In this way, it remotely controls and tests the other subsystems.

4. MOC's Data Distribution System

A radar rain gauge system processes analyzed rainfall information into such forms that are convenient for uses at a river/road management office and sends them to each terminal station. The data distribution system is shown in Figure 4.

(1) Data distribution in regional construction bureaus

As stated before, a radar rain gauge receives radio waves which are reflected from rain drops, obtains their mean power intensity (Pr) of each

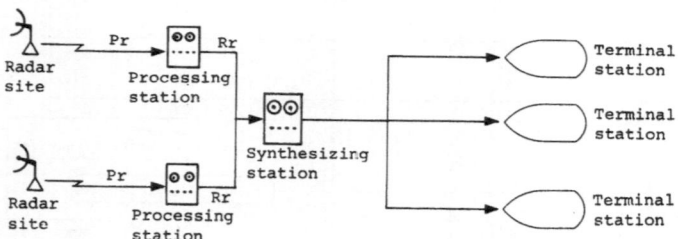

a. Data distribution system in regional construction bureaus

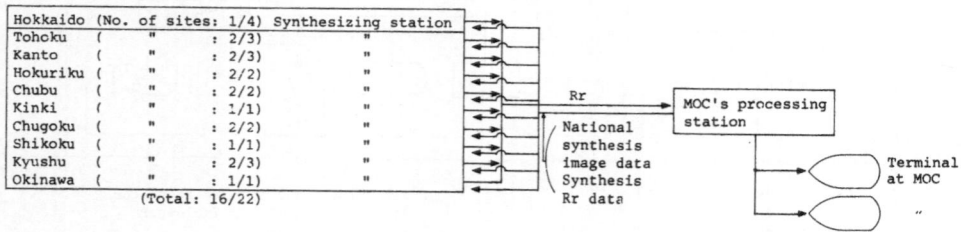

b. National

Figure 4. Data Distribution System

mesh, calculates the 5-minute mean data and determines the rainfall
intensity (Rr) by radar equations. By using the 5-minute mean rainfall
intensity of polar coordinates as basic data, it sends various data to the
terminal at local construction offices.

A radar rain gauge gives the following data to a terminal.
1   Data for qualitative display
2   Data for quantitative display
3   Data for secondary processings (polar coordinates rainfall
    intensity data: Rr)

The data for qualitative display and the data for quantitative display
are used for displaying rainfall areas and the distribution of its rainfall
intensity levels within the qualitative observation range and the
quantitative observation range, respectively. They are displayed on a
terminal color display. They are called qualitative display and
quantitative display, respectively. Since they display the distribution of
the rainfall intensity levels in colors or numerics on a map, observers
instantly can see the appearance or disappearance of a rainfall area, the
size of a rainfall area and the intensity of rainfall. Since the displayed
data are updated at 5-minute intervals, he can see the moving direction and
speed of a rainfall area and make a short-term forecast to some extent.
Since it has MT, it can obtain historical data.

(2)   National data distribution

MOC's processing station receives rainfall intensity data Rr from the
synthesizing stations throughout the country, interpolates Rr data of
overlapped observation areas (including the interpolation of shielded
areas) and distributes synthesized Rr to the synthesizing stations. They
also distribute national synthesis image data (4,800 characteris, 2,000
characters).

(3)   Data transmission time chart and formats

Figure 5 shows the data transmission and processing time chart,
starting from the completion of data collection at a radar site.

Data are transmitted from all the stations in the specified formats.
Especially, the following mode is specified for exchanging data between
synthesizing stations and terminal stations (shown as Table 2).

Table 2.   Specifications of Transmission Mode

(1)	Time interval	5 minutes
(2)	Communication mode	Unidirectional communication
(3)	Modulation	4-phase differential phase modulation
(4)	Communication speed	2400 BPS (or 4800 BPS)
(5)	Synchronization method	Independent synchronization
(6)	Transmission mode	Binary mode (transparent mode)
(7)	Transmission line	MOC's leased line

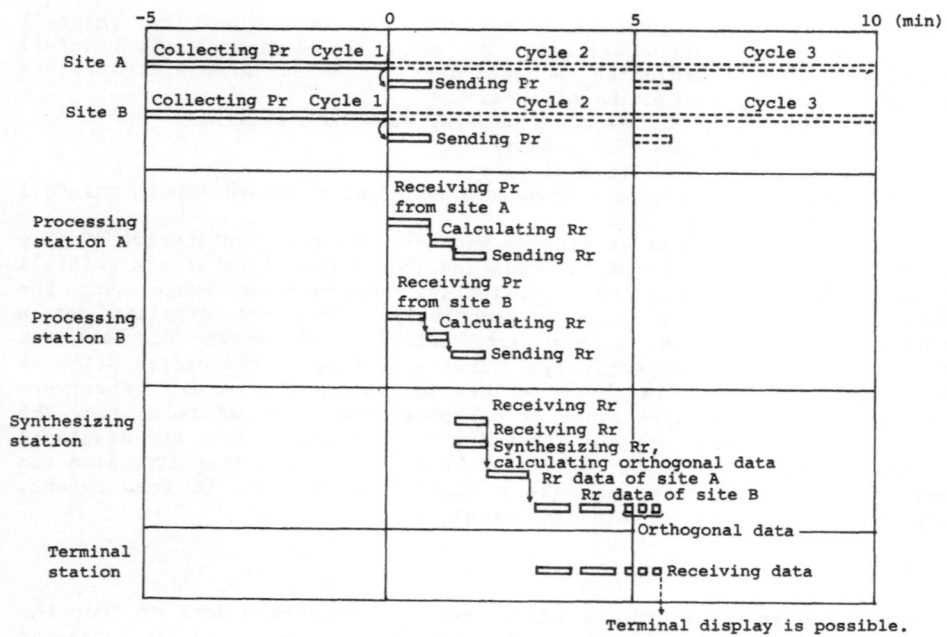

Figure 5.   Standard Time Chart of Radar Rain Gauge

5. FRICS System

MOC has an enormous amount of thorough on-line radar data (described above), telemeter rain and water level data in both the national scale and the local scale (See Table 3.) Foundation of River & Basin Integrated Communications, Japan (FRICS) was established in 1985 in order to disseminate these data to prefectures and local municipalities in on-line mode and at low cost.

As a result of various studies on reliability, speediness and cost etc., a 24-hour private videotex system was adopted as the information media. It was designed for preventing disasters with reference to NTT's CAPTAIN system.

Information users lease receiving terminals to make an access to FRICS.

(1) Method of supplying information

Ordinarily, information is supplied when requested from a receiving terminal. Picture images are displayed on CRT. The Center can send telop messages and turn on buzzers at the terminals when the announcement of such emergency information as flood-fighting warning is issued.

Moreover, if the persons in charge are absent at the terminals, the system will send messages to the homes of the persons in charge through automatic voice generator units. Besides, when the hourly rainfall exceeds 20 mm, the cumulative rainfall exceeds 80 mm or the water level exceeds a certain threshold, this function also works.

Table 3.   Radar Sites, Telemetry Stations (for river administration)

	Radar sites	Rainfall gauges	Water level stations	Water quality stations	Snow-depth sensors
Hokkaido	4*	235	226	8	-
Tohoku	3*	198	181	18	9
Kanto	3*	244	271	39	-
Hokuriku	2*	99	106	7	6
Chubu	2	244	148	10	-
Kinki	1	189	185	4	3
Chugoku	2	158	142	11	1
Shikoku	1	99	61	3	-
Kyushu	3*	225	171	12	-
Okinawa	1	11	10	-	-
TOTAL	22*	1,702	1,501	112	19

In addition, there are many prefectural stations.

* Include those under planning. As of March 31, 1988

(2)   Network and communication processing protocols

In principle, NTT's switched telephone network is used for supplying information. CAPTAIN PLPS is adopted as the protocols.

(3)   Locations of regional centers and range of information supply

A regional center is established for each regional construction bureau etc. of MOC (Japan is divided into 10 regions.). In principle, information is to be used within each region. However, accessing from the users of other region is also possible.

(4)   Types of picture images

Some examples of images are shown in Figure 6. On-line input information based on on-line data and off-line information are available.

(5)   Receiving methods

The following three methods are available for receiving images:
Individual receiving: Required images are received and displayed frame by frame by inputting the image number on the keyboard. This is the ordinary videotex procedure.
Package receiving: Four sets, each containing less than 5 images frequently used, are set in advance and are received continuously at a time. And the images are displayed after the circuit is disconnected. This method is especially developed to be used during busy emergency time.

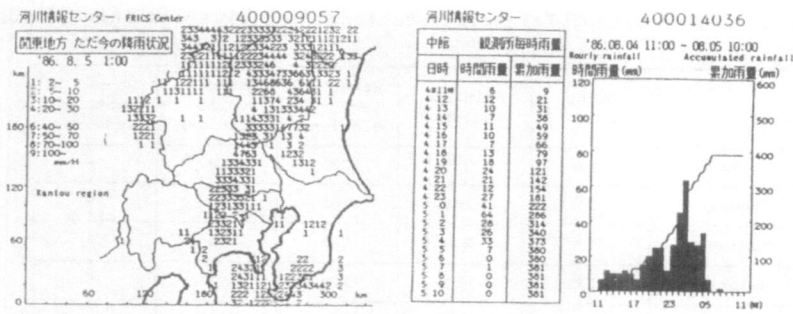

(Example of radar and telemetry raingauge information)

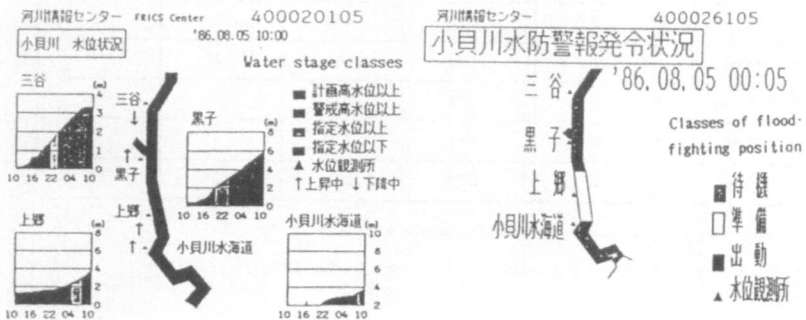

Figure 6   Example of Water Stage and Flood-fighting Information

Hysteresis playback: Radar rainfall images are stored in the Center.
7 pictures in the past 6 hours are received successively at a time by
single keyboard operation.   They are then stored in the memory as in the
case of package receiving and continuously played back as required.

(6) Automatic telephone circuit shut-off function

The automatic disconnection function is provided for the telephone
circuit connected to the Centers in order to prevent the useless occupancy
of the circuit and save the telephone service charges.

(7) Memory function

For the same purpose above, ordinary maps and other background images
are stored in the floppy disks at the terminal in order to reduce the
quantity of information (number of bits) to be received from the Centers.

(8) Color copy function

Color display (16 colors) is made on the CRT and copies are also made
in color (8 colors, ink jet method or thermal transfer method) to prevent
the decrease in the quantity of information.   The copies may be distributed
within the terminal station (city hall) and used for meetings and so forth.

(9) Simultaneous image receiving and copying functions

It is possible to call the subsequent image while making copies.

(10) General specifications

General specifications are as follows.

Power	:	AC 100 V (single phase) 50-60 Hz
CRT display unig	:	14 inches or 37 inches, 16 colors
Control system	:	16 bit processor, 1 MB/diskette
Color hard copy unit:		8 colors, ink jet type or thermal transfer type
Alarm unit	:	lamp and buzzer

(11) Number of terminals installed

As of March 1989, some 2290 terminals have been installed, and it is scheduled eventually to have 2500 terminals by 1990. Finally installation of 10,000 units is expected.

Acknowledgement

We were able to write this paper with the cooperation of a large number of staffs of MOC's Electronics and Telecommunication Section and FRICS. We had a large help especially from Mr. Yamada of Electronics and Telecommunication Section. We would like to express our gratitude to them.

COMPOSITE RADAR AND SATELLITE DATA PROCESSING IN THE METEOROLOGICAL SYSTEM IN SPAIN
(A brief description of SIRAM project)

Teresa NEVADO
Responsible of the SIRAM project
(Sistema Integrado de Radares Meteorológicos)
at ISEL S.A. company

1. Summary

ISEL, S.A. is a spanish public company working in the meteorological area.

ISEL is involved in two meteorological projects for de National Meteorological Office in Spain (Instituto Nacional de Meteorología).

SIRAM (Meteorological Radar Integrated System), and
SAIDAS (Meteorological Satellite Data System)

ISEL is the software responsible for both projects.

The SIRAM is a radar network architecture expects a centralized processor to integrate the radar data from up to 13 radars. Each radar connected to a processor constitutes a Regional Centre (RC), and the central processor that concentrate all the radar information to generate a national map constitutes the National Centre (NC).

The purpose of this project is the meteorological surveillance from each RC, and the elaboration in the NC, connected with the RC's by a low speed asynchronous link, of pictures composed from the individually collected RC's pictures, and also a set of products coming from the satellite images received by the NC from the SAIDAS system, with which the NC is connected by a high speed channel link.

This system have been developped by ERICSSON RADIO SYSTEM company, the RADAR WEATHER OBSERVATORY of McGill University (Montreal) under the supervision and management of ISEL company. ISEL is doing also the integration of all the components of the system and the communications software of SIRAM itself and between SAIDAS and SIRAM.

2. GENERAL DESCRIPTION OF REGIONAL CENTRE

The application program system, developped in colaboration with ERICSSON RADIO SYSTEM and McGill University, consists of a generation and utility part and a operational part.

The utility part consists of programs used to set and change parameters and tables used by the operational part.

The operational part is continuosly running when the system is in operational use.

This part consists of a number of separate processes. One proces handles the user communications and in the background there are continuosly running programs for the radar input, polar to CAPPI conversion, automatic picture preparation, forecasting and picture distribution.

2.1. OPERATIONAL ENVIRONMENT

Main hardware items:

Processor MICROVAX II
        2 disk driver (71 MBy each)
        1 magnetic cartridge

Meteorological terminal
        2 Raster Technologies One/75

Hardcopy Equipment
        1 Colour CANON printer PJ-1080A

Printer
        DEC LA210

User terminal
        2 DEC VT220

The system use VMS operating system, Vax DECNET, and the programs are writing in PASCAL language.

2.2. NORMAL AND DOPPLER MODE

The radar is equipped with doppler hardware and software. Doppler gives measurement of wind velocity and wind turbulence. With doppler, complete volumes are received in the same way as during normal operation. However three diferent rectangular volumes exist: Reflectivity as the normal mode, wind velocity, and turbulence data. All of these volumes are used to prepare pictures.

Reception of doppler volumes are interlaced with non doppler receptions in arbitrary sequences user controllable.

The sizes of volumes are set independently for doppler and non doppler reception.

The SIRAM system is able to obtain every 10 minutes one complete normal volume with 20 elevations and 12 CAPPIS, and a doppler volume with a PPI and one height plane.

In doppler mode, the radius will be 120 km, giving a picture 240 x 240 km. The resolution is 1 km x 1 km and the antenna speed is 2 rpm.

In normal mode the radius will be 120 km, giving a picture 480 km x 480 km. The resolution is 2 km x 2 km, and the antenna speed is 6 rpm.

In normal and doppler way can be make the suppression of ground clutter existing in the area.

2.3. PICTURE PREPARATION

The stored rectangular volume contains in itself the CAPPI pictures, ready to display.

The system can generate the more complicated pictures: Maximum intensity in any column, height of maximum intensity, echo top, and the vertical picture with diferents sectional view.

The pictures are displayed in 16 colours user selected. The levels represented by those 16 colours are also user selectable.

With every picture there are 3 overlays: Border and provinces of Spain, river of Spain, and aeroways.

2.4. USER COMMUNICATION

The operator controls the system through menues. When the program is started the system presents a head menu.

In this system, the users are divided into two groups, CONTROLLERS and USERS, of which the controllers are allowed to control the system, change parameters, etc., while the users only are allowed to load and display pictures.

2.5. SPECIAL PRODUCTS

In this system are included special programs developped in colaboration with McGill University, to obtain special products (Accumulation, Forecast, and VPI).

The Forecast module computes the velocity vector of precipitation pattern over the entire CAPPI, comparing two CAPPIS with a different time of 30 minutes, and produces the point forecast for up to 16 user-selected stations.

The accumulation products indicate the rainfall or snowfall accumulation in millimetres in each cartesian gread in the next periods: From 00 h to 06 h, from 06 h to 12 h, from 12 h to 18 h, from 18 h to 24 h, daily from 07 h to 07 h, and daily from 00 h to 24 h, and the accumulations in the time requested by user.

This kind of accumulation can be calculated also in the user selected sub-areas up to 200.

The VPI generates a vertical profile indicator map obtaining the division between the reflectivity of two cappis at the same time, but at two differents user selected heights.

3. REGIONAL CENTRE SURVEILLANCE

Some Regional centres are predefined as surveillance RC (RC/S). These RC/S can from 1 or 2 adjacent RC by telephone line, receive and make a composite picture presentation of 240 x 240 km pixels, 4 x 4 km each, total area 960 x 960 km.

This composite picture can be presented in the RC/S and can also back to the adjacent RC.

The RC/S presents the normal picture received from his radar, and permits to enter and handle the RC adjacent. RC/S can control the radars of the adjacent RC and perform system management for the region.

All RC in a surveillance region have the same regional map.

4. GENERAL DESCRIPTION OF NATIONAL CENTRE

The NC, developped in colaboration with McGill University (Radar Weather Observatory) have the responsability of collecting information of RC's and the satellite data system (SAIDAS), generating new data from the merging of that, displaying that new products to the operators and distributing it to all possible users (RC's, SAIDAS, etc.).

All this tasks will be carried on real time.

Apart from that, the NC will archive historical data and will allow the simultaneous working of several operators working either with real time data or with historical data coming from tape.

The main functions can be summarised as follows:

- Collection of satellite sectors and synoptic data from SAIDAS.
- Collection of CAPPIs, ECHOTOPS, Accumulation an Doppler products, and request for "window" products from R.C.
- Generation of RAINSAT products and forecasts.
- Generation of composite images of satellite and radar data sets.
- Automatic display of products generated.
- Storage of products in product file.
- Transmission of RAINSAT and composite products to SAIDAS and R.C.s
- Archival of user selected products on tape.

4.1. OPERATIONAL ENVIRONMENT

Main hardware items:

Processor VAX 11/785
        2 disk driver (205 MBy each)
        1 magnetic tape

Meteorological terminal
        2 Raster Technologies One/80

Hardcopy Equipment
        1 TEKTRONIX 4693 RGB Colour Screen printer

Printer
        DEC LA210

User terminal
        2 DEC VT220

The system use VMS operating system, Vax DECNET, and the programs are writing in FORTRAN language.

4.2. SOFTWARE DESIGN

The software is designed to run on the DEC VMS operating system in a multiprocess, multitasking mode. All the processes are created and initialized by SIRAM_MASTER. This module creates in memory all the global data areas that are necessary to maintain the operational system parameters indicating to the different modules their status. In these global data areas is also dumped the products directory. After that the module will create the aboved mentiones processes:

— **RAINSAT_MASTER**

This module is the heart of the NC. Its purpose is to elaborate every new product from the original data. These new products are tipically images obtained by the combination and process of different input products.

This process starts every time there is a new satellite sector coming every half an hour.

— **COMP_PRO_MASTER**

This process is in charge of compositing radar and radar with satellite images from 13 Regional Centres. It works every time there is new radar data to compose.

— **NC_RC_COM_MASTER**

This process is in charge of comunicating and handling the data between the National Center and the Regional Centers. It starts the 26 following tasks:

    TASK_RCn   n = 1,13 tasks
    COPY_RCn   n = 1,13 tasks

These tasks have the following functions:

TASKRCi establish the logical link with the regional center and wake up the SENDNC process in the regional center.

COPYRCi reads the products from the regional center.

- **SIRAM_SAIDAS_COM_MASTER** with 2 tasks:

RECEIVE
SEND

This process is in charge to receive and send the meteosat picture from/to SAIDAS system.

- **AUTOMATIC_DISPLAY**

displays automatically every time a new product is generated.

- **TAPE_ARCHIVE**

Process lets the archiving on tape of the required products. You can recover affterwards from the tape any 10 minutes products slot and loaded on the disk, summing up to 6 hours, by using TAPE_RAINSAT.

- **TAPE_RAINSAT**

This module must work with same capabilities as RAINSAT bu using tape data stored by ARCHIVE module. This module must let the user do an off-line research work.

In adition to these processes, SIRAM system has the following commands available to the user:

* **DISPLAY**

To enable him to interactively display the products in the product file. The product file contains all the product generated for the past 48 hours.

* **DIALOGUE**

To enable him to change the schedule of products to be generated and other importan parameters wich control the operation of SIRAM system.

* **STATUS**

Which enables him to view the status of the operation of SIRAM system: Regional Centers to National Center conection status.

* **TAPERAINSAT**

To generate RAINSAT products of satellite sectors on tape.

4.3. DATA INPUTS FROM RADAR AND SATELLITE

The data input to National Centre are:

- Satellite images from SAIDAS covering the whole observed area every 30 minutes:

    VISIBLE (VIS)
    INFRARED (IR)
    WATERVAPOUR (WV)

- Radar images from RCs every 10 minutes:

    Precipitation CAPPI
    ETOP image
    Hourly accumulation image (every an hour)

- Synoptic data coming from SAIDAS twice a day of three
different sources:

    Radiosounds (from baloon experimets)
    TVOS (Temperature vertical soundings from Tiros
    satellite)
    Numerical model
    Climatological Algorithms

The information contained in the satellite images comes
in satellite projection format with a pixel definition of
2.5 Km in the VIS band and 5 Km in the Ir an Wv bands.

In the case of the radar information, the input to this
modules is the image composed from the RCs individual
images.

This composite image have a pixel size of 4 x 4 kms
covering an area of 512 x 512 pixels in Lambert projection. The composite precipitation CAPPI contains informa-
tin about the rain rate in mm/h, the composite ETOP has
information about echoes maximun height in km and the
composite accumulation comes as the precipitation accumulated during a given period in mm.

The precipitation CAPPIs from radar contain information
about the rain rate in mm/h, the ETOP'S comes in km and
the accumulation in mm, coming all of them with a pixel
resolution of 4 km in a format of 120 x 120 pixels.

The synoptic data comes with information about temperature (C degrees), pressure (mbars), wind strength (Kms/h),
wind direction (degress from North clockwise) and dew
point at several heights. These data sets will come with
with their corresponding lattitude and longitude in a
predefined format.

4.4. GENERATED PRODUCTS

The major objetive of the processing methodology is the generation of the following products:

* Remapped satellite images to the Conical Lambert projection (every 30 minutes)

   - Visible
   - Infrared
   - Watervapour

   The visible images is corrected for sun angle (this is diferent correction for each latitude and longitud pixel, date and time of day).

* Probability of rain image (every 30 minutes) generated using climatological rainfall algorithm.

   This map is compared with teh rainning areas shown in the radar composite map and iterated if necessary (ie. if areas of satellite detected rain >> or << than the radar area).

* Rainfall intensity image (every 30 minutes)

* Probability of rain forecast image (every 30 minutes)

* Rainfall intensity forecast image (every 30 minutes)

* Composited radar rain images with satellite information in the pixels lacking radar information (every 10 minutes).

* Composited echotop map with satellite information (every 10 minutes).

   For that purpose, the IR images is converted to hight map using representative sounding data entered by the operator.

   A default climatological sounding is available.

* Composited accumulation map covering all the country, composed from the hourly accumulation sent by the RC's (every 1 hour).

   For the forecast version of these images is used a cross-correlation algorithm for 16 subareas. The system provides also forecast for 16 selected user station.

These products cover all the national map of Spain, in conical Lambert conformed projection and are displayed locally and retransmitted back to the regional offices.

These images are stored on disk during 48 last hours and they may view either the whole area 2048 x 2048Km at 4Km resolution.

The system can display over these pictures four overlays: geographical limits, lattitude and longitude grid, radar coverage limits, and the wind field calculated directly by th RAINSAT module in real time.

4.5. RAINSAT MODULE FUNCIONALITIES

We explain here the RAINSAT funcionalities, because is the heart of the system, and this module has the respon-sability to elaborate the above mentioned list of output products in displyable format from the input data.

Once the module is created by RAINSAT_MASTER, it remains hibernated until het SIRAM_SAIDAS_COM reports that a new set of satellite images has been received; then it starts up remapping these images to the Lambert projection covering the 512 x 512 pixels (4 x 4 Kms) area that the system is able to display. Once the images habe been remapped, the module have to normalize the VIS image, correcting the image brightness in function of the sun angle (day of the year and daytime) and calibrate the IR image converting the level coming from the satellite to a temperature scale. These prepared images are deposited in the main output products file to allow any user to dis-play them.

Later the module elaborates the probability of rain and the rainfall intensity images from the VIS and IR images supposing classitication schemes previously defined in the system and after, it elaborates any other image according to the possible user defined classification schemes.

Following these tasks, the module dumps the satellite information over the composite radar images in the areas not covered by any radar site.

Once generated every nowcast product, the module begins to generate the forecast's ones. It splits the overall image in a matrix of areas (5 x 5) in order to evaluate the correlation between two 30 minutes delayed consecuti-ves images. It evalues every single area separately (25 areas). From this computation, it obtains a culculated wind field covering the whole 2084 x 2084 Km area. The initial estimated wind field can be obtained from the synoptic products obtained from the SAIDAS and from the wind stimation obtained in the RCs with their own FORE-CAST module. With the calculated wind field, the module shifts every pixel in the input image the calculated amount to its target position in the output forecast image. The system elaborates one forecast image for the probability of rain image and other for the rainfall intensity image.

Every output image (nowcast and forecast) are deposited in the main product file, as soon as available, in order the user to be able to display them. Also the RAINSAT module communicates to any other interested module the availability of the products, in order that to distribute the elaborated products to their corresponding target system (SAIDAS and RCs).

Apart from the images, the RAINSAT module is also responsible of sending to the DISPLAY module the calculated wind field for displaying it in the third overlay.

THE UNITED KINGDOM WEATHER RADAR NETWORK

C A Fair, P K James and P Larke

Operational Instrumentation Branch,
Meteorological Office, Bracknell, UK

SUMMARY

Twenty years of research and development have led to the
establishment of an integrated operational weather radar network
in the United Kingdom. The system produces quantitative
measurements of surface precipitation which are used for real-
time flood prediction and short period weather forecasting.

1. INTRODUCTION

The UK weather radar network currently contains eight operational
radar systems with a ninth system due to be added in the next year. Plans
are in hand to introduce three further systems to provide coverage of
Scotland and by 1993 the network will contain twelve radars supplemented
by one in Jersey.

2. DEVELOPMENT OF THE NETWORK

The current network of radars has developed as a result of over 20
years research and development performed by the Meteorological Office in
conjunction with the Royal Signals and Radar Establishment at Malvern.
Almost all of the systems have been installed as part of co-operative
ventures: the partners include nine of the ten Water Authorities in
England and Wales, Devon County Council, the Directorate of Naval
Oceanography and Meteorology, the Ministry of Agriculture, Fisheries and
Food, and the Departments of the Environment and of Agriculture for
Northern Ireland.

From the earliest days of radar it was realised that they were able
to detect precipitation but effective exploitation of the fact awaited
the development of computers able to automatically reduce the large
volumes of data coupled with an understanding of the factors affecting
the relationship between the return signals and rainfall rate. The first
attempt to assess the utility of such measurements was made in the Dee
Weather Radar Project (1968-1977). A prototype Plessey weather radar
system was installed in North Wales and the output was assessed both by
the Water Industry and the Meteorological Office. The Dee Project
demonstrated that the operational use of radar was practicable, that
precipitation measurements were effective in hilly country as well as
over lowland areas, that the rainfall measurements could in turn be used
for river flow forecasts. It was also clear that radar was the only
practicable means of measuring snow fall as it occurred. In parallel with
the Dee Project a method was developed for accepting the output from a
number of overlapping radars and integrating these into a composite

database. This showed that coverage by a network of radars was feasible.

Developments of technique and understanding continued throughout the North West Radar Project (1977-1985) (Collier et al. 1980) The aims of this project were to establish and evaluate an unmanned operational weather radar station at Hameldon Hill integrated with Water Authority communications, to utilise the radar-defined data in producing quantitative forecasts of rainfall and river flow, and to assess the cost benefits of this approach. The system proved to be highly reliable and the project showed that radar data could be used together with hydrological models developed during the project, to provide forecasts of floods. In some risk zones this had not been possible previously and in other areas the prospect of floods could be detected earlier thus allowing more warning to be given. During the project methods of calibrating radar by the use of a small number of telemetering raingauges were devised so that improved estimates of surface rainfall could be obtained (Collier et al. 1983, Collier 1987).

Thus it was demonstrated that radar could satisfy certain requirements of the water industry but the Meteorological Office was keen to exploit the opportunities presented by the availability of both radar and satellite data for weather forecasting purposes. The ability of the radar systems to present a 'picture' of the changing distribution and intensity of rainfall over a wide area was considered to be a powerful aid to meteorologists preparing short-period forecasts.

The Short Period Weather Forecasting Pilot Project (1978-1982) was aimed at developing the observational and processing facilities necessary to improve the ability of the Office to forecast for the period one to six hours ahead. Radar data were available from old radars installed at Camborne and Upavon and also from new radars on Clee Hill and Hameldon Hill. In addition Meteosat visible and infra-red images were available half-hourly. Although there is no simple relationship between cloud imagery from satellites and surface rainfall it was hoped that the project would identify procedures which would provide some useful information in this respect.

The project showed that the production of short-period forecasts based on radar and satellite data was feasible but that subjective intervention by a forecaster with ready access to conventional meteorological observations was vital.

During 1981-83 a Meteorological Office/National Water Council Joint Working Group carried out a detailed study of the potential benefits of a national radar network, particularly for the Water Authorities. In its final report the group confirmed that operational experience had demonstrated significant advantages in flood prediction and warning, and endorsed the development of a network of unmanned radars. A network of 11 or 12 radars was envisaged to cover catchments in England and Wales that were at risk from short-period flooding and where warnings might be expected to reduce damage. The group identified relevant catchments and calculated the likely benefits. They assumed that quantitative precipitation forecasts out to 4 hours ahead would be achieved through the use of network data by FRONTIERS (Section 5). Overall, a benefit-cost ratio of the order of 3:1 was estimated for this network.

On the basis of this report, which noted the substantial benefits of a radar covering the London area, by 1985 a further radar had been established at Chenies. At this stage the network of five radars (Camborne, Upavon, Clee, Hameldon and Chenies) were declared operational and data from each site together with those from a radar at Shannon in the Republic of Ireland, were being relayed to a central network computer (RADARNET, Section 4) where composite images covering most of the southern half of the UK were formed.

By then it had become clear that it was not economic to install radars in a piecemeal fashion and the older, technically inferior radars at Camborne and Upavon were demonstrating their limitations. Therefore, in collaboration with the Water Authorities and others, a number of consortia were formed and six new radars purchased. Those at Castor Bay, Predannack, Ingham and Crug-y-gorllwyn are now operational and their data is being included in the composite image. Further radars in Devon and Dorset should be completed next year.

The Meteorological Office is also co-operating with the States of Jersey to install a weather radar system on the island. In addition plans are in hand to extend the network into Scotland where the Office is collaborating with the Scottish Development Department to install systems near Glasgow and Aberdeen and in the Western Isles.

Figure 1 shows the coverage achieved at present and the coverage which will be achieved by 1992.

3. INDIVIDUAL RADAR SITES

Each radar site is equipped with a radar system together with a computer system which is used both for radar control and for data processing (Collier and James 1986).

The most recently installed radars are Plessey type 45C with the following characteristics:-

radar frequency	5.650 GHz (C-band)
antenna type	parabolic 3.7 metre diameter
antenna gain	43 dB
antenna beamwidth	1.0 degrees
radome diameter	6.0 metre
polarization	linear vertical
transmitter	magnetron, non-coherent
peak power	250 kW
PRF	300 Hz
pulse length	2.0 μsec
receiver dynamic range	70 dB
minimum detectable signal	-108 dBm
hardware range correction	$1/R^2$ to 100 km

The antennas are enclosed in radomes to protect them from the effects of wind and icing. The on-site processing is performed by a DEC PDP 11/84 computer. Both control and data signals are passed between the radar and computer via a custom designed interface box which, if necessary, can also generate test signals. Initial processing of the radar video signal is handled by an Analogic AP400 array processor.

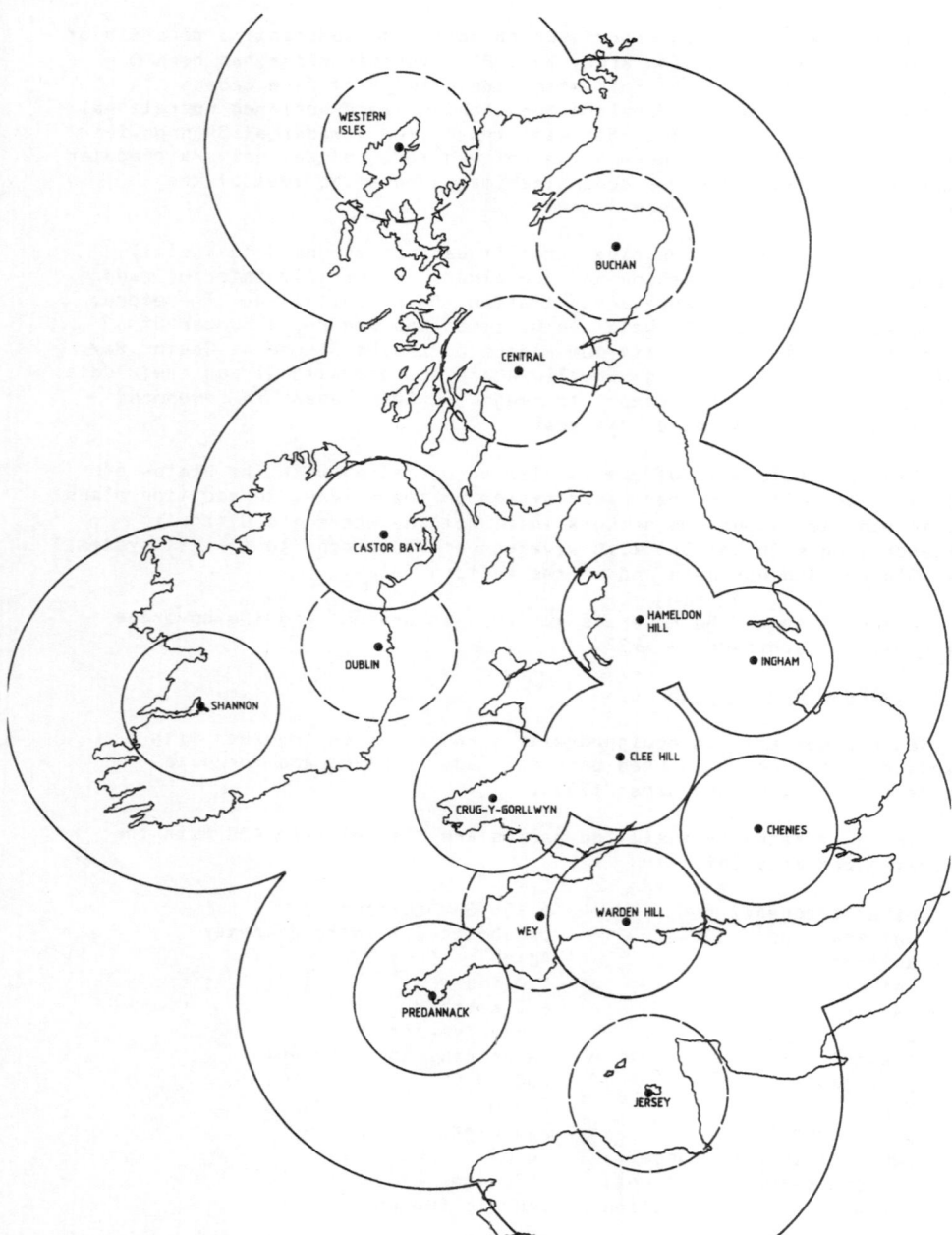

Figure 1: The coverage achieved at present by the UK Weather Radar
Network (solid lines) and by 1992 (dotted lines). The inner circles, of
radius 75 km, show the regions where quantitative precipitation
measurements are achieved and the outer circles, of radius 200 km, show
the areas where qualitative coverage is available. The location of radars
in the Republic of Ireland at Shannon and Dublin are also shown.

The individual radar systems operate routinely on a cycle which repeats every 5 minutes. Data are collected in PPI (Plan Position Indicator) scans at 4 elevations (nominally 0.5°, 1.5°, 2.5° and 4.0°) and are then processed and reduced to measurements of rainfall on both 2 and 5 km mesh Cart·ssian grids. As well as producing grid data the on-site computer also calculates the total rainfall in up to 200 different river subcatchments. Selected local users can also request RHI (Range Height Indicator) scans at chosen azimuths. The grid and subcatchment data from individual radar sites are then transmitted to local users such as Water Authorities and Meteorological forecast offices and each quarter hour the grid data are transmitted to the central radar network computer.

Radar site software runs under the real-time RSX11-M operating system and is mainly written in FORTRAN-77 with a few time-critical tasks coded in MACRO-11 assembler language.

The main tasks carried out by the software are:

- radar antenna control and sequencing,
- equipment monitoring and error detection,
- digitisation, input and averaging of radar video data,
- ground clutter removal using clutter map,
- occultation, range and attenuation corrections,
- conversion of radar reflectivity to rainfall rate,
- bright band detection,
- conversion from polar to Cartesian coordinates,
- infilling (insertion of higher elevation data into the lowest, elevation image)
- raingauge interrogation,
- adjustment using raingauge data,
- subcatchment integration,
- data archiving to magnetic tape,
- data transmission to local users and to the central network computer.

The radar program has been designed using MASCOT techniques. Two kinds of tasks (or programs) exist, a single master task and a number of slave tasks. The master task is initiated automatically on start up and this task runs a sequence of slave tasks as and when required. The various tasks communicate with each other via common data areas. Some of these common data areas are site dependent, containing such information as site location rainfall catchment areas etc., and others such as areas for holding arrays of grid data are identical at all sites.

4. THE RADAR NETWORK CENTRE (RADARNET)

RADARNET is the name given to the radar network central computer facility situated in the main telecommunications centre at the Meteorological Office headquarters at Bracknell. Every 15 minutes it receives data from each radar site and produces a composite image covering England, Wales and Ireland. This composite and products derived from it, are then transmitted by various means to users throughout the UK and to the Belgian, Irish and French Meteorological services. It is also used as the processing centre for the COST-73 project (Section 6).

RADARNET consists of two PDP 11/44 processors each with 2 Mbytes of memory and twin 10 Mbyte disks, one of the machines acts as a warm standby ready to take over in the event of the other failing. Each system has a 16 channel synchronous communications pre-processor to receive data from the UK radar sites and a total of 22 asynchronous channels and 8 ports using the X.25 protocol which are used for data reception from other European countries and for product dissemination. The systems are also connected into a DECNET network using Ethernet which connects most computers within the Meteorological Office. Magnetic tape decks are available for data archiving.

The software runs under the real-time RSX11-M operating system and is written in CORAL-66 and FORTRAN-77 though a few machine dependent routines are coded in MACRO-11 assembler language. The system also runs DECNET which is used for communication with systems, such as FRONTIERS, which are located at Bracknell.

Almost all of the operational software on RADARNET is run to a strict time schedule by running tasks from a clock queue. This allows tight but flexible control over the whole system and makes it much easier to prevent over-loading at busy times. It also facilitates the inclusion of new tasks into the overall system. However it does have the disadvantage of increasing the time taken to produce and disseminate the radar images as each task must be scheduled on the assumption that all previous tasks in the cycle have taken their maximum time to run.

Every 15 minutes, during a predefined temporal window, each radar transmits its latest image to RADARNET. When data from all the radars in the network have been received, or when the cut-off time is reached, a composite 256x256 image of 5 km grid squares covering England, Wales and Ireland is produced. Each square in the composite image contains data from just one radar, no attempt is made to average data from overlapping radars. For each 5 km square a hierarchy map indicates the radars which are able to provide coverage ranked according to 'quality-of-coverage'. Experience has shown that normally the radar whose beam is the lowest over a particular point will give the best coverage. Data from the 'best' radar is used in the composite but in the event of that data not being available data from the next best available radar is used. The hierarchy maps are generated automatically by calculating the centre-of-gravity of all the radar beams over each 5 km square taking into account the detailed radar horizons, radar beamwidths etc.

Having produced the 256x256 composite image it is archived on magnetic tape and disseminated in various forms to many Meteorological Offices, Water Authorities and other users throughout the UK and abroad. The complete image is transmitted via computer-to-computer links to the Belgian and French Meteorological services, also to the main UK Meteorological Office computer complex (COSMOS) and to outstation computer systems (OASYS) using X.25 protocols. It is also transmitted hourly on a nationwide broadcast communications network for the benefit of those users that have display systems with sufficient memory to store and display them. Smaller 128x128 regional segments are prepared and transmitted every 15 minutes on this broadcast network for those users with less powerful displays. Currently over 80 users receive these radar images via this dedicated network. The images will soon be disseminated over the Meteorological Office's new all digital Weather Information System (WIS) which is superseding the existing analogue facsimile and teleprinter network.

When the initial single site radar images are received they are forwarded to the interactive FRONTIERS computer system where quality controlled composite images and forecasts are prepared and sent back to RADARNET for dissemination to COSMOS, OASYS and, in the form of regional 126x128 segments, on the nationwide broadcast network. The quality controlled images being disseminated to COSMOS are used as part of the input to a mesoscale numerical forecast model, combined with raingauge data in a climatological rainfall database and re-transmitted to the BBC for use in the daily TV weather forecasts.

5. FRONTIERS

The concept of FRONTIERS (1981-present) (Forecasting Rain Optimised using New Techniques of Interactively Enhanced Radar and Satellite data) arose out of the Short Period Forecasting Pilot Project (Section 2) (Browning 1986). FRONTIERS is a computer system designed to provide a forecaster with all the necessary information to produce short period precipitation forecasts. Information is presented on colour displays and the system allows the operating forecaster to manipulate data interactively and quality control the radar images prior to the generation of a forecast. At present forecasts are produced using simple linear extrapolation but improvements are now in hand which will utilise wind fields derived from numerical weather prediction models. The quality of forecast images for up to three hours ahead is being assessed before they can be disseminated operationally.

6. EUROPEAN ACTIVITIES

Under the auspices of COST-73 the UK Meteorological Office acts as a European Weather Radar Networking Centre, receiving data in real time from Belgium, Denmark, France, the Netherlands and Switzerland (Collier 1988). These data are combined with UK and Irish data and infra-red satellite data to produce the COST-73 European radar /satellite composite image. This image is sent via computer-to-computer links to Belgium, Denmark and the Netherlands and via a GTS broadcast to Finland and Ireland. The area covered by the composite is currently being expanded in anticipation of data becoming available from other European countries.

7. FUTURE DEVELOPMENTS

With the addition of the new radars in Scotland and Jersey new wider area composite products will be introduced. Similarly the area covered by the COST-73 image will be extended as data becomes available from more European countries.

A trial will be conducted using one of the Plessey radars converted to coherent (Doppler) operation. Depending on the outcome of this trial consideration will be given to the use of Doppler at other radar sites.

Work is continuing to improving the telecommunication links within the network. DECNET is to be introduced between RADARNET and the radar sites. This will allow the two-way transmission of real-time information. The height of the freezing level, for example, could be used by the radar site software to assist in the identification of bright bands.

8. REFERENCES

Browning, K.A.   1986 Weather radar and Frontiers. Weather, **41**, 9-16.

Collier, C.G.,   1980 The North West Weather Radar Project: the
Cole, J.A. and   establishment of a weather system for
Robertson, R.B.   hydrological forecasting. In Proceedings of the
                              Oxford symposium on hydrological forecasting,
                              April 1980. UGGI-IAHS Pub. No. 129, 31-40.

Collier, C.G.,   1983 A weather radar correction procedure for real-
Larke, P.R. and   time estimation of surface rainfall. Q J R
May, B.R.   Meteorol Soc, **109**, 589-608.

Collier, C.G. and   1986 On the development of an integrated weather
James, P.K.   radar data processing system. In 23rd
                              Conference on radar meteorology, Snowmass,
                              Colorado. 22-26 September 1986, Vol 3. Boston,
                              American Meteorological Society, JP95-JP97.

Collier, C.G.   1987 Accuracy of real-time radar measurements. in
                              Weather Radar Flood Forecasting (ed Collinge,
                              V.K., Kirby, C.), J. Wiley, 71-95.

Collier, C.G.,   1988 International weather-radar networking in
Fair, C.A. and   Western Europe. Bull Am Met Soc, **69**, 16-21.
Newsome, D.H.

National Water   1983 Report of the Working Group on National Weather
Council -   Radar Coverage. June 1983. pp 31.
Meteorological
Office

North West Weather   1985 North West Weather Radar Project - Report of
Radar Project   the Steering Group.North West Water,
                              Meteorological Office, Water Research Centre,
                              Department of the Environment and the Ministry
                              of Agriculture Fisheries and Food. September
                              1985.

COMPOSITE WEATHER RADAR DISPLAYS

R. L. Durand
Technology Service Corporation

Summary

A number of physical and environmental constraints limit the
usefulness of conventional weather radar displays, often leading to
misinterpreted weather situations. However, the availability of
geographically overlapping radars networked together by a common data
transmission format, as currently exists in the United States,
facilitates the generation of a composite, or mosaic, image that
provides a more accurate weather presentation. This paper addresses
three major limitations of the conventional weather radar display and
describes how integrated returns from multiple weather radars can
significantly improve display quality and usefulness. An
introduction to the technical aspects of generating a composite
display is provided, along with an example to illustrate the
composite image concept.

1. INTRODUCTION

Radar has been used to detect precipitation since the early 1950s,
when radar equipment developed during World War II first became commercially available. Radars now supply valuable real-time weather information to a wide spectrum of government and private users throughout the
world. Moreover, they are no longer considered an exotic and expensive
meteorological tool, thanks to access provided by government agencies,
technological advances, and competition among radar-display equipment
manufacturers.

The U.S. National Weather Service (NWS) operates a network of 128
weather radars (see Figure 1). Since the late 1970s, the NWS has allowed
organizations with compatible display equipment to directly access these
radars via conventional telephone lines. Many aircraft surveillance
radars operated by the U.S. Federal Aviation Administration (FAA) can also
operate in a weather detection mode, but access to these radars is limited
to government users only.

In the early 1980s, the NWS and FAA established the Remote Radar
Weather Distribution System (RRWDS), which linked many of their radars
together using a common transmission format. Access to this network,
coupled with technological advances and competitive display-equipment
suppliers, has drastically reduced costs, making the use of weather radars
practical and affordable.

A recent development in the use of remote weather radar display
equipment is the composite weather radar display. This display takes
advantage of the overlapping coverage provided by the RRWDS network radars
by combining their data to produce a composite image that is superior to
current single-radar images. This paper describes the practical
limitations of weather radar systems and addresses the technical issues
associated with using multiple radars to form a composite image.

2. WEATHER RADAR SHORTCOMINGS

To understand the improvements offered by the composite image, it is first necessary to understand the shortcomings of conventional weather radars and how these shortcomings affect weather radar displays. There are three major problems: the attenuation and penetration limitations associated with storm activity, the range limitations caused by the earth's curvature, and the effects of ground clutter.

To detect storm activity, NWS weather radars broadcast a burst of electrical energy that is focused by the transmitting antenna into a 1 to 2 degree beam. This burst has been tuned to achieve the maximum detection possible while also providing the energy needed to penetrate and see through the intervening layers of precipitation commonly found in storm activity. However, the further the energy travels, the more it is subject to absorption, scattering, and other attenuation effects that limit a radar's ability to provide full detail for a large area. The small, low-power marine and aviation radars operating at X band (3.5 cm) are particularly sensitive to attenuation. The powerful NWS 250 to 500 kW C and S band radars (6 to 11 cm) offer significantly better performance, but even these do not completely eliminate the need to consider attenuation when interpreting displays.

The second shortcoming of conventional weather radars is the limited distance over which they can provide meaningful data because of the earth's curvature. As the radar beam projects outward from the transmitting antenna, it travels essentially in a straight line (refraction effects cause it to curve slightly downward). Thus, the further the beam travels, the greater its distance above the earth. This increasing altitude causes a problem, since precipitation tends to stay at elevations below 12,000 to 15,000 ft, and tops rarely exceed 50,000 ft. Table I shows radar beam altitude as a function of range and antenna elevation angle. As can be seen, the earth's curvature limits the radar's ability to detect normal precipitation activity to approximately 100 to 125 nmi, which is in contrast to the 200 to 240 nmi display presentations commonly used.

Altitude (ft)

Range (nmi)	At 0° Elevation	At 0.5° Elevation	At 1.0° Elevation
20	265	1,325	2,385
50	1,654	4,305	6,955
100	6,615	11,916	17,215
125	10,336	16,960	23,583
150	14,882	22,830	30,776
200	26,452	37,044	47,632
240	38,082	50,788	63,486

TABLE I. RADAR BEAM ALTITUDE VS. RANGE AND ELEVATION ANGLE

The third shortcoming is the radar's inability to differentiate ground clutter from weather data. Weather radars are specifically tuned to detect precipitation, but they also pick up echoes from the ground, mountains, buildings, and other objects in the environment. These unwanted echoes show up as an annoying pattern on the radar display, often reducing the display's usefulness, particularly for areas near the radar site. The only practical solution to this problem for remote weather radar displays is to blank out all data at the display's center, the area typically containing the most ground clutter. Unfortunately, this solution eliminates valid weather echoes as well.

3. COMPOSITE IMAGE

As stated in the Introduction, a new weather radar image capability is now available, a capability based on using data from multiple radar sites that, together, provide overlapping coverage for a given geographic area. The overlap is key to this concept, since it offers a way to minimize the three shortcomings of conventional radar that cause the major radar display problems.

The storm-activity penetration problem is lessened when more than one radar is reporting on a given area, contributing to a process that leads to an augmented, composite image. Moreover, multiple radars scanning the same area from different locations offer reduced attenuation and range limitations. Finally, with the right geographic distribution of radar sites, ground clutter can be reduced or eliminated by having the radars fill in the blanked areas of adjacent radars.

The typical 200 nmi radius display of a weather radar image covers a geographic area of about 400 x 400 nmi. Computer analysis of location data for the 128 NWS radars reveals that, on the average, nine radar sites surround the central site within each such area, all of which can contribute information to the display.

4. TECHNICAL ISSUES

Two principal technical issues need to be addressed when generating an image that uses data from multiple radars: how to treat overlapping areas that contain conflicting levels of precipitation, and how to accurately position each radar on the composite image.

One method of resolving conflicting levels in overlapping cells is to select the level reported by the closest radar. A more conservative method, and one that is relatively easy to implement, is to simply select the highest of the conflicting levels. This latter method provides a worst-case presentation, using the peak precipitation levels seen by a combination of radars in the same six-level format generally used for individual radar displays.

Accurate positioning of each radar return on the display is, however, a more difficult task, requiring more than just horizontal and vertical alignment. To assure that weather returns are positioned correctly and register with a common map background, complex geometric calculations must be made to account for earth curvature effects.

The necessity for these calculations stems from the fact that a weather radar scans a 360° two-dimensional area above the spherical earth. From the radar's perspective at the center of the display, the earth's imaginary latitude and longitude lines are curved, not straight. That is why geographic boundaries aligned to latitude or longitude often appear curved on weather radar displays.

All NWS and FAA radars are aligned to true north--that is, each site is oriented so that the resulting weather radar displays are presented with true north directly at the top. To correctly combine radar data from multiple locations on a common map, azimuth rotation corrections must be made to compensate for each radar's alignment to true north. Fortunately, data that comes through the NWS/FAA RRWDS system is transmitted in a radial format, which means it can be easily rotated to correct for the alignment problem. (This correction is significantly more difficult for older dial-up systems that use a raster-scan transmission format.)

The offset transformations just described are adequate for compositing weather radars for the relatively small geographic area typically covered by a single radar. For larger areas, however, further processing is required to compensate for the distortions introduced by the earth's geometry. To accurately position weather radar data for a large geographic area, individual radar images should be generated in a projection that facilitates radar compositing. The polar stereographic projection is a good example, since the returns from each radar are projected on a plane referenced to a common geographic location. Images created via this projection can be used to generate a mosaic by simply determining the range scale and horizontal and vertical positioning within the base map. Unfortunately, this projection requires more complex computations than the simple polar-to-cartesian coordinate transformation used to generate conventional radar image displays.

5. FAA METEOROLOGIST WEATHER PROCESSOR

The proposed FAA Meteorologist Weather Processor (MWP) is one element of the FAA's modernization program, whose goal is to enhance the safety and efficiency of air traffic operations in the United States. The MWP will provide meteorologists in Air Route Traffic Control Centers (ARTCCs) with individual and composite displays of up to 17 of the NWS/FAA radars located within its jurisdiction.

Data from individual radars will be continually ingested using dedicated land lines connected to each radar. In addition to generating conventional radar-centered displays, the MWP will generate images in a polar stereographic projection that will then be used to assemble a composite image encompassing the entire geographic area covered by each ARTCC. This area can be as large as 350,000 square nautical miles, as in the case of the Salt Lake City ARTCC. Other MWP products, such as satellite images and lightning data, will also be generated in the polar stereographic presentation, facilitating the generation of combined radar/satellite/lightning data displays.

6. SAMPLE COMPOSITE IMAGE

Figure 2 shows a composite weather radar image made with data from only three NWS radars. This image is centered at Patuxent River, MD. The three radars used to make the composite image are situated at Patuxent River, MD, Atlantic City, NJ, and Cape Hatteras, NC. The individual images for these radars, which were processed in parallel with the composite image, are shown in Figures 3, 4, and 5, respectively.

As can be seen in the composite image, precipitation was detected along the entire coastal area. This large detection area is made up of the individual contributions from the three radars.

Comparison of this coastal area in the images shows how compositing alleviates the penetration, attenuation, and range limitations of weather radars. Take, for example, the precipitation activity off the coast near the Virginia-North Carolina border. As can be seen in Figure 3, the Patuxent River radar detected only minimal activity past this border, so its image shows isolated level 1 and 2 precipitation in this area. It was left to the closer, Cape Hatteras radar (Figure 5) to detect that this area actually held larger precipitation areas containing level 3 and 4 activity. The Atlantic City radar (Figure 4) detected the outer fringe of this same localized activity off the Maryland coast; it also penetrated through the fringe to reveal what it saw as a small area of level 1

precipitation.   The Patuxent River and Cape Hatteras radars, however,
being closer, detected that this small level 1 area was actually a large
area containing level 2, 3, and 4 precipitation.

Two other examples of how compositing eliminates range limitations
are quite obvious.   The Patuxent River radar did not detect precipitation
in Pennsylvania and northern New Jersey, nor to the south in North Carolina.   The Cape Hatteras radar missed the precipitation north of Richmond,
VA.   These individual range limitations were overcome by using the composite image capability.

The ability to reduce clutter is also evident in the composite image.
The blanking value used for this image was 20 nmi for all three radars.
Notice how the precipitation detected by the Patuxent River radar almost
completely fills in the area surrounding the Atlantic City radar.

7. SUMMARY
Composite weather radar displays are significantly better than
conventional single-radar displays, since their use of data from multiple,
nearby radars solves many of the inherent physical limitations of weather
radars.   Moreover, the accessibility of the NWS network of radars and the
availability of low-cost display equipment make composite displays feasible not only for government users but for the general public as well.

**NATIONAL WEATHER SERVICE
WEATHER RADAR NETWORK**

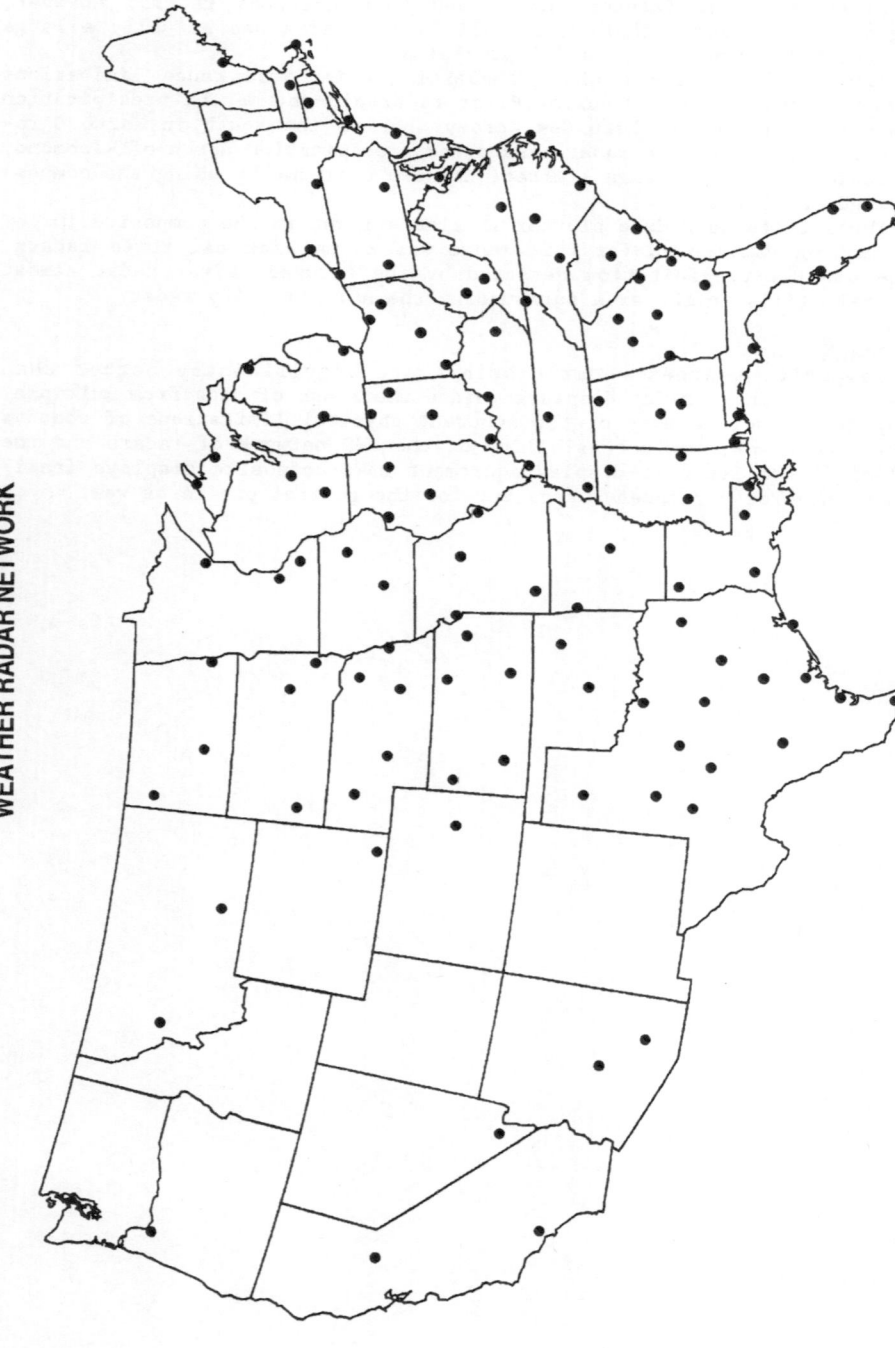

Figure 1: National Weather Service Weather Radar Network

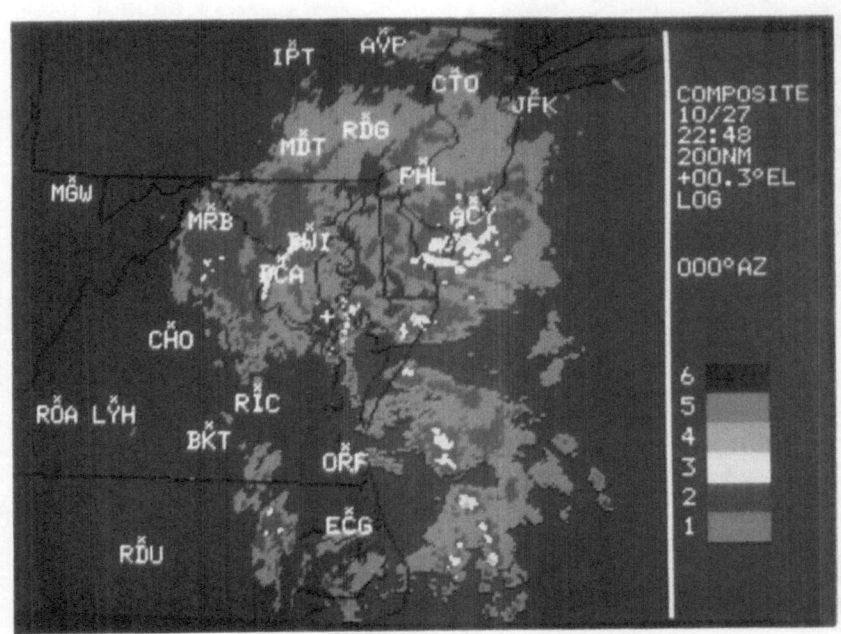

Figure 2: Composite Weather Radar Example

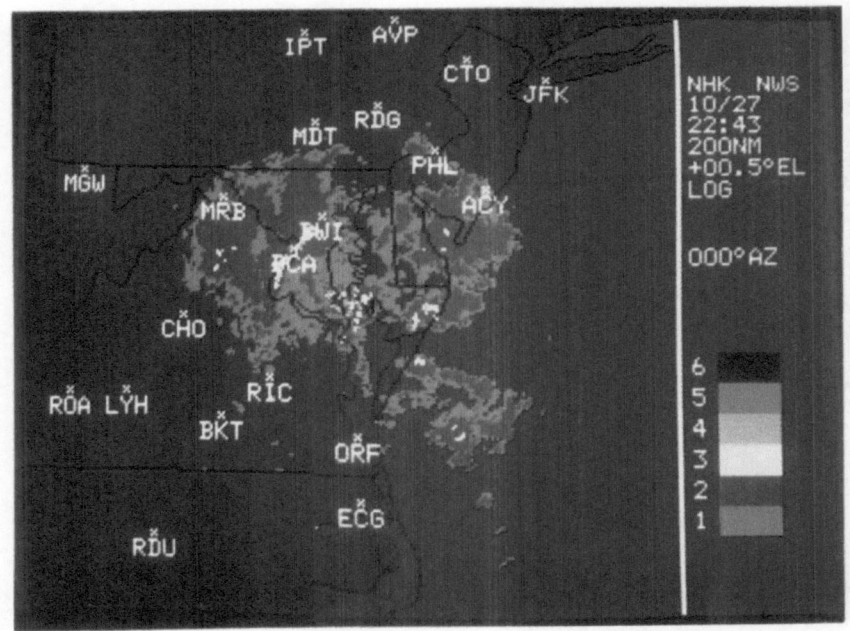

Figure 3: Patuxent River Weather Radar

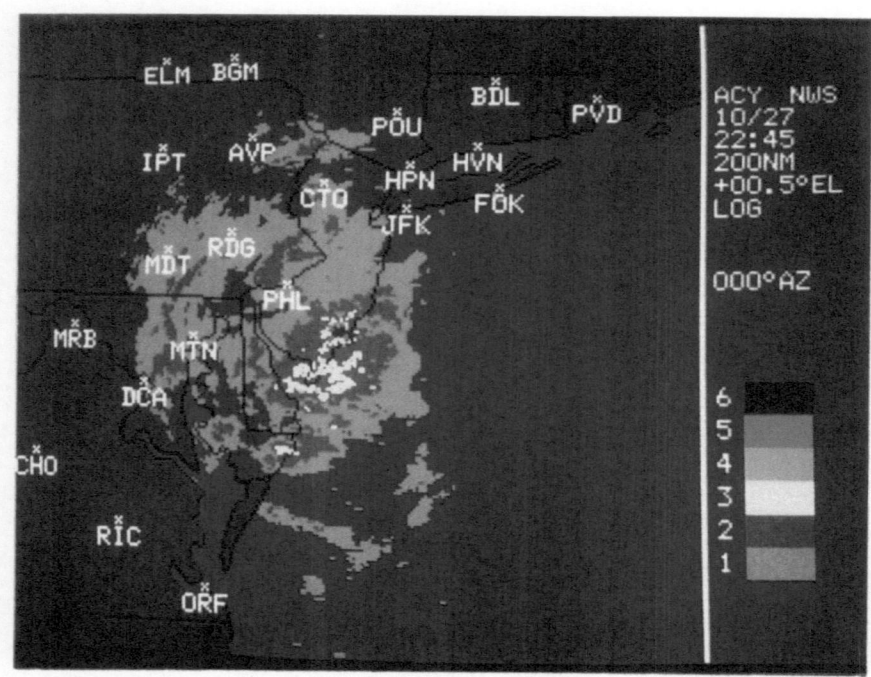

Figure 4: Atlantic City Weather Radar

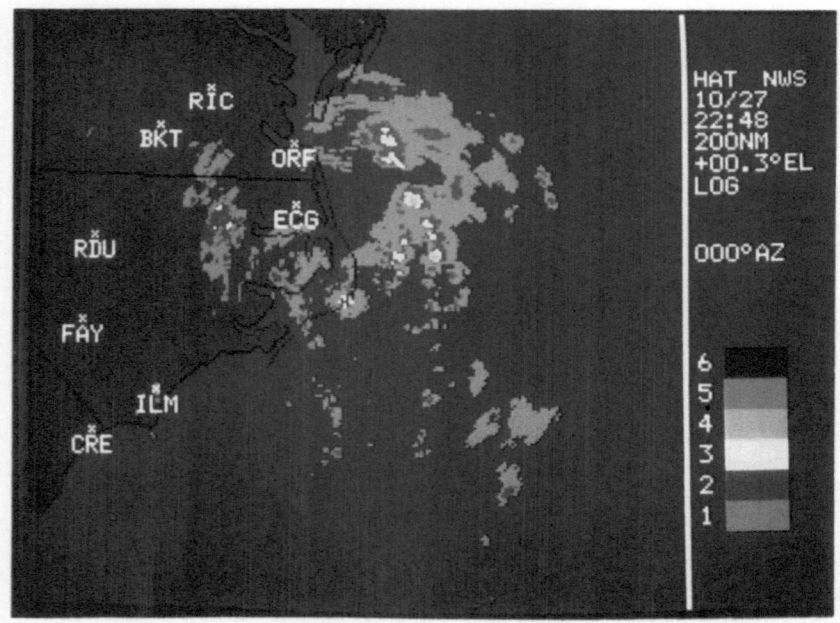

Figure 5: Cape Hatteras Weather Radar

ATC RADAR DATA NETWORKING
IN THE BENELUX AND NORTHERN GERMANY

H-J. BATZER and J.C. SHULSTAD
Task Force 2
Four States
Integration Project

Summary

Under the auspices of a Four States Integration Project managed by a Group
of Senior Officials representing the Air Traffic Control (ATC)
administrations of the Benelux and the Federal Republic of Germany plus
Eurocontrol, an effort is underway to realize a data communications network
to support the sharing of radar data by the early 1990's. The data currently
planned to be transported by the network includes ATC target data, radar-
derived weather messages, station monitoring and control data, and processed
radar data. The sharing of data will enhance both the coverage and quality of
the radar information available to the several Air Traffic Control Centres in
the included region, allowing extension of the geographical area being
monitored by each centre to 30 nm beyond the boundaries of the assigned
area of control, and will support the adoption of a uniform 5 nm enroute
separation standard in the region. The initial network configuration is
foreseen to have five nodes which are interconnected by eight 64 Kbps
digital links leased from the national communications agencies (PTT).
Future expansion to other areas such as Southern Germany is anticipated. In
consideration of the perishable operational value of radar data, the driving
requirement imposed on the network, which is to be based on packet
switching, is a very low transit time characteristic.

1. INTRODUCTION

There is an ongoing initiative to provide broad access to the plot data
measurements available from the radar sites in the area of the Benelux and
Northern Germany to the ATC radar data users for that area. This report provides
a summary description of that initiative and the current status of the effort. The
report is based on the work of Task Force 2 of the Four States Integration Project,
which is chartered by a Group of Senior Officials representing the four national
ATC administrations plus Eurocontrol with the responsibility to plan for the
implementation.

This report provides background information which relates the technical endeavor to the current approach to the coordination of ATC improvements in the Four States area. It delineates the objectives of the initiative to share radar data and provides some history of the development of the network concept. The topics of network topology, performance requirements, and cost tradeoff analysis are also treated.

2. BACKGROUND

The responsibility for Air Traffic Control of the geographical region of the Benelux and Northern Germany is divided among the four national administrations plus Eurocontrol. In general, the division of responsibility is that Eurocontrol, through its control centre at Maastricht, has operational responsibility for control of the upper airspace and the operational responsibility for the control of the lower airspace is divided geographically among the control centres at Amsterdam, Bremen, Brussels, and Duesseldorf. Due to its operational support role in the Four States area, the funding to support the Eurocontrol operations at Maastricht is provided by the Four States. Management of the budget for Maastricht is performed by the Preliminary Maastricht Coordination Group PMCG) with membership representing the Four States plus Eurocontrol.

The PMCG established a subordinate group called the Group of Senior Officials charged with coordination of technical requirements for the Four States Area. High level officials of the ATC administrations comprise the membership of the GSO. The GSO, in turn, established technical subgroups to study particular technical issues. One key technical issue has been the harmonization of ATC requirements among the ATC administrations in the Four States Area. To this end, a Master Plan was developed to provide direction to upgrade programs consistent with the overall objective of harmonization. The current efforts in support of the Masterplan comprise the Four States Integration Project.

Currently there are two active Task Forces chartered by the GSO. Task Force 1 is charged with the planning for the INTNET which is to be a data communications capability for the interchange of flight plan and coordination data. Task Force 2 has been assigned in its Terms of Reference with the responsibility to investigate several topics in the general area of radar as it relates to the support of ATC functions, but the current main thrust of the group's activity is the planning for the implementation of a radar data network, the topic of this paper.

3. OBJECTIVES OF THE RADAR DATA SHARING CONCEPT

The objectives of the radar data sharing concept are several. In relative order of importance, first of all is the augmentation of the data available to the ATC control centres to eliminate current gaps in radar coverage. Second is the improvement of the quality of the coverage data to promote the adoption of a uniform 5 nm. separation standard for enroute traffic throughout the region. Third is to support airspace monitoring to 30 nm. beyond geographical boundaries of responsibility. All of these objectives plus support to the development of new "value added" functions for the ATC community can be achieved by providing increased access to radar data currently available in the region by the ATC control centres and radar data processing systems.

4. ALTERNATIVE APPROACHES TO THE REALIZATION OF RADAR DATA SHARING

There are currently 17 radar sites available to support the operational requirements for data on aircraft and weather in the Four States area. Today, in general, the digital data from radar sites is communicated to local ATC centres through the use of modems over dual 4800 Baud lines. The dual lines provide a capability for load sharing in addition to serving reliability considerations. The data from selected radar sites is similarly communicated both directly to the Eurocontrol ATC centre at Maastricht to support its operational responsibility for the upper airspace and to some military users. Processed radar data is communicated from data processing centres to military air traffic control centres and air defense systems also over 4800 Baud lines.

The concept of radar data sharing requires providing the means to transport radar data from the radar sites to additional destinations. The two ways considered to accomplish this were to expand the current structure of point-to-point direct interconnections or to replace it with a data communications network linking the ATC centres and radar sites. For reasons primarily of flexibility and extensibility, a network approach is preferred, but not so much so that it need not be cost competitive. Therefore, a cost tradeoff analysis has been performed comparing the two alternate approaches.

At present there are 26 leased single or dual point-to-point connections in the area of interest; 27 additional point-to-point connections have been determined by the Task Force to be operationally necessary to close coverage gaps. The actual cost of the communications links presently being leased tallies to approximately 75,000 ECU per month; adding the communications links needed to support the additional connections would increase the monthly leased line cost by approximately 81,000 ECU per month. The total cost for the leased communications links then would be 156,000 ECU per month.

By comparison, a network approach has costs associated with access lines to the network plus the switching nodes and inter-switch links. The monthly cost of the required access lines is estimated at approximately 68,000 ECU and that of the eight 64 Kbps inter-switch links required for the initial network configuration, at approximately 60,000 ECU, applying current tariffs. The backbone nodes are assumed to be purchased and depreciated over an eight year period. A net cost of 5,000 ECU per month for five nodes has been estimated. These figures total to approximately 132,000 ECU per month.

The net result of this simplified cost comparison is a savings of approximately 24,000 ECU per month associated with a network solution versus extension of the current point-to-point approach to radar data communications. Added to the numerous other advantages, a favorable cost comparison solidifies the choice of a networking approach.

5. NETWORK TOPOLOGY

The network topology as currently planned is depicted in the accompanying figure. The switching nodes are placed at the location of the ATC centres at Amsterdam, Bremen, Brussels, Duesseldorf, and Maastricht. Maastricht is directly connected to each of the other four nodes in support of its operational need for radar data from throughout the region. Each other node is connected to two others in consideration of routing flexibility with the particular interconnections being selected considering the potential for mutual interests in radar data sharing.

This initial topology is foreseen to be extended in the near-term to Southern Germany after an initial period of evaluation and achievement of operational stability. Candidate sites for the additional nodes are Frankfurt, Munich, and Karlsruhe.

6. NETWORK ENGINEERING

The special application environment of this data communications network calls for the use of nontraditional planning considerations from the viewpoint of communications network engineering. The engineering of the network is not aimed at maximizing the utilization of limited resources. The inter-switch connectivity is provided by 64 Kbps digital links leased from the public communications agencies (PTT's). The planned use of such links is based on considerations both of 64 Kbps links becoming a basic standard PTT leased service offering and of providing sufficient throughput capacity to avoid significant queuing for the communications links. The engineering analysis has verified the ability of these planned inter-switch links to readily handle a conservatively estimated data traffic load. With only eight inter-switch trunks in the network and with the data load highly dependent on the particular user requests, such conservative planning appears to be called for. The intended objective of the network engineering is to achieve a very responsive network with provisions for queuing and data traffic control only intended to be invoked under unusual circumstances. Even alternate routing of data is not foreseen to be a routine occurrence. A transit time requirement for data which both is very stringent and must be very consistently achieved cannot routinely tolerate bottleneck situations.

PROPOSED RADNET TOPOLOGY

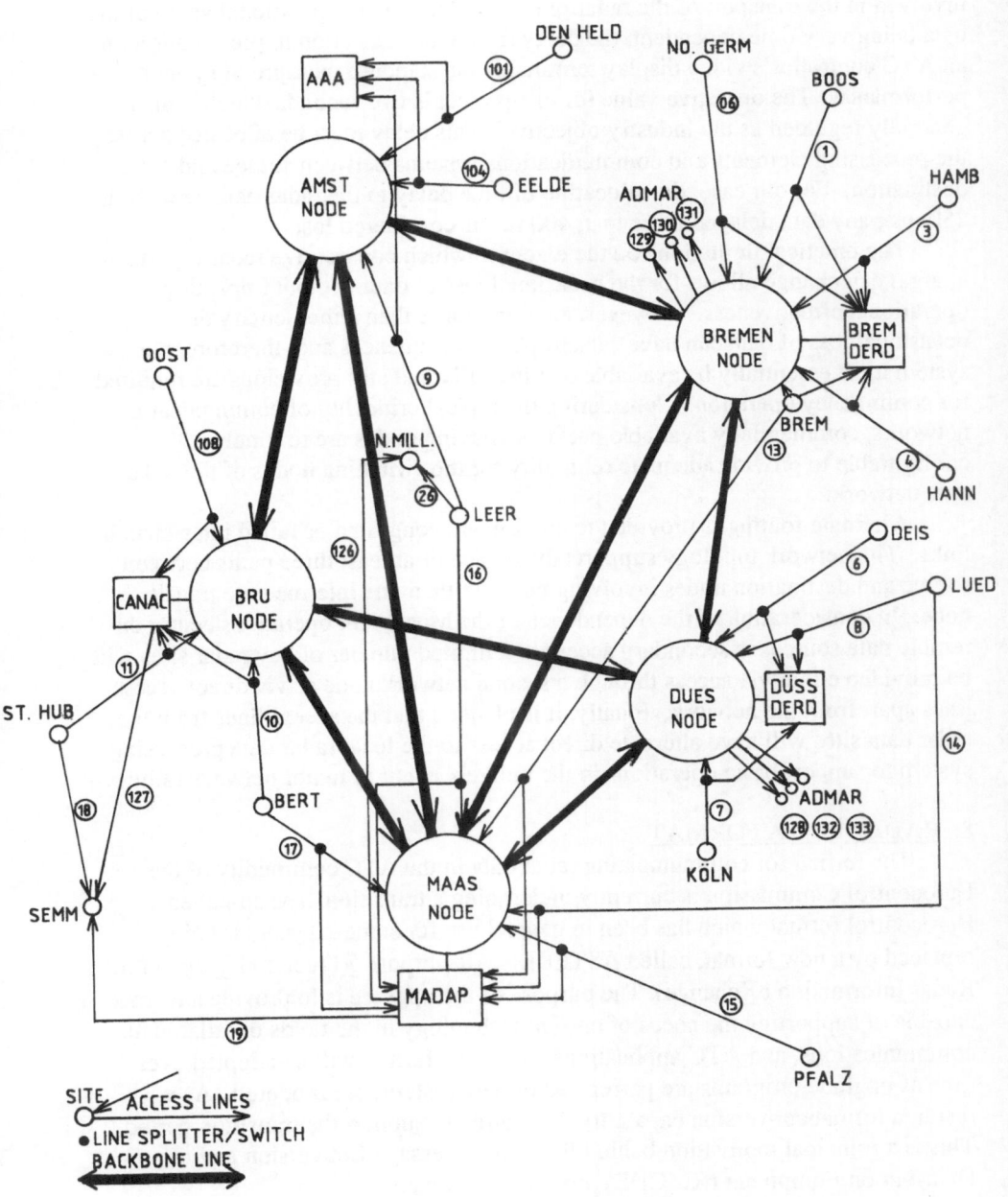

SITE ↗ ACCESS LINES
● LINE SPLITTER/SWITCH
◀▬▬▶ BACKBONE LINE

7. NETWORK PERFORMANCE

The driving requirement on network performance relates to the time delay involved in the transport of the radar plot data. Due to the operational value of the data being very time dependent, the delay from radar detection to presentation on an ATC controller's video display terminal is the standard measure of system performance. The operative value for this project is two seconds which is also generally regarded as the industry objective. This delay must be allocated across the processing elements and communications systems between source and destination. For our case, the allocation of time delay to the radar data network is 250 ms.; any data delayed more than 400 ms. is considered lost.

The practical limitations on the extent to which consecutive radar reports on a target can change allows for the occasional loss of data without impacting operational effectiveness. However, anything more than a momentary or occasional loss of data can have catastrophic consequences and, therefore, the system must essentially be available continuously and still provisions are required for contingency operation. Considering the typical criticality of communications networks, commercially available packet switching nodes are routinely configurable to provide adequate reliability for the switching nodes of the radar data network.

Alternate routing is provided to circumvent congested or failed inter-switch links. The network topology supports the choice of at least three paths between source and destination nodes involving no more than one intermediate transit node. In consideration of the dependence of the Maastricht operations centre on remote data sources, a secondary access to a limited number of key radar sites will be provided either via access through a second network node or via direct access lines apart from the network. Finally, it is planned that the access lines from the radar data sites will have alternate direct access to the local radar data processing system to support basic operations in the unlikely event of major network failure.

8. RADAR DATA FORMAT

The format for communicating radar data in the ATC community of the Eurocontrol Commission is currently undergoing a transition. The so-called Eurocontrol format which has been in general use for some 20 years is being replaced by a new format, called ASTERIX (All-purpose STructured Eurocontrol Radar Information eXchange). The purpose of the upgrade is to provide a format capable of supporting the needs of current technology in the fields of radar, data communications, and ATC applications. The new format will be adopted over time as upgrade programs are performed or new systems are procured. As a result, a format conversion capability is required to support the transition period. This is a principal motivation behind the Radar Message Conversion and Distribution Equipment (RMCDE) program initiative.

9. NETWORK ACCESS FUNCTION

The concept of the RMCDE is also linked very closely to the concept of the radar data network. In fact, the two are viewed to have a synergistic relationship in which the combination of the two enhances their values taken separately. Eurocontrol is currently developing a prototype of a redundant RMCDE unit to be completed in the second quarter of 1990. The prototype unit includes functional capabilities for format conversion between Eurocontrol and ASTERIX formats, for filtering of radar plot data, and for distribution of data via several communications output ports. These functions are classified as access node functions, viewed as a sort of preprocessor to a communications function associated with transfer of data via a network. The prototype RMCDE will provide a basic capability to interconnect RMCDE units directly in a networked configuration in addition to via a standard X.25 packet switching network.

10. RADAR DATA FILTERING

The concept of radar data filtering is key to the planning for the radar data network. Fundamentally, the concept involves remoting user radar data requirements through the network to the access nodes serving the radar data sources and then only distributing data as specifically requested through the network. This allows users with an effective means of access to the best data available to support their operational requirements. The effect is to greatly reduce the network data load at a minor expense in terms of data processing and network administration.

Step-by-step, the process works as follows. A user performs a planning function based on a master grid of radar data sort boxes relative to each radar data site from which he wishes to receive radar plot data. The master grid for each radar site is aligned to true North and then divided into sectors formed by concentric circles 16 nautical miles (nm.) deep and approximately .1 radians wide. This scheme has been chosen to simplify processing requirements on the data such as coordinate translation. The result of the planning is a message to the host RMCDE of the user detailing what data is desired from which radar data sources. The host RMCDE merges the radar data requirements of all of its users and then communicates the radar data requirements to the host RMCDE's of the radar data sources via the data communications network.

The radar data sources send all radar plot data to their host RMCDE's in polar coordinates. The RMCDE at the data source end then filters and sorts the data according to the current active set of data requests on file. The buffered data sets are then presented to the communications network for transfer to the appropriate destination RMCDE's on a one-to-one basis.

When a destination RMCDE receives filtered radar plot data, it sorts the data according to the requests it has on file from its set of users. Once sorted, the data is delivered directly to the appropriate users. The user then can perform whatever additional data filtering he may desire to arrive at the final data set.

11. NETWORK MANAGEMENT

The network management required to support the transport of radar plot data has the objective of ensuring that the network consistently meets the transit time requirement and that there is no problem such that data is systematically being lost due to a fault in the network. In the event of network failure or congestion, network management messages are required to be sent to the access nodes where control or contingency actions can be effected. The RMCDE is specified to selectively delete data types from being offered to the network in response to indications of performance degradation. Also, indicators of the relative priority of radar plot data as specified in the data request from the user can be used to further selectively reduce the data load offered to the network.

Due to the limited scope of the radar data network application and the operational nature of that application, it is the opinion of the Task Force that the network management capabilities could be adequately implemented in a decentralized manner within the network nodes and communicated over the normal data communications links of the network. This is, in fact, the approach being taken in the prototype RMCDE implementation. However, it is recognized that providers of packet switching equipment generally employ a centralized network management approach in consideration of potentially complex application environments and, therefore, there may not really be a choice in the matter.

12. SCHEDULE

The goal is to have the data communications capability operational in early 1992. To achieve this, all preparations are to be completed in 1990 and the implementation contract performed in 1991.

13. RECENT DEVELOPMENTS

Most recently, a special subgroup of Task Forces 1 and 2 has been convened to pursue the concept of using a shared, common network to transport both the radar data and the flight plan data. This development is due to the coincidence of the planned nodes of the two network concepts which should translate directly into cost savings and the fact that the communications requirements for the flight plan data can be met without compromising the performance required in the communication of the radar data. The current forecast is that such a common network will result. The basic work documented in this paper for the radar data network remains valid, but an analysis task is required to verify the engineering and performance of the network operating under a combined data load before the implementation can proceed.

Weather Radar Networking:

Computing versus Communications

by

Richard E. Passarelli, Ph.D.

SIGMET, Inc.,

2 Park Drive

Westford, Massachusetts 01886 USA

FAX (508) 692-9575

1. Introduction

In developing an effective network strategy for weather radar, a fundamental decision is how to allocate computing and communication resources. Fundamentally, a radar network must accomplish the following tasks:

- Acquisition of raw data.
- Processing of the raw data to generate single-radar output "products".
- Generation of multi-sensor "composite" products.
- Communication of single-radar and composite products to users.

The first step, namely data acquisition, is tied to the radar site. However, there are several options for the subsequent processing steps. Examples of relevant questions are:

- Is it better to generate single-radar products at the radar site and transmit them to users, or is it more economical to transmit raw data to a central processing and product distribution facility?
- Is it better to perform compositing at a central facility or at each individual radar site?
- What picture product resolution is best for different types of users and communications capabilities?
- Should end users have access to raw data so that product generation can be done on local workstations?

In evaluating these types of questions, one must take into account the expanding inventory of low-cost, high-powered workstations which make a distributed processing approach much more feasible today than was possible only a few years ago. Indeed, workstations now have the capability to generate virtually any type of radar product as well as a wide variety of other meteorological products. This means that the crucial issue is not the availability of computer power, but rather the bandwidth of the communications that must supply raw data and products to the new generation of workstations.

2. Products and Product Resolution

A "product" is generated from the "raw" spherical coordinate radar data. Products can be either pictures, text or data, although generally users receive picture products. The number of output products that can be made from radar data has expanded very rapidly. This expansion is driven by new nowcasting and forecasting applications and the availability of low-cost computing to perform the product generation. The primary radar product was and still is a two dimensional PPI presentation of weather radar echo intensities. However, most new systems require volume scan products for Doppler and intensity data, e.g.,

- PPI (available at all elevations for intensity, and Doppler information).
- CAPPI (e.g., 12 heights).
- Precipitation accumulation maps (e.g., 1, 3, 6, 12 and 24 hours)
- Echo Tops Maps
- Cross-sections
- Vertically Integrated Liquid (different layers)
- Ad hoc RHI's
- Storm Tracking
- VVP or VAD wind analysis
- Wind Shear Display
- Hail Detection

The number of products that are actually computed depends on the needs of the users, the available computer power and the bandwidth that is available to transmit them. As we shall see, in some cases it is more cost effective to transmit the raw data than to transmit a full complement of products.

Most products are in the form of pictures and these usually have the greatest impact on the communications bandwidth. TABLE 1 summarizes the applications, storage requirements and transmission speeds for three common picture resolutions. Low resolution pictures (320 by 240 pixels) have the advantage that they are very economical to transmit. Medium resolution pictures (640 by 480 pixels) offer better resolution and can be used for split screen displays. High resolution pictures (1024 by 800 pixels) require considerably more storage and communication.

TABLE 1. Typical Picture Product Resolution

Low Resolution

320 by 240 by 4 bits

These picture products provide excellent full screen images however 3X and 4X zoomed images appear "blocky". 2X zooming is acceptable. The advantage of these products is that they are economical to generate, store and transmit. Suitable for most dial-up remote display applications.

Storage per Image: 38.4 KBytes

9600 Baud Update Rate: 16 seconds*

Medium Resolution

640 by 480 by 4 bits

These picture products provide excellent resolution for full or split screen displays. Zoomed pictures are also excellent. This is the highest nominal resolution that can be resolved on PAL or NTSC monitors. Suitable for dial-up remote displays and most other workstations.

Storage per Image: 153.6 KBytes

9600 Baud Update Rate: 64 seconds*

High Resolution

1024 by 800 by 8 bits

These picture products are designed for high-end graphics workstations. The appearance is superb., but they require substantial resources to generate, transmit and display. The images cannot be converted to PAL or NTSC without a special "frame-grabber" hardware. Because of the communications bandwidth required to send these products, they are best suited for workstations on dedicated lines, Ethernet LAN's or microwave/optical-based communications.

Storage per Image: 819.2 KBytes

9600 Baud Update Rate: 5 minutes 40 seconds*

* 9600 Baud update rates calculated assuming 50% image compression efficiency.

3. System Users

Although there are many different types of users, they generally fall into three different categories with regard to their communications, i.e.,

- Local Office Users located at the radar. These can be connected directly to the radar computer or LAN for very high speed communication.

- Local Dial-Up Users who access via standard phone lines, either single-radar or composite products that are stored at the local office.

- Network Users who access a network (e.g., X.25) to obtain products or raw data from any radar or to perform control/maintenance functions.

TABLE 2 summarizes the different types of users and their display and communications requirements.

Local Office Users will often have a workstation with ample computing power and a high speed link via an Ethernet LAN to the radar host computer. This makes it possible for these users to request raw data and perform custom product generation locally at their workstations. Any custom products that are generated in this manner can become part of the data base at the local office so that they can be accessed by other users. This type of distributed processing for custom product generation at the local office serves to reduce the amount of processing that must be performed by the radar host computer.

Local Dial-up Users are typically end users such as local power, water and transportation authorities. These users usually have a low cost display (low or medium resolution). By accessing the local office directly as opposed to a nationwide network, they save on communication costs and do not place any burden on the nationwide network. The maximum communication bandwidth for these users is typically 9600 baud using packet encoding modems over standard phone lines.

Network Users access the nationwide network directly to receive products or raw data. Control, calibration and maintenance functions can also be accessed via the network. Network users have a variety of powers ranging from "observers" who are permitted only to receive product output, to operators who can take full control of the radar. The level of privilege is granted by a "system manager". The hardware capabilities of network users can range from high-powered workstations with high resolution displays, to inexpensive low resolution display-only devices. The communications bandwidth available to Network Users depends primarily on the choice for the national wide area network (WAN), (e.g., satellite, microwave or fiber optic, leased lines, standard phone lines).

TABLE 2. The System Users and Communications

Local Office Users

Many installations are manned for all or part of the day or are associated with a local forecast office. These users require control and display capability for both Local and Network products.

- **Examples**: Local forecasters, operators, maintenance personnel.
- **Workstation Displays**: Medium or high resolution
- **Workstation Communications**: Ethernet LAN or direct high-speed parallel or serial lines.
- **Typical Workstation Bandwidth**: 1 Mbaud.

Local Dial-up Users

These are users who require access to both Local and Network Products stored at the radar host computer. The advantage of the local dial-up is that it reduces the Network communications traffic.

- **Examples**: Local power, water, surface transport, and aviation authorities. Individuals in agriculture, recreation, construction, etc.
- **Workstation Displays**: Low or medium resolution.
- **Workstation Communications**: Standard phone lines and modem.
- **Typical Workstation Bandwidth**: 1,200 to 9600 baud.

Network Users

These are users who have direct access to the network. They have the capability to request products directly from radar sites or the central facility.

- **Examples**: Forecasters, maintenance personnel, operators, system managers.
- **Workstation Displays**: Low or medium resolution.
- **Workstation Communications**: Standard or dedicated phone lines and modems.
- **Typical Workstation Bandwidth**: 1,200 baud to 38.4 Kbaud depending on line quality.

4. Example of a Network Architecture

FIGURE 1 shows an example of the network architecture used for SIGMET's IRIS system (Interactive Radar Information System). TABLE 1 summarizes the types of system users and their display and communications capabilities. The design of this architecture is based on our experience with a wide variety of customer requirements and applications.

4.1 The Radar Site and Local Office

A common requirement is to locate the radar at or near a local forecast office. The local office is usually manned for either all or part of the day, during which time local office personnel usually have responsibility for radar operation, monitoring and maintenance. In some cases, there is no local office associated with a radar installation and control is from a central facility. This is also the case "after hours" when there are no qualified operators at the local office. In these cases, control is performed via a direct telephone line or the Wide Area Network.

4.1.1 The Radar Host Computer

The radar host computer handles the radar/user interface, control, system monitoring and data acquisition for the system. This is an automatic procedure except for occasional ad hoc scans such as miscellaneous RHI's requested by an operator. The radar host computer usually performs all or part of the single-radar product generation.

4.1.2 Local Office LAN

If there is a local office associated with the radar, there may be a local area network (LAN) for linking workstations, computers and peripherals. The LAN will often provide access to other data bases, such as satellite or upper air, as well as radar. An Ethernet LAN has a bandwidth of (10 MBits/sec) which is usually adequate to service personnel at a local office. In addition, the LAN serves to connect the radar host computer to the wide area network (WAN). In the absence of a LAN, the radar host computer can perform the functions of the LAN. However, there is a performance impact when the radar host computer also handles the low-level communications tasks.

Depending on the number of products to be generated and the speed requirements of the system, the LAN may also provide connection to a "compute server" -- a dedicated processor which takes data acquired by the radar host computer and performs single-radar product generation.

The LAN also connects the local office user workstations. If compatible workstations are used, they can also perform product generation from raw data transmitted over Ethernet. This allows the local office users to generate custom products, without effecting other users by burdening the radar host computer.

In summary, the communications bandwidth of the LAN provides a great deal of flexibility in deciding which networked processor actually performs the product generation. To have this flexibility, the radar host computer, compute server and local user

FIGURE 1. Network Architecture Example

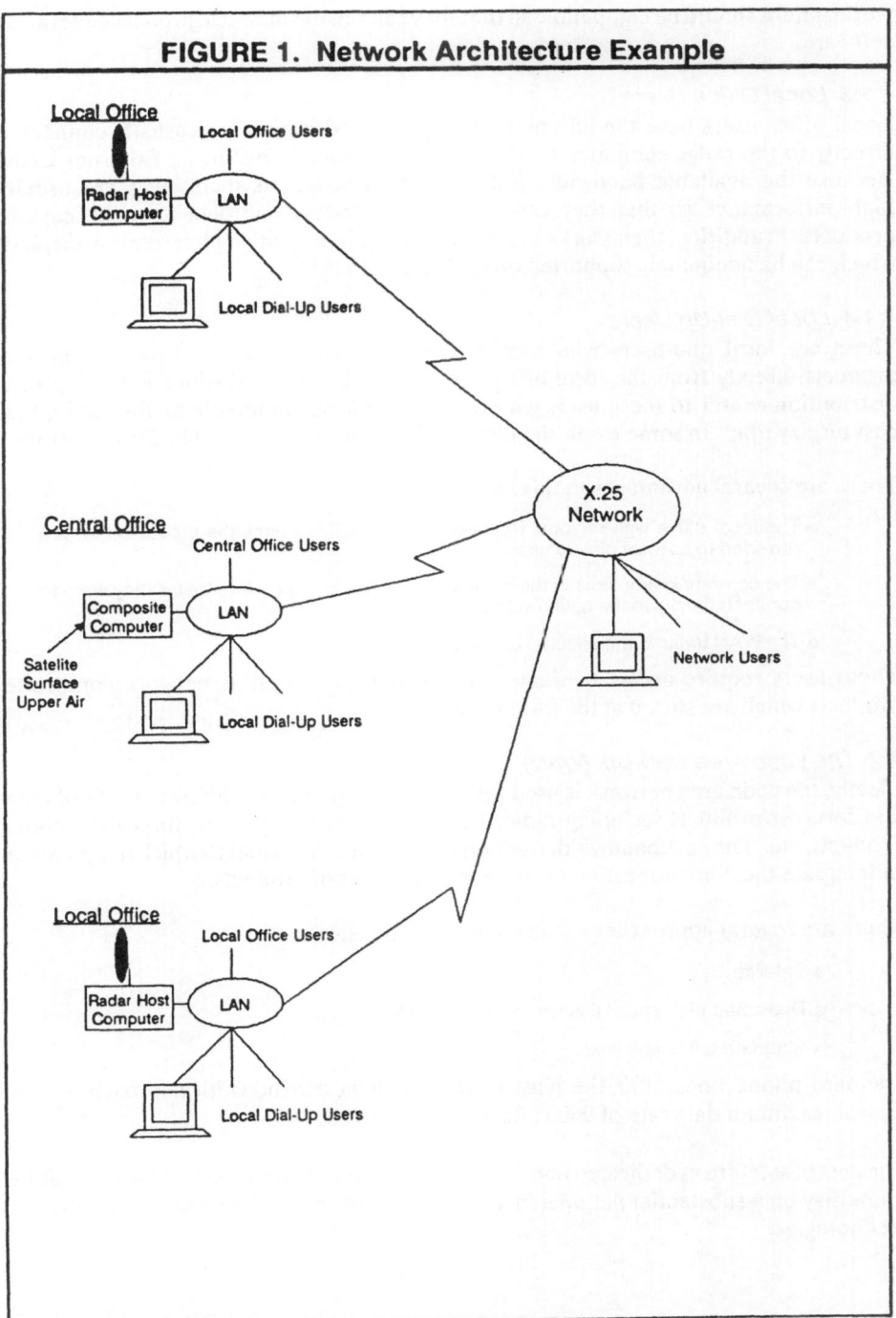

workstations should be compatible so that they can run the identical product generation software.

4.1.3 Local Office Users

Local office users have the advantage that their workstations are usually connected directly to the radar computer via dedicated high-speed lines or an Ethernet LAN. Because the available bandwidth is high, the workstations themselves are usually high-performance so that they can access and process raw data to make custom products. In addition, their workstations may be equipped with high resolution displays which can be adequately supported on an Ethernet LAN.

4.1.4 Local Dial-Up Users

These are local end-users who receive both single-radar and network composite products directly from the local office. In this case, the local office is serving as a distribution center to these users who receive products via telephone line and a low cost display unit. In some cases, the distribution of products via cable TV is feasible.

There are several advantages to this approach:

- The local office with regional forecasting responsibility edits the products that are provided to its local dial-up users.
- The communication cost to the local dial-up users is usually less than if they were to connect directly to the nationwide WAN.
- The WAN traffic is not effected by these users.

These users require access to single radar products as well as network composite products which are stored at the local office.

4.2 The Wide Area Network (WAN)

Ideally, the wide area network is used for communicating many different kinds of data and forecast products including radar, satellite, surface, upper air, forecasts, model products, etc. The cost/bandwidth tradeoff determines the extent to which it is possible to integrate the communication of these various types of products.

There are several approaches to the communications:

- Satellite
- Dedicated high-speed lines or microwave links.
- Standard telephone lines.

Standard phone lines offer the least expensive, lowest bandwidth approach with a typical maximum data rate of 9600 bits per second.

The cost of satellite vs dedicated lines should be evaluated on a case-by-case basis since there may be a substantial national investment already made for one or both of these technologies.

Later in this paper we present a discussion of the WAN bandwidth requirements for radar products and data. The example shown in FIGURE 1 is for an X.25 packet switching network. However, other protocols can be used as well.

There is an important tradeoff between WAN bandwidth versus total traffic volume. There are two approaches to providing products over the WAN, i.e.,

- Regularly scheduled standard transmissions where a fixed set of products is routinely supplied to local offices or other network users.

- Ad hoc transmissions of products that are made only when a user requests a product.

An advantage of regularly scheduled product transmissions is that the WAN traffic can be very highly regulated to stay within the bandwidth limitations of the WAN. This is important if the bandwidth is limited. Unfortunately, at night or during quiet meteorological periods, many of the products that are transmitted will not be used so that the bandwidth required to send the products is actually wasted. The ad hoc product request approach assures that a product will be used if it is transmitted, so that the total traffic on the WAN may be reduced. However, for an ad hoc approach to be successful, the bandwidth must be relatively high so that the response to the request is timely. Potentially, an ad hoc product request approach on a high-bandwidth WAN can result in less total traffic than a regular transmission approach on a relatively low-bandwidth WAN. To calculate the most cost effective approach requires a detailed system use and cost analysis. However, a flexible system should permit either approach so that the system can be changed "on-the-fly" in response to changing traffic patterns or evolving communications capabilities.

4.3 Compositing at the Central Office
Most countries have a central forecast office which often has its own radar as well. The Composite Computer, shown in FIGURE 1, has the task of collecting single-radar products and/or raw data from all of the radar sensors, generating composite products and distributing them back to the local offices. The primary reason for generating a set of standard composites at a central office is to minimize the traffic on the WAN, i.e.,

- If the central office has a radar it requires less communication to transmit the data to a central site and then retransmit a composite, than to have each radar site or local office request data from all of the other sites.

- If the central office has nationwide forecasting responsibility, then it must always receive products from all of the sites on a routine basis to fulfill this mission.

- If the central office has network maintenance and control responsibility for all or part of the day, then it must always receive products from all of the sites for quality control.

After a composite product is made, then it can be transmitted to local offices either automatically or on request.

As discussed in the preceding section it is usually more economical to transmit a composite product on request since bandwidth is wasted when a product is transmitted automatically and no one ever looks at it. This can be made very transparent to a user at a local office-- the user simply makes a request for a product; if it is at the local office already he gets it immediately and, if not, he gets it as quickly as the WAN can send it.

5. Examples of Product and Raw Data Transmission and Storage Requirements

In this section we present calculations of the data storage and transmission requirements for two typical operational scenarios. The two examples presented here can be used as a starting point for specifying the performance of a system. To provide a rational comparison between the communications requirements required for raw data as opposed to products, we first introduce the concept of "matched resolution sampling", described below.

5.1 Matched Resolution Sampling

To compare the data transmission required to send a full complement of products as compared to the resolution required to send "raw" data, we define the concept of "matched resolution sampling", i.e., to make a rational comparison, the sampling resolution of the raw data in spherical coordinates should be roughly equivalent to the required picture product resolution. For example, a low resolution PPI or CAPPI product (320 by 240) requires 120 range bins of raw data to match the number of pixels to the number of range bins that are sampled, Similarly, 240 range bins are required for a medium resolution product (640 by 480) and 400 bins for a high resolution product (1024 by 800).

5.2 Example 1: 15 Minute Update Doppler Case

The first example is summarized in TABLE 4. The top of the table shows the basic sampling assumptions. A PPI volume scan consisting of 20 elevation angles serves as the basis for data collection. For convenience, a range of 240 km is assumed (PRF of 625 Hz). A scan rate of 1.3 RPM is capable of completing the entire volume scan in 15 minutes. For 1 degree resolution in azimuth, this permits 78 pulses to be averaged each degree or "ray" and 8 rays per second (1.3 RPM). For the Doppler example it is assumed that the intensity, velocity and width are each stored in 1 byte (8 bits).

Note that a raw data compression of 50% is assumed. The actual data compression that can be achieved is highly dependent on the extent of the echo coverage as a function of elevation angle. At very high elevation angles, there is less echo at the far ranges because the beam is above the weather. An important point is that to transmit compressed data, there must be adequate buffering to handle the case when the low elevation angles are completely filled with data. This is usually not a problem if there is a disk on the system.

The bottom part of TABLE 3 shows the data volumes and transmission rates required for 4 different range resolutions. Each of the resolutions is matched to a picture product resolution (low, medium or high). In addition, an over-sampling case is presented. Note that the high resolution case is slightly mismatched (480 bins as opposed to 400 bins).

Of primary interest is the transmission rate required to send all of the compressed raw data in 15 minutes. In all cases, the raw data transmission rate exceeds what is typically

TABLE 3. 15-Minute Update Examples of Matched Sampling: Doppler Case (dBZ, Velocity and Width)

Basic Volume Scan Parameters

Update Time (minutes)	15.0
Maximum Range (km)	240
Nominal PRF (Hz)	625
Pulses per "Ray"	78
Minimum Antenna Scan Rate (rpm)	1.3
Azimuth Resolution (degrees/ray)	1.0
Number of Elevation Tilts	20
Bytes per Range Bin	3
Rays per second	8
Raw Data Compression Efficiency	50%

	Range Sampling Matched to Product Resolutions			
Data Volumes and Rates	Low Res	Med Res	HI Res	Over Sampled
Range Resolution (km)	2	1	.5	.25
Number of Range Bins	120	240	480	960
Range Bins per Second	960	1,920	3,840	7,680
Un-compressed Raw Data (MBytes)	2.6	5.2	10.4	20.7
Un-compressed Raw Data Rate (Kbaud)	23.0	46.1	92.2	184.3
* Compressed Raw Data Rate (Kbaud)	11.5	23.0	46.1	92.2
Equivalent Low Res Products	68	135	270	540
Equivalent Medium Res Products	---	34	68	135
Equivalent High Res Products	---	---	13	25

* Assumes that buffering is adequate for compression.

available on a standard phone line with packet encoding modems (e.g., 9600 baud), although the low resolution sampling scenario is close. Both the medium and high resolution cases can be achieved only with dedicated high speed lines.

The bottom of the table shows the equivalent number of products corresponding to the raw data. The low resolution sampling scenario requires the same amount of storage (or transmission bandwidth) as 68 low resolution products. In other words, if fewer than 68 product images are to be generated, it is better to calculate the products at the radar and then transmit them. For the case of medium resolution sampling vs medium resolution products, the "break even" point is at 34 products. Finally for the case of high resolution products and high resolution sampling, it is better to send raw data if more than 13 products are to be sent.

TABLE 4. 15-Minute Update Examples of Matched Sampling: dBZ Only Case

Basic Volume Scan Parameters

Update Time (minutes)	15.0
Maximum Range (km)	240
Nominal PRF (Hz)	625
Pulses per "Ray"	78
Minimum Antenna Scan Rate (rpm)	1.3
Azimuth Resolution (degrees/ray)	1.0
Number of Elevation Tilts	20
Bytes per Range Bin	1
Rays per second	8
Raw Data Compression Efficiency	50%

Data Volumes and Rates	Range Sampling Matched to Product Resolutions			
	Low Res	Med Res	Hi Res	Over Sampled
Range Resolution (km)	2	1	.5	.25
Number of Range Bins	120	240	480	960
Range Bins per Second	960	1,920	3,840	7,680
Un-compressed Raw Data (MBytes)	0.9	1.7	3.5	6.9
Un-compressed Raw Data Rate (Kbaud)	7.7	15.4	30.7	61.4
* Compressed Raw Data Rate (Kbaud)	3.8	7.7	15.4	30.7
Equivalent Low Res Products	23	45	90	180
Equivalent Medium Res Products	---	11	23	45
Equivalent High Res Products	---	---	4	8

* Assumes that buffering is adequate for compression.

This example illustrates that in the case of high resolution products, it is probably more efficient to send raw data than final products. More importantly, it illustrates that flexibility is required so that there is the option of performing product generation at the radar host computer or another computer, depending on the sampling and product mix that are required.

5.3 Example 2: 15 Minute Update Intensity Only Case

The second example shown in TABLE 4 is identical in all respects to the example shown in TABLE 3 except that instead of 3 bytes of Doppler and intensity information for each range bin, there is only 1 byte of intensity information. In this case, the storage and communications requirements are one third of what they were for the Doppler example. The compressed raw data rates for low and medium resolution sampling are within the range of standard phone lines. Also, the "break even" equivalent number of products shifts the advantage to sending raw data as opposed to products. For example, if the composite computer is to generate medium resolution CAPPI's at 1 km intervals up to 11 km, it is equally efficient to send medium resolution raw data as opposed to 11 medium resolution CAPPI's generated at the radar.

6. Summary

The paper describes a flexible architecture for a radar network which permits the computation of products to be performed either at the radar site, local workstations or central computers. By employing a compatible class of computers to serve as radar host computers, composite computers and workstations, identical software can be used to perform product generation or compositing at a number of different sites. This flexibility for performing product generation at different locations provides numerous options for implementing and upgrading the communications network.

Two approaches to product transmission were discussed, i.e., regularly scheduled product transmission and ad hoc product requests. The advantage of ad hoc requests is that virtually all of the products that are requested will be used, while many regularly scheduled products may never be used. The disadvantage of ad hoc requests is that a relatively high bandwidth is required so that requests are fulfilled in a timely manner even during peak use periods. In practice, it is usually possible to analyze the system usage and define a minimal product set for regular transmission and permit more exotic products or raw data to be generated and transmitted on request to network users.

The concept of matched resolution sampling was introduced and used to make comparisons between the efficacy of transmitting raw data versus transmission of radar products. For intensity-only data, or for high resolution displays, it is often more effective to transmit raw data rather than a full set of products. However, the architecture of the network should permit either approach so that the system can be tuned to changing network traffic and evolving communications capabilities.

Acknowledgements
The author thanks the IRIS team at SIGMET, Inc.-- A. Siggia, J. Holmes, T. Huang, W. Czarneicki, J. King and P. Wolf for their creative efforts in designing and implementing the IRIS system, and Mssrs. R. Braswell and L. Collins of Enterprise Electronics Corporation for their support of the IRIS project.

Transmission and visualisation techniques for a remote weather radar display

R. Heylen and M. Van Loey

Royal Meteorological Institute
Ringlaan 3, B1180 Brussels, Belgium

Over the past years, powerful, inexpensive and easy to use computers and telecommunication equipment became commercially available. High speed local area networks enables equipment from different manufacturers to be interconnected and meteorological data resources can be shared throughout the network. At the Royal Meteorological Institute (RMI) of Belgium the implementation of a small weather computer system has started in june 1988. An open architecture concept will allow this system to grow as funding becomes available. As a first step, telecommunication equipment for the exchange of weather radar data and systems for display of data from the local radar (Zaventem), bi-lateral composites (France and UK) and the COST image, were installed.
During the COST73 seminar, a remote radar display was connected to this system. After reception of new images (local data and COST73 composite) at RMI and after data transcoding, radar data was transmitted to a display system installed at the seminar. Techniques used for this live display are explained.

Introduction

An integrated meteorological data processing system has to satisfy five basic functions: data ingest, data processing and analysis, data storage, data presentation and data distribution. To meet these requirements, a plan for a system with distributed processing in a networked environment was made. To reduce cost, it was decided only to use open architecture hardware and to maximize the use of commercially available software packages. As a start, a small system for exchange of radar data was implemented and is operational at the RMI. Future expansions will include satellite processing systems for both METEOSAT and NOAA satellites, integration of radio sounding equipment, mainframe connection for retrieval of graphical products from numerical models and ECMWF products and installation of additional workstations for data processing and data presentation.

System description

A layout for the system is shown in figure 1.
Both workstations are Hewlett Packard 9000/360 computers running the HP-UX operating system. The Unix operating system has evolved to become one of the most popular

computer environments but was originally not written as a real-time system. When real-time enhancements are added to a standard UNIX system, low cost workstations can be used for real-time telecommunication applications. At present, the telecom workstation has interfaces installed for simultaneous data reception and transmission on 12 asynchronous lines and 6 synchronous lines. Spare slots allow for future expansion and a direct X.25 connection.

The graphic workstation is used as a display for both graphics data and image files. The X window system is used as the graphical user interface.

Data transfer between computers is done over an ethernet 802.3 baseband local area network using the TCP/IP protocol.

Radar network

At present time, one weather radar (EEC, C-band with Doppler facilities) is installed at Brussels international airport. The radar is operated by the Met Service of the Belgian airways agency. Transmission of digital data to the RMI started in january 1989 over a leased line (2400 bps channel on a 9600 bps line) using V29 modems at both sides. Two other 2400 bps channels were reserved for data exchange with Bracknell and Paris. Reception of the UK composite and the COST picture together with the transmission of Belgian radar data to Bracknell, was started on a test basis in june 1989. It is expected to start the exchange of radar data with France by the end of 1989.

All software tasks for transmission and transcoding of radar images are loaded as high priority, resident modules on the telecom workstation. During the COST seminar, a personal computer (PC1) was connected to both the local area network using an ethernet interface card and to the public telephone network (PTN) using a V22 Hayes compatible asynchronous modem. The PC1 was programmed to detect if recent COST73 and Belgian radar images were correctly received and if they were coded in BUFR94 format by the telecom workstation. At the COST seminar, a second computer (PC2) was installed and connected to the PTN through a V22 modem. This PC2 was used for ingestion of new radar data and for visualisation of these images on a color screen. For the data transfer over the PTN, a commercially available communications package with a standard file protocol was used (XTALK, XMODEM). In this way, new radar data were downloaded onto the hard disc of the live display. Additional software for data presentation was installed on the PC2. A flow chart for the software installed on the live display is given in figure 2.

Results

During the seminar, the following products were shown:
    . mosaic of the 4 most recent COST73 images
    . full screen display of the most recent COST73 image
    . mosaic of the 4 most recent Belgian images (PPI, 120 km range)
    . full screen display of the most recent Belgian image (PPI,120 km range)
Time and legend for the color scales were added to each image.

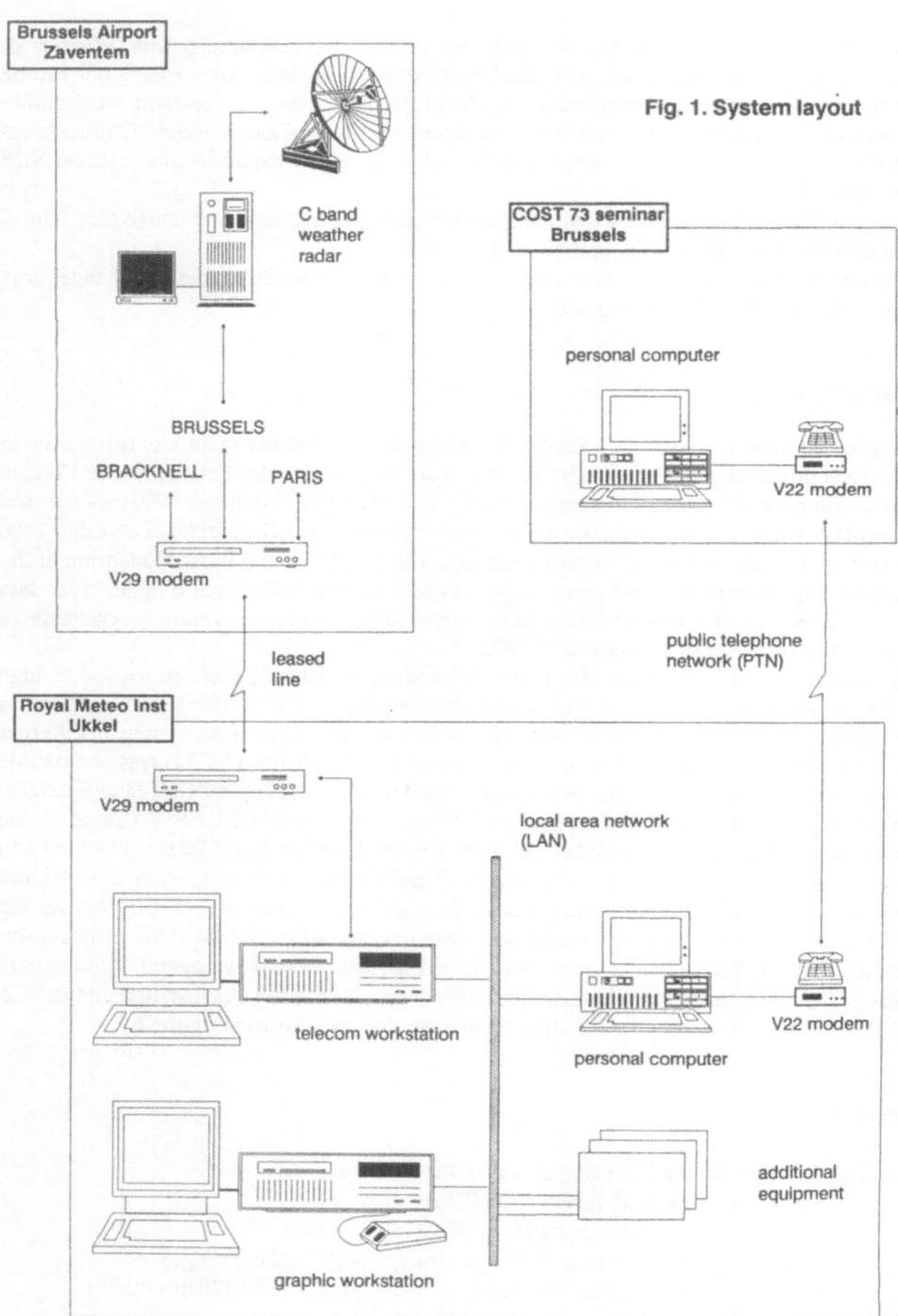

Fig. 1. System layout

As the period during which the seminar was held was characterized by sunny weather, only small precipitation areas could be observed on the display screen.

A very large zone of high pressure from the Urals to the Atlantic Ocean, with centers of 1025 hPa over Central Europe and the south of Ireland, blocked during more than 24 hours a waving cold front over the north of England. A wave was travelling on this front from the west of Ireland (52 °N,13 °W, 89/09/07, 00.00 UT) to the North sea (57 °N, 4° E, 89/09/08, 00.00 UT) and it formed there a low(!) of 1020 hPa. In the mean time, thermic lows developed over the north of Spain.

For this period, COST73 pictures are shown on figure 3.

Figure 2. Flowchart for software installed on COST73 display

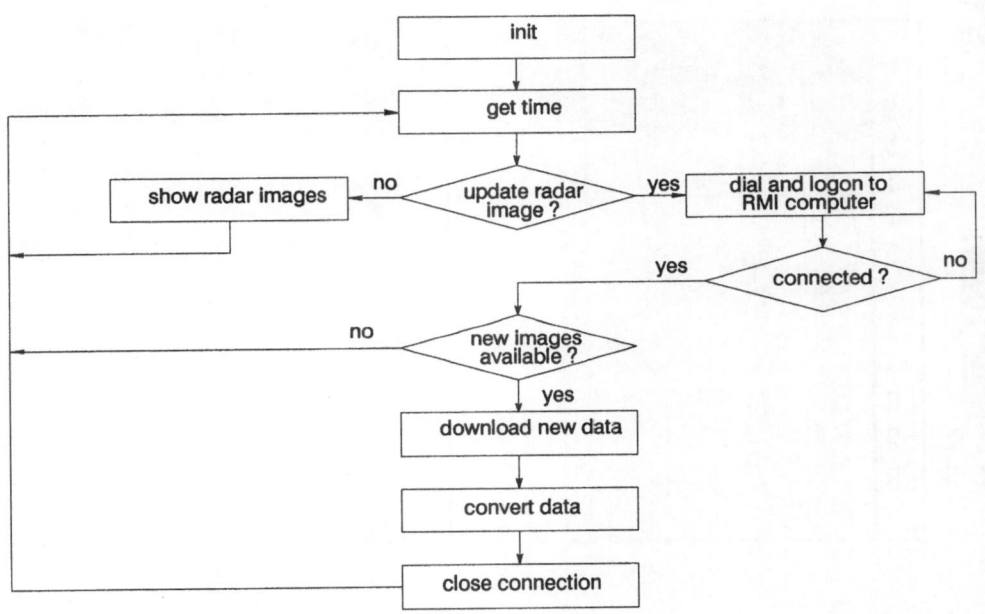

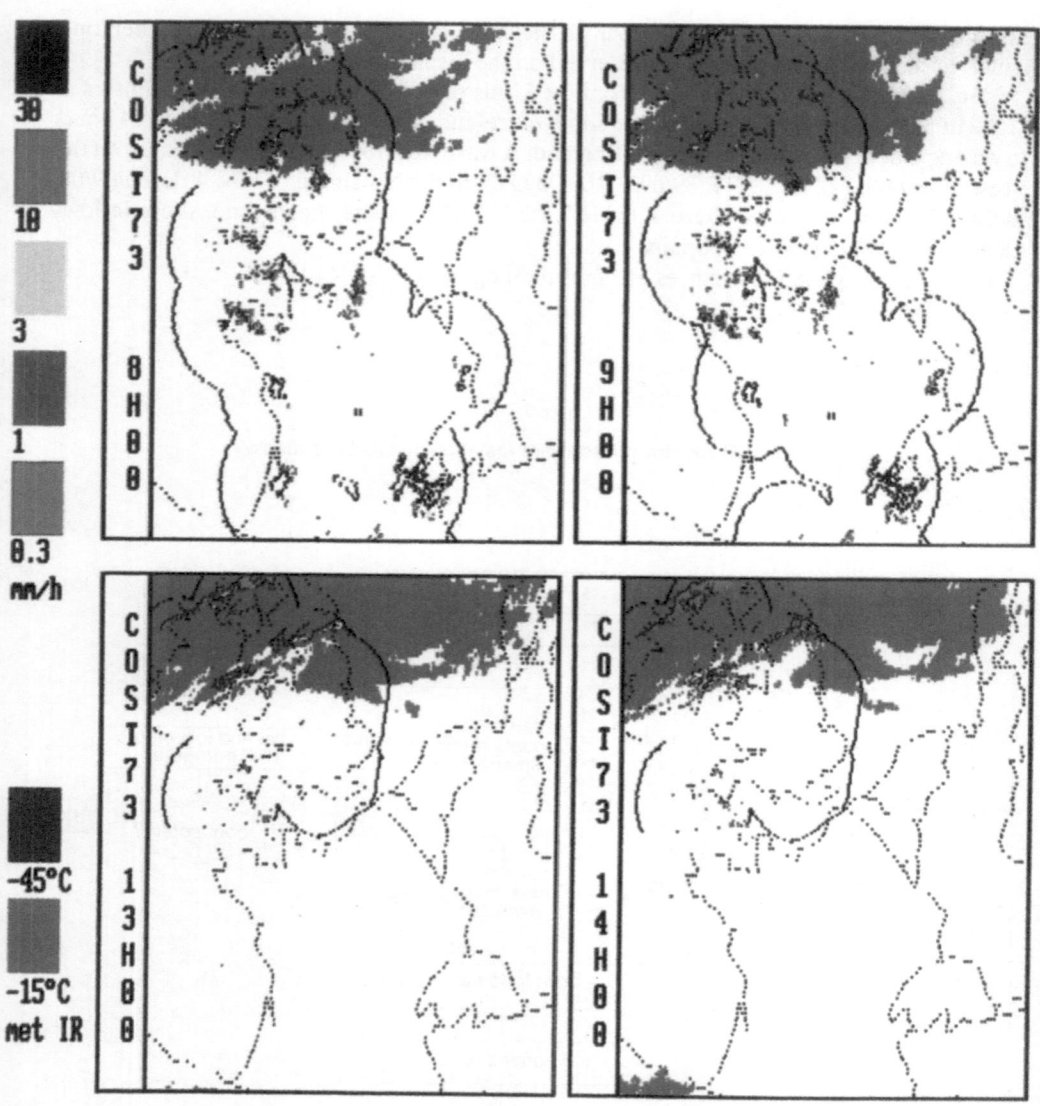

Figure 3. COST73 images taken on september 7th, 1989

SESSION 2

The role of reflectivity-based techniques in operational radar networks

Chairman : C.G. Collier

WAYS OF USING, AND CORRECTING FOR ERRORS IN
CONVENTIONAL RADAR REFLECTIVITY DATA

Jürg JOSS
Swiss Meteorological Institute
Osservatorio Ticinese, CH 6605 Locarno Monti

ABSTRACT

Over the years much research has been directed towards exploring the
potential of radar as an instrument for estimating rain. It is shown that
with the present reflectivity measuring radar, acceptable quantitative
information is already obtained from radar networks in many places and in a
wide range of applications. It is unlikely, however, that radar will ever
replace the raingauge, since gauges are vital as ground truth for adjusting
and/or checking the radar data. On the other hand, as pointed out by many
workers, we would need an extremely dense and costly network of gauges to
obtain a spatial resolution easily attainable with radar.

But radar will only estimate the precipitation it can "see". Even the most
ingenious procedure using sophisticated equipment will not allow us to make
measurements in parts lost because of reduced visibility. Reduced
visibility will occur behind mountains. But also at longer ranges, even in
flat country, we will find increased errors due to losses caused by the
earth curvature and the reduced resolution of the radar beam. Thanks to the
availability of inexpensive, high-speed data processing equipment, it is
possible today to determine the echo distribution in the whole radar
coverage area in three dimensions. This knowledge, together with knowledge
about the position of the radar and the orography around it, allows one to
correct in real time for a large fraction of - or at least to estimate the
magnitude of - the vertical profile problem. This correction allows us to
extend the region in which an accuracy acceptable for many hydrological
applications is obtained. This paper is in part based on the reviews by
Joss and Waldvogel (1989) and Joss and Smith (1989), where the reader may
find more details.

1 INTRODUCTION AND THREE TYPES OF ERRORS OF DIFFERENT MAGNITUDE

The most important advantage of using radar for precipitation measurements
is the coverage of a large area with high spatial and temporal resolution
from a single point and in real time. Furthermore, the three dimensional
picture of the weather situation can be extended over a very large area by
compositing data from several radars. However we have not been able until
recently to make measurements over a large area with an accuracy which is
acceptable for hydrological applications.

Since the quantity Z is by definition equal to the sum of the 6th power of
the particle diameter and the rainfall rate R is roughly proportional to
the 4th power of the particle diameter, there is no unique relationship
between the radar reflectivity and the precipitation rate. The relationship
depends on the particle size distribution. Thus, natural variability in
drop-size distributions is an important source of uncertainty in radar
measurements of precipitation.

Since the precipitation at the ground must be deduced from a measurement sampled aloft, we encounter a second kind of uncertainty which is often not given proper consideration. If there are variations in the vertical reflectivity profile, considerable error - usually underestimation - may result in the rainfall as measured by radar. This problem is especiallly severe in a mountainous country, but also applies to a flat country and may lead to unexpectedly large errors at longer ranges.

Uncertainties of a third type are related to instrumental considerations. Provided that modern radars are well maintained, the stability usually presents no problem. P.M. Austin (1987) discusses and illustrates the relative importance of various kinds of uncertainty using case studies.

On one side, excellent agreement between radar and ground truth is reported from many experiments, on the other side little operational use of radar is made so far in operational hydrology. One of the reasons for this situation probably lies in the large variability of meteorological phenomena and the different types of applications. In other words, the fact that well defined field experiments have shown good agreement between radar and gauge measurements does not necessarily mean that similar results can be obtained in operational applications over large areas. In such limited experiments some errors may be hidden in the radar-raingauge assessment factor (i.e. the ratio of the rain amount obtained with gauges to the corresponding radar amount). When the ideal conditions of such experiments are not fulfilled, the errors may emerge. They might produce unexpected deviations in operational applications and could thereby be responsible for the lack of faith in radar as an operational tool to measure rainfall. Within the last few years the situation has improved significantly. Availability of modern digital techniques for data recording and processing has mitigated many of the problems associated with radar measurements of rainfall, and currently a number of important operational projects are already established or being developed.

2 PARTICLE-SIZE DISTRIBUTIONS AND Z-R RELATIONS

Particle-size distributions of raindrops, snowflakes and hailstones have probably been observed since the beginning of mankind. They are of special interest in radar meteorology because they provide a means of deducing in a direct way the relationship between the radar reflectivity factor, Z, and the associated rainfall rate, R.

Various techniques for using polarization diversity radar to improve rainfall measurements have been proposed. In particular, it has been suggested that the difference between reflectivities measured at horizontal and vertical polarization (ZDR) can provide useful information about the drop-size distributions. The method depends on the hydrodynamic distortions of the shapes of large raindrops, more intense rainfalls with larger drops giving stronger polarisation signatures. There is still considerable controversy, however, as to whether or not this technique has promise for operational use in the measurement of precipitation. At close ranges (with high spatial resolution) polarization diversity radars may give valuable information about precipitation particle distributions and other parameters pertinent to cloud physics. At longer ranges one cannot be sure that the radar beam is filled with a homogeneous distribution of hydrometeors, so the empirical relationship of the polarimetric signature to the drop size distribution has increased uncertainty. Of course knowing more about Z-R will help, but even if multiparameter techniques worked perfectly one could reduce the error caused by Z-R only from 33% to 17% as shown by Ulbrich and Atlas (1984). For the ranges required (area covered) for hydrological applications we also need to correct for other biases which are usually much greater, perhaps by an order of magnitude or more.

3 VERTICAL REFLECTIVITY PROFILE, A BASIC PROBLEM IN RADAR HYDROLOGY

Because of growth or evaporation of precipitation, air motion and change of phase (ice and water in the bright band), highly variable vertical reflectivity profiles are observed, both within a given storm and from storm to storm. With increasing distance from the radar the vertical distance between the sample volume of the radar and the ground usually increases too. Therefore the differences between estimates of rainfall by radar and the rain reaching the surface must also increase.

A frequently observed feature in the vertical reflectivity profile is the "bright band", a layer of enhanced reflectivity where snow is melting into rain. It may lead to a strong overestimate of precipitation by radar. Overestimates of up to a factor of 5 may be observed, provided that the radar has the necessary spatial resolution to resolve the bright-band layer, which has a typical thickness of the order of 300 m. As most weather radars have sufficient spatial resolution only at close ranges, overestimation produced by bright-band enhancement occurs mainly near the radar. At medium ranges the bright band will compensate for the reduced reflectivity aloft and at longer ranges underestimation will dominate.

Above the bright-band (upper level at ca. O degrees C wet-bulb temperatur) vertical reflectivity profiles tend to show a sharp decrease with height, amounting to reflectivity gradients of up to 10 dB/km which corresponds to a change of up to a factor of four in measured rain rate per km. Average profiles for a variety of situations have been presented by Joss et al (1970), Rogers and Yau (1981) and Koistinen (1986). The reflectivity profiles in these references represent averages. Instantaneous values may deviate strongly in both directions. Although the sharp decrease generally starts at the level of the bright band, there are some situations where the decrease may start near the surface. Low-level growth in a moist atmosphere or evaporation in a relatively dry one can bring about significant changes in reflectivity with height. P.M.Austin (1987) has made some calculations of the probable magnitude of these effects.

In summary differences between radar indications and rain reaching the surface consist mainly of underestimation by radar. At long ranges, for low level storms, and especially when low antenna elevations are blocked by obstacles, such as mountains, the underestimate may be severe. This type of error often tends to dominate all others, a fact easily overlooked when observing storms at close ranges only, or when analysing storms which are all located at roughly the same range.

4 GROUND CLUTTER AND SHIELDING

When the radar beam or its sidelobes encounter ground targets, strong
persistent echoes occur and add up in time to appear as large rain amounts
if no precautions are taken in the data analysis. A method for eliminating
them is to use a clutter map in the computer memory and block contaminated
pixels out. But this procedure may leave blind zones in badly cluttered
areas and thereby lead to a loss of data. Because clutter, and especially
that caused by anomalous propagation, is variable in time the clutter map
has to be made up from many situations without rain, including ones with
anomalous propagation. As a consequence of this, the loss in any given,
single weather situation is often larger than it need be. Doppler
techniques hold promise for reducing these difficulties (Passarelli et al.
1981) but some problems are likely to remain in cases of stationary
rainfall patterns. Golden et al, (1986) expect that radars in the new
network being developed in the U.S.A. will have clutter suppression
capabilities of at least 30 dB in the reflectivity channel.

Topography and the curvature of the earth may hinder the detection of
important parts of the precipitation. This difficulty usually increases
with increasing distance of the illuminated volume from the radar and is
more severe in winter than in the warm season when the melting level is
high. For example at 200 km with an elevation angle of 0.5 deg, the average
height of the radar beam is at an altitude of 4.25 km. Fig.1 illustrates
what the radar sees at various ranges in the absence of hills and
obstructions. Even in this ideal situation the profile is considerably
distorted except for convective rain close to the radar. Tab.I demonstrates
that we have to work at longer ranges.

Country	No of radars		Area per radar 1000 km2		Average range km		Ave Beam Dia km	
	1989	1993	1989	1993	1989	1993	1989	1993
Austria	3	4	28	22	92	80	1.8	1.5
Belgium	1		32		96		1.8	
Denemark	1		43		113		1.8	
Finland	3		112		83		1.6	
Germany	3	10	83	25	157	86	3.0	1.7
France	20		28		92		2.4	
Ireland	1		70		145		2.5	
Italy	12		25		87		2.3	
Netherlands	2		20		77		1.5	
Norway	1		386		340		5.0	
Portugal	1		92		166		3.2	
Spain	1	15	500	33	387	100	6.1	1.6
Sweden	3		133		200		3.0	
Switzerland	2	3	20	13	77	63	1.5	1.2
United Kingdom	8	13	13	8	61	48	1.1	0.8
Yugoslavia	4	5	64	51	138	124	2.4	2.2

Tab.I: Number, Area per radar, average distance (=0.543 SQRT(Area)) from
the radar and horizontal extension of the radar beam at the average
distance, for existing and planned radars as reviewed in COST-73.

Errors caused by attenuation, considerations when choosing a wavelength and a radar site for hydrological applications, hardware calibration of radar systems, problems related to sampling, averaging and meteorological adjustment of radar data are discussed in Joss and Waldvogel 1989.

Fig.1: Vertical profiles seen by the radar at various ranges in convective and widespread rain and in low-level rain or snow. The number in each figure gives the percentage (referred to the true, melted water value which we would measure at ground level) in rain rate deduced from the maximum reflectivity of the measured profile. A radar with a 1 degree beam is assumed, in a flat country, which means that obstacles and radar horizon are of the same height as the radar itself (of the order of 100m). Of course putting the radar on a high tower or on a mountain would change the situation. But the change would be not only in the desired direction of reducing shielding but also would increase clutter, thus reducing visibility as well.

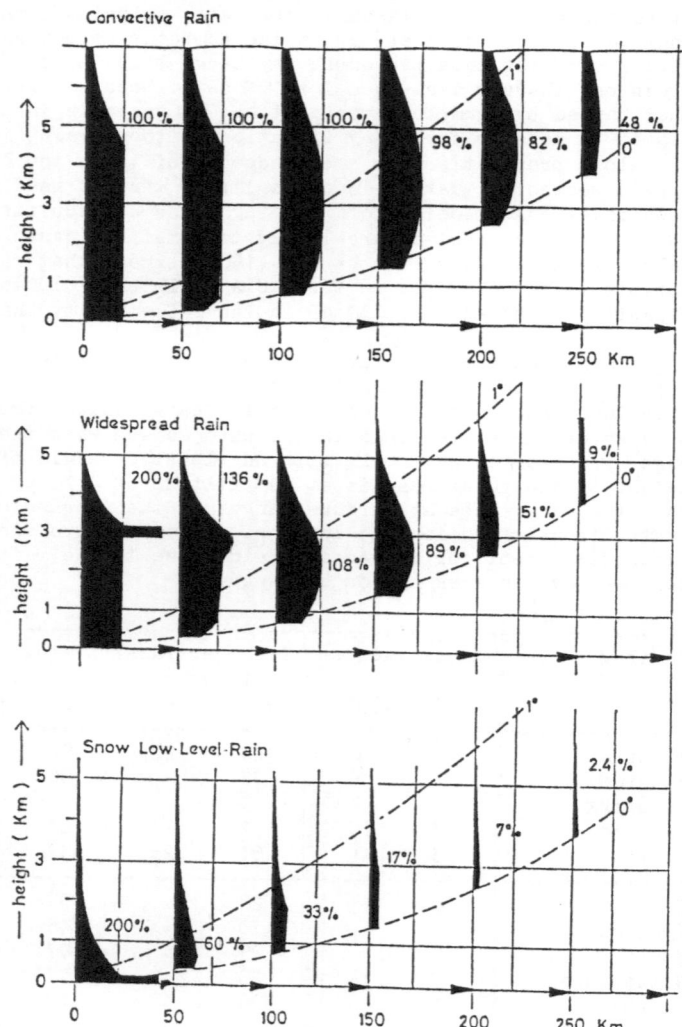

5 MEASUREMENT PROCEDURES

The basic procedure of deducing rainfall rates from measured radar reflectivities for hydrological applications requires the following steps:

1. Making sure that the hardware is stable by calibration and maintenance.

2. Correcting for errors of the vertical reflectivity profile and orographic enhancement of precipitation.

3. Taking into account all the information about the Ze-R relationship and deducing the rainfall.

4. Adjustment with raingauges.

The first three parts are based on known physical laws and only the last one uses a statistical approach to compensate for residual errors. This allows the statistical methods to work most efficiently. In the past, a major limitation in carrying out these steps was caused by analog circuitry and photographic techniques for data recording and analyses. It was therefore extremely difficult to determine and make the necessary adjustments, and certainly not in real time. Today the data may be obtained in three dimensions in a manageable form and the computing power is available for accomplishing these tasks. Much of the current research is directed towards developing techniques for doing so on an operational basis.

The method of approach for steps 2 to 4 and the adequacy of results obtained from radar precipitation measurement depend very much on the situation. This can include the specific objective, the geographic region to be covered, the details of the application and other factors. In certain situations an interactive process is desirable such as developed for FRONTIERS and described by Collier in Section 3.3.2 of Appendix 1 in Joss and Waldvogel (1989). It makes use of all pertinent information available in modern weather data centers.

As yet no one method of compensating for the effects of the vertical reflectivity profile in real time is widely accepted (step 2). However, three degrees of compensation methods can be identified:

1. Range-dependent correction: The effect of the vertical profile is associated with the combination of increasing height of the beam axis and spreading of the beam with range. Consequently, a climatological mean range-dependent factor can be applied to obtain a first-order correction. Different factors may be appropriate for different storm categories, as for example convective versus stratiform.

2. Spatially-varying adjustment: Where the precipitation characteristics vary systematically over the surveillance area, or where the radar coverage is non-uniform because of topography or local obstructions, corrections varying with both azimuth and range may be useful. If sufficient background information is available, mean adjustment factors can be incorporated in suitable look-up tables. Otherwise, the corrections have to be deduced from the reflectivity data themselves or from comparisons with gauge data (a difficult proposition in either case).

3. Full vertical profiles: The vertical profiles in storms vary with location and time, and the lowest level visible to the radar usually varies because of irregularities in the radar horizon. Consequently a pixel-by-pixel correction process using a representative vertical profile for each zone of concern may be needed to obtain the best results. Representative profiles can be obtained from the radar volume scan data themselves, climatological summaries, or storm models. This is the most complex approach but can be implemented with modern data systems.

After making the profile corrections, one should use a reflectivity –
rainrate relationship appropriate to the situation, geography as well as
season, to deduce the value of R (step 3). There is general agreement that
comparisons with gauges should be made routinely, as a check on the radar
performance, and that appropriate adjustments should be made if a radar
bias is clearly indicated (step 4). In situations where the radar estimates
are far from the mark due to radar calibration or other problems, such
adjustments can bring about significant improvements. However, the
adjustments do not automatically assure improvements in the radar
estimates, and sometimes the adjusted estimates are poorer than the
original ones. This is especially true for convective rainfall where the
vertical extent of echo mitigates the difficulties associated with the
vertical profile, and the gauge data are suspect because of
unrepresentative sampling. A general guideline is that the adjustments will
produce consistent improvements only when the systematic differences (i.e.
the bias) between the gauge and radar rainfall estimates are larger than
the standard deviation of the random scatter of the gauge versus radar
comparisons. That guideline allows one to judge whether gauge data should
be used to make adjustments, and leads to the idea that the available data
should be tested before any adjustment is actually applied. Various methods
for accomplishing this have been explored, but at this time there is no
widely accepted approach.

6 TYPES OF APPLICATIONS

6.1 Short Term Forecasting

Probably the most attractive features of weather radar in an operational
context is that it provides an overview of the actual weather situation
(Nowcasting) and the possibility for making short term forecasts. Most of
the systems currently in operation have the description of current weather
as their primary output. The information is disseminated as a rainfall or
reflectivity map. Most of them are working on development of computerized
short term forecasts but many are still in a research stage.

6.2 Hydrological Forecasts, Streamflow And Urban Hydrology

Comparing radar data with river flow brings in a new variable. In other
words, it liberates our radar-verification task from the problem of point
measurement and representativeness of single gauges but adds a new problem
to the complexity of the quantitative radar measurement, the one of fluxes,
storage and delays in a river basin.

Collier (1985) reviews requirements for hydrological forecasting; he
considers groundbased radar techniques, the future potential of remote
sensing and the need for a total system approach in the way which was
proposed by Browning (1979) and which is being discussed and developed with
different degrees of complexity in various countries.

When applying radar to predict the flow of water in the sewers of a town,
variables and equations are similar to those for the flow of a river in a
large basin, but the analysis has to be done on a much finer time scale.

6.3 Flood And Avalanche Forecasting

One of the most important operational hydrological applications is flash flood warnings. Cluckie et al. (1987) investigated flood-producing storms and concluded that the use of real-time hydrological forecasting models 'has now been clearly established' and 'efforts are now being directed towards the further improvement of the operational use of such data'.

Kappenberger and Joss (1986), in an exploratory analysis of four storms which led to a catastrophic avalanche situation, found that after correcting by hand for the vertical reflectivity profile, there was reasonable agreement between radar-estimated melted snow and raingauge-measured melted snow. They conclude that radar together with automatic weather stations and manual observations can give valuable information about potential avalanche activity.

6.4 Combination Of Radar Data With Data From Space

It is generally recognized that on a global scale the areas which can be adequately covered by networks of gauges or radars are limited. Therefore for hydrological measurements on a near-global scale, it is necessary to resort to satellite observations.

Collier (1986) concludes that satellite and radar techniques for measuring rainfall for hydrological applications are complementary and that radars are more appropriate for small basins (of the order of 10^4 km^2 or less, a number which may be increased if correction for the reduced visibility at longer ranges proves possible). Satellites, on the other hand are useful for larger basins only and in regions where no radar information is available.

It appears that the reasonable approach will consist of adjusting radar with gauges and using radar in turn to adjust satellite data. Whether this is best done with human interaction as used successfully in FRONTIERS (Conway 1987) or by automatic ways of data analysis as used in RAINSAT (King and Yip 1987) remains to be seen. The choice for a given application will probably depend strongly on the availability of qualified personnel.

7 STATE OF THE ART AND SUMMARY

Over the years much research has been directed towards exploring the potential of radar as an instrument for measuring rain. In general, radar measurements of rain, deduced from measured reflectivity by using an empirical Z-R relation, agree well with gauge measurements for the area close to the radar but increased variability and underestimation by the radar occur at longer ranges. For example, the Swiss radar indicates at a range of 100 km on the average only 25% of the actual rain amount, in spite of the fact that it measures 100% at close ranges. Similar, but not quite so dramatic variations are found in flat country or in convective rain. The reasons are earth curvature, shielding by topography and the spread of the radar beam with range. Thus the main problem in using radar for precipitation measurements and hydrology in operational applications comes from the inability to measure precipitation close enough to the ground over the desired range of coverage. Because this problem often does not arise in well defined experiments, it has not received the attention which it deserves as a dominant problem in operational applications.

Thanks to the availability of inexpensive, high-speed data processing equipment, it is possible today to determine the echo distribution in the whole radar coverage area in three dimensions. This knowledge, together with knowledge about the position of the radar and the orography around it, allows one to correct in real time for a large fraction of - or at least to estimate the magnitude of - the vertical profile problem. This correction allows us to extend the region in which an accuracy acceptable for many hydrological applications is obtained.

To make the best possible use of radar, we offer the following rules:

1. The radar site should be chosen such that precipitation is seen by the radar as close as possible to the ground. "Seen" means here that there is no shielding or clutter echoes, or that the influence of clutter can be eliminated, for instance by Doppler analysis. This condition may frequently restrict the useful range of radar for quantitative work to the nearest 50-100 km.

2. Wavelength and antenna size should be chosen such that a suitable compromise between attenuation caused by precipitation and good spatial resolution is achieved. At longer ranges this may require a shorter wavelength to achieve a sufficiently narrow beam.

3. Systems should be rigorously controlled (sufficient stability and calibration of equipment).

4. Unless measurements of reflectivity are made immediately over the ground, they should be corrected for errors originating from the vertical profile of reflectivity. As this profile changes with time it should be monitored continuously by the radar. The correction may need to be calculated for each pixel, as it depends on the height of the lowest visible volume above the ground. It is important that the correction for the vertical reflectivity profile, as it is the dominant one at longer ranges, be done before any other adjustments, such as using information about Z-R relationships or adjusting to agree with raingauges.

5. The sample size must be adequate for the application. For hydrological applications, and especially when adjusting radar with gauges, it is desirable to integrate the data over a number of hours and/or square kilometers. The integration has to be done over the desired quantity, i.e. the linear rain rate R, not the $\log(Z)$, in order to avoid any bias caused by this integration.

Even a crude estimate of the actual vertical reflectivity profile can produce an important improvement. Polarimetric measurements may provide some further improvement, but it is questionable whether the additional cost, complexity and risk of misinterpreting polarization measurements can be justified for operational applications in hydrology.

The main advantages of radar are its high spatial and temporal resolution,

wide area coverage and immediacy (real-time data). Radar also has the
capability to follow a "floating target" or a "convective complex" in a
real-time sequence, for instance to make a short term forecast. Although it
is only to a lesser degree suited to give absolute accuracy in measuring
rain amounts, good quantitative information is already obtained from radar
networks in many places. It is unlikely that radar will ever replace the
raingauge, since gauges provide additional information and are essential
for adjusting and/or checking the radar indications. On the other hand, as
pointed out by many workers, we would need an extremely dense and costly
network of gauges to obtain a resolution easily attainable with radar.

REFERENCES

Austin, P.M. 1987: Relation between measured radar reflectivity and surface
    rainfall. Monthly Weather Review, pp 1053-1070.
Browning, K.A., 1979: The FRONTIERS plan: a strategy for using radar and
    satellite imagery for very-short-range precipitation forecasting,
    Met.Mag, 108, pp 161-184.
Cluckie, I.D., P.F.Ede, M.D.Owens, A.C.Bailey and C.G.Collier, 1987: Some
    hydrological aspects of weather radar research in the United Kingdom.
    Hydrological Sciences Journal, 32, 3, 9/1987.
Collier, C.G., 1985: Remote sensing for Hydrology. Facets of Hydrology,
    Volume II, Chapter 1, John Wiley Sons Ltd.
Collier, C.G., 1986: Accuracy of rainfall estimates by radar, Part I:
    Calibration by telemetering raingauges. Journal of Hydrology, 83
    (1986) 207-223.
Conway, J.B., 1987: FRONTIERS: An operational system for nowcasting
    precipitation. Proc.Symp.Mesoscale Analysis and Forecasting,
    Vancouver, ESA SP-282, pp 233-239.
Golden, J., E.D.Sarreals and F.Toepfer, 1986: NEXRAD products and
    algorithms, Part II: Operational impact of NEXRAD technology on NWS
    operations. Preprints 23rd Conference on Radar Meteorology, Vol. 3,
    pp JP87-90, AMS, Boston.
Joss, J. and A.Waldvogel, 1989: Precipitation measurements. To appear in
    the CIMO-Guide, Editor: Marc Gilet.
Joss, J. and A.Waldvogel, 1989: Precipitation measurements and Hydrology, a
    Review. To appear in the Battan Memorial and 40th Anniversary Radar
    Meteorology Volume, Editor: David Atlas.
Kappenberger, G. and J.Joss, 1986: Use of radar and automatic weather
    stations in avalanche forecasting. Proceedings of the International
    Conference on Formation, Mouvement and Avalanches, IAHS No. 162,
    Wallingford, GB, pp 305-310.
King, P. and T.C.Yip, 1987: Evaluation of the rainsat precipitation
    analysis system in real-time use. Proc.Symp.Mesosocale Analysis and
    Forecasting, Vancouver, ESA SP-282, pp 263-268
Koistinen, J., 1986: The Effect of some Measurement Errors on Radar-Derived
    Z-R Relationships. Preprints 23rd Conference on Radar Meteorology,
    Vol. 3, pp JP50-53, AMS, Boston.
Rogers, R.R. and M.K.Yau, 1981: Summertime radar echo coverage at Montreal.
    Preprints 20th Conference on Radar Meteorology Boston, Mass. Published
    by the Am.Met.Soc., Boston, 469-475.
Passarelli, R.E., P.Romanik, S.G.Geotis and A.D.Siggia, 1981: Ground
    clutter rejection in the frequency domain. Preprints 20th Conference
    on Radar Meteorology Boston, Mass. Published by the Am.Met.Soc.,
    Boston, 295-300.
Ulbrich, C.W. and D.Atlas, 1984: Assessment of the contribution of
    differential polarization to improve rainfall. Radio Science, Vol. 19,
    Nr. 1, Jan-Feb 1984, pp 49-57.

REVIEW OF THE TELECOMMUNICATIONS WORK OF COST 73 PROJECT

B. BERINGUER
Direction de la Météorologie Nationale.
Service des Equipements et des Techniques
Instrumentales de la Météorologie.

Abstract

The data supplied by weather radar allows us to see what the weather is like at every moment and to estimate likely weather changes in the very short term - renown as "nowcasting".
However, it is necessary to note that weather radar can only produce information relating to a delimited area.
For this reason, it is necessary to link several radar images concerning neighbouring areas.
Therefore it is necessary to arrange not only means of telecommunications but also a common format for weather radar data.
Concerning the telecommunications, several technical possibilities are offered : specialized lines, network, micro-waves, satellites, etc.
The choice is generally made according to the financial conditions in every country.
So it appears clearly that it is the disparity of the radar image formats which reduces or sometimes completely stops the possibilities of exchanges of radar data between the countries.
This situation has been analysed by the COST 73 Working Group Telecommunications. Its purpose has been to built a proposal of common format concerning radar images available for all - weather radars and whatever the mode of telecommunications means may be.
Initially, the working group studied the possibilities offered by the WMO formats and more particularly "FM 94 BUFR".
It appears that concerning the radar data, the parameters were not yet defined in the edition 0.
In these conditions the COST 73 Working Group Telecommunications has been obliged to complete the radar parameters tables, to built a radar data format and to propose it to WMO.
This project is now being studied by the WMO experts for integration in the "FM 94 BUFR".
Exchanges of radar data using this format have given good results in several countries.

Résumé

Les données fournies par les radars météorologiques permettent de montrer le temps qu'il fait à chaque instant et de prévoir différentes informations météorologiques en particulier la prévision à courte échéance.
Ces données sont d'autant plus utiles qu'elles sont disponibles rapidement pour les différents usagers.
Par ailleurs, il faut noter que chaque radar fournit seulement des informations sur une zone bien délimitée. C'est pourquoi il est nécessaire de disposer non seulement de moyens de télécommunications efficaces, mais aussi un format commun de données de radars

météorologiques.
En ce qui concerne les télécommunications, plusieurs techniques sont offertes : lignes spécialisées, réseau, micro-ondes, satellites ...
Un choix est fait généralement selon les conditions financières pratiquées dans chaque pays.
Ainsi, il apparaît clairement que c'est la disparité des formats d'images radar qui réduit ou même qui empêche les échanges de données radar entre les pays.
Cette situation a été analysée par le Groupe de Travail du COST 73.
La première action a consisté à établir un projet de format commun d'images radar valable pour toutes sortes de radars et quelque soit le mode de télécommunication employé.
En un premier temps, le Groupe de Travail a étudié les possibilités offertes par les formats de l'OMM et plus particulièrement le FM 94 BUFR.
Il est apparu que les données radar n'étaient pas encore définies dans l'Edition 0.
Dans ces conditions, le Groupe de Travail COST 73 a été amené à établir les tables de paramètres adéquat et à proposé à l'OMM un format de données radar conforme au FM 94 BUFR.
Ce projet est maintenant à l'étude par les spécialistes de l'OMM pour intégration dans le FM 94 BUFR.
Les échanges de données radar utilisant ce format ont jusqu'à présent donnés entière satisfaction dans les différents pays.

Weather radar is the preferred system for real-time evaluation of precipitation intensity and distribution over wide areas of the Earth's surface.
It can detect dangerous atmospheric phenomena and monitor their development in time and space.
The quality of the observations provided by weather radar is now such that the needs of both meteorology and hydrology can be catered for.
In particular, more and more users want to know what the weather conditions will be in the short term (i.e. in a few hours'time) in specific locations (known as "nowcasting").
The observations provided by weather radar are capable of answering these questions, provided that the information reaches the recipients in good time and that they are able to interpret it without difficulty.
It was against this background therefore that COST 73 Working Group on Telecommunications looked into the possibilities of exchanging radar information between countries.
From earliest stages of its work, the Working Group concluded that the means of telecommunications ought to be dealt with separately from the formats of the radar data proper.
First, the Working Group studied the various possibilities of using transmission for the exchange of radar data between the countries taking part in the COST 73 project.
This study showed that each country has access to a large number of different means of telecommunication, including specialized lines, microwaves, multiple telecommunications networks, satellites etc. In many cases, there are points of interconnection between these various means of data transmission.
Moreover, excellent results are obtained using telecommunications procedures, which in turn means far fewer errors due to data transmission.

The rates charged for the use of these various means of telecommunication vary not only according to the lines used, but also from one country to another.

Alongside these investigations, the Working Group looked into the possibility of using the existing Global Telecommunication System (GTS) circuits. The meteorological services of the member countries of the World Meteorological Organization (WMO) exchange weather data by GTS on a daily basis.

The observations provided by radar could be considered as meteorological information and could be transmitted using GTS channels.

The members of the Working Group identified two areas of difficulty:

1) The volume of radar data to be transmitted is large and therefore likely to lead to serious overloading of the existing GTS circuits, as well as long transmission times.
2) No common format for radar data has yet been approved by the World Meteorological Organization.

The members of the Working Group therefore decided to devise a common format for radar data which would be used by the fifteen countries belonging to the COST 73 project and, what is more, would be endorsed by the WMO.

The elaboration of this common radar data format came up against a number of constraints.

- Definition of an appropriate reception structure capable of expressing the various data provided by radar.
- Definition of the parameters for identifying the type of radar information.
- Reduction in the volume of information taking account of the specific characteristics of radar data.

The COST 73 Working Group on Telecommunications looked first at the existing possibilities offered by the data formats proposed by the WMO.

After study and consultation, it was felt that the FM 94 BUFR format (Binary Universal Form for Data Representation) was likely to meet the radar information requirements.

However, the first editions of FM 94 BUFR did not include a chapter on radar data.

Confronted with this situation the Working Group on Telecommunications decided to rectify this omission, at least for the purposes of meeting the needs expressed by the sixteen countries taking part in COST 73.

It should be recalled at this point that the FM 94 BUFR format can be used not only to express the data themselves but also to provide information on the data. This is because the FM 94 BUFR format is divided into six sections :

Section 0 : Indicator Section
Section 1 : Identification Section
Section 2 : Optional Section
Section 3 : Data Description Section
Section 4 : Data Section
Section 5 : End

Sections 0, 1, 2 and 5 are functional sections which contain only a limited number of data. Sections 3 and 4, on the other hand, are the principal parts of the format. Section 3 is used to describe the data actually appearing in Section 4. For this purpose, the FM 94 BUFR format makes available to users a very full set of semantic tools to facilitate the description and expression of a wide range of information.

For example, the meaning of a radar image can differ according to the type of radar or the algorithms used.

Within the FM 94 BUFR framework, therefore, the Working Group concentrated on two specific areas :

1) Definition of entities for weather radar.
2) Creation of a compacted image suitable for weather radar, making it possible to reduce the volume of data transmitted and transmission time.

Under point 1, the Working Group identified a number of parameters necessary for radar information. This led, among other things, to the inclusion of a class 21 for radar data within FM 94 BUFR.

Under point 2 a simple effective method for expressing radar data in reduced form was proposed.

All these proposals were submitted to the Working Group on Data Management : Sub-group on Data Representation set up by the WMO's Committee on Basic Systems (CBS). This working group endorsed the proposal to add the existing specifications of FM 94 BUFR. The CBS has now to give a final ruling on this proposal.

Alongside this process of integration of WMO texts, the Working Group has also organized practical trials with this format.

For this purpose software was written in several different advanced languages, this enabling each of the sixteen countries taking part in the COST 73 project to carry out long-term tests.

Initial results from these tests appear extremely encouraging in terms of the operational applications of this format for the representation of radar data.

Left to right :
Mr B. BERINGUER (French Delegate -
   Chairman of Session 3)
Mr L. JONES (Plessey Radar)
Mr W. RANDEU (Austrian Delegate)

ESTIMATION OF THE AREAL COVERAGE OF RADARS AND RADAR NETWORKS FROM RADAR SITE HORIZON DATA

H.R.A.WESSELS
Royal Netherlands Meteorological Institute

Summary

The range of weather radars is restricted by the curvature of the earth and by blocking of the radar beam behind obstacles. Each radar has its own typical horizon, which may have a large influence on the quality of the data, especially at far range. Radar horizon data are important for the planning of new sites and also for the interpretation and the presentation of measurements by single radars or radar networks.
The COST-73 Management Committee has agreed on a standard method for obtaining and archiving radar horizon data in the participating countries. As a result the production of a radar coverage map of a large part of Europe is scheduled. This map will show the areas where operational digitized radars may detect precipitation reaching to a level of 1500 m above the ground. This information is useful for the development of merging algorithms in overlap regions between radars. The map will also indicate where the bilateral exchange of data is beneficial for improved detection or for back-up purposes. Finally such a map may serve to assess the local quality of COST radar composites.

1. INTRODUCTION

The range of a weather radar depends on observational requirements as well as physical restrictions. The operational demands are either quantitative (accurate precipitation measurement) or qualitative (e.g. warning for a certain precipitation threshold). Important physical restrictions are the earth' curvature and the lowest elevation allowed by obstacles surrounding the radar.

As a result the limiting radar range will depend strongly on the application: e.g. 250 km for thunderstorm warning, about 100 km for quantitative use and perhaps only 50 km for drizzle detection. These large differences are not caused by an insufficient radar sensitivity, but primarily by the fact that weaker precipitation systems remain below the radar horizon due to their smaller vertical extent.

Local obstacles with an angular dimension as small as 1 deg. can reduce the radar range with as much as 50 percent. Many obstacles are larger. The arrangement of obstacles around a radar is therefore an important factor for the evaluation and the use of the measurements. Each radar has its own typical horizon with hills, trees and buidings. The site-specific situation can be plotted in a so-called occultation diagram. These diagrams show the limiting range (or the lowest usable elevation) as a function of azimuth.

Horizon data can be collected in various ways: radar observations, optical theodolite measurements in clear weather or computations with a topographical data base. Site survey procedures are summarized by Clift (1). For the planning and operation of radar networks it is important to use a uniform site evaluation for the radar sites involved: Larke and Collier (2), Martinez (3) and Leone et al.(4).

The following chapters will present a discussion on the definition and the measurement of the 'range' of the weather radars contributing to the COST composite.

2. USEFUL RANGE OF WEATHER RADARS

Weather radars are operated for two main tasks: early warning for (e.g. severe) weather phenomena and quantitative precipitation measurements. Both types of observation suffer from measurement errors. Important among these errors are geometrical sampling problems: the radar beam cross-section increases with range and the target may have an inhomogeneous horizontal and vertical structure. Targets vary between thunderstorms reaching to heights of more than 10 km, and drizzle restricted to the lowest km. Horizontal dimensions of showers may be less than 1 km. Moreover there exist elevated echoes: initiating showers, approaching warm frontal clouds or distant melting bands.

Under these circumstances the scanning strategy will inevitably be a compromise between the necessary accuracy, the desired temporal and spatial resolution and the data handling capabilities. For warning purposes most of the radar information is obtained from one or two PPI (plan position indicator) scans at low elevation. The most complete scanning method, favoured for quantitative work, is the 'volume scan': from pseudo-PPI-scans at about 20 elevations a series of CAPPI (constant altitude PPI) levels is constructed, each representing the horizontal echo distribution in a slice with a typical depth of 1 km. The observation results in a regular three-dimensional field cf 'pixels'. For various reasons some of these pixels will never contain precipitation information, especially in the lower CAPPI levels. As described by Joss and Waldvogel (5) pixels can be below ground, shielded by obstacles, only partially visible, or pixels can receive severe ground clutter from the main beam or the side lobe.

If the range exceeds about 100 km, the data in the lowest CAPPI level will be missing due to the curvature of the earth and due to the blocking of the beam by obstacles like hills and buildings. The single low-elevation beam may still contain data up to e.g. 250 km, but by then the precipitation systems must reach into the upper half of the troposphere to become observable. For qualitative use of weather radar this horizon effect is the main limitation to the observing range. For quantitative use the influence of obstacles is just one of many errors, which together make the measurements unreliable at a range beyond, say, 100 km.

As effective range for qualitative use one could e.g. define the distance up to which at least 80% of the precipitation occurrences (excluding drizzle) are correctly detected. This demands that, at its limiting range, the radar should be able to receive an echo from a precipitation system with a certain minimum thickness. According to Stewart (6) an accepted minimum cloud thickness for 20% probability of producing rain is about 1500 m. Taking into account the difference between 'cloud' and 'echo' and assuming the cloud base at 300 m, we may consider 1500 m as a suitable lowest altitude for the limiting radar beam axis. This means that a sufficient concentration of rain drops must be present slightly above the 1500 m level, so that an echo can be detected by the upper half of the beam (Fig.1). Of course this critical height has to be measured above the local terrain and not with respect to the sea level or the site level, because precipitation formation needs a certain depth above the ground. This means that clouds above high ground have a better 'exposure' to the radar and will be observable at larger range.

Of course the critical height depends on the observing task: if the only interest is detecting precipitation of at least 10 mm/hr, the critical height might be as large as 4000 m. The level of 1500 m may be too high in situations with maritime showers or with precipitation caused mainly by orographic ascent.

Although for the COST-area the coverage for compositing and other quantitative use will be presented at the 1500 m level, the primary data (i.e. the elevations of obstacles) remain available, and coverage data can easily be recomputed for a different critical height.

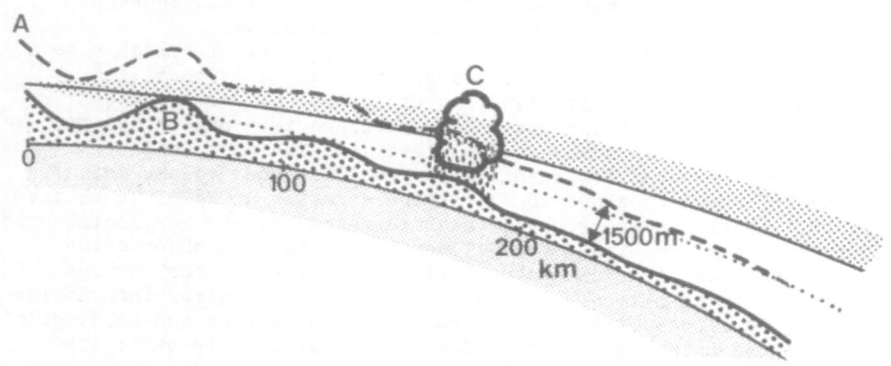

Fig. 1. Radar beam at the limiting elevation. Blocking occurs by the obstacle B, but the upper half of the beam can still detect cloud C, if the precipitation extends just above the 1500 m level.

As was noted, data at short range may be missing due to permanent ground clutter. Distant mountains can cause fixed echoes to occur at ranges exceeding 50 km even under normal propagation conditions. If no ground clutter removal is applied, data from such areas would be equally unavailable as in the case of beam blocking. Information on this type of reduction of the useful coverage (over areas larger than e.g. 5x5 km) is also included in the present COST survey.

3. GEOMETRY OF THE LOWEST USABLE BEAM

In the following computations the curvature of rays (radio or optical) is accounted for by means of the "effective earth's radius model": the rays are straight lines on a fictitious earth with radius R'. An appropriate value for R' is derived in Section 4. Elevations ϕ' measured with radar or (optical) theodolite are therefore different from the values ϕ found without atmosphere.

To describe the geometry of the observation three points along the beam axis are important: the radar A, the target C (e.g. cloud), and the lowest point T of the beam (Fig. 2). The height of these points above mean sea level is indicated by h_A etc. The height of the ground under C is designated by h_G.

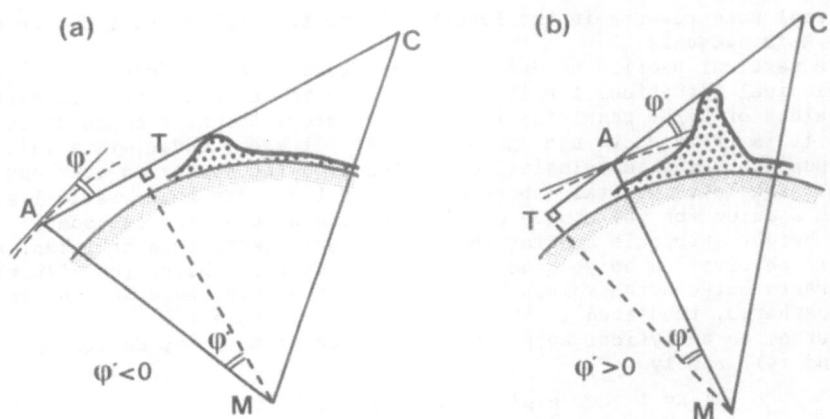

Fig.2. Geometry of the lowest usable radar beam: a) and b) illustrate situations with descending and rising limiting beam respectively. The radar rays are presented as straight lines, because an 'effective earth radius' has been used.

For small angles $\tan(\phi/2)$ is replaced by $(\tan \phi)/2$. The heights h_A and h_T are neglected compared to R'. From inspection of triangle AMT follows

$$h_A - h_T = MA - MT = AT/\sin \phi' - AT \cot an \phi' = AT ((1- \cos \phi')/\sin \phi') =$$
$$= AT \tan(\phi'/2) = AT (\tan \phi')/2 = AT (AT/R')/2 = AT^2/(2R') . \quad (1)$$

An analogous computation in the triangle CMT gives

$$h_C - h_T = CT^2/(2R') , \quad (2)$$

from which (1) can be subtracted:

$$h_C - h_A = (CT^2 - AT^2)/(2R') = (CT+AT)(CT-AT)/(2R') = d (d+2R'\sin \phi')/(2R') ,$$

or

$$h_C - h_A = d^2/(2R') + d \sin \phi' . \quad (3)$$

where d is the (slant) distance between radar and target. If the elevation is known this quadratic equation can be solved

$$d = -R' \sin \phi' + \sqrt{(R'\sin \phi')^2 + 2R'(h_C - h_A)} \quad (4)$$

4. EFFECTIVE EARTH RADIUS

The radio refractive index depends on air density, temperature and vapour pressure (e.g. Turton et al.(7), Bean and Dutton (8), also explaining ray tracing with a modified earth radius). The effective earth radius R' is found from the mean vertical gradient of the refractive index n for the height range of the radar beam. For small elevations the multiplication factor is

$$R'/R = 1/ (1+ R/n (dn/dh)) \tag{5}$$

For optical measurements in the lowest 4 km of the atmosphere a choice of R'=1.15 R is adequate.

The vertical profile of dn/dh may vary considerably, depending on the meteorological situation. A well-known extreme is the occurrence of very large values of R'/R, resulting in anomalous propagation. For the present purpose it is prefered to use one average R' for every radar. This value will depend on the climatological situation over the observing area and also on the height-interval (above m.s.l.) of the radar beam. Table I may serve as a guide for the choice of R'. The values shown are for some typical height intervals and for three (average) temperature profiles. For a certain observation height and temperature the best choice for R'/R will be somewhere between the values specified for the very humid and the very dry atmospheres, tabulated in the two sections of Table I.
If prefered, an analytical approximation can be used in the Equations (3.a) and (4), namely

$$R'/R= 1 + c_1 \exp(-((h_c+h_A)/2 /c_2) \tag{6}$$

where the constants (c_1,c_2) depend on the sea-level temperature: (0.50, 5.1), (0.38, 5.9) and (0.30, 6.8) for temperatures 20, 10 and 0 deg respectively, and (0.27, 10) for the dry cases of Table I.

Table I. Factors R'/R (radar wavelengths, wet-adiabatic temperature profiles)

Height(km,m.s.l.)	0-1	0-2	0-3	0-4	1-2	1-3	1-4	2-3	2-4	3-4
temperature 20	1.45	1.41	1.37	1.34	1.37	1.34	1.31	1.31	1.28	1.25
at 0 km (deg) 10	1.35	1.32	1.30	1.27	1.29	1.27	1.25	1.25	1.23	1.21
(rel.hum. 100%)0	1.28	1.26	1.25	1.23	1.24	1.23	1.21	1.21	1.20	1.18
(deg) 20	1.24	1.22	1.20	1.19	1.20	1.19	1.18	1.18	1.16	1.15
(rel.hum. 0%) 10	1.24	1.23	1.21	1.20	1.21	1.20	1.18	1.18	1.17	1.16
0	1.26	1.24	1.22	1.21	1.22	1.21	1.19	1.19	1.18	1.17

5. METHODS FOR DETERMINING THE LOWEST USABLE ELEVATION

In principle a careful study of radar measurements can provide this information (e.g. by reducing the elevation until a precipitation echo weakens - to about 3 dB reduction on the A-scope). However, as follows from substituting typical values in Equation (3), an inaccuracy of 0.1 deg (about 0.1 beamwidth!) in ϕ' will cause an uncertainty of about 16 km in the range. Therefore, unless the radar location is unaccessible, optical measurements must be prefered. Due to the different propagation characteristics blocking for radar rays may occasionally occur at more distant obstacles than a theodolite may observe. However, the resulting limiting elevation will not be significantly different.

Among the optical tools the theodolite is to be prefered above the camera. Horizon photographs with sufficient angular resolution and with little or no distortion can in principle be obtained with a tele-objective, but then many photographs must be taken and on each an object of known elevation must be identified.

Especially in surroundings with distant mountains and insufficient visibility a map can be a suitable alternative. A disadvantage of maps, however, is the absence of altitude data for buildings and trees, so a visual correction will be necessary.

Every few years the horizon survey should be checked, because buildings and trees may come, grow and disappear! Perhaps for this check only, photographs will be usefull.

5.1. Use of a theodolite

For the desired accuracy of about 0.02 deg., carefull leveling of the instrument is necessary. Calibration of the elevation and azimuth scales can be done by sighting objects of known height and position. If, e.g. due to the presence of a radome, the instrument is located a few meters away from the radar, corrections in azimuth and elevation have to be applied. A displacement of 1m necessitates such corrections for obstacles closer than 2000 m.

It should be noted, that for the same target, optical elevations (R'=1.15) are slightly smaller than radar elevations (R'=1.30 or so). The angular difference follows from Equation (3) applied to a blocking feature located at point C with range d_B (km): $d_B/2230$ (deg). So this difference is smaller than observational inaccuracies for all practical optical targets ($d_B < 40$ km).

Theodolite readings should be taken at 1 deg azimuth intervals and also at the lower and upper corners of any obstacles with angular dimensions exceeding 0.2 deg. The result of the survey is a list of azimuths and limiting elevations, both in units of 0.01 deg.

5.2. Use of a geographical map or a geographical data base

These tools are of course aequivalent. The digitized map offers the advantage of computer processing. If a map is used the inspection may be restricted to pronounced ridges and tops.

The computations are carried out along distance-height cross-sections through the radar, preferably at azimuth intervals of 1 or 2 deg. For ranges of 100 km and more differences between projections become rather serious. Care must be taken not to measure cross-sections along straight lines on a map but to use great-circle computations. The same applies to the determination of azimuth angles.

The horizontal resolution should allow computing with steps of about 0.25 km. The vertical accuracy depends strongly on the distance to the radar: terrain heights must be known within 1 m at a range of 2 km, within 10 m at 20 km, and so on. For some radars, part of the horizon may be formed by rather distant mountains. Up to the distance of the blocking obstacle we need a high accuracy for h_G and consequently an adequate horizontal resolution of the data base.

The necessary (stepwise) calculations are as follows:
- Determine for increasing range d the height h_G of the ground;
- Compute the elevation of the ray pointing at the ground (at distance d) from Equation (3) with h_C replaced by h_G

$$\sin \phi' = (h_G - h_A)/d - d/(2R') \quad . \tag{7}$$

- The maximum of the series of values ϕ' determines the limiting ray ϕ_1'.

6. COMPUTATION OF THE RADAR RANGE

For the remaining part of the calculation an uncertainty of 100 m in h_G is acceptable. The value of h_G is only needed to account for the large-scale elevation of the terrain during the determination of the critical height. So a simplified data set with average values for h_G (e.g. 10 deg sectors azimuth, 20 km range intervals) can be used.

The computation starts with the limiting elevation found at 5.1 for nearby or 5.2 for more distant obstacles and repeats the following steps for increasing range d:
- Determine the free height of the limiting ray from Equation (3).
- Note the last occurrence of h_C-h_G <1500 m, which determines the range of the radar. Here the search is explicitely for the 'last' occurrence, because with h_A > 1500 m the beam (in Fig.2 e.g. at point T) may temporarily descend below 1500 m, while points up to the range of C still offer reasonable observing possibilities.

7. COMPARISON WITH OBSERVATIONS

The coverage of the radar for warning purposes may also be quantified by interpretation of a large number of radar observations. From a digital archive of PPI data, collected during e.g. a year, the percentage of echo occurrences for every pixel can be found (Fig.3a). Here the polar picture is prefered, because the conversion to a rectangular picture may cause additional uncertainties.

The resulting pattern shows usually an artificial increase of echo detection at ranges below 70 km, due to permanent ground clutter and bright band echoes. To prevent the inclusion of anaprop echoes, it is advisable to remove observations under clear skies from the data set.

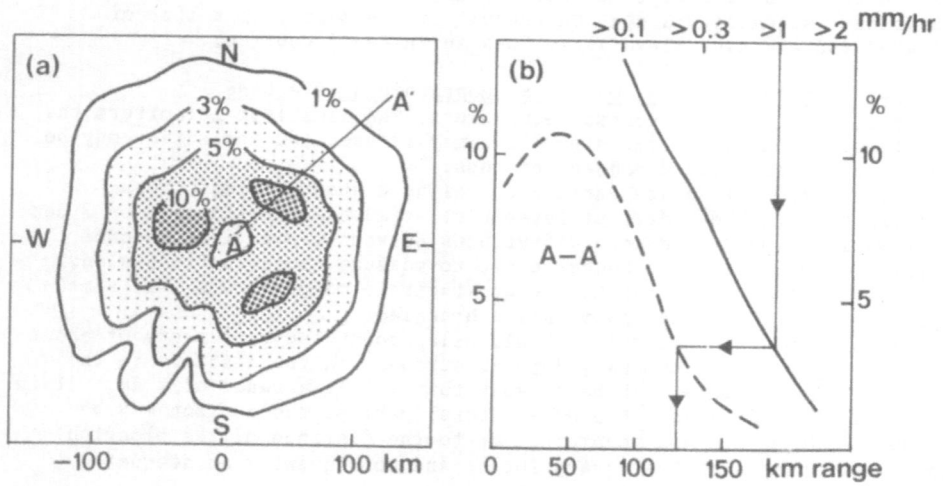

Fig.3. <u>a</u>. Sketch of the distribution of the echo detection frequency around a (hypothetical) radar. <u>b</u>. Dependence with range of echo frequency (along azimuth A-A'), compared with climatological precipitation exceedance.

The resulting angular pattern will reflect the influence of the most important blocking obstacles and may be compared with the coverage map obtained with the procedure described in the preceding chapters. At far range the echo occurrence may be combined with the actual frequency distribution of the rainfall intensity, measured with one or more recording raingauges on representative positions. In fig.3b the dependence of echo-occurrence on range is compared with the frequencies on the right scale. So the radar range for detection of different precipitation intensities may be found (e.g. 1mm/hr at 125 km in Fig.3b). The procedure suggested here offers an independent check of the coverage map and may ultimately lead to a better motivated choice of the critical precipitation height (Chapter 2.). This experiment has not yet been performed with real data, because an echo archive was not available for the present study.

8. APPPLICATIONS

A coverage diagram can contribute significantly to the value of radar data. These benefits can be gained at a negligible fraction of the total cost of the radar installation.

The assistance of coverage diagrams for the site selection of single radars or networks is well accepted. They are useful for evaluating the benefits of exchanging data between services or countries.

Coverage data can also be used for data processing and presentation of single radar data. An example is the correction of data in the shadow of small obstacles or the optimum choice of elevations (depending on azimuth). A further possibility is the improvement of weather echo interpretation by informing the user about the actual coverage pattern or by optimizing the display area.

Even more applications emerge for the compositing of data from different radars. Algorithms for merging or back-up can be based on quantitative coverage data of the contributing radars.

Hopefully the present collective effort to obtain a uniform set of coverage data will add to the quality and use of both national radar data and COST-73 products.

REFERENCES

(1) CLIFT, G.A. (1985). Use of radar in meteorology. WMO Technical Note Nr.181., Geneva, 90 pp.
(2) LARKE, P.R. and COLLIER, C.G. (1981). Merging data from several weather radars. in: Workshop/Seminar on the European weather radar project, Reading, Malvern (UK), 141-158.
(3) MARTINEZ, C. (1985). The Spanish radar network project. In: Proceedings COST-72 Final Seminar, Erice (It), Measurement of precipitation by radar, Report EUR 10353, 191-198.
(4) LEONE, D.A., ENDLICH, R.M., PETRICEKS, J., COLLIS, R.T.H. and PORTER, J.R. (1989). Meteorological considerations used in planning the NEXRAD network. Bull. of the American Meteor. Soc., 70, 4-13.
(5) JOSS, J. and WALDVOGEL, A. (1987). Precipitation measurement and hydrology. Arbeitsber. Schweiz. Meteor. Zentr. Anstalt
(6) STEWART, J.B. (1964). Precipitation from layer cloud. Quarterly Journal of the Roy. Met. Soc., 90, 287-297.
(7) TURTON, J.D., BENNETTS, D.A. and FARMER, S.F.G. (1988). An introduction to radio ducting. Meteorological Magazine, 117, 245-254
(8) BEAN, B.R. and DUTTON, E.J. (1968). Radio Meteorology. Dover, New York, 435 pp.

AN OPERATIONAL SYSTEM FOR DISPLAY AND ANALYSIS OF
HYDROMETEOROLOGICAL RADAR DATA

M. ROSA DIAS
Instituto Nacional de Meteorologia e Geofísica,
Rua C do Aeroporto - 1700 Lisboa - Portugal
J. BIOUCAS DIAS, J. CUNHA SANGUINO, J. NUNES LEITAO
Centro de Análise e Processamento de Sinais,
Complexo I do INIC, Instituto Superior Técnico,
Av. Rovisco Pais -1000 Lisboa - Portugal

Summary
The paper describes the general features of a system designed for
remote exploration and diffusion of information from a
Meteorological and Hydrological Radar Network. If has been
conceived as a versatile and low cost workstation based on
inexpensive and general purpose equipment and it can work as a host
or a node of a point-to-point computer network.
Based on the services provided by the communication and archiving
subsytem, the user can construct his application. The exploitation
benefits of the weather radar equipment can in this way be improved
by tailoring the hardware and software configurations according to
the users requirements. The presently operating units, designed
mainly for hydrological purposes, receive and process information on
the nominal surface precipitation field from a radar station and
some telemetering raingauges.

1. INTRODUCTION
     The system herein described, named TELERAD, is a partial result of a
cooperative project involving the National Institute of Meteorology and
Geophysics, the Technical University of Lisbon and the General Directorate
for Natural Resources. The purpose of this project is to develop tools,
equipment, software and concepts to be applied as guidelines and building
blocks of the future National Meteorological and Hydrological Radar
Network. The radar and the on-site processing system installed at the
Aeronautic Meteorology Center (Lisbon Airport) has been developed
according to this criterion and works as a pilot station.
     The TELERAD has been conceived as a versatile and low cost
workstation based on inexpensive and general purpose equipment. It can
work as a host or a node of a point-to-point computer network, thus
serving as a remote terminal of the radar network and as an information
diffusion element. As shown in figure 2, it can also be directly
connected to a rain-gauge network.

2. SYSTEM DESCRIPTION
2.1. General architecture
     The general structure of TELERAD is represented in figure 1. The
hardware configuration depends on the user needs. A compatible PC AT with
hard disk and graphics monitor (EGA, CGA, HERCULES, ...) is however the
main kernel. For high data volumes and long term archiving a streaming
tape should be used.
     If wanted, an image processing board (RGB 256x256 pixels/8bits)
compatible with the AT bus makes available the same information at

different video monitors connected in parallel. A memory expansion board gives the possibility of displaying stored images at a rate of up to 4 images/second thus creating the possibility of animation.

When transmitting via a Public Switched Telephone Network (PSTN) a modem V.21/22 is needed. A leased line demands a modem V.29. Base Band transmission between close units is of course also possible. These connections are supported in each PC by the UART (Universal Assynchronous Receiver Transmitter).

If the system is to be connected to a Packet Switched Data Network (PSDN) a Wide Area Network (WAN) board and a modem V.29 are needed. Figure 2 shows a possible network configuration.

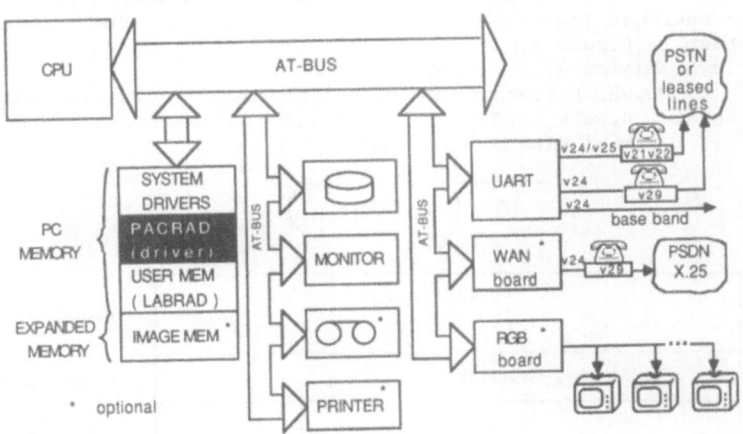

Figure 1.

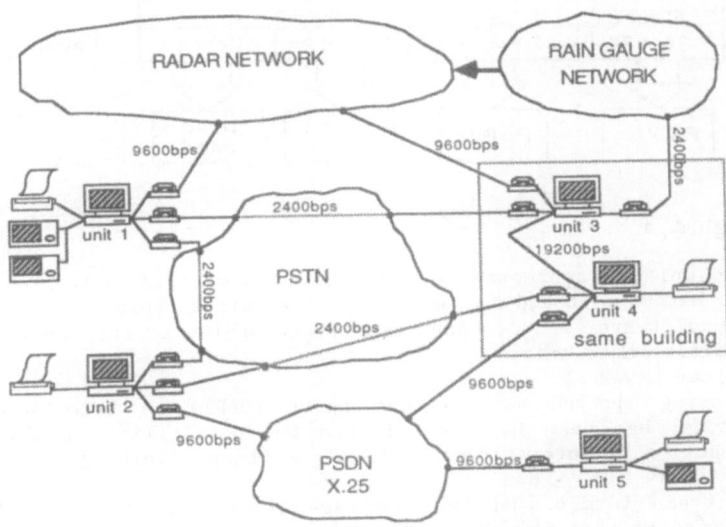

Figure 2.

The two main software packages of TELERAD are :
  i) PACRAD, the communications and archiving subsystem. It behaves
     as a resident driver, being transparent to the user of the
     operating system;
  ii) LABRAD, the display and analysis subsystem. It is an
      application program running on the services provided by PACRAD.

2.2. PACRAD

The communications architecture has been designed assuming :

  i) Open System Interconnection (OSI) Reference Model of
     International Standard Organization (ISO), (1), until level 4,
     as seen in figure 3.
  ii) Wide Area Network (WAN) topology;
  iii) Use of standard communications systems such as Public Switched
       Telephone Network (PSTN), leased telephone lines, Base Band and
       Packet Switched Data Networks (PSDN X.25).

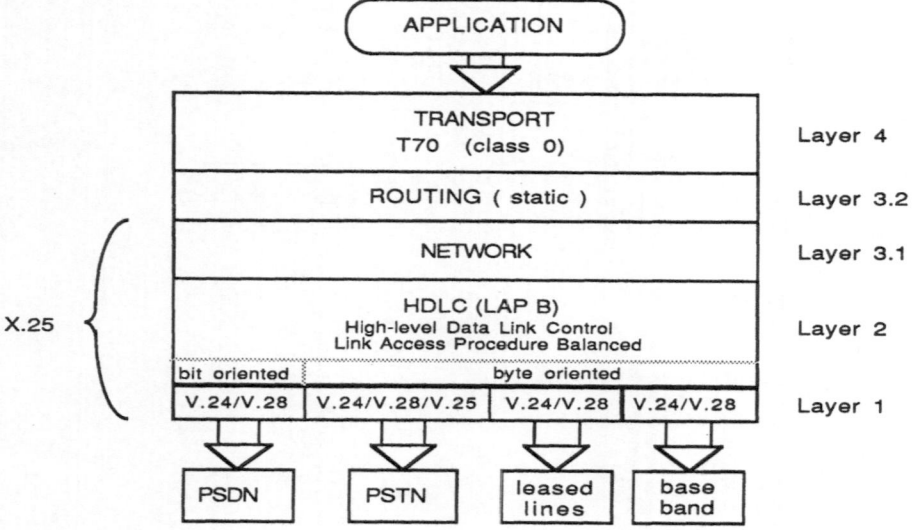

Figure 3.

When the unit is connected to a PSDN, protocol X.25 is implemented
by available hardware board. Otherwise the first three layers are
implemented by software, and designed closely according to the model X.25.
However some differences exist :
  i) Physical Layer
     By using the PC asynchronous ports (UART), a byte oriented
     protocol has been implemented, and the possibility of physical
     connection by automatic call (v.25) has been created.;
  ii) Link Layer
      The frame format, namely the Frame Check Sum (FCS), has been
      modified so that it may support the physical layer changes;

iii) Network Layer
     A routing algorithm (Layer 3.2) added to the normal Network
     Layer provides the possibility to connect A with C through B
     (see figure 4).   At present a static routing algorithm is
     installed.

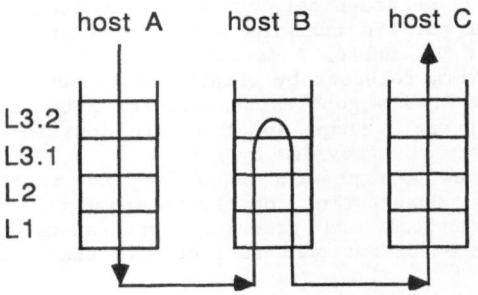

Figure 4.

     As a driver, PACRAD offers to any application the following
services:
          - selective information and data retrieving;
          - virtual circuit establishment;
          - virtual circuit disconnection;
          - file transfer.

     An important feature of PACRAD is that data representation in
transmission and archiving meets the requirements of FM 94 BUFR as
suggested by COST 73 (2).   The data format to the user is however
independent of that representation. This feature gives the TELERAD a high
modularity.

2.3. LABRAD
     Based on the services provided by PACRAD the user can construct his
applications, LABRAD being one.
     The presently operating units, designed mainly for hydrological
purposes, receive and process information on the nominal surface
precipitation field from the radar station and some telemetering
raingauges.
     The system receives high-resolution rainfall information (8 bits on
a 2Km square grid) from the radar station and has the capability to
process it both numerically and graphically, thus enabling the rainfall
observation, measurement, analysis and very-short-range forecasting.
     The numerical processing of the high-resolution radar and
telemetering raingauge information enables one to obtain adjusted rainfall
estimates, as well as short-period forecasts, in both cases according with
algorithms programmable by the user.

The system has also the capability to feed other computing systems running hydrological forecasting models with numerical precipitation data.

The graphical processing provides radar imagery display in up to sixteen colours (combined with alphanumerical displaying), including capabilities for animation at variable speeds, zoom, off-centering, visualization of the pixel location in various systems of co-ordinates, use of several geographical overlays (including river basins, towns, etc.), rainfall pattern tracking and determination of their motion speed, etc. Therefore it enables a detailed analysis of the precipitation field and its short-term forecast by simple linear advection.

The system has a good capability for data storage and enables image colour hard-copying. Phone and Mail are additional application programs based on the services provided by PACRAD.

With a view to a greater capability for weather watch, analysis and very-short-range forecasting, namely in convective storm conditions, units capable of numerical and graphical processing of radar upper levels information (three-dimentional reflectivity field) are under development.

3. CONCLUSIONS

The TELERAD remote processing and display system was designed to solve a key problem associated with the operational exploration of a weather radar network, namely to achieve rapid dissemination of user oriented products.

The system pays particular attention to some important aspects of such an user-oriented exploitation :

i)   Ready access to the radar network information, either directly or through other units;

ii) Good programmability for both numerical and graphical applications so as to optimize the use of the radar network information to meet the local needs;

iii) Compatibility with the usual telecommunication media formats and protocols;

iv) Good graphical capabilities and user-friendly operation at low cost.

TELERAD aims at improving the benefits of the exploitation of weather radar equipment by better tayloring the hardware and software configurations according to the users requirements.

REFERENCES
(1) Fred Halsall, "Data Communications, Computer Networks and OSI", Second Edition, Addison Wesley, 1989.
(2) "Binary Universal Form for Data Representation - FM 94 BUFR - Collected papers and specification", European Centre for Medium-Range Weather Forecasts, February 1988.

CASTOR PROJECT
IMPROVED PROCESSING OF RADAR DATA IN FRANCE

Bernard Beringuer
Jean-Louis Maridet

DIRECTION DE LA METEOROLOGIE NATIONALE
SERVICE DES EQUIPEMENTS ET TECHNIQUES INSTRUMENTALES DE LA
METEOROLOGIE

Summary

Although little used before 1982 radar pictures nowadays are
increasingly being supplied to various users: weather forecasting
services, flood warning services, road traffic monitoring,
hydrological services etc. ...

The first generation computers associated with radar have now
exhausted their potential. The National Meteorology Directorate
therefore decided to design a new radar computer. The CASTOR project
(system for the acquisition, monitoring and processing of radar
observations) is its name.

The first operational version of CASTOR, which became available in
mid-1989, performs the following functions:

- Preparation of the base image from radar video.
- Preparation and transmission of pictures adapted to the various
  users.
- Radar control and monitoring.
- Automatic electronic calibration.
- Maintenance assistance.

Subsequent developments including the preparation of cumulative
rainfall pictures are planned.

1. INTRODUCTION
Since 1982 the Meteorological Directorate has been developing a radar
network as part of the ARAMIS project (1).

At the moment 10 radar stations are in operation (including 3 different models). The computers used in the network are of two different types (SAPHYR, developed by the NMD and the THOMSON MT 750). They were designed in the early eighties around 8-bit microprocessors. Developments in data processing are currently allowing more powerful computers to be used (32-bits, 25 MHz, for example) which are available at a reasonable price.

The main restrictions on the first generation emerging in our experience are: digitalization of the radar video on 6-bits (1 dBZ coding), data integration in dBZ, picture generation at a single elevation, the availability of a single type of picture downstream of processing, the need for human presence to calibrate the radar, difficulties in guaranteeing the development of two different types of computer (SAPHYR and MT 750).

In addition, owing to the exclusive use of the assembler language, the development and maintenance of the software are difficult.

In order to overcome these difficulties the NMD decided to launch a design, within its SETIM technical department, of a new radar computer in 1986. This project was named CASTOR (System for the Acquisition, Monitoring and Processing of Radar Observations).

2. CHARACTERISTICS AND OPERATION OF CASTOR
2.1 GENERAL
The National Meteorological Directorate opted for automatic radar operation: certain radar stations (Nancy) are located far from weather forecasting stations and are unattended. Others (Bourges, Toulouse ...) are unattended around the clock.

This choice and this desire for continuous operation led us to the configuration described in Figure I.

The computer is made up of industrial maps (apart from the signal processing part) implanted on a VME bus. The central (processing) unit is organized around a 32-bit 68020 microprocessor working at 25 MHz and it has 8 megaoctets of RAM. The local console, which dialogues with the maintenance technician, is a PC-compatible microcomputer, like the service terminal.

The main functions formed by the CASTOR computer are as follows:

- Preparation of the base image using the radar video,
- Preparation and transmission of pictures adapted to the various users,
- Radar control and monitoring,
- Automatic electronic calibration,
- Maintenance assistance.

We decided to use an industrial, multi-task, real-time software system in order to concentrate our activities on its application as such. The programmes are written in PASCAL language apart from the basic picture generation part, which processes the integrated data and for which the assembler has been used.

In general terms the CASTOR specifications as far as possible take account, of the recommendations contained in COST 72 document no.(2).

2.2 PREPARATION OF THE BASE IMAGE
The characteristics of the radar stations used in the network are set out in Figure II.

2.2.1 Characteristics of the base image

We decided to set its range at 256 km as a result of a compromise between the limits to the radar measuring distance, the density of the network and the resolution selected (pixel size - 1 km per side).

The data are thus on a 512 x 512 grid in 1 km increments.

They are expressed in tenths of mm/h in order to maintain a resolution lower than the uncertainty of the radar measurement and to work only in integer numbers.

The base picture constitution period is 5 mn. Its preparation may include 1-3 sweeps at different points.

The base image is prepared in two parts:

- via processing using a hard-wired logic;
- via processing by software.

Various stages in this processing are described in Figure III.

2.2.2 Processing via hard-wired logic
The hard-wired logic (developed by SETIM, the NMD's technical division) carries out the repetitive processing operations on the

video signal: digitalization, Z/R conversion, integration. The video received from the radar is logarithmic. It has already been corrected into 1/R2 up to 100 km and the set echoes have been removed.

Digitalization is on 8-bits or a 0.25 dBZ increment for a coding dynamic of 64 dBZ and a frequency of 600 KHz, leading to a depth of 250 m.

These digitalized data are presented during the distance integration processing input. This processing selects one out of four gates. This value is then converted into tenths of a mm/h by a table stored in PROM. The contents of that table were calculated by the Marshall Palmer ratio Z = (dBZ) =200 R **1.6 with R in mm/h.

Where: R (1/10 mm/h) = 10* (10** Z/10)/200) ** (1/1.6).

This early conversion into rainfall intensity has two advantages:

- Where there is an identical number of samples N the standard deviation types of values obtained in mm/h is less than that in values of R deducted from Z. It is shown (3) that integration on R reduces the standard deviation from the average (expressed in mm/h) by a factor of 1.26.

- In the major reflectivity gradient zones integration on log Z (dBZ) causes significant local errors (3).
  Integration on R enables this cause of error to be removed.

The values for R thus obtained are azimuth (time) integrated by means of a first-order recursive filter - a method described by Sirmans and Doviak (4). 30 shots were retained since this figure was compatible with the resolution and accuracy desired and with the balance of the digital filter. This value is programmable in order to enable CASTOR to be adapted to other radar models (different speeds of rotation or PRF).

Every 30 shots, i.e. 90ms for the RODIN radar, a set of 256 integrated values is available at the output of the hard-wired logic. It is transmitted by DMA to the CPU after the azimuth, location, date and time have been added.

2.2.3 Processing by software
A first series of processings is carried out on each integrated shot:

- range correction between 100 and 256 km,
- correction of partial masks: we have scope for applying a multiplying coefficient to gates where it has been established that the beam has been partly masked. We are awaiting the result of statistical studies of radar installations in order to instigate that option.

- conversion of polar coordinates into cartesian coordinates. This conversion is reflected by a table in order to limit computing time.

A second series of processings is carried out after generation of the raw base image (every five minutes):

- removal of isolated pixels;
- processing of the masked sectors or of residual set echoes is possible at this level: an "indicator" picture can contain these data and is stored in the memory.

Outside the time needed to prepare the base image (between 1 and 3 minutes depending upon the number of elevations selected), the central unit is free for computing the radar data.

2.3 PREPARATION AND TRANSMISSION OF PICTURES ADAPTED TO THE VARIOUS USERS

The various users of radar pictures (forecasters, hydrologists, road traffic departments ...) do not all have the same needs.

A distinction is made between their products, in particular via the number and the values of the coding thresholds, the zone covered, the resolution of the picture and the preparation time.

CASTOR enables the treatment of various products deriving from the base image to be converted into parameters.

For example, the picture sent to the NMD's concentrator is a 256 x 256 x 4 - bit picture.

A pixel corresponds to an area of 2 x 2 km. This picture is obtained by assembling together 4 of those base picture pixels. It is emitted every quarter of an hour and it is planned to reduce this to five minutes. The pictures are transmitted to the National Meteorological Directorate's concentrator and towards users equipped with a Meteotel terminal via specialized lines at 4 800 b/s. The standard version of CASTOR manages four lines of this type, which may simultaneously emit four different pictures.

The existence on the market of industrial boards and X25 softwares
will enable us, where appropriate, to use this protocol in order to
transmit pictures.

2.4 RADAR CONTROL AND MONITORING

The fact that the radar site is not attended round the clock means
that automated monitoring is required. Thus the CASTOR computer
acquires different parameters:

- logical parameters (present/absent) providing information on
  the presence of various voltages or alarms: the presence sector,
  the EHT presence, incident modulator, locking of the AFC ...

- analog parameters to indicate an anomaly in a supply voltage, the
  magnetron current and the transmitted power.

Any change in state (logic parameter) or abnormal value (analog
parameter) is detected by CASTOR, stored in a disc file and
transmitted by a dial up telephone line to the departmental terminal.

The disc file thus set up may be consulted locally or from the remote
control terminal. It is the radar operating log.

Where there is a particularly serious alarm (pedestal incident, fire,
unauthorized entry), CASTOR shuts down the radar and maintenance staff
are required to restart operation.

The meteorological station responsible for monitoring has a reduced
set of controls: radar stop/start, 90° positioning of the dish,
stop/start of the generating set ...

A larger set of controls is accessible from the local console and this
enables the maintenance staff to test the various functions of the
radar.

2.5 AUTOMATIC ELECTRONIC CALIBRATION

A quantitative measurement of rainfall intensity is increasingly being
sought after by radar picture users. It is therefore essential
regularly to check the validity of the correspondence between the
receiver input levels and the values available after integration.

$$Z \text{ (dBZ)} = P \text{ (dBm)} + C$$
C : Radar constant

This operation currently requires the services of two qualified technicians and shutdown of radar operation for at least one hour. We have therefore decided to automate this operation. The computer checks and corrects, if necessary, the correspondence between the known hyperfrequency levels injected by a generator into the receiver input and the associated rainfall intensity values.

The generator is controlled by CASTOR via a GPIB bus. CASTOR regulates the process by adjusting the reference levels (gain and offset) of the analog/digital conversion.

This operation is carried out automatically once a week.

2.6 MAINTENANCE ASSISTANCE

The local console has three operating modes:

- the "telephone" mode which is the standard operating mode. The changes and incidents described in 2.4 are sent to the departmental terminal.

- the "visualization" mode: the maintenance technician present at the radar site may request the display of two pictures according to choice:

  - a raw PPI picture (512 x 512 on 16 colours)
  - base picture, line-by-line.

- The "VT 100 emulation" mode. This mode enables local orders to the input into CASTOR: aerial controls, calibration request, issue of the operating log ... In this mode the maintenance technician may access a disc file receiving the following information: breakdowns requiring his intervention, changing of subassemblies. This file must permit better coordination between the various departments (local, regional, national maintenance) required to work on the radar.

3. FUTURE DEVELOPMENTS

The first examples of CASTOR must be included in the network before the end of 1989, together with the scope described above. The medium-term developments planned are as follows:

- the processing of partial beam masking zones,
- the preparation and transmission of cumulative rainfall pictures.

4. BIBLIOGRAPHY

(1)   Gilet, M. (1985). Etat d'avancement du réseau français de radars météorologiques COST 72 Report.

(2)   COST 72. Operational requirements specifications for a European weather radar system.

(3)   David, P., Musiedlak, J.P., Bissonnier, P. (1984). Note de travail du SETIM sur le radar RODIN.

(4)   Sirmans/Doviak (1973). Meteorological Radar signal intensity estimation NOAA Technical Memorandum ERL NSSL 64.

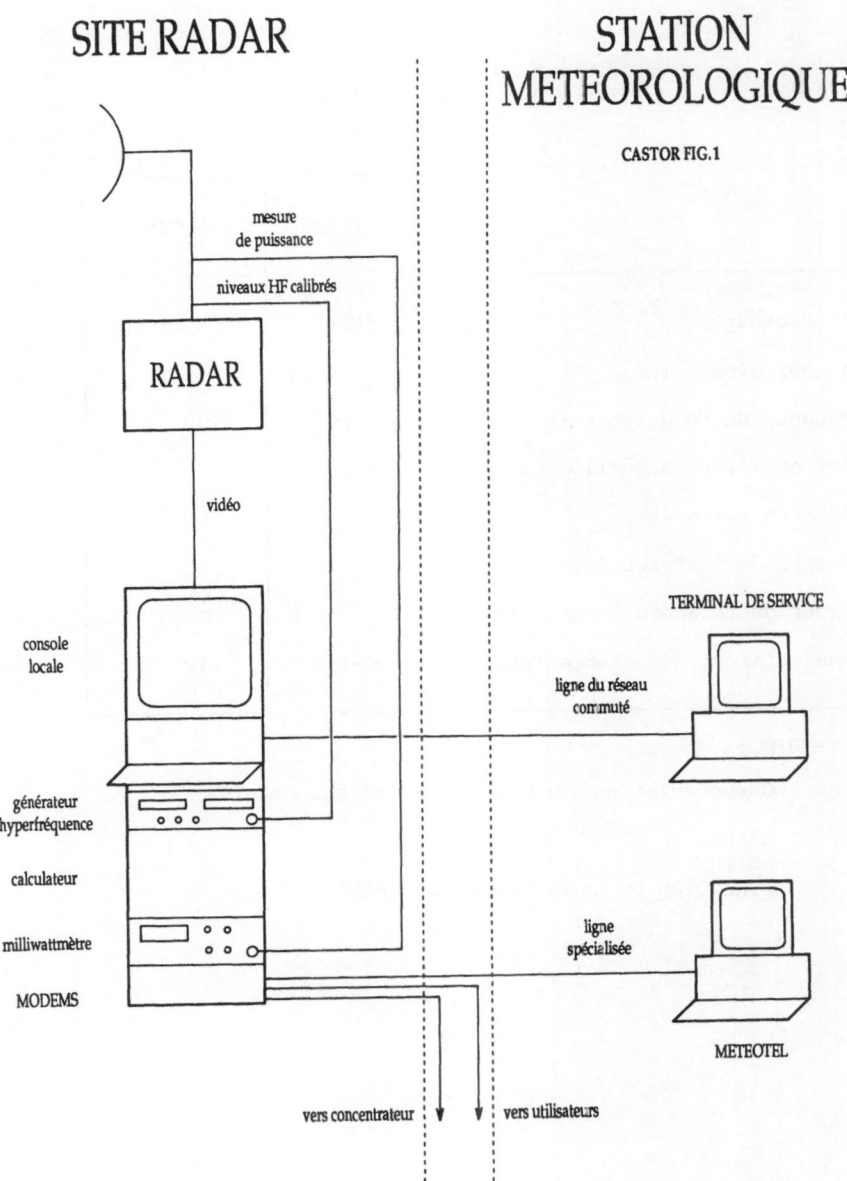

SITE RADAR

STATION METEOROLOGIQUE

CASTOR FIG.1

RADAR

mesure
de puissance

niveaux HF calibrés

vidéo

console
locale

générateur
hyperfréquence

calculateur

milliwattmètre

MODEMS

TERMINAL DE SERVICE

ligne du réseau
commuté

ligne
spécialisée

METEOTEL

vers concentrateur   vers utilisateurs

Figure I

    Configuration d'installation de CASTOR

    BERINGUER, MUSIEDLAK, MARIDET, BARDY, BARTHEZ
    SETIM
    DIRECTION DE LA METEOROLOGIE FRANCE

	MELODI	RODIN
Constructeur	OMERA	THOMSON-CSF
Longueur d'onde (cm)	10,7	5,3
Fréquence de récurrence (Hz)	250	330
Durée de l'impulsion (10 s)	2	2
Puissance émise (KW)	700	250
Diamètre de l'aérien (n)	4	3
Largeur du faisceau à -3 dB (degrés)	1,8	1,3
Signal minimum détectable (dBm)	-106	-112

Figure II

Caractéristique des radars du réseau Aramis

GILET
SETIM
DIRECTION DE LA METEOROLOGIE FRANCE

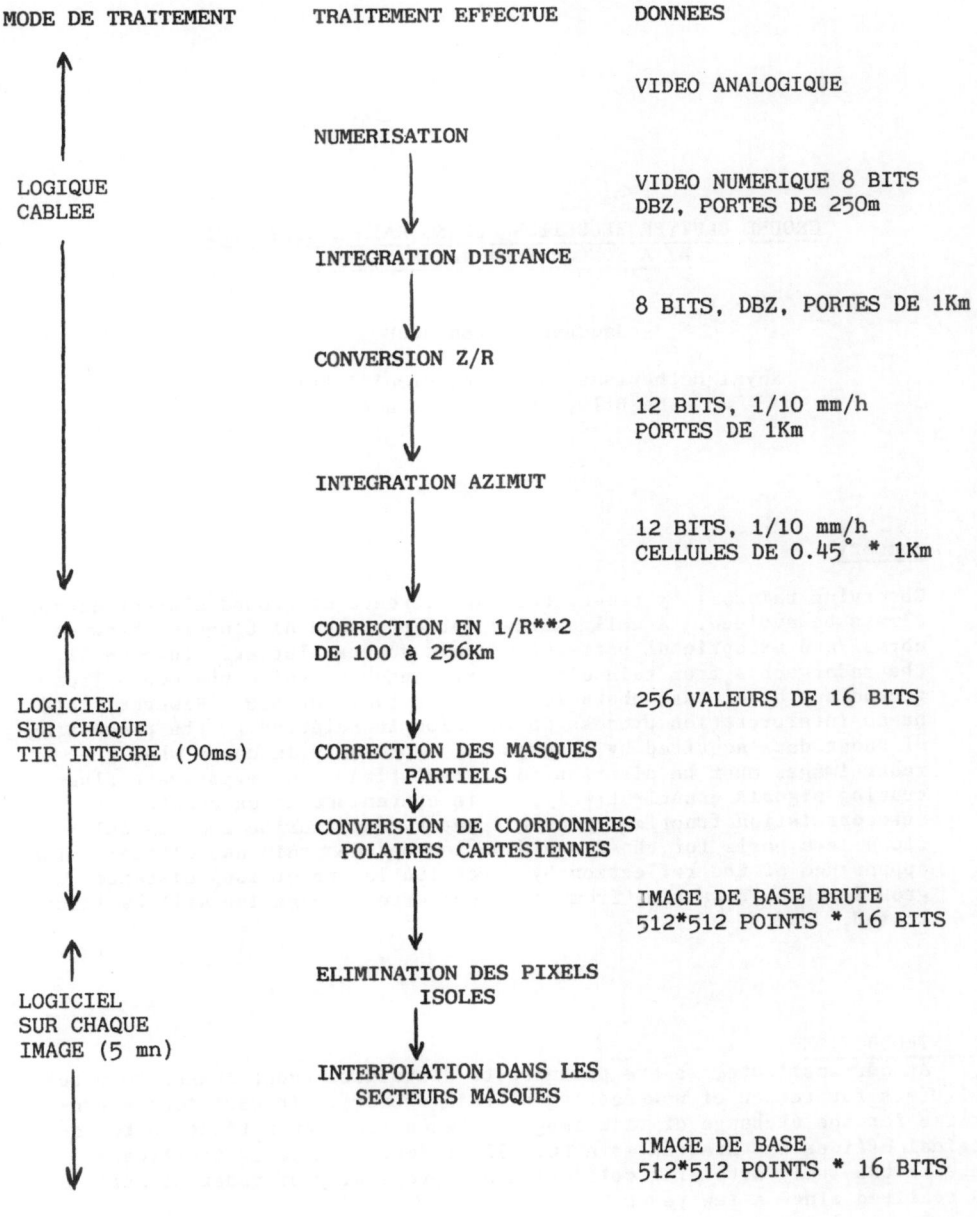

MODE DE TRAITEMENT   TRAITEMENT EFFECTUE   DONNEES

                                                   VIDEO ANALOGIQUE

                         NUMERISATION

LOGIQUE
CABLEE   VIDEO NUMERIQUE 8 BITS
                                                   DBZ, PORTES DE 250m

                         INTEGRATION DISTANCE

                                                   8 BITS, DBZ, PORTES DE 1Km

                         CONVERSION Z/R

                                                   12 BITS, 1/10 mm/h
                                                   PORTES DE 1Km

                         INTEGRATION AZIMUT

                                                   12 BITS, 1/10 mm/h
                                                   CELLULES DE 0.45° * 1Km

                         CORRECTION EN 1/R**2
                         DE 100 à 256Km

LOGICIEL   256 VALEURS DE 16 BITS
SUR CHAQUE
TIR INTEGRE (90ms)   CORRECTION DES MASQUES
                               PARTIELS
                         CONVERSION DE COORDONNEES
                         POLAIRES CARTESIENNES

                                                   IMAGE DE BASE BRUTE
                                                   512*512 POINTS * 16 BITS

                         ELIMINATION DES PIXELS
                                ISOLES
LOGICIEL
SUR CHAQUE
IMAGE (5 mn)   INTERPOLATION DANS LES
                            SECTEURS MASQUES

                                                   IMAGE DE BASE
                                                   512*512 POINTS * 16 BITS

Figure III

    Etapes de l'élaboration de l'image de base CASTOR

    BERINGUER, MUSIEDLAK, MARIDET, BARDY, BARTHEZ
    SETIM
    DIRECTION DE LA METEOROLOGIE FRANCE

GROUND CLUTTER REDUCTION DURING RAIN MEASUREMENTS
BY A NONCOHERENT RADAR SYSTEM

Jacques J. van GORP

Royal Netherlands Meteorological Institute,
De Bilt, The Netherlands

Summary

Observing rainfall by radar, the interference of ground clutter cannot
always be avoided. A well skilled observer can distinguish between
normal and exceptional patterns of rain versus clutter. In general,
the radar echos from rain clouds vary randomly, while the echos from
ground and other vast obstacles are relatively stable. However, this
human interpretation process is too slow in relation to the processing
of radar data acquired by a computer, especially in cases when the
radar images must be distributed in real time. To investigate fluctuating signals quantitatively, it is convenient to calculate its
autocorrelation function. In this paper, we describe such calculations as a basis for the discrimination between rain and clutter. The
appearance of the reflection by inversion layers of long distance
ground echos, resulting from anomalous wave propagation will be illustrated.

1. INTRODUCTION

At our institute, we are planning to distribute radar images to external users for reason of now-casting of rain showers. In addition, a programme for the exchange of rain images between European national meteorological offices was started as a COST-73 project. These initiatives
require the fully automatic collection and processing of radar signals, as
is realised since a few years.

Presently, radar information is accompanied by the relevant remarks
given by the meteorologist on duty, and upon request additional information
is provided. This additional information is needed for a reliable
interpretation of the resulting radar images. In this paper, we describe
an automatic validation process which can be incorporated with the image
processing. Some disturbances are detectable from the behaviour of the raw
radar signals, while others can be determined only by taking into account
the prevailing meteorological situation.

2. CAUSES AND EFFECTS OF RADIO WAVE REFLECTIONS

All kinds of obstacles besides raindrops reflect radio waves, e.g., buildings, trees and hills. Additionally, the earth's surface can reflect radio waves too, especially under certain meteorological conditions. Because these obstacles appear as large bright areas on the radar display, the obstacle reflection is named "ground clutter". Ground clutter from obstacles at close range dominants the radar image, making the simultaneous appearance of clutter and rain difficult to distinguish. A method for separating the effects can be provided by ignoring the reflections of fixed obstacles by means of a geographical mapping during dry weather. A disadvantage of this technique is that objects moistened by rain produce different reflections then when dry. In addition, the wave propagation and thus the ground clutter is highly variable, especially over flat land.

Another solution can be found by manipulating images within the mixing process of two different, but simultaneous radar observations. In our case, we use the radar systems at our institute in De Bilt and at Schiphol airport, which are about 37 km apart. In the nearest 10 km around each radar system, the information from the other system is used. Although this results in a reduction of the clutter, but it is difficult to compose a well fit image.

More serious, then the previous ones, are ground clutters caused by the so called "anomalous propagation". In the next section, an explanation for this phenomena will be given, with special emphasis to the meteorological effects. An alternative method to reduce the effects of ground clutter, may be possible by an analyses of the specific signal pattern of rain versus clutter echos and by taken into account the prevalent meteorological situation.

3. ANOMALOUS PROPAGATION OF RADAR WAVES

The atmosphere's refractive index is vertically stratified, hence radar waves travelling through the atmosphere follow curved rather than straight paths. Often, the radar beam is refracted back to the earth's surface, causing distant ground clutter to be sometimes falsely interpreted as rain cloud echos. This downward curving of radar waves is called an "anomalous propagation" (AnaProp). The curvature of a ray depends upon the rate of change of the refractive index (n) of the air with height (z), expressed as $dn(z)/dz$. The refractive index is related to the pressure, the temperature, and the water vapour content. Given a vertical profile of these meteorological variables, one can determine the path of the radar beam through the atmosphere.

In the atmosphere the refractive index is so near unity at microwave frequencies, it becomes convenient to introduce a different measure of the refractive properties of air. The refractivity N is defined as,

$$N = (n - 1) * 10^{6}. \tag{1}$$

In a wave front, the phase velocity is inversely proportional to the refraction index ($v_p = c/n$, c = speed of light in vacuum = constant). Therefore, the speed of the signal components increase with height, resulting in a downward bending of the radar wave path. In the ideal case (when $dN/dz = 0$) the radar beam is traveling straight line, so that the distance to the earth's surface rapidly increases. This effect is illustrated in Fig. 1, in which the radar beam is initial horizontally. In the standard atmosphere $dN/dz < 0$, but since the value is rather small in magnitude, the resulting path is little curved.

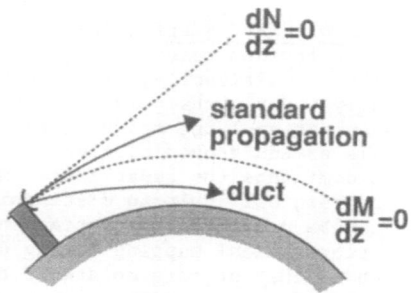

Fig. 1. Refractive atmospheric propagation categories.

With higher negative values of dN/dz, the curvature of the radar waves increases, and in extreme conditions it becomes comparable with the curvature of the earth. This special situation is indicated in Fig. 1. We introduce a second differential, dM/dz = 0, where M is the "modified refractivity". The modified refractivity accounts for the curvature of the earth and can be obtained by adding a term to the refraction index,

$$M = N + 157 \ z. \tag{2}$$

where z is the height above the surface in km, and M in N-units. By this addition, the earth "appears" flat with respect to the propagation paths of radar waves. In the case where dM/dz < 0, the radar waves are bent toward the surface of the earth and a so called "duct" is formed. In such a region, the radar beams collide with the surface at large distance, and after reflection lead to ground clutter. The inspection of the equation for dM/dz provides an indication for the appearance of this type of "inversion ground clutter" (described in following section).

Note: A similar phenomena in acoustics exists. Sound waves carried with the wind bend towards the earth's surface when in an inversion layer and when the wind speed increases with height, see Wessels and Velds (5).

4. WAVE PROPAGATION IN A STANDARD ATMOSPHERE

The refractivity is dependent on the prevailing meteorological situation as given by Bean and Dutton (2):

$$N = 77.6 * p \ / \ T + 373256 * e \ / \ T^2. \tag{3}$$

where p is the atmospheric pressure in hPa, T the temperature in $^{\circ}$K, and e the vapour pressure also in hPa. For a well-mixed atmosphere with a relative humidity of 60 % and a temperature of 15 $^{\circ}$C at sea level, characteristic values for N(z) are given in Table I.

Height [m]	Temperature [$^{\circ}$C]	Atm. Pressure [hPa]	Vap. Pressure [hPa]	Refractivity [N-units]
0	15	1013.3	10.2	319
100	14.4	1001.5	9.6	314
500	11.8	954.6	8.3	298
1000	8.5	898.6	6.7	279

Table I. Refractivity N(z) of a well-mixed atmosphere.

According to the values in the table the refractivity decreases linearly with height ($N = 319 - 40z[km]$) in a "standard" atmosphere at low altitudes. To derive a practically expression for the sensitivity of dM/dz, equations (2) and (3) must be manipulated. First we rename the expressions in equation (3) as $N_p = N(p,T)$ and $N_e = N(e,T)$, respectively. Substitution of (3) in (2) and partial differentiation leads to the differential equation,

$$\frac{dM}{dz} = N_p \left[\frac{1}{p} \frac{dp}{dz} - \frac{1}{T} \frac{dT}{dz} \right] + N_e \left[\frac{1}{e} \frac{de}{dz} - \frac{2}{T} \frac{dT}{dz} \right] + 157 \qquad (4)$$

Substituting the ground based numerical values from table I for $z = 0$ and rearranging the result gives,

$$dM/dz = 0.27 \ dp/dz + 4.5 \ de/dz - 1.27 \ dT/dz + 157. \qquad (5a)$$

In the case of a well-mixed atmosphere to 1 km (using values from table I),

$$dM/dz = -31 -16 + 8 + 157 = 118 \ [N\text{-units/km}]. \qquad (5b)$$

The sum of the first three terms ($dN/dz = -39$) are characteristic of a nearly well-mixed atmosphere, see Fig. 1.

5. WAVE PROPAGATION IN AN INVERSION LAYER

The sensitivity of the modified reflectivity M to changes in the meteorological variables p, e and T provide an indication for the appearance of ground clutter at large distances. Ground clutters from distance sources is usually brought about by inversion layers, i.e. an atmosphere in which the temperature increases with height (z). In a standard atmosphere the temperature decreases with height.

The gradient of the refractive index is not always a constant, and we have particularly severe departures from linearity when strong temperature inversions exist and are accompanied by large moisture gradients. Furthermore, the refractive index cannot decrease without bound, because at large heights (30 km) it must asymptotically approach unity. The nearly horizontal radar transmissions can only rebound back to the radar receiver when the inclination angle of the collision of the radar waves with the earth's surface is very small. Consequently the refractions from distant ground clutter are limited to low heights.

The meteorological tower at Cabauw (height 213m, 25 km from our radar, see Monna and vd Vliet (4)) can provide the pertinent meteorological information necessary to determine if, strong inversions exist. We can use the data (received at KNMI every 10 minutes via direct link) to determine the received signals are contaminated by ground clutter. For example, on 9-2-1989 a possible severe ground-clutter map over Belgium existed. Using the Cabauw temperature and humidity data, an inversion layer with strong vapour pressure gradient was found between 80 and 140m. With the measured values, eqn (4) becomes,

$$dM/dz = 0.28 \ dp/dz + 4.84 \ de/dz - 1.26 \ dT/dz + 157. \qquad (6a)$$

With numerical values substituted for the gradients, between 80 and 140m,

$$dM/dz = -27 - 89 - 65 + 157 = -24 \ [N\text{-units/km}]. \qquad (6b)$$

The negative value for the modified refractivity gradient indicates the possible appearance of a duct, which can generate ground clutter maps

on the radar display (see Fig. 2).
By comparison of equations (5) and (6), the following conclusions about
whether ground clutter exists can be made:
 - large moisture gradients are the dominating factor,
 - the contribution of a marked temperature inversion is relatively
smaller,
 - and the pressure gradient is least important.

6. OBSERVATIONS OF INVERSION GROUND CLUTTER
     Persistent ground clutter was observed on 9-2-1989 between 9:00 and
10:15 GMT. On the radar signals frequency analyses is performed, see following section. The clutter image was located above Belgium (AZ=170)
detected with an antenna elevation angle of 0.2 degree, and a beam width
1.05 of degrees. The results are given in Fig. 2a.

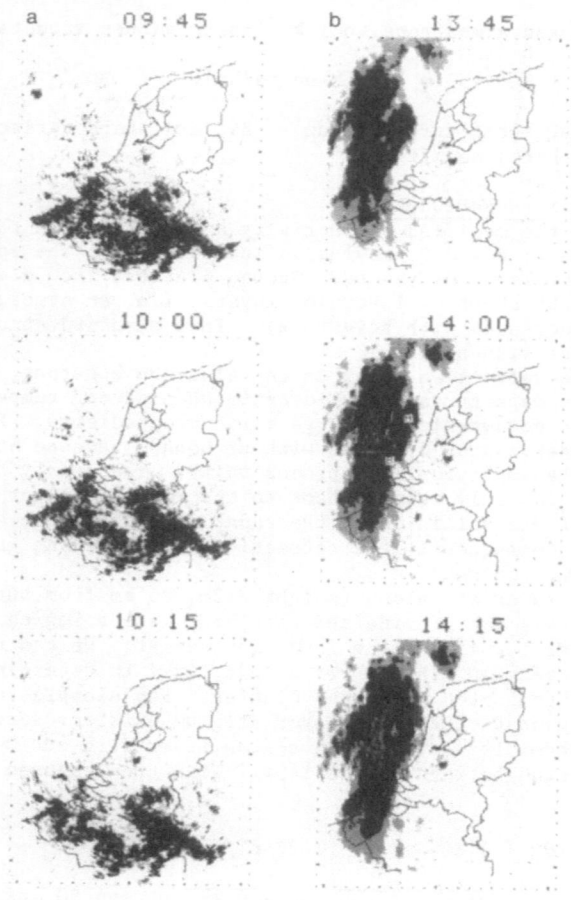

Fig. 2. Radar images of ground clutter (a), versus rain clouds (b).

For comparison, a heavy rain cloud observation above the North sea on

13-2-1989 is included in Fig. 2b. Striking are the differences between the two types of pictures.

The rain cloud images are clearly characterised by coherent (broad) regions of intensity. From the outside of the cloud to the centre, the rain intensity is increasing, leading to an observation of high cloud tops. In the middle picture of Fig. 2b, two tops are indicated by means of the characters A and B. Furthermore, the rain clouds are moving eastward.

In contrast, the pictures for the inversion clutters (Fig. 2a) are more of a mosaic structure, with no explicit distribution of regions of high intensity. Furthermore, ground clutter pictures are stationary and display a peculiar circular pattern. A reason for the circular pattern may be that the radar waves graze the ground surface at about same distances from the radar system. Using additional meteorological information has proved the real existence of rain on 13-2-1989 together with the complete absence of rain on 9-2-1989.

7. DATA ACQUISITION OF THE RADAR SIGNALS

With a digital storage oscilloscope (Philips PM 3320) series of finite time period from the direct output of the logarithmic receiver (A-scope) are acquired. A high sample rate (f_s) of 2.5 Mhz is selected to eliminate the aliasing of the signal, leading to space samples at every 60 m. On the trigger of the transmit pulse of the radar, 512 successive samples taken every 200 micro-sec represent an echo distance interval of 30 km in radial direction. Our radar system has a pulse repetition frequency (prf) of 250 Hz, thus the receiver collects radar echos over a distance of 600 km during each scan. Adjusting the time delay (t_d) between 0 and 4 msec, the measurement window (30 km) can be moved over the full radar echo scan of 600 km. When echos of high strength are found t_d is fixed, e.g., with $t_d = 1$ msec an echo interval from 150 to 180 km will be sampled.

A maximum of four such radar echos can be archived into the oscilloscope memory for "off-line" processing. Signal analysis is done on an IBM-PC/AT with of an interactive programme (Asystant). The transfer of data between oscilloscope and PC, together with the archiving process of the sampled data in the PC, is rather slow (the manual control takes about 10 sec). The antenna of a weather radar makes a full rotation in 20 sec to acquire data for the radar images. For that reason, we can only measure four successive scans. Or by locking the antenna we can make a time history of echo signals of one specific obstacle.

8. ANALYSIS OF RADAR ECHO SIGNALS

The radar echo signal belongs to the category of non-deterministic signals and requires special precautions. For reason of the limited time records, signals are enhanced by averaging radar echos of the same obstacles. To eliminate abrupt changes at the borders of the data set, a smoothing process is applied by multiplication of a gradual lapsed, tapered rectangular- or Blackman Window, see Harris (3). Another pre-processing procedure is the removal of the steady state portion of the echo signal.

The behaviour of the random radar echos can be characterised by correlation functions. Specifically, we use correlation to measure the dependence of the value of a signal at one instant of time with its value at another instant of time. The auto-correlation of N discrete samples is approximated using,

$$R_{xx}(k) = \frac{1}{N-k+1} \sum_{n=0}^{N-k} x(n) * x(n+k) \tag{7}$$

where x(n) is the value of the signal at the discrete time n/f_s, in which f_s is the sample rate. The value for k is usually less than N/10 to achieve a good accuracy in the calculation of the auto-correlation. For k=0, the correlation takes its maximum value R(0) = square sum, which represents the energy content of the original signal. The value of R(k) decreases with increasing k. In the case of a rapid decrease, the dependency between the time-lagged signals is low, indicating a highly variable signal. Conversely, a slowly decreasing R(k) indicates a less variable signal.

Cross-correlation is a measure of the dependence of one signal to another. as a function of time-lag. When the cross-correlation of signals x and y from two successive radar scans are computed, a measure of the ensemble correlation results.

The auto- and cross-correlations are normalised in order to compare radar echos,

$$r_{xx}(k) = \frac{R_{xx}(k)}{R_{xx}(0)} \quad \text{and} \quad r_{xy}(k) = \frac{R_{xy}(k)}{\sqrt{R_{xx}(0) * R_{yy}(0)}} \tag{8}$$

To conclude, the slope of a limited number (e.g. 10) of correlation functions allows for comparison of the time variability of radar echos. Examples will be given in the following section.

9. AUTO-CORRELATION FUNCTIONS OF RAIN AND CLUTTER

Fig. 3 shows a set of observed radar echo signals from rain, clutter, and a sunrise ("white noise").

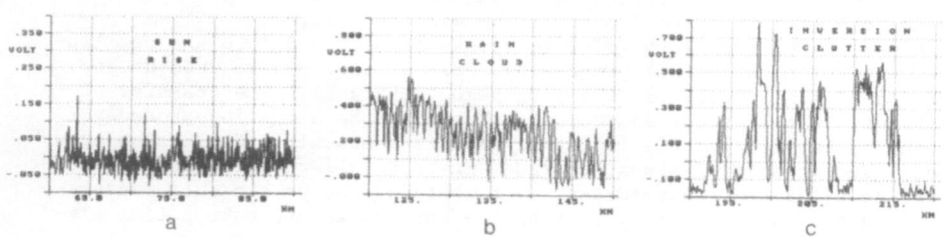

Fig. 3. Radar echos from the sun (a), rain (b), and clutter (c).

The rain-induced echo (Fig. 3b) originates from the same rain cloud as shown in Fig 2b. It illustrates the highly fluctuating signal produced by rain drops. In Fig 3c, an echo signal is represented from the same ground clutter occasion shown in Fig. 2a. Clearly, the ground clutter signal displays much less fluctuations.

The signals are converted from logarithmic to linear values and the deviation to the mean value is computed. After a smoothing procedure (see previous section), the normalised auto-correlation functions for the three types of signals shown in Fig. 3 are computed and the results are displayed in Fig. 4,

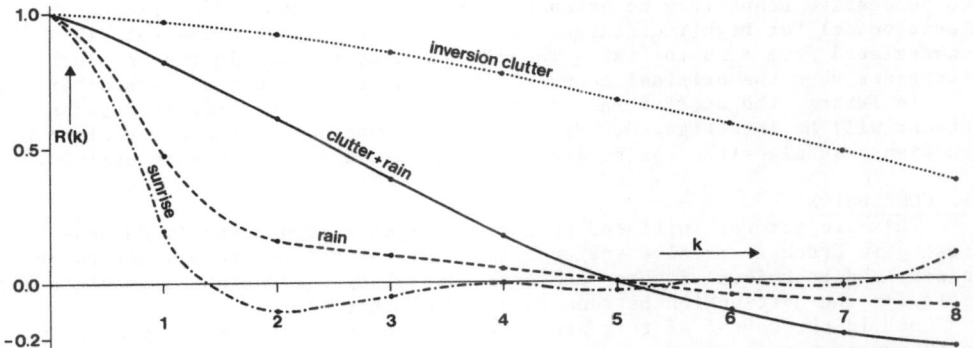

Fig. 4. Normalised auto-correlation functions for radar echos types.

The radar echos from the sun results in a sharp decreasing auto-correlation
function. The signals reflected by the raindrops cause comparable fluc-
tuatings and are characterised by a somewhat milder decrease in the auto-
correlation function. The slope of the auto-correlation function of
clutter echos is nearly horizontal, illustrating its stable behaviour. The
auto-correlation function in between originates from simultaneous clutter
and rain at the same place (rain at nearby obstacles, within 30 km distance
of the radar system in De Bilt, on 1-3-1989). Moreover, the energy of the
clutter echos are remarkably higher than those of the rain echos, e.g.
R(0)=4.000 versus 5 [volt^2], respectively. These results are in close
agreement with Aoyagi (1).

10. COMPARISON OF THE AUTO- AND CROSS CORRELATIONS
The individual autocorrelation functions (r_xx) and the cross-
correlation functions (r_xy) are computed from four successive rain and
clutter signals. Typical functions are displayed in Fig. 5,

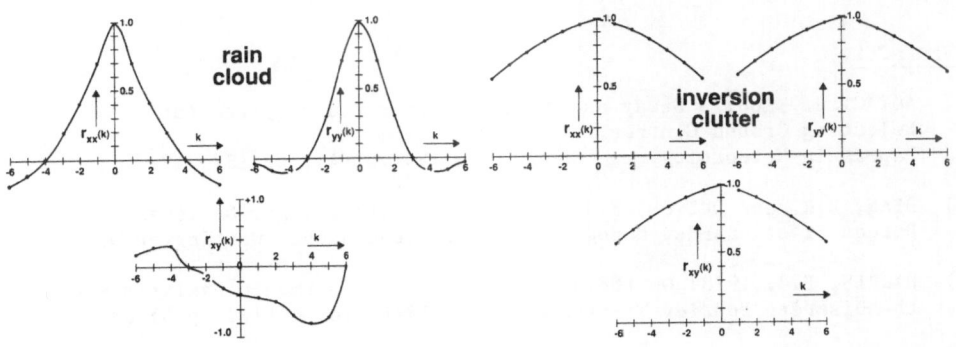

Fig. 5. The auto- and cross correlation for rain and clutter.

For the rain echo, the correlation functions are nearly identical,
while those for the clutter echo are different.

Two successive scans from an antenna at the same azimuth and at the same elevation deliver highly correlated clutter observations, while rain is uncorrelated from scan to scan. The clutter contribution in the rain echo disappears when the original scans x and y are subtracted from each other.

In future, the acquisition of successive scans with a rotating radar antenna will be investigated. By a similar comparison of their correlation functions, an algorithm for separating of rain and clutter may be derived.

11. CONCLUSIONS

This project was initiated to exploit digital signal and image processing in order to resolve the problem of ground clutter in weather radar. Indeed, the technique of auto- and cross-correlation computation provides a basis for the distinction between rain and clutter.

But in the course of this study, we became convinced of the necessity to integrate a variety of meteorological measurements since individual meteorological parameters are dependent on others. In the case of rain measurements, useful information can be added from radiosonde launches (high altitude temperature and humidity profiles), meteorological towers (low altitude temperature and humidity profiles), satellite images (cloud cover) and lastly with ground base rain-gauge measurements for validation. The evolution in the field of high technology, together with the already existent basic theory, support the realisation of this integration process.

This approach is similar to what is done in daily practise, in which an experienced meteorologist observes the weather charts together with looking outside to observe the current weather. This process may be automatised by an artificial intelligence system. It is in this context that we shall proceed with this study, continuing the development of an algorithm for separating rain and clutter.

ACKNOWLEDGEMENTS

The author wishes to thank Dr. L.Vogten from the Institute for Perception Research in Eindhoven for his advice on the signal analysis, and Dr. M.H.A.J. Herben from the University of Eindhoven for his suggestions on the wave propagation.

REFERENCES

(1) AOYAGI,J.,1983: A Study on the MTI Weather Radar System for Rejecting Ground Clutter. Papers in Meteorology and Geophysics, Vol.33(4), pp 187-243.

(2) BEAN, B.R. and DUTTON, E.J.,1966: Radio Meteorology. National Bureau of standards, Monograph 92, U.S.Government, Washington D.C.

(3) HARRIS, F.J.,1978: On the Use of Windows for Harmonic Analysis with the Discrete Fourier Transform. Proc. IEEE, Vol.66(1), pp 51-83.

(4) MONNA, W.A.A., and van der VLIET, J.G.,1987: Facilities for research and weather observations at Cabauw and at remote locations. KNMI Scientific Report 87-5.

(5) WESSELS, H.R.A., and VELDS, C.A., 1983: Sound propagation in the surface layer of the atmosphere. J. Acoust. Soc. Am. 74(1), pp 275-280.

THE OPERATIONAL USE OF A HIGH SPEED INTERACTIVE
RADAR DATA PROCESSING SYSTEM (RDPS)
PART 1 : A TECHNICAL DESCRIPTION

G.L. AUSTIN, A. KILAMBI,
McGill Radar Weather Observatory
Ste Anne de Bellevue Quebec
Canada H9X 1CO

H.P. BIRON
Quebec Weather Centre
Environment Canada
St. Laurent, Quebec
Canada H4M 2N8

Summary
     The RDP system at the Centre Météorologique du Québec receives
a complete 3-D volume scan data every five minutes from a radar
operated by McGill University about 24 km away and generates rapidly
a large number of products from the volume scan based on user
selected schedules and interactive requests.   The system's
repertoire of products include horizontal sections (CAPPI's) at any
of 18 pre-defined heights, vertical cross-sections along any user-
selected corridor, Echotops of selectable reflectivity threshold,
rainfall accumulations, rainfall forecast maps based on image cross-
correlation and extrapolation.   The products generated are stored in
a disk file for up to 24 hours and these may be viewed in a single
image or animation loop modes.   The interactive display options such
as the BLINK and ZOOM facilities and instant colour schemes changes
enable the user to enhance and study in detail desired signatures
and features in the radar imagery.   In addition, RAINSAT, a system
which combines precipitation and echotop information from a network
of weather radars and Geostationary satellite imagery, is described.

1. INTRODUCTION

     While the potential of meteorological radars to aid in the
identification and forecasting of meteorological events involving
precipitation has been apparent for many years, the actual achievement of
these objectives in the operational environment has been less than
spectacular.
     In many countries around the world a great deal of money has been
spent on purchasing, installing and operating radars where the only
display system is a PPI either of the conventioanl fading screen type or a
TV scan converter.   The difficulties with interpretation of PPI displays
have been well known for years, the differing height of beam as a function
of range and the large areas of ground echo being but two of the more
obvious problems.   This dismal state of display technology is in sharp
contrast to the situation with satellite image processing systems, where
interactive techniques have been pioneered most noticeably by a group at
the University of Wisconsin, and with the elaborate plans proposed for the
NEXRAD system.   The latter suffers from the disadvantage that it is not
currently available and may well be expensive in comparison with existing
radar networks.

It has been our goal to devise a sophisticated interactive
processing system which can be attached to a conventional radar so that
the measurement and forecasting potential of the radar may be realized in
real time operational environments. It is fair to say that this activity
parallels some of the activities of the RADAP group in the U.S. (McGoven
and Saffle (1984)). (5)

2. HARDWARE

The arrangement of the equipment used is illustrated in Figure 1.

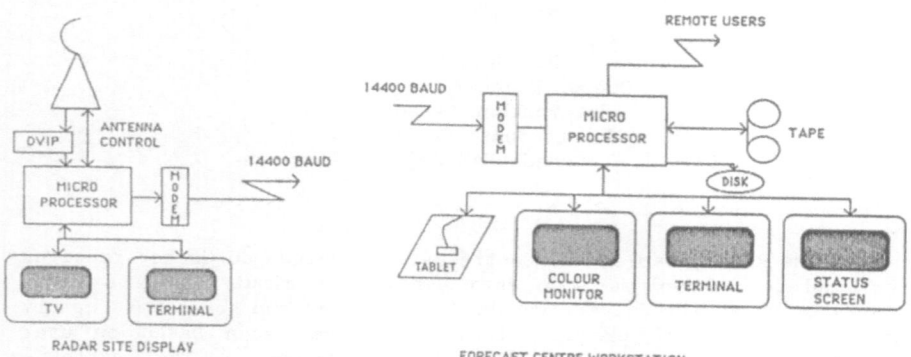

Figure 1

At the radar site a microprocessor controls the antenna elevation angle
program so that in 5 minutes the antenna is elevated to 24 different
angles at a rotation rate of 6 rpm. The video from these different angles
is digitized and then compressed using zero suppression encoded strategy
(Austin and Riley (1983)). (1) It is then sent on a 14400 baud line to
the regional forecast office where it may be recorded on tape and be
available for processing in the interactive display system (RDP) which is
the major subject of this paper.

3. DESCRIPTION

The RDP system is capable of generating in real time, from the 3-D
volume scan data, the following products :

- CAPPI at any of the 18 pre-defined altitudes.
- Echotop for any reflectivity thershold.
- Accumulations over various time intervals.
- Forecast Accumulations
- Vertical section over various time intervals.
- Forecast Accumulations
- Vertical section along any selected corridor
- Forecast rainfall intensities at selected points and over the field of
  view.
- Maps of highest rainfall rate over selected time inteval.
- Maps indicating the highest rainfall in the vertical extent and the
  heights at which this maximum occurs.

The user can create a library of schedules, which specify the combination of products to be generated for each 5 minute interval and control parameters such as the Z-R relationship, suitable for different seasons and meteorological conditions. He can thus call upon the schedule appropriate for the current situation quickly.

The meteorologist's workstation consists of :

- A high resolution colour graphic system for display of radar imagery.
- A terminal for the selection of system parameters and interactive image display options.
- A terminal for the continuous display of upto date meteorological and system status information.

A considerable effort has been devoted into making the user interface for the selection and display of the large number of products involved, as simple as possible. Thus, the conventional menu system is replaced by a tree structure which groups together a great number of commands and their objects in one single screen of the console. The user can select the desired option from among the available ones using only a few arrow keys. The status terminal alerts the meteorologist (with bells etc.) to severe weather, heavy rain, electrical and flood situations when pre-defined thresholds are exceeded in the generated products and to the radar or the communication line malfunctions.

4. METEOROLOGY

The foregoing description is primarily of a processing system. In order to proceed beyond this engineering aspect of the problem to allow the machine to show some "meteorological artificial intelligence" it is necessary to develop some algorithms capable of deducing the presence of operationally significant events indirectly from the radar imagery and other available information.

The techniques implemented presently in this system are a thunderstorm algorithm (35 dBZ at a CAPPI height of 6 km (Marshall and Radhakant (1978)) (3) and the extrapolation forecasting scheme described in Bellon and Austin (1978). (2) The detection of regions showing well defined bright band anf the automatic rejection of ground echo using the relative intensity of the radar echo at two heights as described by Collier et al (1980) (4) is being implemented. Algorithms which allow calibration against raingauge data for verification of the accumulated rainfall products are also available.

An important area of ongoing research is the effective combination of radar data with other meteorological fields and imagery. It is important not just to overlay them and then do intuitive meteorology but to develop operational, meteorologically significant and statistically verified algorithms for important meteorological parameters and their forecasts.

5. RAINSAT

The RAINSAT system combines precipitation and echotop information from several weather radars with Geostationary (GOES, GMS or METEOSAT) satellite imagery to generate composite images at synoptic scale (2000 x 2000 km) for real time operational monitoring and nowcasting. Tha RAINSAT procedure remaps the visible and IR imagery onto a Lambert conic or Polar steregraphic projection, normalizes the visible data for sun angle variations and then combines the visible and IR data to generate maps indicating the raining areas or thunderstorms based on statistical classification schemes derived from a sufficiently large number of radar-satellite comparisons as described in Bellon et al (1968) (8), Cherna et al (1985) (7). Radar precipitation data, wherever available, is overlayed onto this RAINSAT rainfall map. The cross-correlation and linear extrapolation methodology described in Bellon et al (1980) (6) is applied to the rainfall map thus obtained to generate short term forecasts in the 1 to 6 hour domain. Synoptic velocities at the steering wind level may also be used in junction with the velocities generated from the cross-correlation technique in generating the forecast maps. A composite echotop map which combines the echotop information from the radar network and IR imagery is also available. An interactive display package similar to the one described for the RDP system enables the meteorologist to take full advantage of the potential of satellite and radar data. Two products a RAINSAT image based on Meteosat VIS and IR images and a 5 hour forecast are illustrated in Figures 2 and 3. A RAINSAT system (SIRAM) operating on a 13 radar network and Meteosat imagery is currently being implemented at Madrid, Spain.

6. CONCLUSIONS

The installation of radar systems having no processing capability beyond PPI is not cost effective. No doubt powerful new systems of the next NEXRAD type will include these processing strategies but our purpose here has been to suggest that even for lower budget and existing installations relatively low cost processing systems can considerably enhance the capabilities of the radar, particularly in producing;
1) clearer, less ambiguous images
2) quantitative rainfall accumulations
3) objective severe weather warnings automatically
4) short range forecasts automatically.
5) network several radars together on an appropriate map background of satellite estimated rainfall rates.

It is possible with these systems meteorological radars will come nearer to realizing the potential recognized more than 30 years ago.

REFERENCES

(1) AUSTIN, G.L., M.M. Riley and E.H. Ballantyne, 1983: 21st Conference on Radar Meteorology.
(2) BELLON, A., and G.L. Austin, 1978: J. App. Met. 17, 1778-1787.
(3) MARSHALL, J.S., and S. Radhakant, 1978: J. App. Met. 17, 206-212.
(4) COLLIER, C.G., S. Lovejoy, and G.L. Austin, 1980 : 19th Radar Conference 44-47.
(5) Mc Goven, W.E., and R.E. Saffle, 1984: 22nd Radar Conference, 188-191.
(6) BELLON, B., S. Lovejoy, and G.L. Austin, 1980 : Combining satellite and radar data for the short-range forecasting of precipitation. Mon. Wea. Rev., 108, 1554-1566.
(7) CHERNA, E., A. Bellon, G.L. Austin, and K. Kilambi, 1985: An objective technique for the delineation and extrapolation of thunderstorms from GOES satellite data. J. of Geophysical Res., Vol. 90, 6203-6210.
(8) BELLON, A., and G.L. Austin, 1986 : On the relative accuracy of satellite and raingauge rainfall measurements over middle latitudes during daylight hours. J.C. and App. Met?; Vol. 25, 1712-1724.

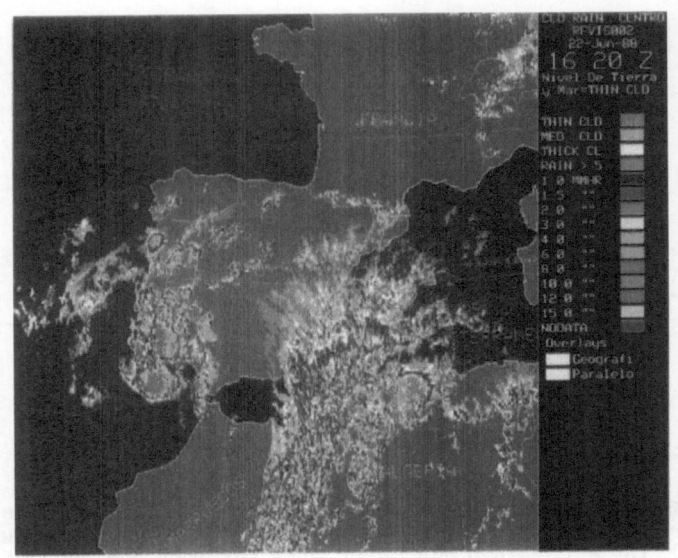

Figure 2

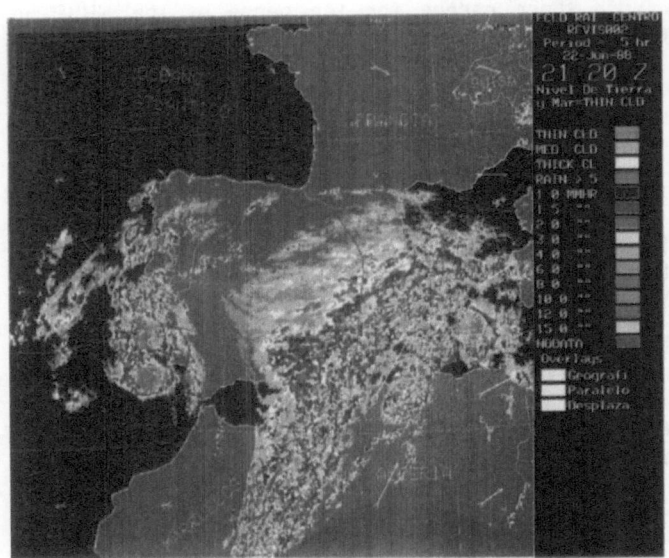

Figure 3

Figures 2 & 3 :   Five hour forecast : RAINSAT image based on Meteosat VIS
and IR images.

SITING OF A WEATHER RADAR NETWORK FOR

OPERATIONAL HAIL SUPPRESSION IN GREECE

N. R. DALEZIOS
ENVIROTECH Ltd. ,132 Olympou str, 54635 Thessaloniki, GREECE

N. PAPAMANOLIS
Greek Agricultural Insurance (ELGA), Thessaloniki, GREECE

P. C. LINARDIS
Dept. of Physics, Aristotle University of Thessaloniki, GREECE

Summary

This paper describes the procedure followed to select a site for a C-band weather radar in the Greek National Hail Suppression Program (NHSP). The established objectives were followed and most of the siting criteria were satisfied. At first, the strategic or regional criteria were fulfilled and then the local siting criteria were satisfied for the selection of the specific site. The C-band weather radar was installed at the site ruins at Ktima near the small town of Filiron in region 3 of Kissos oros. This radar constituted part of an effective weather radar network with three radars for the needs of the NHSP in the three protected areas.

1. INTRODUCTION

The selection of a suitable site for a weather radar presents problems depending on the purpose of use and type of application. In a hail suppression program a site should satisfy operational requirements as well as scientific or program evaluation needs. For the first purpose the equipment is to be used as a meteorological aid in the detection and tracking of storms as well as controlling and directing seeding operations. In these cases the user wishes to obtain information from as great range as possible, which necessitates operating at a low angle of elevation with a good field of view throughout $360°$ in azimouth although for short range information an angle of elevation of about $2°$ may be used (1); besides, information on the vertical structure and evolution of a storm is needed to assess the operational status and decide and control seeding missions.

For the second purpose precipitation intensity measurement and areal rainfall analysis (2,3) is needed, and siting problems are mainly related to the incidence of permanent echoes and obstructions as well as to the maximum range achievable. Intersection of the beam with the melting layer should also be avoided, which suggests that the radar should be operated at the lowest possible angle of elevation having an antenna with a narrow beamwidth.

This paper examines the site selection criteria and the procedure followed in 1986 (6) to install the C-band weather radar in area A2 (Figure

1) of the Greek National Hail Suppression Program (NHSP). A suitable site should be able to meet the following objectives: to constitute an operational center to direct seeding operations in area A2; to track storms and detect their vertical and horizontal structure; to be used as a backup radar for operational seeding in area A1; to consist part of an effective weather radar network for operational hail suppression in central and northern Greece.

The paper is organized as follows: in section 2 background information is provided on the NHSP; in section 3 the selection criteria are described; and in section 4 the selected site is considered.

2. BACKGROUND

The NHSP has been initiated in 1981. A five-year operational cloud seeding program (1984-88) has been designed (5) to reduce hail damage over three distinct agricultural areas, namely two areas A1 and A2 in northern Greece and one area A3 in central Greece (Figure 1). A research component of the program has also been designed in the form of an exploratory experiment in area A1 to conduct a statistical evaluation of the NHSP (4,5). The operational period was from 15 April till 30 September every year.

The anomalous ground relief with high mountains between the areas necessitated three operational centers, one for each area. Cloud seeding operations in areas A1 and A2 have been conducted by three aircrafts based at Thessaloniki operational center (airport), whereas area A3 has been supported by two aircrafts based at Larisa airport. Throughtout the program the Greek Agricultural Insurance Institute (formerly OGA, currently ELGA) has provided two S-band weather radars to direct seeding operations. In particular, area A1 has been covered by one radar located at Thessaloniki airport and area A3 has been supported by the second radar located at Larisa airport. These S-band radars have been recently digitized. Furthermore, ELGA requested that a third C-band radar should be installed for the coverage of area A2 (Figure 1).

3. RADAR SITING CRITERIA

Two types of criteria have been selected to determine optimum location(s) for the position of the C-band weather radar for the control of seeding operations in area A2, namely strategic or regional siting criteria and local or logistical siting criteria (6). Strategic criteria have been selected according to the requirements of effective seeding control. After satisfying the strategic criteria and selecting a region then the local criteria were considered. The following subsections briefly describe the strategic and local siting criteria.

3.1 Strategic or regional criteria

The strategic or regional criteria mainly consist of setback and communication considerations. The setback aspect implies a seeding buffer zone and radar view angle setback (Figure 2). Seeding should be carried out on storms that are within 20 minutes of entering the protected area, which corresponds to a seeding buffer zone of about 15 km for storms moving with about 40 km/hr. Moreover, for storms with significant radar information below 8 km altitude this criterion corresponds to a setback of about 22 km, since effective controlling is difficult for antenna elevations above 20 degrees. The above result in a setback of about 35 km which allows for a seeding buffer zone and radar view angle setback.

The communication considerations require that the radar controller must be in constant communication with seeding aircraft at all times, which suggests the need for a FM relay station. Furthermore, it is desirable to have IFF contact with aircraft at low altitudes (about 700 m). However, this criterion cannot be met at all times, since earth curvature effects limit this IFF contact to 60 to 80 km even for an ideal site.

3.2 Local Criteria

The local radar siting criteria include the area geometry or morphology, as well as local logistics and practical considerations. In particular, a suitable radar site should have minimum ground clutter. Thus hills near to the radar site should not be lower than zero degrees elevation. For a site at the top of a high hill ground clutter exists to large distances in all directions at low elevation angles resulting in difficult identification of stationary storms. Similarly, maximum elevation to nearby hills between the radar and the protected area could be about one degree or less, whereas distant hills should be at a lower elevation than the nearby hills. Otherwise, radar side lobes may cause ground clutter at hights much above hill height.

Moreover, the antenna should be above buildings and trees in the immediate vicinity of the site. In order to reduce effects of side lobes there should be rough vegetation within 0.5 km from the site. Since one of the objectives remains to have backup coverage of area A1 from the area A2 radar site, some occlusion in the form of low hills between the two radars is desirable.

Similarly, logistics and practical aspects should also be considered. Specifically, all-weather access roads are preferred and the site should be convenient for staff accommodation with water and sewer facilities highly desirable. Furthermore, the site should have access to reliable line power with capability of 20 Amps or more at 220 Volts, as well as access to telephone or possibly a radio relay station for communication to Thessaloniki airport (Figure 2). Other factors to be considered included appropriate building for electronics or office, existing tower or roof for radar, size of land required and available, right of way and cost.

4. SITE SELECTION PROCEDURE

The steps to determine acceptable sites include: preparation of ground clutter and area-of-coverage maps for each of the existing radar sites; identification of regions of blockage for each of the existing sites with consideration of project areas, buffer zones and upwind zones, as well as the minimum elevation angle and height to clear obstacles; use of detailed topographical maps and the above described siting criteria to establish potential sites; selection of specific sites within these regions and measurement of obstruction-free azimuths and elevations; and consideration of local logistics and practical aspects. This procedure has been applied to four potential regions (Figure 2), namely Vironia (region 1), Kalabakion (region 2), Kissos oros (region 3) and Arnea (region 4). The following paragraphs describe the four regions and the selected site within region 3.

Region 1 of Vironia has been rejected for the following reasons: the region is located within the buffer zone; there is restricted view of buffer to west of protected area due to terrain with peaks between 500 to 800m within 10 km southwest; there is restricted view of east end of protected area due to higher terrain west of site within 20 to 30 km; there is restricted view of sky to north due to higher terrain north of site within 10 km; there is IFF contact 80 percent in protected area and buffer zone. Similarly, region 2 of Kalabakion has been rejected for the following reasons: the region is located within the buffer zone on east side of protected area; there is restricted view of northern buffer zone due to high elevation angles; there is no backup by Thessaloniki radar in restricted view area; the site is remote to area of approaching storms; and there is IFF contact 80 percent in protected area and buffer zone.

Region 3 of Kissos oros has been accepted for the following reasons: the region is located outside the buffer zone; there is generally unrestricted view of protected area and buffer zone; there is backup coverage of area A1 on ridges northeast of Thessaloniki; the region is near to Thessaloniki airport; and there is IFF contact 80 percent in protected area and buffer zone. Similarly, region 4 of Arnea has the following characteristics: effective view angle 310 degrees through 0 to 70 degrees; the region is located outside of the buffer zone; there is no coverage of area A1; no specific sites have been found; and it is not close to Thessaloniki airport. Region 4 is acceptable in general, but it would require time and effort to identify sites meeting local siting criteria.

Within the acceptable region 3 of Kissos oros (Figure 2), five specific sites have been examined around the small town of Filiron, namely, Phenix, Water Tower, Village, Ruins at Ktima and Saddle between Hortiatis and Exohi (Figure 2). Although most of these sites have met the local siting criteria, the site Ruins at Ktima has been selected to install the C-band weather radar, the main technical characteristics of which are shown in Table 1 (6). The selected site has the following features: altitude 488 m; effective view azimuth 260 through 0 to 100 degrees; dirty road direct to site; distance to power 50 m; distance to nearby building 50 m; land publicly owned; view to area A1 elevation angle 0 degrees; and 40 minute drive to Thessaloniki airport.

TABLE 1: C-band Radar Main Characteristics

1. Band	C
2. Frequency (MHz)	5600
3. Peak Power Output (KW)	250
4. Pulsewidth (μs)	2.0
5. PRF (PPS)	250
6. Beamwidth (deg)	1.65 max
7. Antenna (Reflector)	parabolic
8. Reflector Diameter (ft)	8
9. Antenna Gain (db)	4
10. MDS (dbm)	-108
11. Receiver (75-78 db range)	logarithmic

The selected site satisfied most of the specified criteria, since due to the complex topography and numerous logistics it was not possible to fulfill all the criteria. It has been observed that in certain directions an elevation angle of one degree was required for a good horizon. This drawback has caused some minor difficulties in directing and effectively controlling operational missions in the center of area A2 for cloud-base seeding. However, there were only a limited number of such cases and most of them were overcome by top seeding. The C-band weather radar was successfully used as a backup radar for area A1 in a number of cases. The installed C-band radar in the specific site constituted part of an effective weather radar network for the three areas A1, A2 and A3, respectively, in the operational hail suppression program in central and northern Greece. Moreover, during the 1988 operational period this site was successfully and effectively used as a 20-hour per day radar watch site.

5. CONCLUSIONS

A suitable radar site has been selected to install a C-band weather radar as part of a weather radar network in the Greek National Hail Suppression Program. The selected site satisfied the describrd objectives and most of the established strategic and local siting criteria. It was possible to use this C-band radar as a backup radar for operational missions in area A1 in a number of cases throughout the three year period. During the last season this radar site was used as a radar-watch site for 20 hours daily for the three areas.

ACKNOWLEDGEMENTS

We would like to thank all the ELGA and the contractor (Intera) personnel for contributing to the selection, installation and operation of the weather radar network in the Greek National Hail Suppression Program. We would also like to thank Mr. C. Papageorgiou, ELGA Program manager, for his support. We are also grateful to Dr. T.S. Karacostas of the Arist. Univ. of Thessaloniki for valuable discussions and suggestions. Finally, we would like to thank Mr. D. Skepastianos and C. Anogianakis for their technical assistance.

REFERENCES

(1) CLIFT, G.A., 1985: Use of Radar in Meteorology. Tech. Note No 181, WMO- No 625, 90p.

(2) DALEZIOS, N.R., 1988: Digital Processing of Weather Radar Signals for Rainfall Estimation. Proceedings, Intern. Conf. on Advances in Remote Sensing, Arist. Univ. of Thessaloniki, Greece, 10-12 Oct. (under review).

(3) DALEZIOS, N.R., and N. Kouwen, 1989: Radar Signal Interpretation in Warm Season Rainstorms. Submitted to Nordic Hydrol. for possible publication.

(4) KARACOSTAS, T.S., 1989: The Greek National Hail Suppression Program: Design and Conduct of the Experiment. Proceedings, 5th Scient. Conf. on Weather Mod. and Applied Cloud Physics, WMO, Bejing, China 5-8 May, pp.605-608.

(5) KARACOSTAS, T.S., 1984: The Design of the Greek National Hail Suppression Program. Proceedings, 9th Conf. on Weather Mod., AMS, Park City, Utah.

(6) RUDOLPH, R. et al., 1987: The Greek National Hail Suppression Program: 1986 Annual Report. Intera Technologies Ltd., Rept. No M86-215, Jan., 165p.

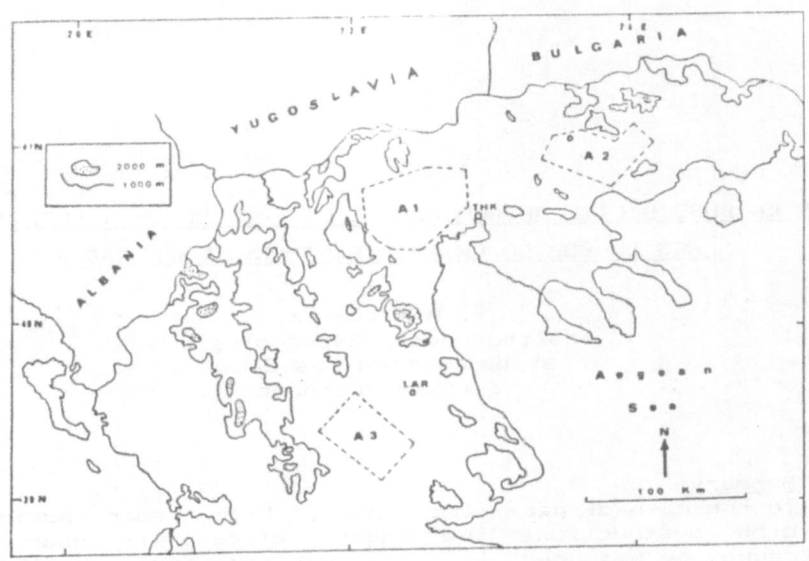

Figure 1. Map of the central and northwestern part of Greece, showing the three targets

Figure 2. Map of the northern part of Greece, showing the proposed sites

SUR LA POSSIBILITE DE REPRODUIRE LE FACTEUR DE REFLECTIVITE PRES DU SOL DU CHAMP TRIDIMENSIONNEL RADAR

S. Moszkowicz
Institut de la Météorologie
et de Gestion des Eaux
Varsovie, Pologne

Summary

In the case of permanent echoes, high radar horizon or other obstructions the radar reflectivity near ground cannot be measured. As it is very important for the rain intensity determination one would want to re — establish it from the radar reflectivity data taken at higher levels.

For that purpose a set of 1200 recorded radar situations containing maximum reflectivities from 0-2, 2-4, 4-6, km layers and radar top hights over squares of 10 x 10 km were examined.

The convective and separately stratiform echoes were divided into classes of different "bright-band" hights. For each case the linear regression and discrimination function were found and the errors of estimation were determined.

1. INTRODUCTION

Il advient bien souvent que le facteur de réflectivité radar près du sol, qui est essentiel pour déterminer l'intensité de précipitation par radar, ne peut pas être mesuré. Les raisons principales de ce fait sont les cibles parasites, les masques et la position trop élevée du faisceau radar. Mais on sait bien que l'information radar est essentiellement tridimensionnelle: on peut mesurer le facteur de réflectivité dans les couches plus élevées et l'altitude du sommet de nuage. Sachant que le développement vertical d'un nuage est lié à l'intensité de précipitation on peut espérer qu'il existe une relation entre le facteur de réflectivité près du sol et celui des couches supérieures et l'altitude du sommet. Dans cet article on cherche telles relations par la méthode de régression et de discrimination.

2. DONNEES

On a pris pour l'analyse environt 1200 situations mesurées par le radar MRL – 2 (3 cm) en 1988 à Legionowo (près de Varsovie) et enregistrées par le Système radar semi-automatique (PSR) composé de Processeur météorologique specialisé (SPM – 1) et micro- ordinateur compatible IBM PC/XT.

SPM – 1 en regime "exploration en volume" divise l'espace en couches de l'épaisseur de 2 km et chaque couche en carrés de 10 x 10 km. Il trouve, dans chaque élément de 10 x 10 x 2 km, le facteur maximal de réflectivité et encore l'altitude du sommet de nuage au-dessu de chaque élément horizontal de 10 x 10 km.

Cette information sert à reconnaître des phénomènes météorologiques (grêle, orage, averse, pluie stratiforme, Cumulus, nuage stratiforme) pour le service d'avertissement de l'aviation, elle est aussi mémorisée sur un disque ou une disquette pour le traitement tardif.

Au premier stade de reconnaissance des signeaux radar, ils sont divisés en échos convectives et stratiformes par la fonction discriminante:

$$g = 0.46 \text{ DHM} + 0.147 \text{ DZM} + 0.346 \text{ ZM} + 0.114 \text{ P} - 14.01$$

ou DHM et DZM sont respectivement pour l'altitude de sommet (HM) et pour le facteur maximal de réflectivité (ZM) – les differences absolues maximales entre élément donné et les éléments voisins contenant écho, ZM – la valeur maximale du facteur de réflectivité dans la couche 0 – 6 km, P – pourcentage des éléments vides autour de l'élément donné. La valeur $g = 0$ divise les échos en convectives $(g >= 0)$ et stratiformes.

La fonction discriminante g était testée sur un échantillon indépendant et on a trouvé que l'erreur de classification ne depasse pas 10.2 %, etant la plus grande pour les échos convectives composés des petites cellules simples, uniformement et tout dru distribuées dans l'espace. Elle est évidement provoquée par la résolution assez basse du SPM – 1 qui "stratiforme" de tels échos.

On utilise ici la fonction g pour diviser les échos en convectives et stratiformes.

On n'a pris pour traitment que le saison d'été déterminé par l'altitude d'isothèrme $0°C$ (HO) supérieure à 1 km. Les échos de la region 40 – 90 km en distance de radar ont été analysées, c – a – d là ou le facteur de réflectivité près du sol pouvait être mesuré.

Les notations sont suivantes: Z1, Z2, Z3 [dBZ] – le facteur maximal de réflectivité radar respectivement pour la couche 0 – 2, 2 – 4 et 4 – 6 km, HM – l'altitude du sommet de nuage en demi – kilomètres.

Les échos radar ont été classés comme suit:

		- existent Z1,Z2,Z3,HM -	C11
	HO < 2 km		
ÉCHOS CONVECTIVES		- existent Z1,Z2,HM -	C12
		- existent Z1,Z2,Z3,HM -	C21
	HO >= 2 km		
		- existent Z1,Z2,HM -	C22

		- existent Z1,Z2,Z3,HM -	S11
	HO < 2 km		
ÉCHOS STRATIFORMES		- existent Z1,Z2,HM -	S12
		- existent Z1,Z2,Z3,HM -	S21
	HO >= 2 km		
		- existent Z1,Z2,HM -	S22

La classe C11 ne contient que 3 échos et la classe S11 — 15 éléments et ensuite ces classes ne sont pas analysées.

3. LA REGRESSION

Les résultats de modèle de la régression linéaire sont présentés sur le tableau I. L'ordre des paramètres est Z1,Z2,Z3,HM, alors p.ex. COV24 n'est que la covariance entre Z2 et HM. A,B,C,D sont les coefficients de la droite de régression cherchée:

Z1 = A Z2 + B Z3 + C HM + D
et σ - l'écart - type residuel de la régression.

On a comparé ensuite les erreurs de l'intensité de précipitation provenant:
- de remplacement de la valeur "vraie" de Z1 par la moyenne de Z1 designée comme E1,
- d'utilisation de la regression pour Z1 - erreur E2.

Les erreurs étaient déterminées comme facteurs par lesquels il faut multiplier et diviser la valeur obtenue pour trouver les limites entre lesquelles se trouve la valeur vraie. On les obtient par transformation de l'equation de Marshall et Palmer en prenant l'écart - type de Z1 pour E1 et l'écart - type residuel de la regression σ pour E2. Le quotient des ces facteurs donne le gain de la précision si on applique la régression au lieu de la moyenne de Z1.

On voit que pour la classe C12 la régression ne donne aucune amélioration de la précision, pour autres classes on a le gain entre 18 et 45 %, le meilleur pour les classes S21 et C21, alors pour les cas ou HO > 2 km et les nuages sont bien développés en verticale - Z3 existe. C'est encourageant car ces sont les classes ou la plus forte précipitation peut avoir lieu.

Tableau I. Analyse de régression

Classes:		S12	S21	S22	C12	C21	C22
Taille d'échant.		549	2674	2692	165	4052	1687
Moyennes:	Z1	17.50	20.04	15.80	21.69	27.85	20.95
	Z2	14.95	21.90	16.67	18.94	28.58	19.79
	Z3	–	13.94	–	–	21.92	–
	HM	5.79	10.94	6.68	5.90	12.93	7.19
Écart-type	Z1	5.24	6.30	5.42	5.00	7.11	6.39
Covariances							
	COV11	27.51	39.63	24.02	24.95	50.56	40.86
	COV12	14.80	29.66	14.81	6.81	34.87	22.60
	COV13	–	16.01	–	–	35.46	–
	COV14	0.762	4.14	2.09	0.78	9.12	1.99
	COV22	31.08	35.24	25.54	33.28	40.91	37.70
	COV23	–	18.67	–	–	39.28	–
	COV24	0.412	3.86	6.89	0.75	8.83	3.44
	COV33	–	31.79	–	–	69.97	–
	COV34	–	10.67	–	–	19.22	–
	COV44	0.669	8.91	6.71	0.61	13.82	3.55
Coefficients							
	A	0.695	0.85	0.665	0.181	0.805	0.60
	B	–	-0.047	–	–	0.024	–
	C	0.384	0.153	0.174	1.064	0.112	-0.025
	D	4.90	0.408	3.55	11.99	2.87	9.20
Écart-type residuel	σ	4.12	3.81	3.87	4.78	4.54	5.23
Facteurs d'erreur	E1	2.13	2.50	2.18	2.05	2.78	2.51
	E2	1.81	1.73	1.74	1.99	1.92	2.12
Quotient	E1/E2	1.18	1.45	1.25	1.03	1.45	1.18

En règle générale les relations obtenues sont bien faibles,
c'est, au moins en partie, le résultat de la faible résolution
du SPM – 1 et de la mesure du facteur maximal de réflectivité
au lieu de sa valeur moyenne.

4. DISCRIMINATION DE PRECIPITATION

Au service d'avertissement de l'aviation nous ne prêtons
pas tellement attention à la mesure précise de l'intensité de
pluie, mais nous voulons plutôt déterminer (reconnaître) si la
pluie existe ou non. Pour le saison d'été nous avons adopté le
seuil de Z1 = 17 dBZ comme la limite inférieure de
précipitation, qui correspond à l'intensité ~ 0.8 mm/h de
l'equation de Marshall et Palmer.

Pour les mêmes classes qu'au chapitre 3 on a cherché une
fonction discriminante linéaire de précipitation, et sur le
même échantillon le pourcentage de classifications erronées a
été testé.

Les tailles d'échantillon étaient ici plus grandes parce
que on a pris aussi en consideration les cas ou Z1 n'existait
pas (la base de nuage supérieure à 2 km). Les résultats
d'analyse sont présentés sur tableau II. L'ordre des paramètres
est le même qu'au chapitre 3, mais en omettant Z1, alors on a
Z2,Z3,HM et p.ex. COV23 c'est la covariace entre Z3 et HM. A,
B, C et D sont les coefficients de la fonction discriminante
linéaire

$$G = A\ Z2 + B\ Z3 + C\ HM + D$$

qui donne la classe de précipitation si G >= 0 et le nuage au cas contraire. Pour les classes S22 et C22, ou la taille d'échantillon est assez grande, on a essayé aussi la fonction discriminante quadratique, mais elle n'a donné aucune amélioration.

Tableau 2. Discrimination de précipitation

Classes:	S12	S21	S22	C12	C21	C22
Taille d'échant.						
total	705	3145	3721	194	4283	2023
pluie	367	1991	1248	144	3776	1287
nuage	338	1154	2473	50	507	736
Moyenne Z2						
pluie	16.50	24.07	19.99	19.10	29.22	21.15
nuage	12.25	14.53	12.76	17.36	19.54	15.13
Moyenne Z3						
pluie	–	14.88	–	–	22.37	–
nuage	–	11.37	–	–	16.32	–
Moyenne HM						
pluie	5,94	11.14	6.98	6.01	13.02	7.32
nuage	5.8J	10.33	6.49	6.08	11.49	6.87
Covariances						
COV11	32.25	43.55	33.53	31.61	45.66	39.87
COV12	–	20.73	–	–	40.86	–
COV13	0.274	4.30	1.596	0.215	9.31	3.42
COV22	–	31.55	–	–	69.75	–
COV23	–	10.29	–	–	19.05	–
COV33	0.982	8.63	2.145	1.108	13.72	3.70
Coefficients						
A	0.131	0.246	0.212	0.055	0.287	0.153
B	–	-0.059	–	–	0.062	–
C	0.092	0.085	0.070	-0.077	0.177	-0.020
D	-2.344	-4.34	-4.63	0.512	-5.96	-2.074
Erreur de classification en %						
totale	34.2	15.8	19.0	25.8	10.5	27.8
pluie	32.5	6.2	43.6	0	2.9	11.9
nuage	36.4	32.4	6.5	100.0	67.1	55.7

Comme on voit au tableau, les erreurs totales de classification se trouvent dans l'intervalle 10 à 30 %, la discrimination n'etant pas suffisement bonne. Il est toujours encourageant que les erreurs de classification sont minimales aux classes S21 et C21, alors pour les cas ou on peut espérer une grande intensité de pluie.

5. CONCLUSIONS

L'information radar provenante de couches supérieures de l'atmosphère permet, à un certain degré, réproduire les valeurs manquantes de facteur de réflectivité près du sol. Les relations obtenues - la régression et la fonction discriminante de précipitation - sont en effet assez faibles, mais elles sont les plus fortes pour les classes les plus interessantes c-a-d quand l'intensité de pluie peut être très grande.

La faiblesse des relations résulte, au moins en partie, de la faible résolution spatiale du SPM -1. En appliquant des appareils de haute résolution on pourrait espérer une correlation plus grande et alors une meilleure réproduction de facteur de réflectivité près du sol.

Reception at Hôtel de Ville (5/09/89)
Right : Dr H. TENT
Deputy Director General
GD XII "Science, Research & Development"
Commission of the European Communities
Left : Dr C.G. COLLIER
COST 73 Chairman

Réception at Hôtel de Ville (5/09/89)
Left to right :
Mr H. WESSELS (NL - Chairman of Session 5)
Mr R. KING (COST 73 Delegate - Finland)
and T.W. SCHLATTER (NOAA Forecast Systems
Lab. Boulder USA)

Reception at Hôtel de Ville (5/09/89)
Left to right :
Dr D.H. NEWSOME (COST 73 Coordinator),
Mr G. McDONALD (Ir) and Mrs C. NIEWHONER
(Gematronik - DE)

Reception at the Hôtel de Ville
Guided tour of the state rooms of the Hôtel de Ville

SESSION 3

The role of new techniques in
operational radar networks

Chairman : B. Beringuer

NEW WEATHER RADAR TECHNIQUES: READY FOR
OPERATIONAL USE ?

Walter L. Randeu
Graz University of Technology, Austria

Summary:

Relevant new weather radar techniques are reviewed with
respect to their operation principles, expected benefits
and acceptance for use in operational networks.

Dual-frequency and frequency-agile radars are found to
be of little value for routine operation, and therefore
might remain restricted to special applications, probably in the research domain.

Doppler weather radars are available off-the-shelf from
all major manufacturers, and can already be found in
several operational networks. Their value lies in the
suppression of ground clutter and the measurement of
precipitation particle velocities.

The improvement of precipitation intensity estimates and
the liquid/solid-phase discrimination are benefits
brought in by polarisation diversity radars. With today's technology to be implemented without problems,
these radars are just on the way to leave the research-
domain and to become accepted for routine operation.

The typical weather radar for installation during the
90ies should be a combination of Doppler and polarisation diversity. Other advanced systems, like electronically steered (multibeam) antennas, are expected as last
step forward at the end of the next decade.

1. Introduction

Since more than 30 years, the application of pulsed radar systems for the qualitative and quantitative detection of precipitation phenomena is worldwide well established. Starting with
relatively simple devices with analogue data outputs, improving technology led to the development of rather sophisticated
systems, including extensive digital signal processing equipment and software. This change from analogue to digital processing and representation techniques took place during the late
70ies, mainly influenced by the improving availability of reliable and powerful mini-computers. Today only few old installations with analogue output are in use, while 100% of recent
or new operational installations are of course of the digital
type.

Influenced by this change in technology, several European
countries started the implementation of national weather radar
networks, utilizing one important property of data in digital
representation, i.e. their communicability to remote processing centres and user facilities.

In December 1979 the CEC-project COST-72 (Measurement of Precipitation by Radar) commenced its activities towards the European co-ordination of related national programmes, to result

in a trial network of weather radars. The follow-on project, COST-73, is continuing these efforts, and shall, within the remaining two years of working time, pave the way to a pre-operational weather radar network throughout Europe.

A weather radar network, producing instantaneous precipitation intensity displays with pan-European coverage, receives its value from the quality of its data output as well as from the interest of groups of users addressed. Therefore not only the communication and networking aspect is being treated by COST-73, but also working groups for the improvement of COST-images and the application of new radar techniques have been established. While the improvement of COST-images is oriented towards the post-processing and presentation of weather radar data to suit (new) users' requirements, new radar data techniques are intended to improve the quality and extend the information contents of the sensor's, i.e. the radar's direct data output.

This paper is to give an overview of "new" techniques available, and to discuss to what extent these techniques are being accepted by every-day operational users. Furthermore the standpoint of radar manufacturers will be illuminated and recommendations for future installations given.

"New" techniques are to be considered as those, which extend a weather radar's capability beyond those of a "conventional" radar. Conventional means, that one single quantity only (in most cases the radar reflectivity for horizontal polarisation, Z_H) is measured by the instrument.

2. The limitations of conventional (single-parameter) weather radars

The basic problem in the interpretation of conventional (or single-parameter) weather radar data is that the radar-measured reflectivity Z is roughly proportional to the sum of the sixth powers of the sphere-equivolumetric particle diameters:

$$Z = \frac{\sum_{\nu=1}^{n} D_i^6}{V} \qquad (m^6/m^3) \qquad (1)$$

(for Rayleigh-scattering, i.e. $D_i < \lambda/10$; applicable to S-band radars)

or

$$Z_{H,V} = \frac{\lambda^4}{\pi^5 |K_0|^2} \frac{\sum_{\nu=1}^{n} \sigma_{H,V}}{V} \qquad (m^6/m^3) \qquad (2)$$

(equivalent reflectivity for non-Rayleigh-scattering and non-spherical drops; applicable to C-band and above radars)

$Z_{H,V}$ ....   radar reflectivity factor (for Horizontal or Vertical polarisation)

D_i ....   equivolumetric sphere diameter of the i-th particle

V ....   tropospheric volume illuminated by the radar beam resp. pulse-cell

λ .... wavelength

$|K_0|^2$ .... factor determined by the refractive index of the particle material (0.93 for water at usual radar frequencies)

$\sigma_{H,V}$ .... rigorously derived radar cross-sections (H- or V-polarisation)

In contrast, the quantities being of main interest to the meteorologist or hydrologist, i.e. the instantaneous rainfall rate R and the rainfall total, are roughly proportional to the sum of the fourth powers of the particle diameters (the water volume).

Interpreting Z in terms of R therefore has to rely on some assumption made on the distribution of drops with size. We all know several different relationships between Z and R having been reported in the literature.

Some examples are (Z in mm^6/m^3, R in mm/h):

$Z = 291 * R^{1.43}$ (C-Band, Attmannspacher & Riedl, 1977)

$Z = 490 * R^{1.43}$ (S-Band, 0°C, R = 9 mm/h, Cherry, 1978)

$Z = 130 * R^{1.43}$ (S-Band, 0°C, R = 23 mm/h, Cherry, 1978)

$Z = 350 * R^{1.6}$ (X-Band; thunderstorm; Dutton, 1967)

$Z = 200 * R^{1.6}$ (X-Band, Attmannspacher & Riedl, 1977; S- and X-Band, Austin & Schaffner, 1970; X-Band, wide-spread rain, Dutton, 1967; X-Band, Ochs & Rücker, 1977; Ka-Band, Reid, 1969)

The coefficients in these Z-R-relationships vary not only from frequency to frequency, but also from author to author. The most often stated equation, $Z = 200 * R^{1.6}$, seems to be the "universal" one, and can be found in many conventional radar processor implementations. The proper choice of the Z-R-conversion law is in some applications aided by data from telemetering rain gauges, but this technique is not suitable for radar-real-time rainfall intensity presentations. So there remains some uncertainty how to convert Z to R, resulting in considerable errors in the radar-derived instantaneous rain fall rate.

Not only the drop-size distribution influences the conversion law, also the material phase (water or ice, or a mixture of both), as well as particle shapes and orientations are of importance. Simply spoken, too many assumptions have to be made in order to derive a hand full of parameters from a single measured quantity.

The way out is pointed by new types of radars, delivering additional independent measurables.

These radars resp. techniques are:

- the dual-(multiple-)frequency radar
- the polarisation diversity (dual-polarisation) radar
- the Doppler radar
- the fast-scanning or multibeam radar
- the application of frequency agility.

Of course any combination of these techniques (except Doppler with frequency agility) is possible.

The first two techniques mentioned deliver information about the micro-structure of precipitation fields (drop shapes and canting, as well as liquid/solid phase discrimination), and therefore support the interpretability of Z in R.

Doppler radars do not improve the Z-R-conversion, but are very useful for two other commonly needed purposes, i.e. ground clutter suppression and particle velocity measurements, from which such important quantities as wind shear and turbulence can be derived.

The last two techniques result in higher measurement speed without loosing accuracy, an aspect very important for near-real-time large volume scanning.

The way to implement, use and interpret these new techniques (except Doppler) has been pointed by research institutes concerned with cloud physics or tropospheric microwave propagation research. Table 2.1 (Bringi & Hendry, 1987; Randeu, 1988) shows a selection of advanced weather radars being in use for research activities mainly.

A quick analysis of this table (which is restricted to installations in North America and Europe) gives the following statistics:

Location: 8 * Europe, 12 * U.S. or Canada;

Frequency: 9 * S, 4 * C, 4 * X, 1 * Ku, 2 * Ka;

Polarisation: 7 * any, 12 * LIN H/V,
                     1 * CIRC LHC/RHC;

Reception: 13 * co- and cross-polar, 6 * co-
                     polar, 1 * cross-polar;

Echo phase: 11 * yes, 9 * no;

Doppler: 14 * yes, 6 * no;

Frequ. agility: 2 * yes, 18 * no;

Measured quantities: 18 * ZDR, 14 * LDR, 9 * CDR,
                     8 * correlation co- vs. cross-polar,
                     14 * Doppler;

The following sections each focus on one of the new techniques, discussing operational principles, implementation problems and benefits in daily routine operation.

3. The dual-frequency weather radar

This instrument simultaneously measures reflectivities at two clearly separated frequencies (e.g. S- and X-band). Two mechanisms provide additional information about the target's physical structure (Cherry, 1978):

a)   The lower frequency has lower attenuation due to rain than the higher one. The differences in echo signal strength at both frequencies can be used to estimate drop size distribution parameters, and hence improve the accuracy when converting Z to R.

Agency	Location	Frequ. (GHz)	Polarisation radiated	Polarisation received	Phase Doppler capability	Pol.Quantities measured	Note
NRC of Canada	Ottawa, Ontario	16.5	any	co+cross	no	CDR,LDR,cross-corr.	1)
Alberta Res. Council	Penhold, Alberta	2.9	LHC,RHC,LIN	co+cross	no	CDR,LDR,cross-corr.	
NOAA, ERL/WPL	Boulder, Colorado	35	LHC,RHC,LIN	co+cross	yes	CDR,LDR,cross-corr.	
VPI & SU	Blacksburg,Va.	2.8	3)	copolar	no	ZDR in 4 planes	2)
CHILL	Urbana,Ill.	2.8	LIN (H + V)	copolar	yes	ZDR	2)
SHAPE Tech. Centre	The Hague, NL	10	4)	co+cross	yes	scatt. matrix, complex cross-corr.	
NSSL	Cimarron,Okl.	2.8	LIN (H + V)	copolar	no	ZDR	2)
CNR/Istit. di Electr.	Italy	5.6	LIN (H + V)	co+cross	yes	ZDR,LDR	
AFGL	Sudbury,Mass.	3	LHC,RHC,LIN	co+cross	yes	ZDR,LDR	
Lab. d'Aero-logie	Lannemezan, France	35	LHC,RHC,LIN	co+cross	yes	ZDR,CDR,LDR	
New Mexico Tech.	Socorro, New Mexico	10	LIN (H + V)	co+cross	yes	ZDR,LDR,cross-corr.	
SPANDAR	Wallops Isl.	2.8 6)	LIN (H + V)	copolar	no	ZDR,Doppler	2)
RAL	Chilbolton,GB	3	LIN (H + V)	5)	no	5)	2)
DFVLR	Oberpfaffen-hofen, FRG	5.6	any	co+cross	yes	ZDR,CDR,LDR, Doppler,cross-corr.	
IAS/FGJ	Graz, Austria	5.6 8)	LIN (H + V)	co+cross	no	ZDR,LDR,cross-corr.	7)
NRC of Canada	Ottawa, Ontario	10	RHC,LHC	co+cross	yes	CDR,LDR,cross-corr.	
NCAR (CP-2)	Boulder, Col.	2.8 10	LIN (H + V) LIN (H)	copolar co+cross	yes no	ZDR,Doppler LDR	9)
TU Delft	Delft, NL	3.3	any linear	co+cross	yes	ZDR,LDR,Doppler	10)
CSIM	Teolo/Padova,I	5.6	LIN (H + V)	copolar	no	ZDR,Doppler	11)

Table 2.1: Summary of polarisation diversity research radars (notes on next page)

Notes to table 2.1:

1) manually operated polarizer
2) single-channel receiver
3) linear, 22.5 deg. change per pulse
4) 28 discrete linear and 137 RH & LH elliptical
5) copolar (Z_H, Z_{DR}) or crosspolar (Z_H, LDR) (not simultaneously)
6) frequency-agile (single-tube electronical tuning)
7) variance of all measured quantities calculated in real time
8) frequency-agile (dual-tube congruent mechanical tuning)
9) dual-frequency radar (S- & X-band)
10) continous motor-driven polarizer; seperate TX- and RX-antenna
11) Z_{DR} in non-Doppler mode only

---

b) For large ice-particles, such as hail, the departure from Rayleigh-scattering at the higher frequency is useful as an indicator for the presence of these particles. Hail detection studies with dual-frequency radars have led to useful results (e.g. Eccles & Atlas, 1970; Carbone, 1972).

There are, however, serious disadvantages associated with a dual-wavelength radar:

- The instrument is actually a combination of two radars (at least in the RF-part), associated with high cost and difficult matching of antenna pointing and pulse shape parameters.

- The impact of systematic and random measurement errors on the evaluation of differential attenuation is too severe to consider such a radar as a reliable measurement tool.

- The maximum range of a dual-wavelength radar is very limited by masking effects at the higher frequency (at least X-band). A single thunderstorm cell may cause more than 30 dB two-way-attenuation, making the radar blind for targets beyond this cell.

As a result, dual-frequency radars are not suitable for operational purposes, where large coverage (at least 100 km radius) and stable measurement results are required. This technique remains restricted to research work within limited areas of special interest, and should not further be discussed in the context of COST-73 and operational applications.

4. Doppler weather radars

Caused by absolute and relative (partially due to the finite beamwidth of a radar antenna) movement of precipitation particles, their echo spectrum shows a spread around a mean value, determined by the mean radial velocity. Involved is the well-known Dopplereffect, giving the name to (weather) radars with spectral resolution capability.

Echoes from the ground, on the other hand, which are always of significant disturbing intensity during weather radar operation (low antenna elevation angles are required to reach the first few kilometers above ground also at long ranges), have their spectral energy in most cases closely centered around zero Doppler frequency. Exceptions may occur for wooded target areas in combination with strong wind, but these are rather infrequent. A good overview of Doppler techniques and target spectral behaviour is given by Doviak and Zrnic (1984).

The value of applying Doppler techniques in a weather radar, causing increased investment costs compared to a conventional system, lies not only in the detectability of particle velocities, but even more in the ground clutter suppression ability. Because the major fraction of clutter echo energy is confined near the origin of the base-band spectral domain, it can be removed with proper digital filters (recursive Infinite-Impulse-Response- or non-recursive Finite-Impulse-Response-filters). The error introduced to rain reflectivity estimation is negligible or can be corrected for.

Today two types of pulsed Doppler radar designs are in use, differing in their MTI (moving target indication) improvement factors, and of course also in their complexities and prices.

The so called "transmitter-coherent" design (see block diagram in Fig 4.1) provides up to 50 dB sub-clutter-visibility, and can usually be found in military applications. One exception is the GPM-500C radar offered by SMA/Florence, intended for use in the civil domain. Furthermore, Raytheon in the U.S. is on the way to apply the transmitter-coherent principle in its TDWR (Terminal Doppler Weather Radar), to be used for the early detection of microbursts close to ground.

There are, however, several disadvantages connected with this Doppler design. Besides the higher cost of implementation, transmitter-coherent systems suffer from the necessary use of high-power pulsed amplifiers (klystrons or travelling wave-tubes), requiring very high voltages and voluminous power supplies. Associated problems are the generation of dangerous X-rays, and reduced practical life time.

More suited for every day's routine operation is the second design, the "receiver-coherent" pulsed Doppler radar. The block diagram in Fig. 4.2 illustrates its operation principle, relying on the locking coherent oscillator (COHO) to preserve the phase of the transmitted carrier-frequency pulse. The transmitted energy is produced by rather cheap coaxial magnetrons, being state-of-the-art devices with high practical lifetimes and simple power supplies. This design may provide subclutter-visibilities in the order of 40 dB, which, in combination with the fixed siting of weather radars, may be considered sufficient to discriminate precipitation from ground clutter echoes.

The majority of Doppler weather radars offered by the industry are receiver-coherent designs. At present EEC, Ericsson and Gematronik offer such radars in different frequency bands, with Plessey to follow soon. Moreover, Ericsson has reported to have developed an advanced design with tight magnetron frequency control, improving the total system's clutter suppression to 45 dB, comparable to a transmitter-coherent radar.

Table 4.1 summarizes the present state of operational and experimental installations in Europe. Devices already ordered, but not yet delivered or installed are included.

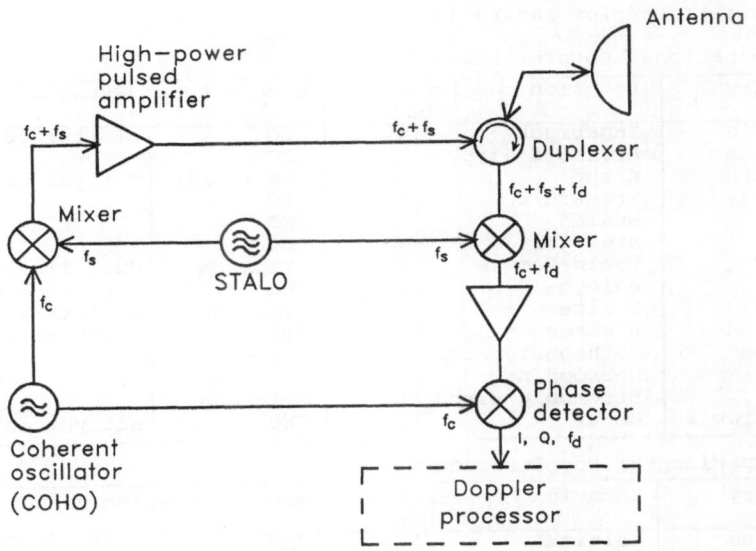

Fig 4.1: Block diagram of a transmitter-coherent pulsed
Doppler radar.

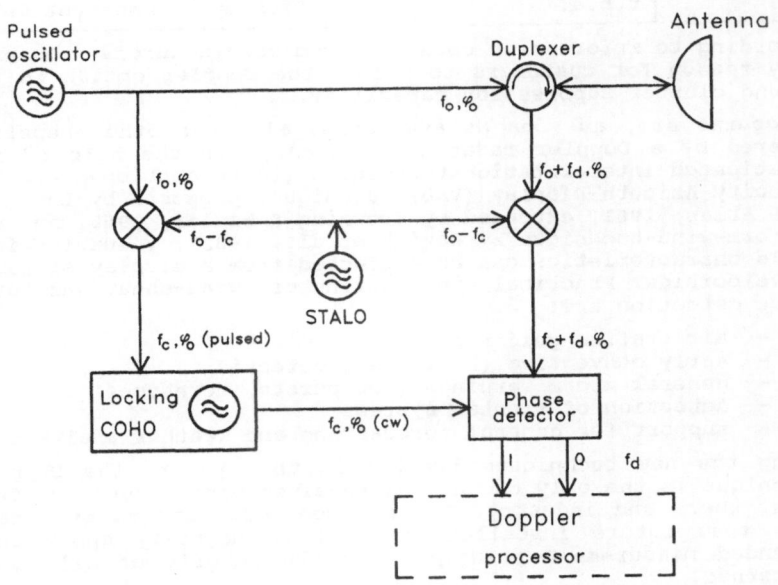

Fig 4.2: Block diagram of a receiver coherent pulsed Doppler
radar

Table 4.1: Doppler radars in Europe:

a) Operational Doppler radars:

Country	Location	Frequency band	Manufacturer	Remarks
Austria	Innsbruck	C	EEC	clutter suppr.
Belgium	Brussels airp.	C	EEC	
Denmark	Karup	S	Gematronik	not yet del.
Germany	Frankfurt	C	EEC	
	Munich	C	EEC	
	Stuttgart	C	EEC	
Italy	Teolo/Padova	C	Ericsson	dual lin. pol.
	Bologna	C	SMA	dual lin. pol.
Spain	9 sites	S	Ericsson	1 installed
	6 sites	C	Ericsson	2 installed
Sweden	Gothenburg airp.	C	Ericsson	
	Norköpping	C	Ericsson	
	Stockholm airp.	C	Ericsson	
Yugoslavia	Koper	C	EEC	not yet del.

b) Experimental Doppler radars:

Country	Location	Frequency band	Manufacturer	Remarks
Finland	Helsinki	C	EEC	
France	Lannemezan	Ka	self	pol. div.
Germany	Hohenpeissenberg	C	EEC	
	Oberpfaffenhofen	C	EEC	pol. div.
Netherlands	Delft	S	self	pol. div.
	The Hague	X	self	pol. div.
Norway	Oslo	C	Ericsson	
U.K.	t.b.d.	C	Plessey	not yet del.

According to information obtained from manufacturers, the primary reason for customers to choose the Doppler option is the ground clutter suppression capability.

Of course are, as soon as available, also the other benefits offered by a Doppler radar being used. With the help of sophisticated interpretation techniques (to be mentioned are the Velocity-Azimuth-Display (VAD) technique, proposed by Lhermitte & Atlas, 1961, extended by Browning & Wexler, 1968, and the uniform-wind-technique by Doviak et al., 1982), valuable windfield characteristics can be extracted from a display of radial velocities. Practical applications of wind-shear and turbulence detection are:

- air traffic guidance
- early convective (hail) cell detection
- general storm warnings (downbursts, tornadoes)
- detection of frontal systems
- support for general forecasting and weather analysis.

Among the new techniques treated in this paper, the Doppler technique is the only one having received wide acceptance from both, users and industry. It is recommended and to be expected, that future installations will in majority apply this extended measurement principle for the benefit of all users concerned.

5. Polarisation Diversity Radars

Proposals to use several polarisation states on transmission, as well as co- and crosspolar echo reception, go back to the early seventies (McCormick & Hendry, 1972; Seliga & Bringi, 1976), and were made mainly by cloud physicists and microwave propagation researchers. These research communities are, dictated by their work, interested in the microstructure of precipitation (particle size and shape, spatial orientation of non-spherical particles, liquid-solid-phase discrimination). High accuracy, when converting radar observables to precipitation characteristics, is especially needed in the microwave propagation domain, where predicted signal attenuation and depolarisation are very sensitive to particle size distribution and type of precipitation. There is a very fruitful co-operation between microwave propagation people and radar meteorologists, including:

- radar meteorologists provide models of precipitation structures (large- and fine-scale), in order to aid the interpretation of complex radar signatures;
- in return they receive tools to correct for the radar signal distortion due to intervening precipitation cells, and can improve the accuracy of radar-measurables and interpreted meteorological parameters.

But now it is time to explain the principle of polarisation diversity radars. This is best done with the help of the so-called "coherency matrix" J.

Assuming complex echo field strengths E_1 and E_2 in two orthogonal polarisations (linear H and V, or circular LH and RH) being received by the radar, the coherency matrix J is defined as:

$$\underline{J} = \begin{pmatrix} \overline{E_1 E_1^*} & \overline{E_1 E_2^*} \\ \overline{E_2 E_1^*} & \overline{E_2 E_2^*} \end{pmatrix} = \begin{pmatrix} J_{11} & J_{12} \\ J_{21} & J_{22} \end{pmatrix}$$

The overbar denotes the short-term average (integration over tens or hundreds of pulses), while the asterisk indicates complex conjugation.

If the echoes are in response to a transmitter polarisation state 1, then the meanings of the matrix elements are:

J_{11} .... prop. to mean power of the copolar echo
J_{22} .... prop. to mean power of the crosspolar echo
J_{12}, J_{21} ... complex correlation between co- and cross-polar echoes.

With a radar, which transmits only one polarisation, but receives co- and cross-polar echoes, the following quantities can be measured (refer also to Table 2.1):

a) radar reflectivity Z (prop. to J_{11})
b) J_{22}/J_{11}, called linear depolarisation ratio (LDR) for linear polarisation, and circular depolarisation ratio (CDR) for circular polarisation.
c) the cross-correlation between co- and cross-polar echoes (J_{12} or J_{21})

For a radar with two transmitted polarisations (in practice the transmission is alternated between the two polarisation states from pulse to pulse), two coherency matrices exist, but are not completely independent of each other. In the case of linear polarisation J_{11} describes Z_H in the first matrix and Z_V in the second matrix, while both matrices show equal values of J_{22}. The ratio Z_H/Z_V is the differential reflectivity Z_{DR}, being the fourth independent measurable for a fully equipped polarisation diversity radar.

The relationships between the four radar-measurables Z_H, Z_{DR}, LDR and Correlation, and precipitation characteristics can best be shown for the case of dual linear polarisation (H & V):

- Increasing particle size delivers increasing Z_H (the extreme case is large hail covered with water),

- Increasing drop size results in increasing drop oblateness for rain; for uncanted raindrops, falling with their minor axes aligned vertically, the differential reflectivity Z_{DR} increases accordingly. In contrast, Z_{DR} remains constantly low for hailstones, or even drops below 1 for the case of conical hailstones or graupel.

- Large oblate raindrops, or ice-needles, being tilted with respect to one of the two orthogonal polarisation axes, cause the generation of cross-polar echoes. Increasing cross-polar echo strength indicates increasing oblateness of canted drops (the maximum is reached at 45° canting angle).

- If all oblate particles show the same canting angle ("common orientation"), then co- and crosspolar echo sequences are correlated well (i.e. J_{12} and J_{21} are high). The reverse is true for random orientation of particles, being an indicator of turbulences.

Table 5.1 tries to present these relationships in a more systematic manner.

RADAR OBSERVABLES				PRECIPITATION CHARACTERISTICS			
Z_H 0dbZ...65dBZ	Z_{DR} −2dB...<LOW 0db...LOW +6dB...HIGH	LDR −∞dB...LOW −10dB...HIGH	Correl 0...LOW 1...HIGH	Particle deviation from sph. shape	Canting angle (0°...±45°)	Degree of common orientation	Remarks
LOW	LOW	LOW	n.a.	LOW	any	n.a.	drizzle
LOW	HIGH	any	HIGH	HIGH	HIGH	HIGH	ice needles, common orientation
LOW	LOW	HIGH	LOW	HIGH	any	LOW	ice needles, snow random orientation
HIGH	LOW	HIGH	LOW	HIGH	any	LOW	strong rainfall random orientation
HIGH	HIGH	LOW	HIGH	HIGH	LOW	HIGH	strong rainfall common orient. 0°
HIGH	LOW	HIGH	HIGH	HIGH	HIGH	HIGH	strong rainfall common orient. 45°
HIGH	<LOW	LOW	LOW	LOW	any	any	large hail, graupel

Table 5.1: Interpretation matrix for linear polarisation diversity radars
(principal states shown, interpolation between cases possible).

A data example obtained from the research radar in Graz/Austria (Riedler & Randeu, 1982; Randeu, 1986) may illustrate the usefulness of dual-polarisation data for the recognition of hail. Fig. 5.1 shows a thunderstorm event on Sept. 5, 1988 in PPI-slant- range presentation with 4,5 elevation angle. Fig. 5.1a presents the absolute reflectivity Z_H, Fig. 5.1b the simultaneously observed differential reflectivity Z_{DR}. It is striking that, within the core of the thunderstorm cell (25 to 30 km south-west of the radar), where Z_H goes up to 60 dBZ, Z_{DR} remains very low, or even shows negative values (in log. representation, see gray "triangle" between 25 and 30 km range, at azimuth = 230). This is a clear indication of the presence of large hailstones aloft, suspected to reach the ground as hail. Would it be strong rain, then Z_{DR} should show values in excess of 3 dB. Hailpad data confirmed the occurrence of hail in this case.

For the purpose of hail suppression, the detection of the presence of hail with a dual-polarisation radar may be too late. It should be combined with Doppler observation, in order to predict the growing cumulonimbus from particle updraft speeds.

For circular-polarisation radars, a similar classification scheme may be used. Figure 5.2 shows the circular polarisation characteristics for various types of precipitation, taken from Hendry & Antar (1984).
Here ORTT stands for $J_{12}/(J_{11} * J_{22})^{1/2}$ (the degree of common orientation), and CAN for $10 * \log (J_{11}/J_{22})$ (the inverse of CDR).

McCormick & Hendry (1975) pioneered the use and interpretation of circular polarisation diversity radars, but it turned out, that the implementation of and interpretation techniques for linear polarisation radars are simpler and more cost effective.

Figure 5.3 shows the simplest form of a dual-linear-polarisation radar, restricted to measure Z_H and Z_{DR}. Three additions have to be made to a conventional radar, to obtain this design:

a) a fast high power switch (ferrite switch), with switching times in the order of microseconds;

b) an antenna feed with dual input and integrated polarisation combiner (the reflector remains unchanged);

c) a second channel in the DVIP to handle the second polarisation's echoes.

It is clear that, having Z_H and Z_{DR} only, the interpretation matrix in Table 5.1 can not fully be utilized. However, as proposed by Seliga & Bringi (1976), good estimates of the parameters N_o and D_o in a two-parametric exponential drop size distribution can be obtained from Z_H and Z_{DR}. The knowledge of these parameters helps to choose the proper Z-R-conversion law for each individual event. A number of authors have performed either simulations or practical measurements with this technique, and obtained clear improvements compared to conventional radar estimates of rainfall rate (Hall et al., 1980, 1984; Atlas, 1984; Rogers, 1984; Bringi et al., 1982; Goddard & Cherry, 1984).

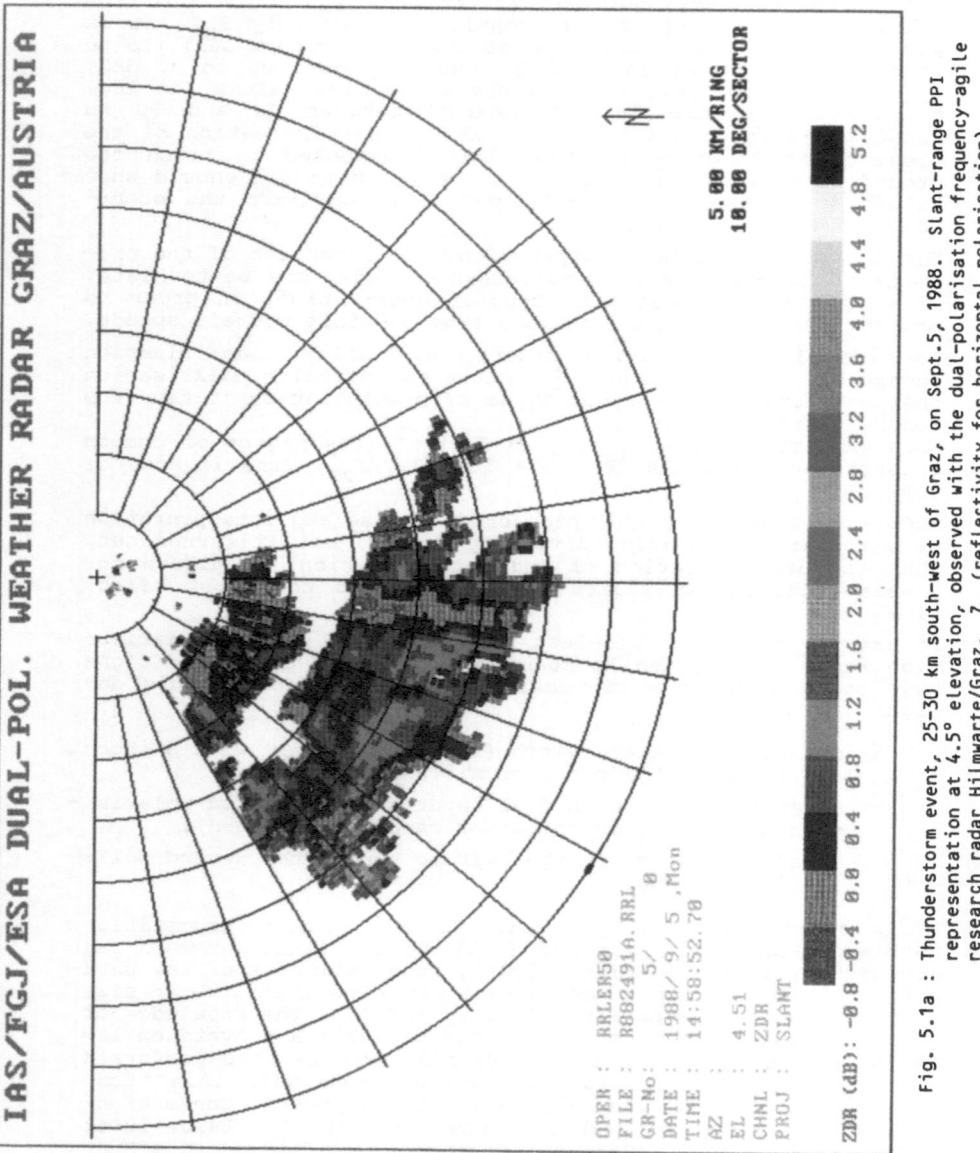

IAS/FGJ/ESA DUAL-POL. WEATHER RADAR GRAZ/AUSTRIA

OPER : RRLER50
FILE : R8024910.RRL
GR-No: 5/ 0
DATE : 1988/ 9/ 5 ,Mon
TIME : 14:58:52.78
AZ : 4.51
EL : ZDR
CHNL : SLANT
PROJ :

5.00 KM/RING
10.00 DEG/SECTOR

ZDR (dB): -0.8 -0.4 0.0 0.4 0.8 1.2 1.6 2.0 2.4 2.8 3.2 3.6 4.0 4.4 4.8 5.2

Fig. 5.1a : Thunderstorm event, 25-30 km south-west of Graz, on Sept.5, 1988. Slant-range PPI
representation at 4.5° elevation, observed with the dual-polarisation frequency-agile
research radar Hilmwarte/Graz. Z_H (reflectivity for horizontal polarisation).
Scan limited to $\pm$ 60° from south.

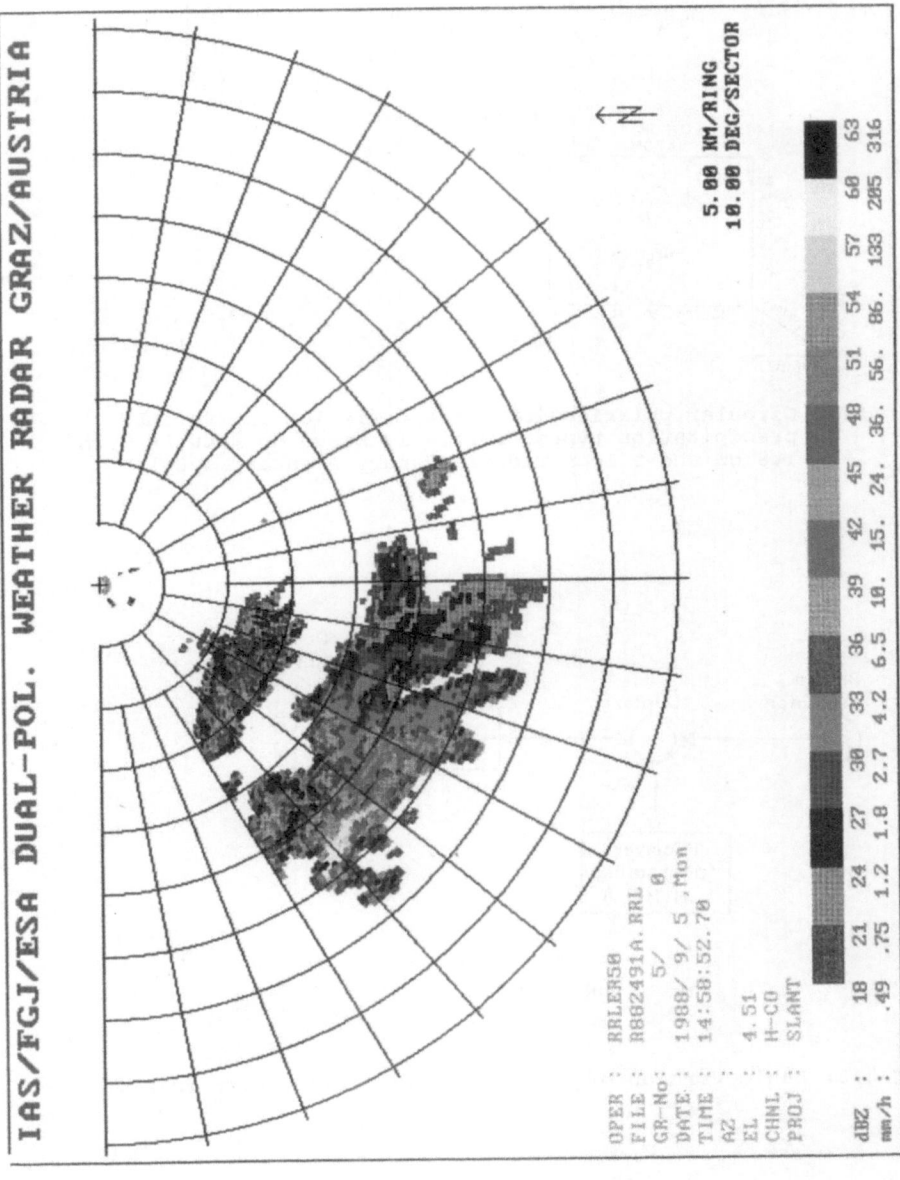

Fig. 5.1b : Thunderstorm event, 25-30 km south-west of Graz, on Sept.5, 1988. Slant-range PPI representation at 4.5° elevation, observed with the dual-polarisation frequency-agile research radar Hilmate/Graz. Z_{DR} (differential reflectivity z_H/Z_V). Scan limited to $\pm$ 60° from south.

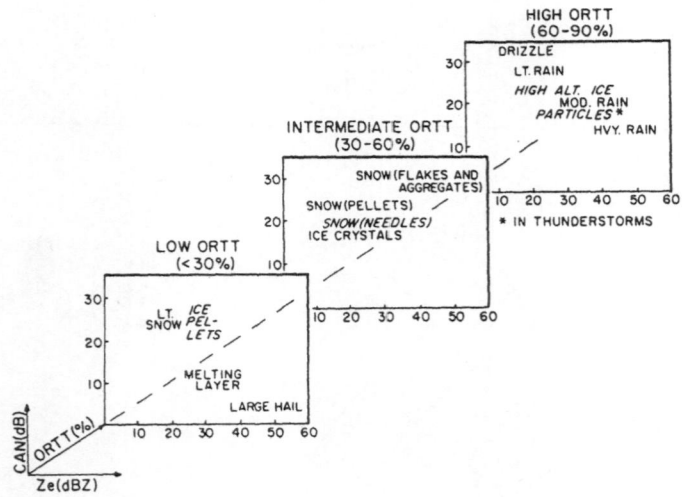

Fig 5.2: Circular polarisation characteristics for various
precipitation types. Partially based on results with
1.8 cm and 3.1 cm radars (Hendry & Antar, 1984).

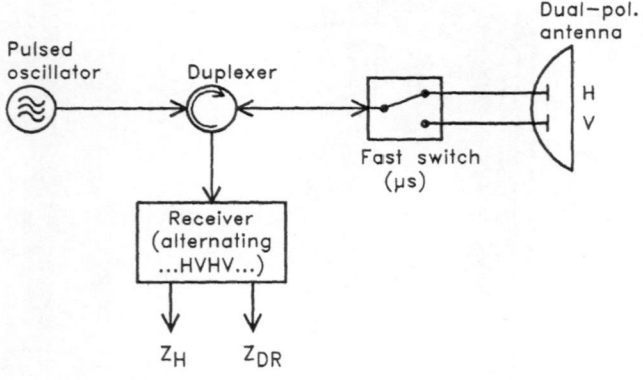

Fig 5.3: Basic version of a dual linear polarisation radar.

Striking examples are reported by Seliga et al. (1981), who found the error between radar-derived and ground-measured rainfall rates to be reduced from 47 % (for the conventional radar) to 22 % (for the Z_H & Z_{DR}-method), and by Cherry et al.(1981), who compared radar-derived with directly measured signal attenuation at 12 GHz. While the Z_H-only-method delivered 2 dB error for the total tropospheric path, the Z_H-Z_{DR}-method reduced this error to 0.5 dB.

But there are a number of drawbacks associated with the simple dual-polarisation radar:

- Raindrop canting may mask the differential reflectivity Z_{DR}, and lead to an overestimation of number of drops per unit volume.

- The recognition of particle canting and mixed-phase hydrometeors is impossible.

These disadvantages recommend the use of fully equipped polarisation diversity radars, receiving also cross-polar echo components and processing the correlation between co- and crosspolar returns. The block diagram of such a system is shown in Fig. 5.4.

Additional hardware needed is:

a) a high-quality dual-polarisation antenna, having at least 35 dB cross-polar discrimination;
b) a second T-R-duplexer;
c) a "slow" high power switch (switching time in the order of the pulse repetition period);
d) a second receiver channel for the cross-polar signal (may have reduced dynamic range), combined with a fast low-power transfer-switch;
e) a second channel in the signal processor.

Extended software is needed for the dual-DVIP, in order to extract Z_H, Z_{DR}, LDR and the correlation from the two received echo trains.

Although this full version of a linear polarisation diversity radar introduces considerable cost compared to the simple version shown in Fig. 5.3, it seems worth to choose it for the purpose of precipitation type discrimination and recognition resp. correction of canting and turbulence effects. Especially at long ranges, as being used in weather radar operation, Z_{DR} and LDR may receive considerable distortion along the radar signal's propagation path, so that careful correction, including that of any cross-polar antenna coupling, has to be applied in order to come close to the true in-situ values of the four radar-observables. A good example for a completely equipped polarisation diversity radar is the DFVLR-Radar "POLDIRAD" in Oberpfaffenhofen/Germany (Schroth et al., 1984). It was manufactured by EEC, showing that the industry is ready to deliver specialised systems.

At present there are only two (semi-) operational polarisation diversity radars, both combined with Doppler capability, installed in Europe. These radars have already been mentioned in table 4.1. The first one, in Teolo/Padova, was produced by Ericsson. It is in principle an off-the-shelf C-band Doppler radar, with a polarisation switch, a dual-channel antenna and special signal processing software added. The device is re-

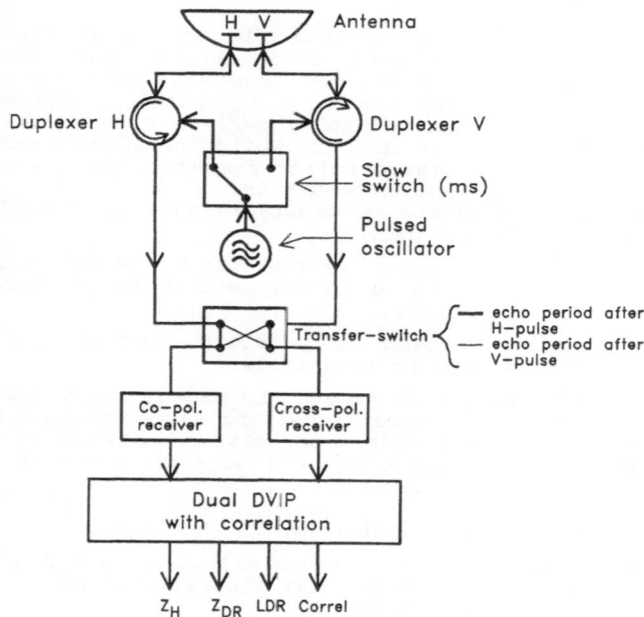

Fig. 5.4: Full version of a dual linear polarisation radar.

stricted to measure Z_H and Z_{DR}, and serves several purposes. First it is to be considered as a part of the (planned) Italian network of digital weather radars. Second, it is being used for local hail detection studies. And third Ericsson is testing different software packages, intended to obtain experience in the interpretation and presentation of extended weather radar data products.

The second operational polarisation diversity radar is near before completion (the inauguration shall take place in October this year) in San Pietro Capofiume (near Bologna). It will be a part of the Italian routine operation network, but serve also research purposes. The device has been developed and manufactured by SMA/Florence (model GPM-500C), and is of the transmitter-coherent Doppler type. It measures co-polar returns only, i.e. it is also restricted to deliver Z_H and Z_{DR}.

So, while the Doppler technique is already well accepted, dual polarisation radars are just making their first steps into practical use. For Italy, being rather late with the buildup of a nation-wide network of weather radars with digital output, it seems, that this lateness results in making two steps forward at one time, i.e. to equip all new installations from the beginning with Doppler and (at least the simple) linear polarisation diversity option.

EEC, Ericsson and SMA have already been mentioned before to be practically concerned with the production and testing of dual-polarisation radars. Gematronik reports to be able to accept orders for specialised systems (incoherent or coherent multi-polarisation), and Plessey is considering polarisation diversity as a useful option for future weather radar products.

The situation with respect to the use of polarisation diversity radars in operational networks for routine operation may be summarized as follows:

- The value of this technique for the improvement of precipitation rate estimation, as well as for precipitation type and hail detection purposes is out of discussion.
- Technology is at a stage to enable the implementation of reliable polarisation diversity radars at reasonable cost, i.e. the industry is ready to deliver such systems.
- There is still a lack of proved correction algorithms needed for long range observations. The same is partially true for the automatic interpretation of observables obtained from such radars. Continuing efforts in the research domain are expected to resolve this problem within the next few years.
- As for every new technique, users and operators show some hesitation to accept new data products offered, or to acknowledge their value for routine warning and prediction tasks. So there should efforts start to make potential users familiar with the features of polarisation diversity techniques and to train them in the use of new data products available.
- Seen from the position of the COST-73 project, provisions are already being made to account for the additional data output of advanced weather radars. This concerns the extended communication capacity needed (the proposed BUFR FM-94 code is flexible enough, and new data transmission links, like high-speed satellite communications channels, provide sufficient capacity), but also the presentation and application of new types of weather radar images (cf. working groups resp. subgroups on the improvement of COST-images, display systems, severe weather algorithms).

The following recommendations may be given:

a) Countries intending to install new radars within the next three years should at least prepare the radars for the later addition of hardware and software to upgrade them to polarisation diversity systems. To be considered as the most important item is the high-quality antenna with excellent polarisation characteristics.

b) Later installations should, from the beginning, be complete polarisation diversity systems, with the interpretation and presentation software to be continually modified according to the state of research and operational experiences.

To my impression, it will be only a matter of time to have the majority of operational weather radars equipped with some version of polarisation diversity, probably in combination with Doppler. The best indicator for the validity of this assumption is that industry has already started to deal with these techniques.

6. Other Techniques

Two other new weather radar techniques concentrate more on the increase in measurement speed than on the improvement of interpretability or provision of additional independent parameters.

These techniques are the application of the frequency agility principle and the use of fast-scanning electronically steered antennas, maybe with multiple beams.

Frequency agility, i.e. the change of carrier frequency by more than the inverse of the pulsewidth from pulse to pulse, provides instantaneous decorrelation of weather echoes. This increases the speed of acquiring a sufficiently high number (usually between 50 and 100) of independent single-pulse echoes, required for the elimination of reflectivity fluctuations. On a theoretical basis, a frequency-agile C-band radar with prf=1000 per second requires only one ninth of the measurement time compared to a fixed-frequency radar. At S-band, the time would even be reduced to one seventeenth. Besides the quicker acquisition of independent echo samples, a frequency-agile radar provides additional benefits. These are the high stability of integrated clutter echoes, being a pre-requisite for clutter map correction (Randeu et al., 1988). The price to be paid for frequency-agility is rather high. Either fast tuning magnetrons, or high-power pulsed amplifier tubes, as needed for the transmitter-coherent Doppler radar, are required as transmit sources. Furthermore, frequency agility cannot be combined with Doppler measurements, because of the intentional decorrelation of amplitude and phase from pulse to pulse.

Presently two research radars apply this principle (SPANDAR/ Wallops Island, U.S.A., and IAS/FGJ/Graz, Austria; see Table 2.1), and the SMA GPM 500C, soon to be put into operation near Bologna, would be able to transmit agile pulse patterns. However, when measuring Doppler shifts, this radar has to be put back to fixed frequency operation.
The result of this short discussion of advantages and disadvantages of frequency-agile weather radars is, that their use will remain restricted to research installations only, and will play no role in operational networks.

Fast-scanning weather radars (also called storm-scan-radars, or multiple-beam radars) are on the way to be prototyped by Ericsson and Plessey.

Especially under severe-weather conditions, cells with significant reflectivity show relatively small spatial extents. The purpose of such a storm-scan-radar with its electronically steered antenna is to locate intensive cells quickly, and restrict scanning to volumes actually filled with precipitation. A multi-beam system could even scan several independent storms at the same time, requiring probably some sort of frequency-diversity to discriminate between the individual echo contributors.

The advantage of such sophisticated radars would primarily be the very early detection of hazardeous weather phenomena, but also the reduced update time during periodical routine scanning. It is, however, expected that the availability and practical use of this type of radar will not become actual before the end of the next decade. Also will its expected cost restrict its use to special situations only (e.g. areas with high thunderstorm and hail activity).

7. Conclusion

A number of new weather radar techniques have been reviewed and discussed, as well as their practical use in operational weather radar networks discussed.

Dual-frequency techniques, because of their high cost and limited usefulness, may remain restricted to special research installations. The same applies to frequency-agile radars.

Doppler radars, with their important abilities to discriminate precipitation echoes from ground clutter, and to deliver particle velocity information, are already state of the art of the industry, and can be found in several national weather radar networks in Europe and elsewhere. The majority of future installations is expected to be of the Doppler type.

Polarisation diversity radars deliver, at rather moderate additional cost, valuable information about the type of precipitation, and help to improve interpretation of radar-observables in terms of precipitation characteristics. These radars are presently doing their first steps out from the research domain to every-day's practical use in operational networks. Combined with Doppler, they are expected to get increasing acceptance during the next decade, and to form then, together with the technique of fast storm scanning, the ultimate tool for accurate remote measurements of precipitation phenomena.

8. Acknowledgement

The author wishes to thank his colleagues in the COST-73 Coordination Committee for providing information about the present and planned use of advanced weather radar techniques throughout Europe. The help of the radar manufacturers EEC, Ericsson, Gematronik, Plessey and SMA, having supplied valuable material about their present and future product lines, is highly appreciated.

9. References

Atlas, D., 1984:
Highlights of the symposium on the multiple-parameter radar measurements of precipitation: personal reflections.
Radio Science, Vol. 19, No. 1, pp. 238 - 242.

Attmannspacher, W., und J. Riedl, 1977:
Gleichzeitige Flächenniederschlagsmessungen mit einem X- und einem C-Band Radargerät.
Kleinheubacher Berichte, Bd. 20, pp. 371-378.

Austin, P.M., and M.R. Schaffner, 1970:
Computations and experiments relevant to digital processing of weather radar echoes.
Proc. 14th Radar Meteorol. Conf., Tucson, Ariz.

Bringi, V.N., T.A. Seliga and E.A. Mueller, 1982:
First comparisons of rainfall rates derived from radar differential reflectivity and distrometer measurements. IEEE Trans. Geoscience and Remote Sensing, GE-20, No.2, pp. 201 - 204.

Bringi, V.N. and A. Hendry, 1987:
Technology of polarisation diversity radars for meteorology.
Battan Memorial and 40th Anniversary Conference on Radar Meteorology; AMS, Boston.

Browning, K.A., and R. Wexler, 1968:
A determination of kinematic properties of a wind field using Doppler radar.
J. Appl. Meteor., 7, pp. 105 - 113.

Carbone, R.E., 1972:
Evaluation of a dual-wavelength hail detector.
Proc. 15th Conference on Radar Meteorology, pp. 7 - 12, AMS, Boston.

Cherry , S.M., 1978:
Dual frequency radars - use for attenuation measurements and solid-liquid phase determination.
Proc. Radar Workshop Graz, Austria, Nov. 1978.

Cherry, S.M., J.W.F. Goddard and M.P.M. Hall, 1981:
Use of dualpolarisation radar data for evaluation of attenuation on a satellite-to-earth path.
Annales des Telecommunications, Vol. 36, No. 1,

Doviak, R.J., et al., 1982:
Pre-storm boundary layer observations with Doppler radar. Preprints 20th Radar Meteorology Conference, AMS, Boston, pp. 546 - 553.

Dutton, E.J., 1967:
Estimation of radio ray attenuation in convective rainfalls.
J. Appl. Meteorology, 6(4), pp. 662 - 668.

Eccles, P.J., and D. Atlas, 1970:
A dual-wavelength radar hail detector.
J. Appl. Meteorol., 12, pp. 847 - 856.

Goddard, J.W.F, and S.M. Cherry, 1984:
The ability of dual-polarization radar (copolar linear) to predict rainfall rate and microwave attenuation.
Radio Science, Vol 19, No. 1, pp. 201 - 208.

Hall, M.P.M., S.M. Cherry, J.W.F. Goddard and G.R. Kennedy, G.R., 1980:
Raindrop sizes and rainfall rate measured by dual-polarisation radar;
Nature, Vol. 285, pp. 195 - 198.

Hall, M.P.M., J.W.F. Goddard and S.M. Cherry, 1984:
Identification of hydrometeors and other targets by dual polarization radar.
Radio Science, Vol. 19, No. 1, pp. 132 - 140.

Hendry, A. and Y.M.M. Antar, 1984:
Precipitation particle identification with centimeter wavelength dual-polarization radars.
Radio Science, Vol. 19, No. 1, pp. 115 - 122.

Lhermitte, R.M., and D. Atlas, 1961:
Precipitation motion by pulse Doppler radar.
Proc. 9th Weather Radar Conference, AMS. Boston, pp. 218 - 223.

Mc. Cormick, G.C., and A. Hendry, 1972:
Results of precipitation backscatter measurements at 1.8 cm with a polarization diversity radar.
Proc. 15th Radar Met. Conf., Champaign Urbana, pp. 35 - 38.

Ochs, A., und F. Ruecker, 1977:
Streuquerschnitte von Niederschlägen bei 11,5 GHz, gemessen mit einem bistatischen Radar.
Kleinheubacher Berichte, 20, pp. 401 - 414.

Randeu, W.L., 1986:
Über die Entwicklung und den Bau eines mehrparametrigen fre-
quenzagilen Wetterradars im Rahmen der Erforschung von Aus-
breitungsstörungen auf Satellitenfunkstrecken.
Habilitationsschrift, Technische Universität Graz, Nov. 1986.

Randeu, W.L., 1988:
COST-73: Review of New Radar Techniques (Dual-frequency, pol-
arisation-diversity and frequency-agile weather radars).
EUCO-COST 73/37/88, Nov. 1988.

Randeu, W.L., W. Riedler, E. Kubista and F. Stampfl, 1988:
The behaviour of ground clutter echoes obtained with a fre-
quency- agile C-band weather radar.
Presentation at ISRP'88, Chinese Institute of Electronics,
Beijing, April 18 - 21, 1988.

Riedler, W., and W.L. Randeu, 1982:
A frequency-agile, dually-polarised C-band weather radar as-
sisting microwave propagation experiments.
Paper presented at URSI/IEE Conf. on Multiple Parameter Radar
Measurements of Precipitation, Bournemouth, UK.

Rogers, R.R., 1984:
A review of multiparameter radar observations of precipita-
tions.
Radio Science, Vol. 19, No. 1, pp. 23 - 36.

Rossettini, A., G. Vezzani and R.W. Lee, 1989:
GPM-500C: A new Italien dual-polarization Doppler weather radar.
Proc. 24th Conf. Radar Meteorology, AMS, Boston, pp. 375 -
379.

Schroth, A., P.F. Meischner and H. Schuster, 1984:
Coherent polarimetric C-band radar for atmospheric research.
Proceedings of IGARSS '84 Symposium, Strasbourg, August 27 -
30, pp. 497 - 505.

Seliga, T.A., and V.N. Bringi, 1976:
Potential use of radar differential reflectivity measurements
at orthogonal polarisations for measuring precipitation.
J.Appl.meteor., Vol. 15, pp. 69 - 76.

Seliga, T.A., V.N. Bringi and H.H. Al-Khatib, 1981:
A preliminary study of comparative measurement of rainfall
rate using the differential reflectivity radar technique and a
raingauge network.
J. Appl. Meteor., Vol. 20, pp. 1362 - 1368.

CANADA'S OPERATIONAL DOPPLER RADAR

T.R. Nichols, P.I. Joe and C.L. Crozier
Atmospheric Environment Service,
King Weather Research Station,
4905 Dufferin St., Toronto, Ont., M3H 5T4

SUMMARY

The King Weather Radar Research Station of the Atmospheric Environment Service (AES), located north of the metropolitan Toronto area in Southern Ontario, is a component of the Canadian weather service radar network. The only Doppler weather radar in the country, it was instituted as part of the research program of the Cloud Physics Division of AES and is a first phase of Canada's Dopplerization plan. The main objectives for this radar were to identify operational uses of Doppler radars in the Canadian climate, to determine and test the functional and technical requirements for an operational Doppler radar, to develop Doppler radar expertise and to initiate the training of operational meteorologists in the use of Doppler radars. This paper will describe how Doppler capability was incorporated with conventional radar observations.

1   The Network Radars

Figure 1 shows the distribution of weather radars within Canada. All of the radars are 5 cm in wavelength with the exception of a 10 cm system in the Montreal region operated by the McGill Weather Observatory. The selection of 5 cm for the network of conventional radars was based on economics and sensitivity. The 5 cm wavelength allows for smaller antennae and lower power for an equivalent beamwidth and sensitivity when compared with 10 cm systems. The wavelength appears as an inverse second power in the radar equation and therefore for equivalent transmit power, the received power is a factor four higher when comparing 5 cm versus 10 cm systems. This was a consideration for radars that would be used for observing winter snow cases. A possible difficulty with the 5 cm wavelength was the increased attenuation. However, climatological studies and operational experience have shown that rarely is a storm so completely attenuated that it is not identifiable qualitatively. The conditions for complete attenuation such as storms lining up along a radial are ephemeral events and storms are only attenuated out momentarily.

The King Doppler radar is also a C-band radar. Its location, highlighted by the solid circle in the figure, provides coverage for over 5 million people.

2   The King Doppler Radar

Doppler radars have proven their worth in the research environment specifically for severe weather identification in climates where tornadoes are a prevalent feature of summer convective weather. Canada has a thunderstorm climatology which is much more benign that those areas in the United States where much of the Doppler radar research has been performed. In terms of mean annual thunderstorm days, the area covered by the King radar has about 25 days per year or half that for Oklahoma and Colorado. In terms of the incidence of tornadoes, measured as the number per year per 10,000 km^2, the number is about 0.8/yr/10,000km^2 or

Figure 1: Radar network in Canada. Highlighted circle represents the King Doppler range of 110 km. The conventional range (250 km with images produced to 220 km) is similar to the other network radars.

one fifth the number in Oklahoma. The Oklahoma and Colorado tornado climatology statistics represent maxima for the United States and results from these areas are somewhat biased towards the extreme cases. What is not indicated by these statistics are the different thunderstorm type climatologies. Supercell or single cell thunderstorms are rare in southern Ontario. Thunderstorms are mostly multi-cellular associated with squall lines or frontal systems. To date, most published results are derived from research projects which are associated with highly trained personnel (2). While these projects showed promising results, the question was asked whether such promising results could be applied both in a weather regime where severe weather is not as dominant and in a fully operational setting. Another question was the investigation of non-summer severe weather uses of a Doppler radar. Most of the Doppler radar research projects in the U.S. are summertime projects with very little experience in the wintertime.

A replacement for an existing Curtis-Wright radar, the King Doppler radar was purchased from Enterprise Electronic Corporation and installed in late 1984. It is a 5 cm system, consistent with the rest of the Canadian network with a magnetron based transmitter. Table I describes the operating characteristics of the radar system. At the time of purchase, this system was the only commercially available equipment.

The choice of a 5 cm wavelength and the coherent on receive transmitter system was somewhat controversial. Not only does the 5 cm suffer from more severe attenuation but also from a smaller Nyquist velocity interval. This latter issue requires a real-time technique to unfold the velocity data and a robust technique to handle velocity derivatives for mesocyclone or shear detection. The magnetron system is a coherent on receive signal; that is, the phase of the transmit pulses vary from pulse to pulse. Therefore, the phase of each transmit pulse must be recorded and compared with the phase of the received pulse in order to determine phase differences and hence target velocity. There is some phase jitter or noise within a pulse which adds some uncertainty to the determination of the target velocity. Questions posed were:

- Is a 5 cm coherent on receive Doppler system operationally adequate?

- Can reliable real-time velocity unfolding be accomplished?

- How does the system perform with respect to shear detection; that is, do errors in the

Table I: Radar General Characteristics

Parameter	Units	Conventional	Doppler
Frequency	MHz	5625	5625
Wavelength	cm	5.33	5.33
Peak Power	kW	260	260
Pulse Duration	μs	2.0	0.5
Pulse Length	m	600	150
Range Resolution	m	300	75
PRF	pps	250	892,1190 (650)*
Scanning Rate	rpm	6.0	0.75 (2.)*
Reflector Dia.	m	6.1 parabolic	
Polarization		linear horizontal	
Gain	dB	48	
Beamwidth		0.65^o	
Radome		Fiberglass laminate	

* long range mode only

unfolding algorithm generate an unacceptably high false alarm rate?

3 Scanning Strategy and Products

The radar has a dual purpose. It is a research as well as an operational facility. The agreement with the regional operational forecast office was that the radar would provide a service at least equivalent to one of the network conventional radars. New conventional or Doppler products would be supplied under a development umbrella. To accomplish this, the scanning strategy consists of alternating 5 minute cycles. Five minutes are devoted to 24 elevation conventional scanning and five minutes are devoted to Doppler scanning. The conventional scan mode is equivalent to the newer conventional network radars where three-dimensional scanning is implemented to produce Constant Altitude Plan Position Indicator products as well as Echo Top products. The older network radar produce low level PPIs only. Nominally, four Doppler scans are made at 0.5^o, 1.4^o and 3.5^o and 0.4^o. These are operator controllable and the fourth scan is sometimes altered to a vertical scan when weather passes overhead.

Table II outlines the products that are generated every ten minutes and transmitted to the regional weather forecast office. The processing system which acquires, processes the data and generates these products is described elsewhere (1).

The Doppler products are displayed at the weather office by an inexpensive high resolution colour display system. The heart of the system is the Number Nine SGT PLUS graphics card sitting in a COMPAQ 386/20 host microcomputer. Display is on a NEC Multisync monitor. The system is capable of storing at least 120 separate images equivalent to 2 hours of 10 products generated every 10 minutes, displaying 4 images simultaneously, zooming and animation. However, the system is capable of displaying any of the ten products easily. Operational experience has demonstrated that it is absolutely essential to display multiple images simultaneously. For example, the radial velocity patterns are ambiguous and require the correlation with the reflectivity images. Animation is required to correlate images from

Product	Comments
1	0.5° reflectivity
2	0.5° radial velocity
3	1.4° radial velocity
4	3.5° radial velocity
5	0.4° reflectivity long range
6	0.4° radial velocity long range
7	automatic mesocyclone detection
8	automatic microburst detection
9	automatic gust front detection
10	0.5° spectral width
11	1.5 km CAPPI reflectivity
12	MAX reflectivity
13	Echo Tops
14	Severe Storm Map

time frame to time frame and for viewing the temporal evolution of the systems.

4 Operational Uses

Although the benefits of Doppler radar as a tool for observing summer severe convective storms are well documented, a meteorologist knowledgeably utilizing Doppler radar imagery will be able to provide better short range forecasts in other seasons and synoptic situations as well. This section will provide three examples of the usefulness of Doppler radar in the Canadian climate regime.

Weak Tornado Case

On September 17, 1988, a small tornado was confirmed to touch down briefly at about 2130Z in a densely populated area near Pearson International Airport, Toronto (4). The importance of this case was the development of a tornado in a situation where from synoptic information alone none was expected and the undetectability of the severity from conventional radar products.

The damage, confined to a few industrial buildings, classified the tornado as F0 on the Fujita scale. There were several sitings and photos of the tornado.

The conventional radar showed small weak isolated storms, occasionally with high reflectivity cores (>50dbZ). The echo tops were relatively low, the maximum observed was in the range 8 to 10 km and high reflectivities (>40dbZ) were confined to below 5.5 km. These two features suggested non- severe weather.

Using Doppler data, a mesocyclone was identified by the detection algorithm beginning at 2030Z and continuing to 2130Z (3). Figure 3 shows the temporal evolution of the mesocyclone as described by the azimuthal shear ($s = v_r/R$) where v_r and R are the estimated rotational velocity and radius of the mesocyclone. Angular momentum ($p = v_r \times R$, not shown) showed a maximum prior to the time of maximum shear and prior to tornado touch down similar to cases from Oklahoma (6). Without this Doppler information, the storm would not have been identified as being severe.

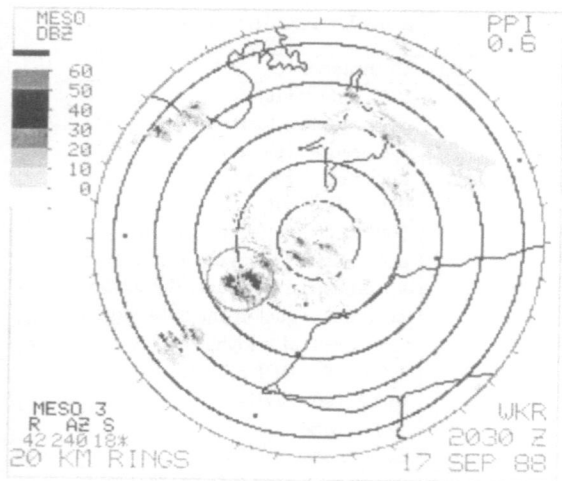

Figure 2: Mesocyclone product indicating the location of a mesocylone that produced an F0 tornado

The importance of this figure is that a mesocyclone producing an F0 tornado can be identified very early in its development. However, many non-mesocyclonic shears can be identified and it is imperative to discriminate these different types of shears. The non-mesocyclonic shears can be due either to noisy data or to other meteorological features such as the inflow-gust front area of a severe storm. Discrimination is done by associating the location of the shear with respect to the thunderstorm cell identified on the reflectivity maps as well as using feature continuity in height and time. For example, shears not associated with any thunderstorm cell or shears located at the edge of a cell and only on the lowest elevation scan are not likely to be mesocyclones. Shears which are correlated with height and time and first observed at mid-levels of the storm likely indicate mesocyclones.

Once mesocyclonic shear has been identified, then the nowcaster must discriminate the tornadic from the non-tornadic mesocyclone. This discrimination is accomplished by quantitative information regarding the shears and angular momentum of the mesocyclone. If thresholds in these values are exceeded then the mesocyclone is classified as tornadic. At the moment, the technique is far from perfect but it is very promising and will improve as more data is collected from tornadic cases.

Low Level Jet

Rear inflow jets have often been observed prior to the bowing out of squall lines and prior to tornado development. One such frequently occurring example is the interaction of a low level wind maximum with a convective line which results in rapid, short- lived changes in the severity of convective weather. In several of the cases studied, a relatively small wind maximum, usually descending, approaches a convective line from the rear. As the higher winds move within a few kilometres of the line, the convection develops rapidly often resulting in heavy showers and damaging winds at the surface. As the wind maximum passes the convective line, the weather weakens, often to the point of all precipitation dissipating.

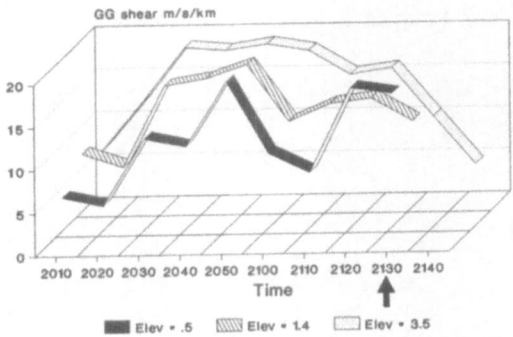

Figure 3: Temporal evolution of a F0 tornadic mesocylone. Arrow indicates tornado touch-down.

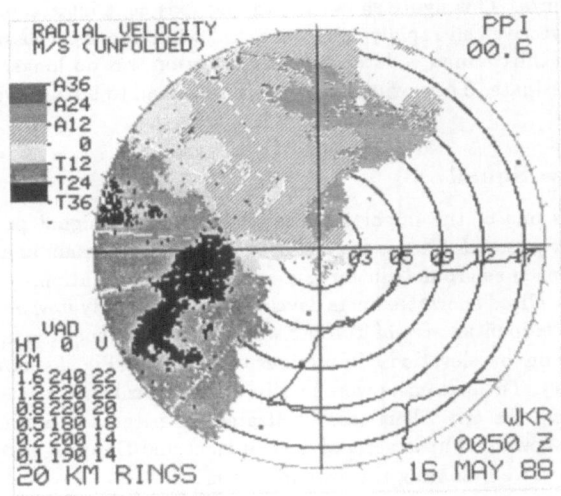

Figure 4: Radial Velocity Map with inflow jets.

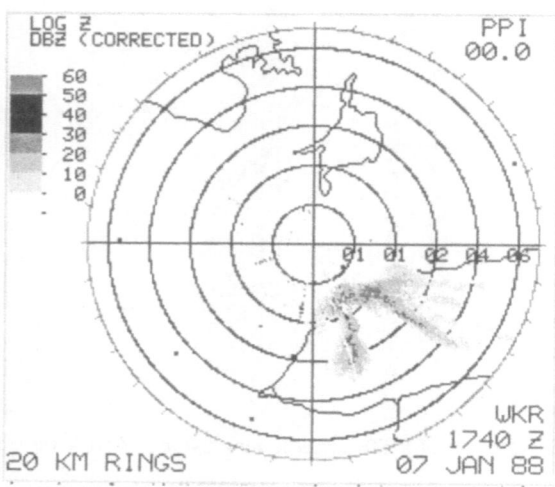

Figure 5: Low level snow storm.

Figure 4 is a radial velocity image portraying such a case. Thirty minutes prior to this image, a wind maximum first appeared on the west-southwestern edge of the image with a speed of at least 30m/s. This figure shows the the jet axis as it intersects a convective line. At this point a convective cell rapidly intensified and wind damage (possibly tornadic) was reported. A further thirty minutes later, severe convection was no longer observed and the weather began to dissipate. These wind features are too small to be observed on the synoptic network.

Lake Effect Snow Squall

This case illustrates one of the important spinoffs of Doppler signal processing for using reflectivity data. A zero notch filter is applied in the frequency domain and is used to remove non-moving targets such as ground clutter and anomalous propagation.

On this day, localized snow streamers developed in a southerly flow off Lake Ontario and travelled over the metropolitan area of Toronto where weather echoes are normally embedded in the ground clutter on low elevation PPI pictures. A 0.0° PPI (Fig. 5) showed two streamers converging on the city. The streamers were localized and contributed to two 50 car accidents on a highway crossing the city. This case illustrates the necessity to scan at low levels to identify heavy snowshowers of little vertical development and Doppler provides the means of isolating the weather by eliminating the ground clutter.

5   SUMMARY AND CONCLUSION

A C-band radar with a magnetron transmitter and a limited Nyquist interval is viable for the detection of severe weather in the existing climatology of the region. Severe high reflectivity storms can cause attenuation that will distort the reflectivity pattern of other storms but good velocity information can still be obtained when the signal is attenuated. There are very few times when the signal is completely missing or reduced to the extent that a second storm would be completely missed. The coherent-on-receive Doppler system has proven to be viable and with real-time velocity unfolding to 48m/s and real-time shear calculations it has been successfully used for severe storm detection. Simultaneous multi-displays are absolutely necessary for the severe storm forecaster .

The identification of the 'weak' F0 to F2 class tornadoes in southern Ontario is important as they represent a significant number of the damaging storms in the region. These can be classified as 'marginal' tornadic storms. Their identification is important in this climate regime in order to maintain public confidence in the weather warning system. Preliminary statistics indicate an improvement in initial forecast accuracy from 30% to 50% (5). Further improvements are occurring with products refinements and forecaster educations and experience.

Doppler radar provides wind data on the mesoscale (10-100km) which have not been routinely available. Rapidly moving jet cores, and other observed cases not portrayed here such as mesoscale convective systems and winter frontal weather have provided examples of complex wind flow patterns that lack an adequate conceptual description and illustrate the new and exciting information that is now routinely collected year round.

The zero velocity notching or the removal of non-moving targets is an important aspect of a Doppler radar. It allows low level weather to be observed in a high clutter environment. Echoes due to anomalous propagation and ground clutter can be removed. This has implications not only for the weather forecaster in enhancing the ability to identify weather but also for the hydrologist who will have the benefit of a clutter free rainfall accumulation map.

There are compromises and limits inherent in any radar and display system that are imposed by the radar, the meteorology, processing technology and costs, and in addition there are limits to the data that can be absorbed by the user in real time. The King Doppler radar system, providing a variety of displays and a modest amount of interaction at reasonable cost, has been used with great enthusiasm by operational forecasters even as it has evolved. The operational forecasters do not want more products to examine or more computer systems generating images with which to interact; they want fewer products that have greater significance. A goal of the King radar research is to summarize the pertinent information from the various algorithms and images in as few products as possible. Artificial intelligence will play a key role in this respect.

6   REFERENCES

(1) Crozier,C.L., P.I. Joe, J.W. Scott, H.N. Herscovitch and T.R. Nichols, 1989. First experiences with an operational Doppler Radar, 14th AMS Radar Conference, Tallahassee, Florida, 179-185.

(2) JDOP Staff, 1979. Final report of the Joint Doppler Operational Project 1976-1978, NOAA Tech. Memo. ERL NSSL-86, 84 pages.

(3) Joe, P.I. and C.L. Crozier, 1988. Evolution of mesocyclonic circulation in severe storms, 10th International Cloud Physics Conference, Bad Homburg, F.R.G., 687-691.

(4) Hogue, R., P.I. Joe, and T.R. Nichols, 1989. The 17 September 1988 Weak Tornadic Storm in Metro Toronto, Ontario Region Technical Notes, Atmospheric Environment Service, 17 pages.

(5) Leduc, M.L. and P.I. Joe, 1987. Use of a 5 cm Doppler radar to detect and forecast severe thunderstorms in a real- time forecast operation, IUGG, Vancouver, 49-54.

(6) Zrnic, D.S., D.W.Burgess and L.Hennington, 1985. Automatic detection of mesocyclonic shear with Doppler radar. J. Atmos. Oceanic. Technol., 2, 425-438.

A DOPPLER APPLIANCE TO MRL-5

R. Petrov
Hydrometeorological Service, Bulgaria
D. Stoyanov, A. Savchenko and L. Mladenov
Institute of Electronics, Bulgaria

Summary

The purpose set in the present study is the development of Doppler
appliance to the existing radars MRL-5. That will enrich the possibilities of those two-wave radars so as to obtain at the same time the
information on the structure of clouds and precipitations and the
information on the dynamics of cloud processes. The Doppler appliance
developed is the basic section of the Doppler channel under elaboration which is to be applied to MRL-5. The meteorological radars MRL-5
are the basic ones in the network of radars in the socialist countries.
The system "STALO-COHO" represents a stable oscillator, a coherent
oscillator and is selected as the most successful variant of Doppler
appliance.
The Doppler appliance design is composed by three sections :
- coherent oscillator
- control phase-shifter
- quadrature phase detector
The general conclusion drawn from the laboratory experiment carried
out indicates that all elements of the Doppler appliance are reliable
enough in the course of their functioning, with high stability of parameters.

1. INTRODUCTION

The use of Doppler (impulse-coherent) radars for meteorological purposes is quite promising as they provide a possibility to define directly
the hydrometeors flight and consequently the horizontal and vertical components of air flows. In particular the medium frequency of the Doppler
spectrum is proportional to the mean radical velocity of scatters in the
range that is specified by the impulse.

The purpose set in the present study is the development of a Doppler
appliance to the existing MRL-5. That will enrich the possibilities of
those modern radars so as to obtain at the same time information on the
structure of clouds and precipitations and information on the dynamics of
the cloud and precipitation processes.

It is obvious from the survey made (1,2,3,4) that the best variant is
that one which is used by the system STALO-COHO. The same could be applied
to all non-coherent radars with magnetronic generators such as MRL-5. As
to memorize the phase of the impulse emitted, the application of a coherent

oscillator (COHO) is needed that functions with the radar intermediate frequency as to get better accuracy and for the sake of convenience. For that reason it is necessary to use also a stable oscillator (STALO) as to set up a background frequency for its mixing with the signal received.

The Doppler frequency processing goes on to the best advantage by means of computer since in this way could be assessed the moments of the Doppler spectrum.

2. DOPPLER APPLIANCE

In Fig.1 is shown the summarized block diagram of the developed Doppler appliance. The main unit in it is the coherent oscillator which is started by the MRL-5 synchronizer and produces a radioimpulse, the phase of which coincide with the phase of the emitted sounding impulse. The latter is derived from the system for automatic coplementary (control) frequency adjustment. The coherent radioimpulse of long duration, worked out by the oscillator, is to be passed to a control phase-shifter which removes in some direction the radioimpulse phase. In that way is introduced an accurate additional frequency shifting that enables to define the direction of motion and to avoid the effect of Gibbs in converting the signals and their analysis into digits in the computer. The frequency and the direction of shifting are set by separate generator.

Further the signal from the phase-shifter is dephased to 90° and along with the non-shifted signal they are passed respectively into two phasing detectors. At their second inputs the radar signal is passed to that is received from the intermediate frequency amplifier. In that way a quadrature phase detector is produced which enables to perform optimum processing of the signals received.

The Doppler appliance design takes form in three sections, installed in a special box. The appliance front and rear panels where are placed all inputs and outputs are shown respectively in Fig. 2a and 2b.

2.1. COHERENT OSCILLATOR

To the "start-pulse input"(SP) (Fig. 2a, pos. 11) an impulse with random polarity and with higher than 1 V amplitude is run into. The start-pulse polarity is selected from pos. 13 (Fig. 2b). The start-pulse level is controlled by resistor - pos. 12 (Fig. 2b).To the input "PhA" (phase adjustment) - pos. 7, a radioimpulse is passed to the coherent oscillator. The radioimpulse amplitude - the signal from the magnetron, should be of 1-3 V. From pos. 8 - "output BS", the background signal is obtained that is submitted to phase modulation in the second section - the control phase-shifter (CPhS) and it is passed to pos. 4 - "input of CPhS". From pos. 3 the modulated background signal (MBS) is got that feeds the phase detector (PhD).

2.2. CONTROL PHASE-SHIFTER (MODULATOR)

The frequency of phase modulation is controlled by pos. 15 within the boundaries as from 205 Hz up to 770 Hz. The direction of phase modulation (the sign of the frequency supplement to BS) is picked out from pos. 16 + ΔF. As to discontinue the phase modulation process , it is necessary to place the key - pos. 17, in state "STOP".

2.3. PHASE DETECTOR

The phase detector has two inputs - a background signal - pos. 1 and a received signal (RS) - pos. 2. To them are connected respectively the background signal from pos. 3 and the received signal from the MRL-5 intermediate frequency amplifier respectively. At the phase detector output two components of the Doppler signal are got (dephased to 90°) - SIN component (pos. 5 and 6) and COS component (pos. 9 and 10).

3. LABORATORY TEST OF THE DOPPLER APPLIANCE

A laboratory test has been carried out to test the normal functioning of the separate units and their reliability during the operation.

3.1. CHECK-UP OF THE COHERENT OSCILLATOR (COHO)

In passing a signal to pos. 11 "input SP" from impulse generator, the level (pos. 12) is selected so as the signal shown in Fig. 3 is observed at the "output BS". The duration of generation of COHO is 700μs (about 100 km). When the COHO operates under continious rate of work (it is not in phased from outside), its stability is measured - a deviation of 10^{-5}. After passing a phasing radioimpulse (PhA) for adjustment to pos. 7, at the "output BS" is obtained a signal which is displayed in Fig. 3d - "BS".

3.2. OPERATION OF THE PHASE MODULATOR AND THE PHASE DETECTOR

The check-up is performed jointly with the control over the coherent oscillator. In this case to pos. 1 and 2 is connected the control line on which runs unmodulated background signal. In case of discontinued modulation (pos. 17 in state "STOP") the signal "output PhD" is under examination at the output of the phase detector (Fig.3e). As the background signal is compared with itself by phase, at the outputs (I and Q) appears a rectangular impulse - "OPhD". In switching on the phase modulation (PhM) by pos. 17, the signal to pos. 2 is phase-modulated (PhM) compared to that one from the "control line", therefore the plateau of the impulse (outputs I and Q) starts to fluctuate according to the law of phase modulation - "output PhD with phase modulation (PhM)" (Fig.3f) - "OPhD with PhM".

Pos. 14 enables the coherent oscillator to be in phased by the signal, received from any selected meteo-target (clouds and precipitations).

The general conclusion drawn from the laboratory experiment performed is that all Doppler appliance units operate reliably with high stability of parameters that corresponds to the needed requirements and the error derived in defining the Doppler frequency up to 100 km do not exceed 0,5 Hz.

4. CONNECTION OF THE DOPPLER APPLIANCE TO MRL-5

The preparation so as to connect the Doppler appliance to MRL-5 is to be effected quickly without accomplishment of any complicated procedures.

The start-pulse impulses to the coherent oscillator are the anticipating impulse from MRL-5 (1') that outdistances the basic impulse (1) by 25μs. As to initiate COHO, a scheme for retention is elaborated which allows (1') to be approximate to 5 s from the basic impulse (1). The amplitude is 3,5 V.

The magnetron phasing impulses to the coherent oscillator "PhA" are picked up from the second cascade of the intermediate frequency amplifier of the system for automatic complementary (control) frequency adjustment. The impulses frequency and amplitude are 30 MHz and 2 V respectively.

The return signals to the phase detector are picked up from the intermediate frequency amplifier of the receiver - 30 MHz with amplitude of 0,8 up to 5,6 V.

The tests showed that the feeding from MRL-5 do not hold out against the capacity of the appliance. For that reason it was necessary to use suitable rectifiers.

Several signals are received from some meteo-targets but as a system for processing is missing, they are not still processed.

5. CONCLUSION

The worked out Doppler appliance has large potentialities and could be switched on optionally to the First or the Second MRL-5 channels.The investigations which have been completed indicated that so as to establish in full a Doppler channel to MRL-5 it is necessary the existing oscillators to be replaced by much more stable oscillators as well as a system for Doppler signal processing to be elaborated.

At present a stable oscillator for the First and the Second MRL-5 channels is already developed - stability of 10^{-7}.The system for Doppler signal processing as well as a synchronizer of repetition frequency of 1000 Hz are under elaboration.

REFERENCES

(1) ABSHAEV, M.T. et al. (1976). Methods and devices of the Doppler measurements of vertical flows in thunderstorm and hailstorm clouds.VGI, Vol. 31.
(2) BATTAM, D. (1973). Radar observation of the atmosphere. The University of Chicago Press.
(3) DOVIAK, R.J. (1979). Doppler Weather radar. Proc. of the IEEE, Vol. 67, No. II.
(4) DOVIAK, R.J. and ZRNIC, D. (1984). Doppler radar and weather observations. Academic Press, Inc, Orlando, Florida 32887.

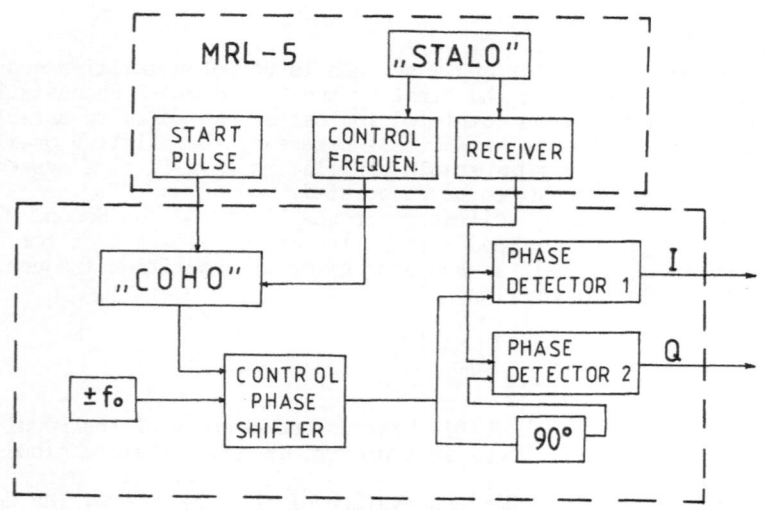

Fig.1. The Doppler appliance block diagram
connected to MRL-5

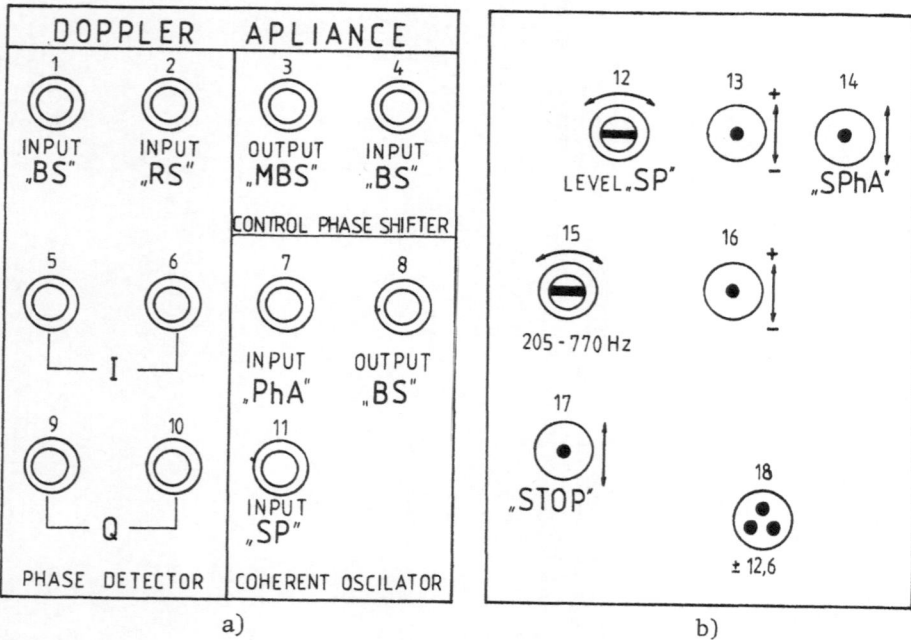

a)   b)

Fig.2. Front and rear panel of the Doppler appliance

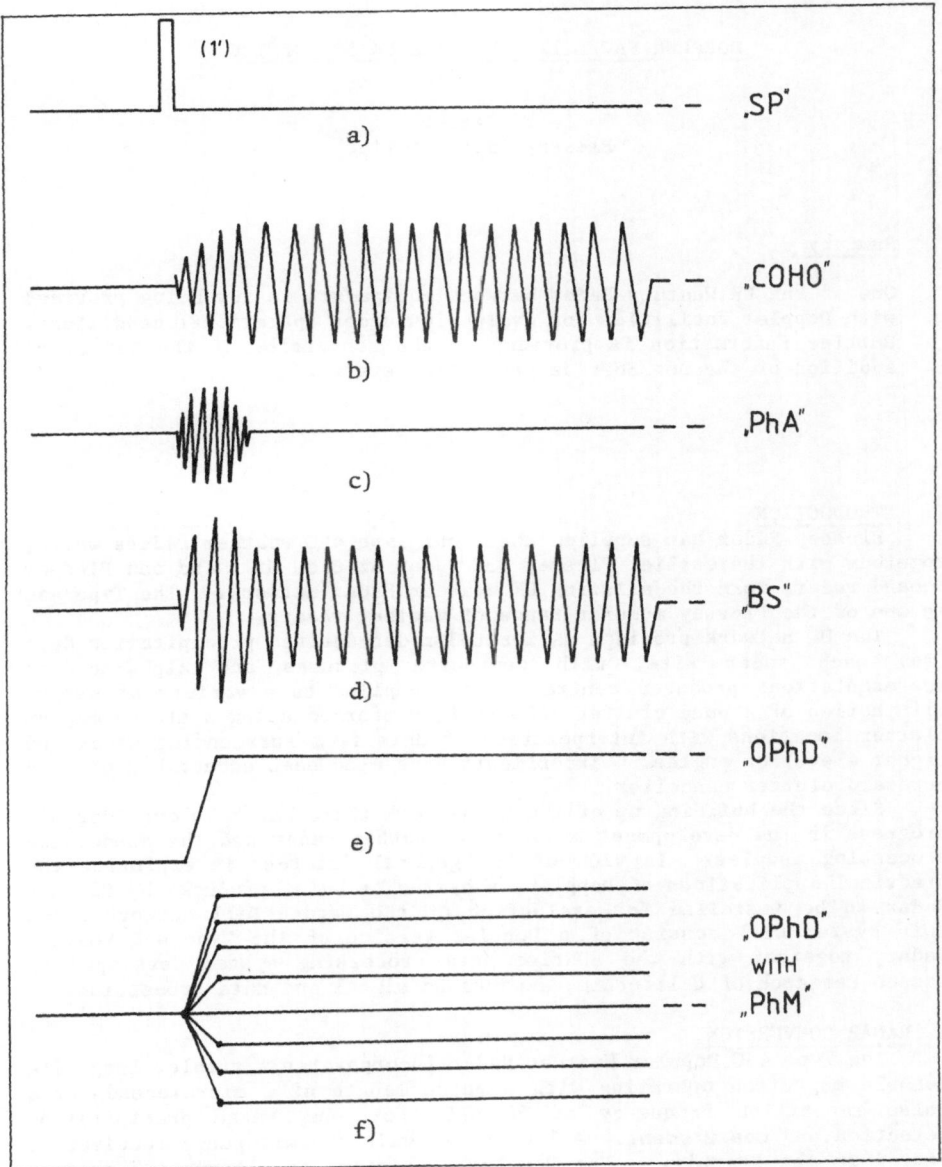

Fig.3. Oscillogram got by laboratory test

DOPPLER FACILITIES FOR THE UK MET OFFICE

P J Bacon
Plessey Radar Limited

Summary

One of the UK Weather Radar Network equipments is now being provided
with Doppler facilities for evaluation under operational conditions.
Outline information is provided on the conversion of the radar and
addition of the new SUNrise processing system.

1. INTRODUCTION

Plessey Radar has supplied the eight Type 45C weather radars which,
together with the earlier Plessey 43C radar at Clee Hill and one Plessey
S-band radar, form the existing UK Weather Radar Network. The Type 45C
is one of the Plessey modular range of weather radars.

The UK network provides calibrated radar-derived precipitation data
from each radar site, with composite pictures and alpha-numeric
representations produced centrally and supplied to a variety of users.
Elimination of ground clutter effects is performed using a stored map of
clutter locations with interpolation of data from surrounding areas and
higher elevation angles. Experiments have also been undertaken using a
hardware clutter canceller.

Since the building up of the UK network there has been considerable
progress in the development of Doppler weather radar and the associated
processing modules. In view of the general interest in exploring the
practical applications of Doppler, a system is being produced by Plessey
Radar to be installed for evaluation at the Warden Hill network site.
This system will consist of a Doppler version of the Type 45C Weather
Radar, together with the SUNrise data processing system developed by
Lassen Research of California, experts in signal and data processing.

2. RADAR CONVERSION

The Type 45C Doppler Weather Radar incorporates a stable, long life
tunable magnetron operating with a pulse length of 2 microseconds at a
pulse repetition frequency of 300 PPS for long range precipitation
detection and measurement. A low noise, wide dynamic range receiver is
fitted in the RF unit. For Doppler operation, an alternative shorter
(0.5 microsec) pulse length at higher (variable) PRF is also provided.
Stable and coherent local oscillators are incorporated. The coherent
local oscillator is connected to a phase sensitive detector which
provides in-phase and quadrature video outputs in addition to the normal
logarithmic video output.

The scanner control unit digitises the azimuth and elevation
resolver outputs and provides an RS232 interface for the transmission of
the angle data to, and receipt of control signals from, the Doppler
processor.

To minimise downtime on site the conversion will be carried out by the provision of factory built replacement units.

3. SUNrise DATA PROCESSING SYSTEM

The core of the SUNrise system is contained in a single rack which includes the host and product computers with 12 MB and 16 MB RAMs, a data bus, disc control units, signal processor and antenna control unit, tape drive, colour display board, and communications interfaces. In addition, the total system includes one or more remote colour workstations.

The SUNrise system is capable of full radar control. During the evaluation programme it will permit the 45C to operate in its normal non-Doppler network mode when required. The system will generate a wide range of data products for immediate use or archiving. Most data products are generated in high resolution cylindrical co-ordinates so that further data products can be derived from them. The product range includes PPI and RHI displays of precipitation intensity, velocity and spectrum width, ground clutter free precipitation rate and accumulation plan displays, precipitation totals for catchment areas, CAPPI type displays of intensity, velocity and spectrum width, echo tops and vertical cross-section displays, vertically integrated reflectivity and velocity-azimuth displays, storm centre tracking, time lapse facilities and a range of standard overlays.

The SUNrise system has the capability of considerable expansion by the addition of extra communication interfaces, colour printer, additional memory and archiving facilities, remote displays, graphics generation and special software developments.

A reduced facility version of SUNrise designated RADEX is also available.

4. THE FUTURE

Following incorporation of the Doppler facility at Warden Hill, an extensive evaluation programme will commence. This is intended to establish the practical value of the Doppler capability and the additional data products now available in an operational environment. Plessey Radar will support the UK Met Office throughout the evaluation period.

BRIGHT BAND ERRORS IN RADAR ESTIMATES OF RAINFALL:
IDENTIFICATION AND CORRECTION USING POLARIZATION DIVERSITY

I J Caylor, J W F Goddard[*], S E Hopper, and A J Illingworth
Dept of Physics, UMIST, Manchester, M60 1QD, UK
[*] RAL, Chilton, Didcot, OXON, OX11 0QX, UK

Summary
    A principle source of error in rainfall rates derived from the
radar reflectivity (Z) is caused by the enhanced radar return due to
melting snowflakes in the bright band. We present observations of
two S-band polarization radar parameters which enable the bright band
to be simply identified. The technique involves transmission of
horizontally and vertically polarized radiation and reception of the
co-polar and cross-polar return signals. The linear depolarization
ratio (LDR) is defined as the ratio of the cross-polar to the co-
polar return and p(H,V) as the correlation of the time series of the
horizontal and vertical co-polar return. The high values of Z in the
bright band are accompanied by values of LDR above -18dB and p(H,V)
below 0.8. In echoes where no bright band is present the values of
LDR are everywhere below -20dB and the correlation is always close to
unity. We discuss potential problems in implementing these
techniques for C-band radars with one degree beamwidths, and also
consider how they could be used to identify spurious echoes from
ground clutter and anomalous propagation.

1. INTRODUCTION
    A major source of error in deriving the rainfall rate (R) from the radar
reflectivity (Z) arises from the enhanced radar return occurring in the
melting layer or 'bright band'. The value of Z typically increases by
10dB when low density snowflakes become wet and scatter microwaves as if
they were giant raindrops. This error in Z in an empirical Z-R
relationship would lead to a fivefold overestimate of the rainfall.

    Most radar networks scan in PPI mode to obtain a complete spatial
coverage, and in truly stratiform rain the bright band should be
recognisable as a concentric ring of enhanced reflectivity centred on the
radar. Smith (1) has suggested an automatic means of identifying the
bright band by comparing the range of the maximum values of Z at a
particular azimuth for two different beam elevations. However, in practice
the height of the bright band can change, the precipitation is never
stratiform and, in the UK at least, quite vigorous showers often have
bright bands. Ground based rain gauges can provide localised real-time
information for correcting bright band errors (2). In this paper we
demonstrate a means of uniquely identifying the bright band using two new
polarization parameters. We shall also consider how such techniques could
be implemented on C-band radars with smaller antennas.

2. THE CHILBOLTON POLARIZATION RADAR

The Chilbolton radar operates at 3-band and, with a 25m dish, is the largest steerable meteorological radar in the world, having a beamwidth of only a quarter of a degree. Earlier reports (3,4,5) have considered the implementation and interpretation of differential reflectivity (ZDR), but here we discuss observations made in 1988 of two new parameters (LDR and p(H,V)) which are described below.

2.1 DIFFERENTIAL REFLECTIVITY

The differential reflectivity (ZDR) provides an estimate of mean hydrometeor shape. It is defined as

$$ZDR = 10 \log(ZH/ZV) \tag{1}$$

where ZH and ZV are the radar reflectivities measured at horizontal and vertical polarizations respectively. For small raindrops or tumbling ice particles, ZH and ZV are equal and ZDR is zero. Positive values of ZDR occur for oblate particles of high dielectric constant when ZH exceeds ZV. In heavier rain ZDR is positive and reflects the mean shape (and hence the size) of the raindrops. The ZDR of ice is more complex (5). Because of the low dielectric constant, dry snowflakes have a ZDR close to zero, but wet snowflakes can have high positive values. Graupel tends to tumble and so be associated with a zero ZDR value.

2.2 LINEAR DEPOLARIZATION RATIO

The linear depolarization ratio, LDR, is a measure of the hydrometeor fall mode and appears to be an excellent indicator of wet ice. LDR is defined as:

$$LDR = 10 \log(ZVH/ZH) \tag{2}$$

where ZVH is the (horizontal) cross-polar return from a vertically polarised transmitted pulse, and ZH (as in Equation 1) is the co-polar (horizontal) return for horizontally polarised transmission.

A cross polar return occurs only when oblate hydrometeors fall with their major or minor axis at an angle to the vertical. Computations of LDR for tumbling oblate spheroids are plotted in Figure 1 and are found to be consistent with the Chilbolton observations (6). Snowflakes have such a low dielectric constant that even if they are very oblate their LDR is below the antenna limit of -32dB, oblate dry hail or graupel could have a value up to -20dB if the axial ratio were as low as 0.5, but LDR values above -20dB can only realistically occur for wet tumbling ice particles. Such high values are restricted to the bright band. Raindrops give rise to a very low cross-polar return.

2.3 CO-POLAR CROSS CORRELATION

The estimates of ZH and ZV in equation (1) are made from the true linear average (over 210msec) of the return at one 75m gate from 64 successive pulse pairs transmitted with alternate horizontal and vertical polarization. The standard error in ZDR is reduced to 0.1dB by spatial averaging over four adjacent 75m gates. The co-polar cross correlation (p(H,V)) is the correlation of these two time series in ZH and ZV.

Observations (7) show that the correlation is generally close to unity in rain and dry ice, with low values being confined to the bright band. Low values of correlation are thought to indicate that a variety of hydrometeor shapes is present. We believe that the low values in the bright band are caused by the coexistence of oblate half-melted snowflakes with nearly spherical raindrops.

3. COMPARISON OF ECHOES WITH AND WITHOUT BRIGHT-BANDS

An RHI through stratiform precipitation is displayed in Figure 2; where the reflectivity does not exceed 30dBZ in the rain but reaches 40dBZ at 2km altitude in the bright band. At this height the oblate melting snowflakes give a clearly visible bright band in ZDR with values reaching 2dB. However automatic recognition of the ZDR bright band can be difficult. ZDR values in the rain at 10-20km range reach 0.5dB, and in heavier rain can be much higher. We also note in Figure 2 the positive ZDR values above 3km altitude; this low Z region presumably containing aligned high density ice crystals.

It is much easier to identify the Z bright band from the LDR data. The maximum values of Z coincide with the peak LDR of about -15dB, which is consistent with wet tumbling snowflakes having an axial ratio of about 0.5 (Figure 1). In contrast LDR values in the rain are near the antenna limit of -32dB, and reach about -27dB in the low Z ice region above 3km where the ZDR indicated high density crystals. Figure 2 also shows that ground clutter results in LDR values above -10dB near to the ground. Because LDR involves measuring the low-power cross-polar return it is much more susceptible to ground clutter than is Z or ZDR. The bright band can also be identified via the correlation parameter. Although data is limited to a 5km range window values of correlation below 90% are restricted to the bright band.

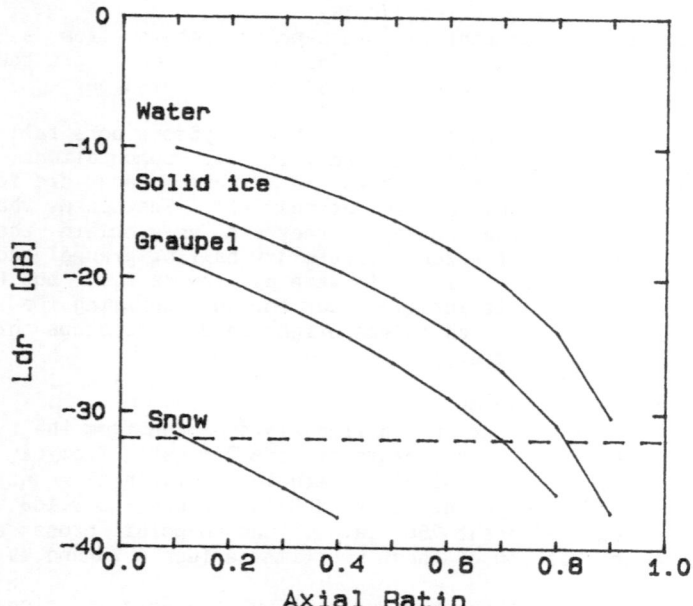

Figure 1. LDR values for randomly tumbling particles as a function of axial ratio. Raindrops do not tumble, but the "water" curve is applicable to wet ice particles.

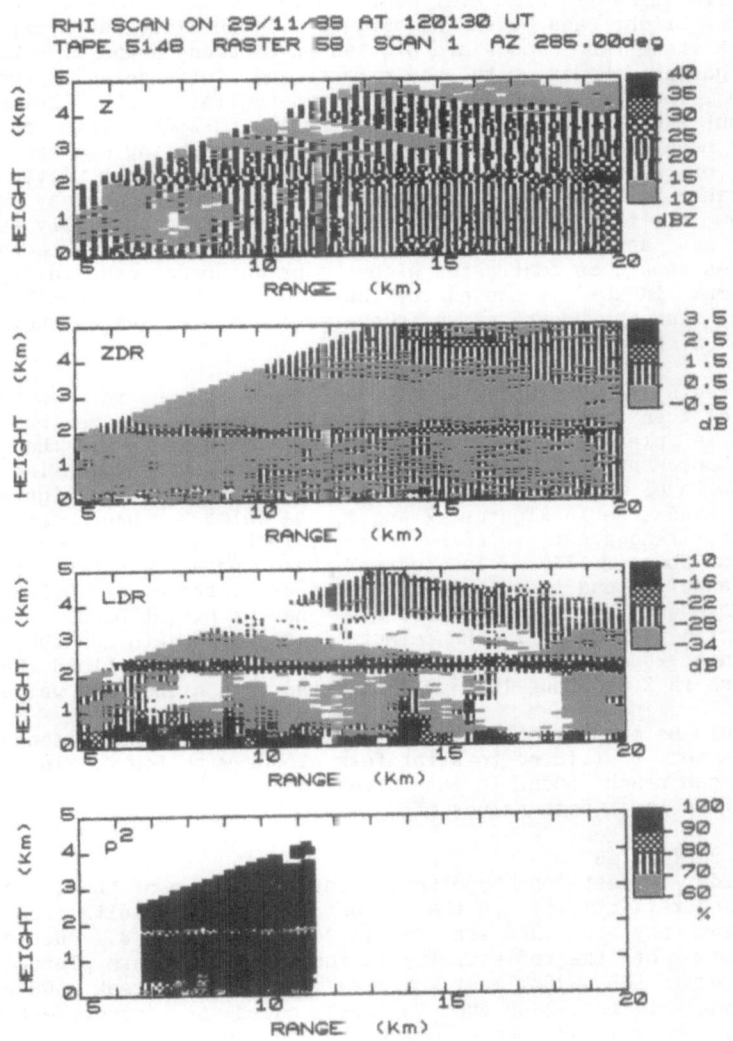

Figure 2. A typical RHI scan through stratiform cloud on 29 November 1988 with a bright band for Z, ZDR, LDR, and p(H,V).

A vigorous shower with no bright band is depicted in Figure 3 and it is clear that the polarization parameters have a quite different character: the correlation is high everywhere while the LDR values are much lower than for the bright band case in Figure 2. The different vertical profiles through these two clouds are plotted in Figures 4 and 5. We believe that the vigorous shower with no bright band in Figures 3 and 5 contains graupel. The dry tumbling graupel gives negligible LDR, but melting occurs at about 2km altitude and LDR rises to about -25dB. This weak "LDR graupel bright band" is consistent (Figure 1) with tumbling wet ice with an axial ratio of 0.8. In the heavy rain the LDR is just detectable and is explicable in terms of a canting angle of about 5 degrees. Values of ZDR are low for the tumbling dry ice, but rise monotonically as the graupel melts and assumes the equilibrium shape of the large raindrops. These profiles should be contrasted with the bright band case in Figure 4, where a maximum in ZDR is caused by the low density wet snowflakes. In the graupel case the correlation is everywhere above 90% as there is no great variety of shapes present.

Figures 6 and 7 display how the value of LDR is related to the enhanced value of Z in the bright band. The altitude of the maximum value in LDR (which is present in all types of cloud) is used to fix the melting level. The enhancement of Z (delta Z) is then estimated by comparing the Z at the melting level with the Z in the rain 500m below. In Figure 6, for the bright band case in Figures 2 and 4, the delta Z enhancement in the bright band for ranges out to 60km is about 10dB, and is accompanied by an LDR value of about -15dB. In Figure 7 these parameters are plotted for both the graupel cloud in Figures 3 and 5 at a range of 10-20km, and for a second shower beyond 40km which does have a bright band. In the graupel clouds the mean delta Z enhancement is close to zero and LDR values are in the range -20 to -25dB, while for the more distant cloud the bright band increase in Z of about 10dB is associated with higher LDR values of -15dB.

It should be emphasised that, in the UK at least, the presence of a bright band is not restricted to stratiform clouds with low Z. In some showers Z values can reach 50dBZ in the bright band, while others, with no bright band, have lower peak values of Z.

4. LDR STATISTICS

In order to test our hypothesis that the value of LDR is related to the increased reflectivity in the bright band, the results from scans on 11 different days in 1988 are summarised in Figure 8. Histograms of the enhancement of the reflectivity in the bright band are plotted for each 4dB increment in LDR. For most vertical profiles the peak LDR values are in the range -14 to -18dB and in these cases the enhancement of Z is, on average, about 10dB. Less common, in this UK sample, are the peak values of LDR in the range -18 to -22dB and -22 to -26dB where the Z enhancement is essentially zero and no bright band is present.

It should be stressed that in 1988 there were no cases of very deep vigorous convection. Measurements of LDR at 3cm (8) suggest that hail in wet growth can give high LDR values, although the measurements at this wavelength are affected by propagation problems. We hope to examine such cases in the future. For the observations discussed in this paper the depth of the bright band is greater than the beamwidth of the Chilbolton radar. In future we will analyze the effect on LDR and correlation

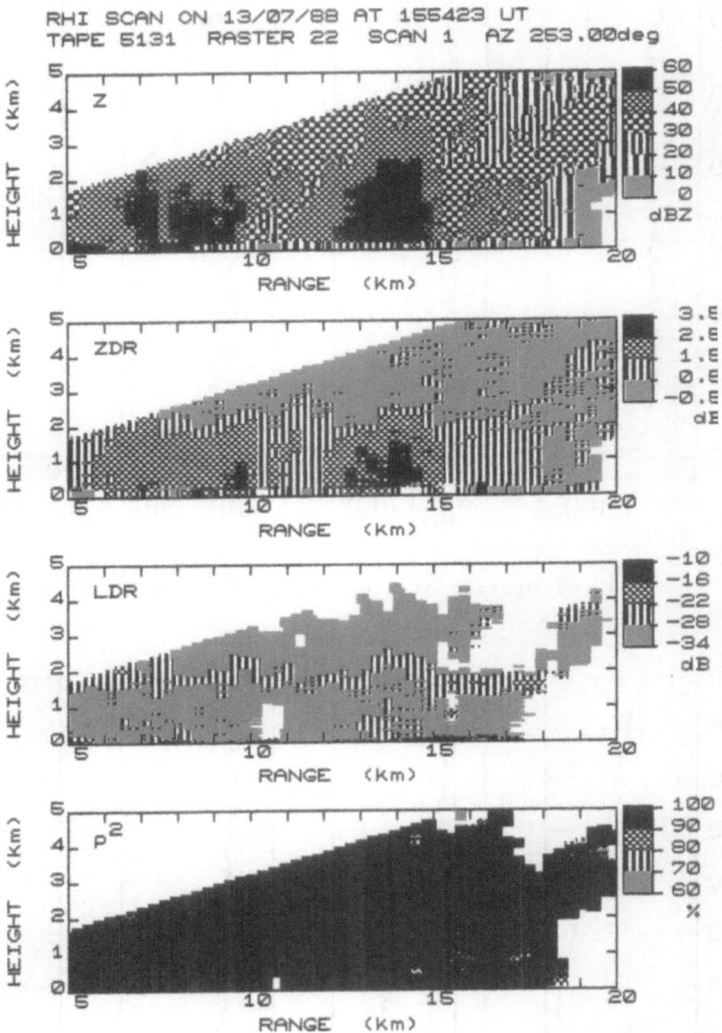

RHI SCAN ON 13/07/88 AT 155423 UT
TAPE 5131   RASTER 22   SCAN 1   AZ 253.00deg

Figure 3. A typical RHI scan through a convective cloud on 13 July 1988
with no bright band for Z, ZDR, LDR and p(H,V).

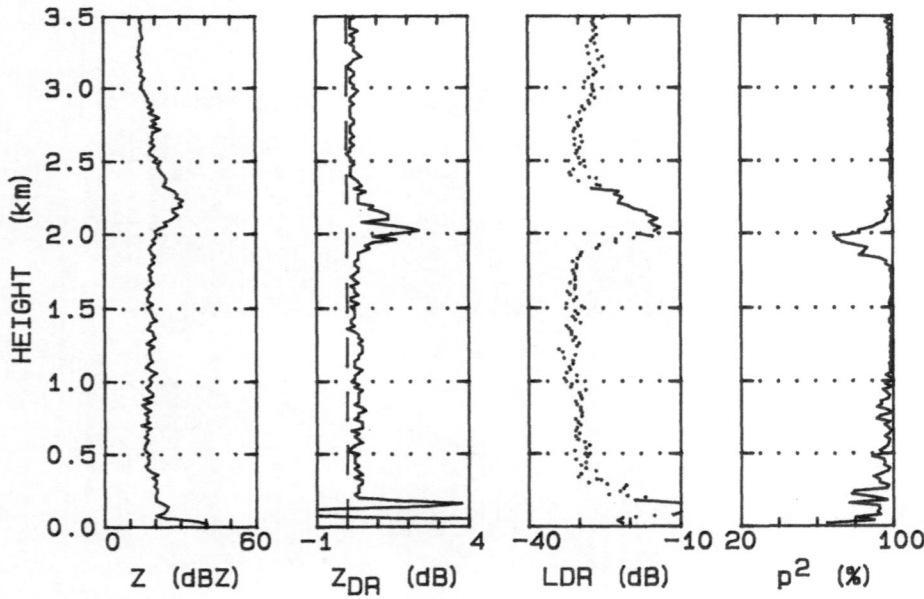

Figure 4. Vertical profile at 9.6km range for the RHI in Figure 2.

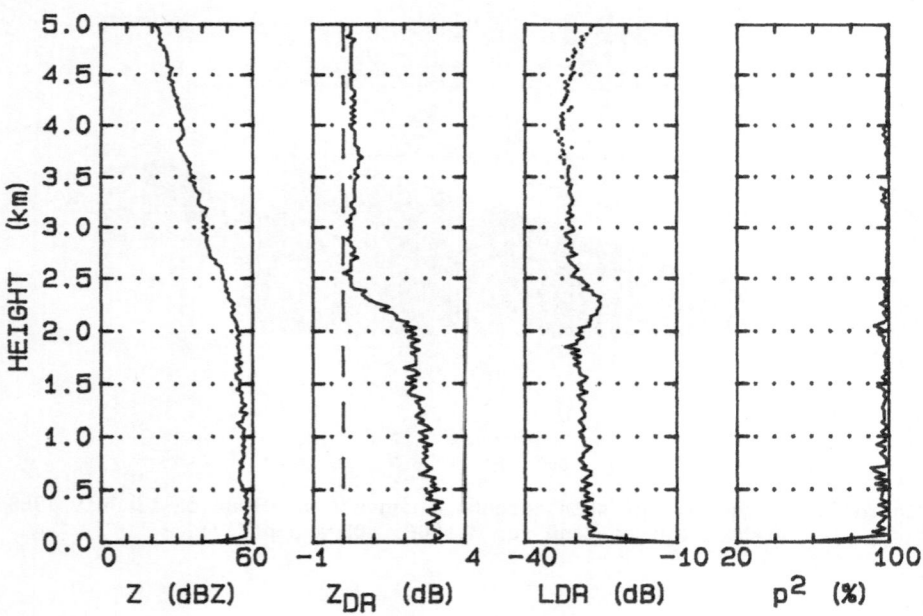

Figure 5. Vertical profile at 13.8km range for the RHI in Figure 3.

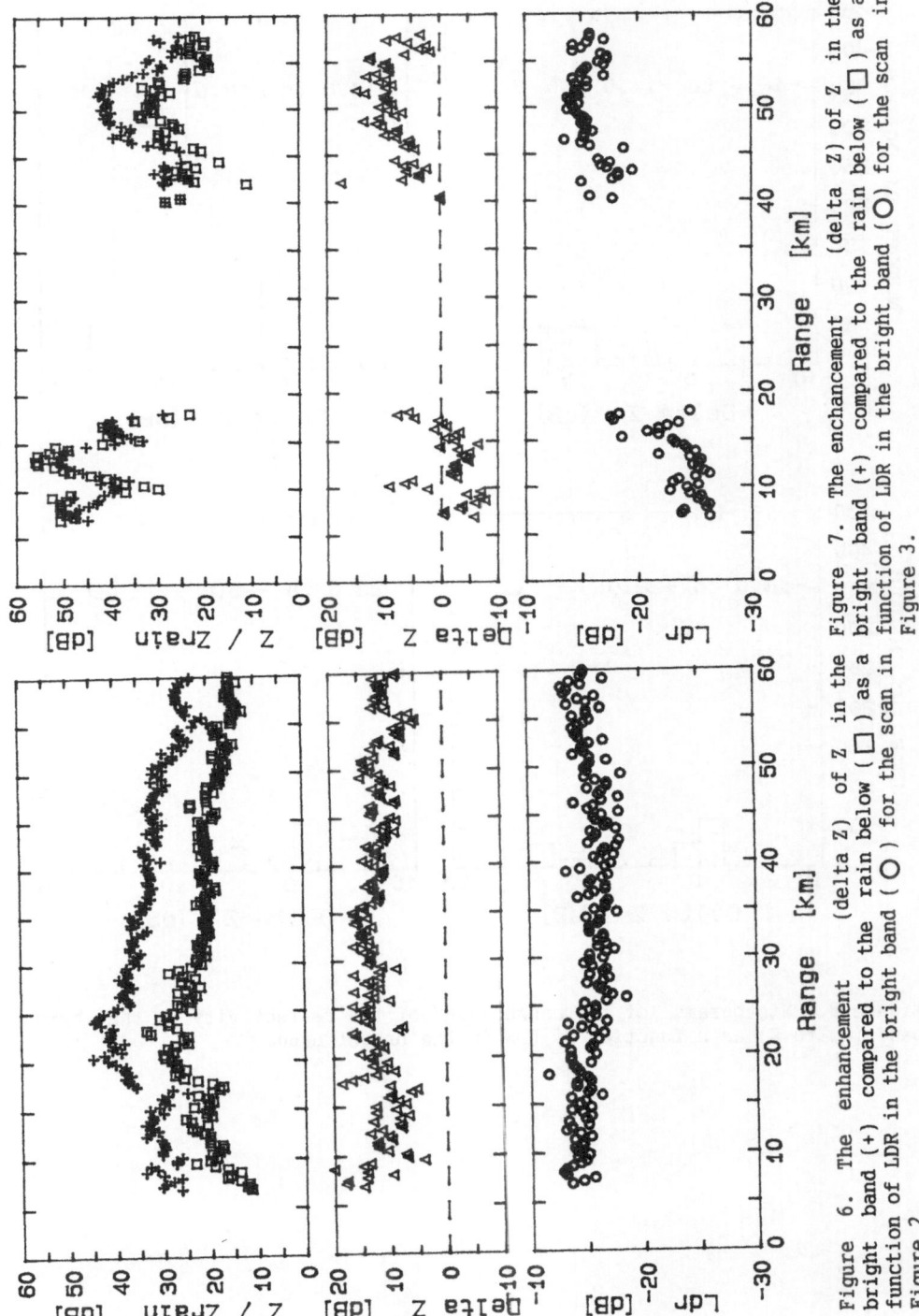

Figure 6. The enhancement (delta Z) of Z in the bright band (□) as a function of LDR in the bright band (○) for the scan in Figure 2.

Figure 7. The enchancement (delta Z) of Z in the bright band (+) compared to the rain below (□) as a function of LDR in the bright band (○) for the scan in Figure 3.

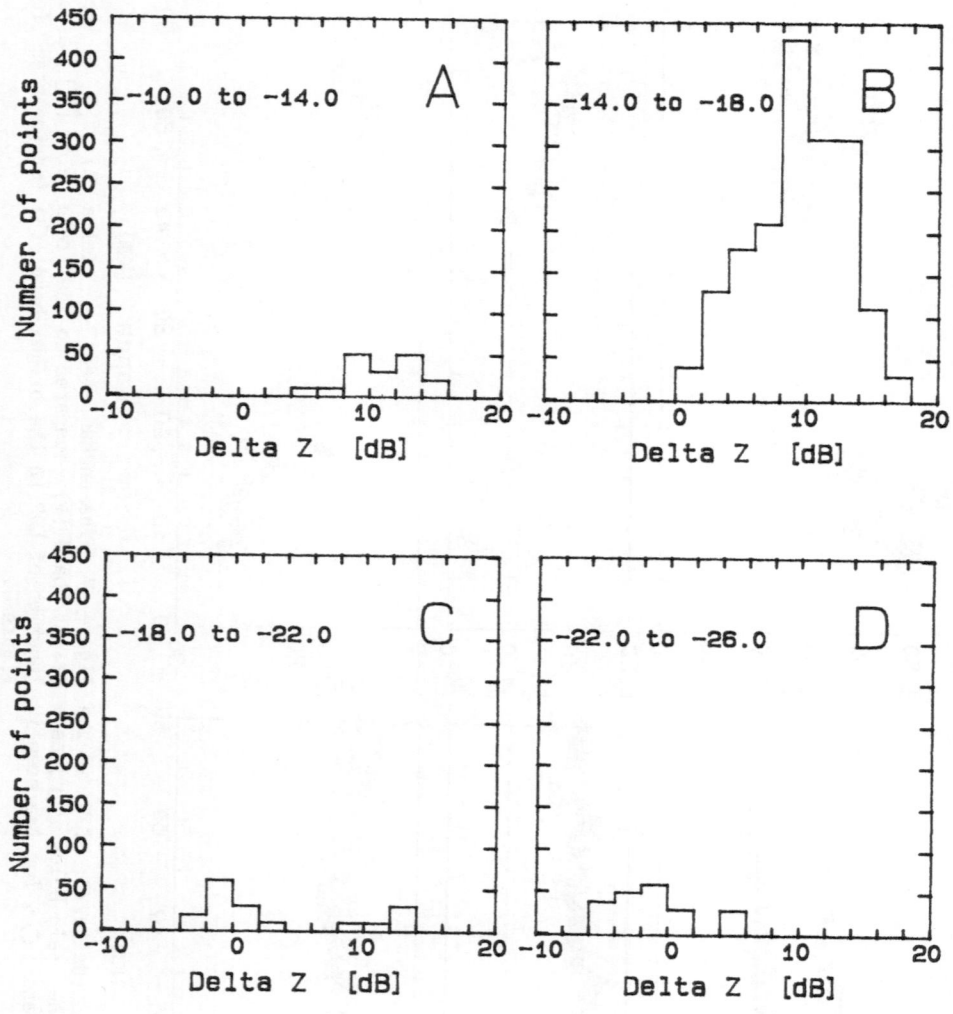

Figure 8. Histograms of the enhancement of the reflectivity in the bright band (delta Z) as a function of LDR in the bright band.

measurements for a one degree beamwidth radar which is only partially filled by the bright band.

5. TECHNICAL ASPECTS OF IMPLEMENTATION
The Chilbolton radar is a narrow beam research radar which can make measurements of unrivalled polarization purity. Several factors need to be considered if the techniques are to be implemented on conventional C-band radars.

The LDR measurement has the advantage that no fast switching of the transmitted signal is needed. However, the cross-polar signal power is very low, and a signal to noise ratio of more than 20dB is required if an LDR of -20dB is to be detected. We also note that the low power cross-polar signal is much more sensitive than Z and ZDR to ground clutter contamination. Finally, and most importantly, the LDR signal is affected by propagation and attenuation problems at shorter wavelengths. Correction algorithms may be possible at C-band.

If the co-polar cross correlation is to be measured then rapid pule-to-pulse switching of the polarization of the transmitted signal is required. However, the correlation is unaffected by attenuation and propagation, provided there is sufficient signal level. A signal to noise ratio of 20dB is needed if the correlation is to be measured to 1%. At shorter wavelengths the correlation measurement will be degraded as the time between transmitted pulses approaches the decorrelation time of the return signal. ZDR measurements generally need long dwell times, but it is possible to estimate the correlation from shorter time series and so scan more rapidly.

6. CONCLUSION
These observations suggest that the linear depolarization ratio and/or the co-polar cross correlation can be used to identify the bright band. These parameters may also be of use in identifying anomalous propagation and ground clutter, both of which should have high values of LDR and low correlations.

ACKNOWLEDGEMENTS
This work was supported by AFOSR 89-0121, NERC GR3/5896 and the Meteorological Office.

REFERENCES
(1) Smith C J (1986) J Atmos Ocean Tech, 3, 129-141.
(2) Collier C G (1986) J Hydrol, 83, 207-223.
(3) Cherry S M and Goddard J W F (1982) URSI Symposium on Multiple Parametr Radar Measurement of Precipitation, Bournemouth, UK.
(4) Hall M P M, Goddard J W F, and Cherry S M (1984) Radio Sci, 29, 132-140.
(5) Illingworth A J, Goddard J W F and Cherry S M (1987) Q J Roy Meteorol Soc, 113, 469-489.
(6) Illingworth A J and Caylor I J (1989) Preprints 24th Radar Meteorol Conf, Amer Meteor Soc, Boston, 323-327.
(7) Caylor I J and Illingworth A J (1989) Preprints 24th Radar Meteorol Conf, Amer Meteor Soc, Boston, 9-12.
(8) Herzegh P H and Jameson A R (1989) Preprints 24th Radar Meteorol Conf, Amer Meteor Soc, Boston, 315-317.

RAINRATE MEASUREMENTS USING A CIRCULAR POLARISATION-DIVERSITY RADAR

A.R. HOLT and R. McGUINNESS
Department of Mathematics, University of Essex,
Wivenhoe Park, Colchester CO4 3SQ.

Summary
Improvements in rainrate predictions using ground based polarisation
diversity weather radars over those provided by conventional weather
radars have been suggested for a number of years. Dual linear
polarisation S-band radar has shown improvements are possible but
their operational speed and the extra costs of processing mean their
use as operational radars is unlikely. Circular polarisation S-band
radars have suffered significantly from on-path propagation effects.
Attempts to correct these effects to obtain rainrate predictions have
proved to be unstable. An alternative method for obtaining rainrate is
considered which adopts a different approach. At S-band the
propagation effects that corrupt the mesurements are dominated by
differential phase. Using the assumption that the canting angles of the
raindrops in the pulse volume are close to zero the differential
propagation phase shift can be extracted independently from each
range-gate. It has been found that differential propagation phase is
almost linearly related to the rainrate. It has also been found to be
less sensitive to the type of dropsize distribution compared with Z and
ZDR. The method therefore uses backscatter measurements to extract
forward scatter differential propagation phase shift, and uses the latter
to predict the ground rainrate.
We present results from an operational radar in Alberta, taken
over a four year period.

1. INTRODUCTION

The method which has been generally adopted for obtaining a
quantitative measurement of rainfall by radar, is to assume a relationship
between radar reflectivity factor Z and rainfall rate R of the form $Z = aR^b$
(cf Browning (1)). Unfortunately there is no unique relationship. Indeed
Battan (2) lists numerous such pairs of parameters a,b which have been
empirically derived by fitting data. In a given situation there is no
a-priori information on the parameters to be used, and hence operational
radar networks have resorted to normalising the rainrate by reference to
measurements from a network of raingauges (Collier et al (3)).

The reason for the failure to obtain a unique relationship is not hard
to find. In using a measurement of reflectivity, a single item of data is
being used to infer rainrate in a volume. The unknowns are many.
Firstly, the nature of the hydrometeors composing the volume being sensed.
These may be rain, ice, melting particles, or some combination of them.
Also unknown are their physical characteristics such as shape, orientation,
size distribution and temperature. Secondly the distribution of
hydrometeors throughout the target volume. Large gradients of density
across a volume may not give the same reflectivity as the same number of
hydrometeors evenly spaced throughout the target volume, due to the
nature of the antenna beam pattern. It is also possible in regions of high
gradient to obtain errors due to sidelobe coupling. The third area of

uncertainty relates to propagation effects along the radar path. Where propagation effects are significant, Hitschfeld and Bordan (4) have shown that a gate-by-gate correction process, using the reflectivities along the radar path, is unstable. In this paper we are limiting our attention to wavelengths of around 10cm where the differential attenuation is very small, but differential phase can be large. Thus there can be significant problems with circularly polarised waves (McGuinness et al (5)). Although for linear polarisation attenuation can be neglected in moderate rainfall, in heavy rain it cannot be neglected.

Variations in drop-size distribution play an important role in the lack-of-uniqueness of the Z-R relationship. The reflectivity is the integral of the drop-size distribution weighted by a factor D^6, (where D is the drop diameter) whereas rainrates can be approximated by the integral of the drop-size distribution times $D^{3.67}$ (Atlas and Ulbrich (6)). Recently Grosh (7) has shown that a true representation of raindrop terminal velocity results in the log-log plot of Z against R being distinctly non-linear. For high rainrates this means that the use of a $Z = aR^b$ relationship will cause the rainrates to be overestimated. Further problems with using reflectivity to estimate rainrate have been discussed by Browning (1).

In an attempt to improve the measurement of rainrate, Seliga and Bringi (8) proposed the use of differential reflectivity (ZDR) in addition to Z. In principle the knowledge of both ZDR and Z enables a two-parameter drop-size distribution to be used to model the rain in each range-gate, and this should therefore improve the estimate of rainrate. ZDR can be obtained by alternatively transmitting vertical (V) and horizontal (H) linear polarisation, and measuring the copolar received power. Therein lies one of the problems since, operationally, the use of switching must increase the scan time by a factor of 2. Since narrow beamwidths are required in order to measure ZDR sufficiently accurately, this will further impact the scan speed. Further, although Z is not particularly sensitive to drop shape, ZDR is (cf Goddard and Cherry (9)). Nevertheless in some situations the use of ZDR has been shown to improve rainrate estimates (Seliga et al (10)).

More recently Seliga and Bringi (11), Jameson (12) proposed the use of differential propagation phase shift Φ_{DP} to estimate rainrate. It more closely resembles rainrate in that it weights the drop-size distribution by a factor of approximately $D^{4.24}$ (Sachidananda and Zrnic (13)). These authors showed that it is possible to extract Φ_{DP} from a V/H switched coherent radar and using this method Golestani et al (14) have given examples of Φ_{DP} in storms.

For more than a decade, the Alberta Research Council (ARC) was operating a circularly polarised S-band radar to monitor intense summer convective storms. The radar emitted left-hand circularly polarised waves (LHC), and measured the powers W_2, W_1 of the main (RHC) and orthogonal (LHC) return channels respectively, as well as the complex correlation W of the LHC, RHC return signals. Holt (15) has shown that Φ_{DP} may be extracted from the data. In addition to a large radar dataset ARC also collected a "ground-truth" database of rain and hail reports. Some of this ground-truth data was obtained by storm-chase vehicles, but most came from a network of farmers.

2. THEORY

The theory of the circular polarisation system used in Alberta was intially developed by McCormick and Hendry (16). We presume the target volume is rainfilled, and that the number of drops with diameter in the range D, D + dD is given by the dropsize distribution N(D)dD. We shall further presume that the forward and backward scattering amplitudes for vertical polarisation (with incidence in a principal plane) are $f_{VV}(D)$, $S_{VV}(D)$

respectively, with subscript H describing horizontal polarisation. We shall assume that the drops are axially symmetric, and hence that the off-diagonal elements f_{VH} etc. vanish. The wavenumber of the incident radiation is k_O (= $2\pi/\lambda$) and the refractive index of water for this wavelength is n_O. For the purpose of this study we shall assume that the drops are all-aligned, since Holt (17) showed that allowing for a narrow spread in canting angles makes little difference.

For a radar transmitting LHC, and receiving the powers W_2, W_1 in the RHC, LHC channels respectively we find that, neglecting propagation effects,

$$W_2 = \frac{1}{4} |C|^2 \int N(D) \ |S_{VV} + S_{HH}|^2 \ dD \qquad [1]$$

and

$$W_1 = \frac{1}{4} |C|^2 \int N(D) \ |S_{VV} - S_{HH}|^2 \ dD \qquad [2]$$

where $|C|^2 = 10^6 \ \lambda^5/\pi^5 \ |(n_O^2 - 1)/(n_O^2 + 2)|^2$. $\qquad [3]$

In practice, however, the radar beam may pass through precipitation on path to the volume being sensed, and this precipitation will effect the signal. The vertical and horizontal components will be effected respectively by a factor

$$d_{V,H} = \exp[i \ \beta_{V,H} - \gamma_{V,H}] \qquad [4]$$

which represents the total on-path phase change and attenuation. Thus the powers actually measured are given by (Holt (17)) as

$$W_2 = \frac{1}{4} |C|^2 \int N(D) \ |d_V S_{VV} + d_H S_{HH}|^2 \ dD \qquad [5]$$

and

$$W_1 = \frac{1}{4} |C|^2 \int N(D) \ |d_V S_{VV} - d_H S_{HH}|^2 \ dD, \qquad [6]$$

whilst the complex correlation is

$$W = \frac{1}{4} |C|^2 \int N(D) \ \{\overline{d}_V \overline{S}_{VV} + \overline{d}_H \overline{S}_{HH}\}\{d_V S_{VV} - d_H S_{HH}\} \ e^{2i\alpha} \ dD \qquad [7]$$

where α is the canting angle in the plane normal to the direction of propagation. It should be noted that d_V, d_H are only dependent on the rain on path, whereas S_{VV}, S_{HH} are depend only on the rain in the volume being sensed.

Following Holt (15), McGuinness and Holt (18), we assume that α may be assumed zero. Then

$$\frac{W_2 - W_1}{2} - i \ ImW = \frac{1}{2} |C|^2 \int N(D) \ dD \ \overline{d}_V^2 d_H^2 \ \overline{S}_{VV} S_{HH} \qquad [8]$$

We note that $d_{V,H}$ is independent of the volume being sensed and may thus be taken outside the integral. Further, $\arg(\overline{S}_{VV} S_{HH}) < 1°$ and hence $(\overline{S}_{VV} S_{HH})$ may be assumed real. Thus

$$\arg \left\{ \frac{W_2 - W_1}{2} - i \ ImW \right\} = \arg \{\overline{d}_V^2 d_H^2 \}$$

$$= 2(\beta_H - \beta_H) = \Phi_{DP} \qquad [9]$$

We have thus extracted the two-way differential propagation phase shift.

It must be emphasised that the property of negligible differential phase on backscatter holds only at S-band and then only for rain, or for small hail. For medium to large hail, $\arg(S_{VV}S_{HH})$ may be significant.

We will assume that the gates being sensed are of length L. The two-way differential propagation constant for gate n is defined as

$$K_{DP}^n = \left[\Phi_{DP}^{n+1} - \Phi_{DP}^n \right] / L \qquad [10]$$

where Φ_{DP}^n is the total two-way differential propagation phase up to gate n.

We may relate K_{DP} to rainrate R using a formula derived from Sachidananda and Zrnic (19) as

$$K_{DP} = R^{1.15}/3.172\lambda \qquad [11]$$

In Table 1 we give values of rainrate for various K_{DP} at a frequency of 2.88GHz. It will be seen that for rainrates < 20mm/hr the value of K_{DP} is small. Therefore due to the difficulty of measuring to this accuracy, it is unlikely that K_{DP} will be a useful parameter in light rain. It was for just this range, (R < 30mm/hr) that the Marshall/Palmer Z-R relationship was designed. However in regions of intense rain, K_{DP} may be 5° or more and measurement errors should not be troublesome.

The differential reflectivity ZDR may be extracted from the circular polarisation data, under the assumption of negligible canting angle. Indeed using the differential propagation phase we may correct for differential attenuation (15).

3. EVENT ANALYSIS

The ARC radar transmitted a left hand circular signal and on backscatter measured four parameters as described in §1,2. The radar performed a helical volume scan rotating at 48°/sec in azimuth (7.5 seconds/revolution) and increasing 1°/revolution in elevation. It measures the 4 channel data every 1° in azimuth along a scan line of 147 range gates each 1.05km long. The radar operated up to 8° elevation or up to 20° elevation depending on the proximity of the storms to the radar. This resulted in a cycle time of either 90 seconds or 180 seconds.

A comprehensive ground truth network collected data on rainfall and hailfall. Storm chase vans collected time resolved hail samples and rainfall rate measurements. Telephone surveys of the 1300 farmers in the project area were also carried out. These gave information on the start of the hail (rain), and the duration and the size of hail (amount of rain). Caution is required when examining the farmer reports as they contain inaccuracies such as the start of the events and their duration.

In this paper we present results from a total of 31 events. Of these 31 events 13 have been classified as rain only events. Some of these 13 events do contain a very small amount of hail, but it does not significantly affect the results. The remaining 18 events contain varying degrees of rain and hail mixed together. The results that are presented are the ground mesured total rainfall against the radar estimated total rainfall. This radar estimate is obtained by calculating the average rainrate for the entire event. The rainrate of each sample point during the event is obtained from the differential propagation phase shift using eqn [11]. For each sample point the phase shift K_{DP} (deg/km) is obtained by averaging the shift over the 3 range gates, along the radial, closest to the sample point. The number of sample points per event is dependent on whether the radar is operating with a 90 second cycle time or 180 second cycle time. The 31 events examined range in duration from 8 minutes to 50 minutes with an average of 17 minutes. The number of sample points range from 3 to 17 with an average of 8 per event.

Figure 1 presents the radar estimated total rainfall against the measured rainfall for the 13 rain events. The average rainrate for these 13 rain events ranged from 12mm/hr up to 66mm/hr. These 13 events contain 4 events sampled by the storm chase vehicles. Points marked with a cross indicate the radar was sampling the event every 90 seconds while those marked with a circle are sampled every 180 seconds. There is generally very good agreement between the radar estimated and observed rainfall. All but one event falls within the factor of 2 error limits. This single event has the lowest average rainrate of 12mm/hr and is sampled every 180 seconds. For the events sampled at 90 second intervals it was generally the case that the higher the rainrate the better the radar predictions become. Because of the near linear relationship between the rainrate and K_{DP} all the sample point rainrates produced realistic values of less then 200mm/hr. This is not true if a standard Z-R relationship is used as shown in figure 2. This plots the same 13 rain events using the Marshall-Palmer relationship $Z = 200R^{1.6}$. This Z-R relationship tends to overestimate the total rainfall. The Joss Thunderstorm model (1) would further overestimate the total rainfall.

Figure 3 presents the 18 events containing a mixture of rain and hail. A significant number of these events give good agreement between the radar estimate of rainfall and the recorded rainfall. During most of the 18 events the derived ZDR is very low which identifies the presence of hail as the dominant factor. However K_{DP} (the slope of Φ_{DP}) is often positive suggesting that the rain is the dominant factor. An example of this is shown in figure 4. This displays Z, ZDR and Φ_{DP} along a radial line during an intense hail storm on 11th July 1985. At the sample point the ZDR falls to zero whereas Φ_{DP} is increasing. The ground observation recorded 5mm hail falling at this time. However when the hail becomes larger then both ZDR and Φ_{DP} remain low as shown in the radial plot in figure 5. Even with $Z \sim 72dB$ the Φ_{DP} is not increasing significantly. The ground observation recorded 12mm hail falling at this time.

The conclusion we draw from these results is that ZDR and Φ_{DP} behave differently for rain and hail mixtures. ZDR is a good parameter for identifying the presence of hail even when rain is present. Φ_{DP} and hence K_{DP} will remain positive in the presence of small hail and K_{DP} becomes negative only in the presence of very large hail. For events that only contain rain K_{DP} is unreliable below rainrates of 20-30mm/hr and in this region a Z-R relationship is more reliable.

To be able to measure the differential propagation phase shift requires a more complex receiver system than a conventional single polarisation radar system. There are however certain advantages associated with the receiver system. The transmitter would not require any enhancements except to transmit a left hand circular signal. The receiver system does not require fast switching but does have to measure the right hand circular (co-polar) and left hand circular (cross-polar) signals as well as the correlation between these two signals. The ARC radar achieved this using a stable magnetron. During the transmit pulse a stable local oscillator was locked onto the pulse phase and frequency, allowing the phase shift of the receive signal to be measured. Processing of these signals is required to obtain Φ_{DP} as described in §2. The one important advantage of this system over the conventional radar system is that the derived rainrate R is not as dependent on calibration errors in the reflectivity Z as the conventional Z-R radars. Also bias errors in measuring Φ_{DP} are not significant as the method requires the change in Φ_{DP} across the three range gates and not the absolute values.

4. CONCLUSIONS

In this paper we have shown that differential propagation phase shift and hence rainrate can be obtained from an S-band circular polarisation radar. The method has proved to be very stable in very heavy rainrate events due to the near linear Φ_{DP} - R relationship. In rain events with rainrate below ~ 25mm/hr the propagation phase shift is small and less reliable. In events where hail is present within a storm the derived ZDR identifies the hail clearly whereas Φ_{DP} can still give reasonable estimates of rainrate. Only when the hail size and hail rate becomes large does Φ_{DP} become an unreliable estimator. Compared with methods that use backscatter parameters Z or Z, ZDR to extract rainrate this new method has been found to be more stable. Bearing in mind that the ground-truth data was recorded mainly for the purposes of the investigation of hail, and that, at the time, the polarisation information was recorded but rarely investigated, we believe that the agreement we have achieved suggests that even better agreement is achievable. Clearly further experiments with this radar in conjuntion with a ground-truth network, are needed.

5. ACKNOWLEDGEMENTS

We would like to thank the Alberta Research Council for the use of their radar data.

6. REFERENCES

(1) Browning K.A. (1978). Meteorological applications of radar. Rep. Progress in Physics 41 761-806.
(2) Battan L.J. (1973). Radar Observation of the Atmosphere. (Chicago: University of Chicago Press).
(3) Collier C.G., Larke P.R. and May B.R. (1983). A weather radar correction procedure for real-time estimation of surface rainfall. Quart. J.R. Met. Soc. 109 589-608.
(4) Hitschfeld W. and Bordan J. (1954). J. Meteor. 11 58-67.
(5) McGuinness R., Holt A.R. and Humphries B.G. (1984). The interpretaion of CDR radar data to obtain rainrates in storms. Conference volume: 22nd Conf. on Radar Met. (Boston: AMS) 276-80.
(6) Atlas D & Ulbrich C.W. (1977). Path and area integrated rainfall measurement by microwave attenuation in the 1-3cm band. J. Appl. Meteorol. 16 1322-31.
(7) Grosh R.C. (1989). The highly non-linear aspects of Z-R and VT-Z caused by raindrop terminal fall speeds. Conference volumse: 24th Conf. on Radr Met. (Boston: AMS) 676-81.
(8) Seliga T.A. and Bringi V.N. (1976). Potential use of radar differential reflectivity measurements at orthogonal polarisations for measuring precipitation. J. Appl. Meteorol. 15 69-76.
(9) Goddard J.W.F. and Cherry S.M. (1984). The ability of dual-polarisation radar (co-polar linear) to predict rainfall rate and microwave attenuation. Radio Science 19 201-8.
(10) Seliga T.A., Bringi V.N. and Al-Khatib H.H. (1981). A preliminary study of comparative measurements of rainfall rate using the differential reflectivity radr technique and a raingauge network. J. Appl. Meteorol. 26 1362-8.
(11) Seliga T.A. and Bringi V.N. (1978). Differential reflectivity and differential phase shift: Applications in radar meteorology. Radio Science 13 271-5.
(12) Jameson A.R. (1985). Microphysical interpretation of multi-parameter radar measurements in rain Part III: interpretation and measurement in rain of propagation differential phase shift between orthogonal linear polarisations. J. Atmos. Sci. 42 607-614.

(13) Sachidananda M. and Zrnic D.S. (1986). Differential propagation phase shift and rainfall rate estimation. Radio Sci., 21 235-247.
(14) Golestani Y, Chandrasekar V and Bringi V.N. (1989). Intercomparison of multi-parameter radar measurements. Conference volume: 24th Conf. on Radar Met. (Boston: AMS) 309-14.
(15) Holt A.R. (1988). Extraction of differential propagation phase shift from data from S-band circularly polarised radars. Electronics Letters 24 1241-2.
(16) McCormick G.C. and Hendry A. (1975). Principles for the radar determination of the polarisation properties of precipitation. Radio Sci. 10 421-434.
(17) Holt A.R. (1984). Some factors affecting the remote sensing of rain by polarisation diversity radar in the 3-35GHz frequency range. Radio Sci. 19 1399-1412.
(18) McGuinness R. and Holt A.R. (1989). The extraction of rain-rates from CDR data. Conference volume: 24th Conf. on Radar Met. (Boston: AMS) 338-341.
(19) Sachidananda M. and Zrnic D.S. (1987). Rainrate estimates from differential polarisation measurements. J. Atmos. and Oceanic Tech. 4 588-598.

TABLE 1

Relationship between differential propagation phase shift / km and rainrate

K_{DP} (deg/km)	Rainrate (mm/hr)
0.5	11.3
1.0	20.7
2.0	37.7
4.0	68.6
7.0	111.5
10.0	151.8
15.0	215.6
20.0	276.7

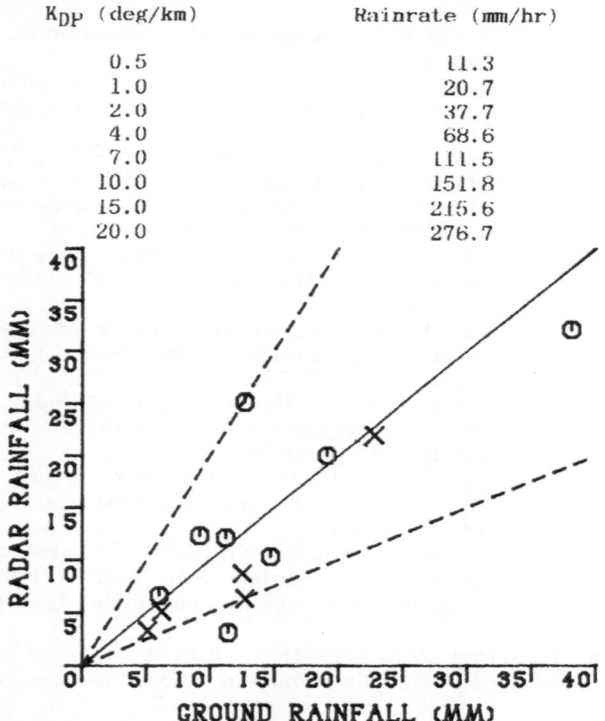

Fig. 1. Comparison between measured rainfall and the radar derived rainfall using differential propagation phase shift for 13 rain events. Dotted lines mark the factor of 2 error limits x = 90 sec. samples   0=180 sec. samples.

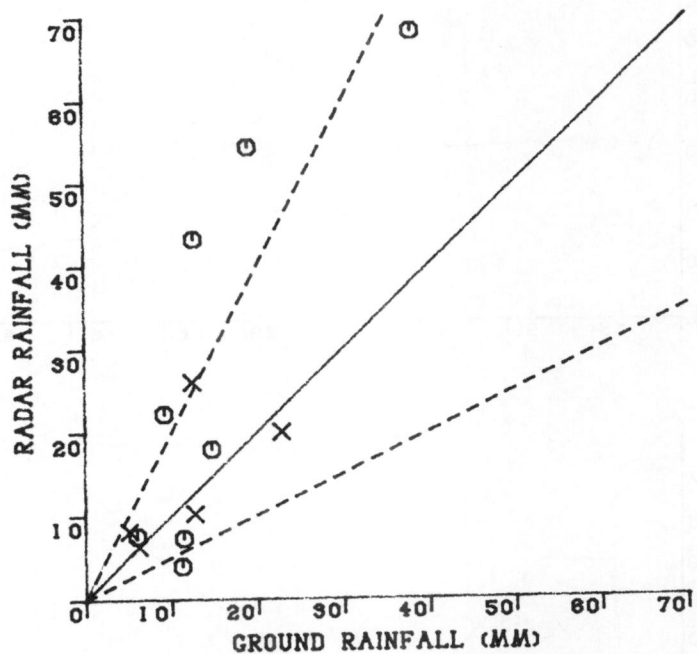

Fig. 2. As Fig. 1 for radar derived rainfall using Marshall-Palmer Z-R method.

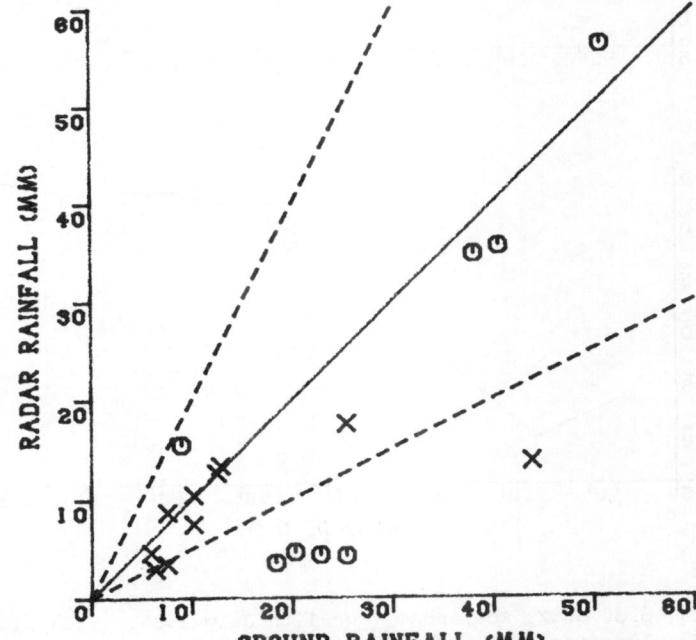

Fig. 3. Comparison between measured rainfall and the radar derived rainfall using differential propagation phase shift for 18 rain/hail events. Dotted lines mark the factor of 2 error limits.

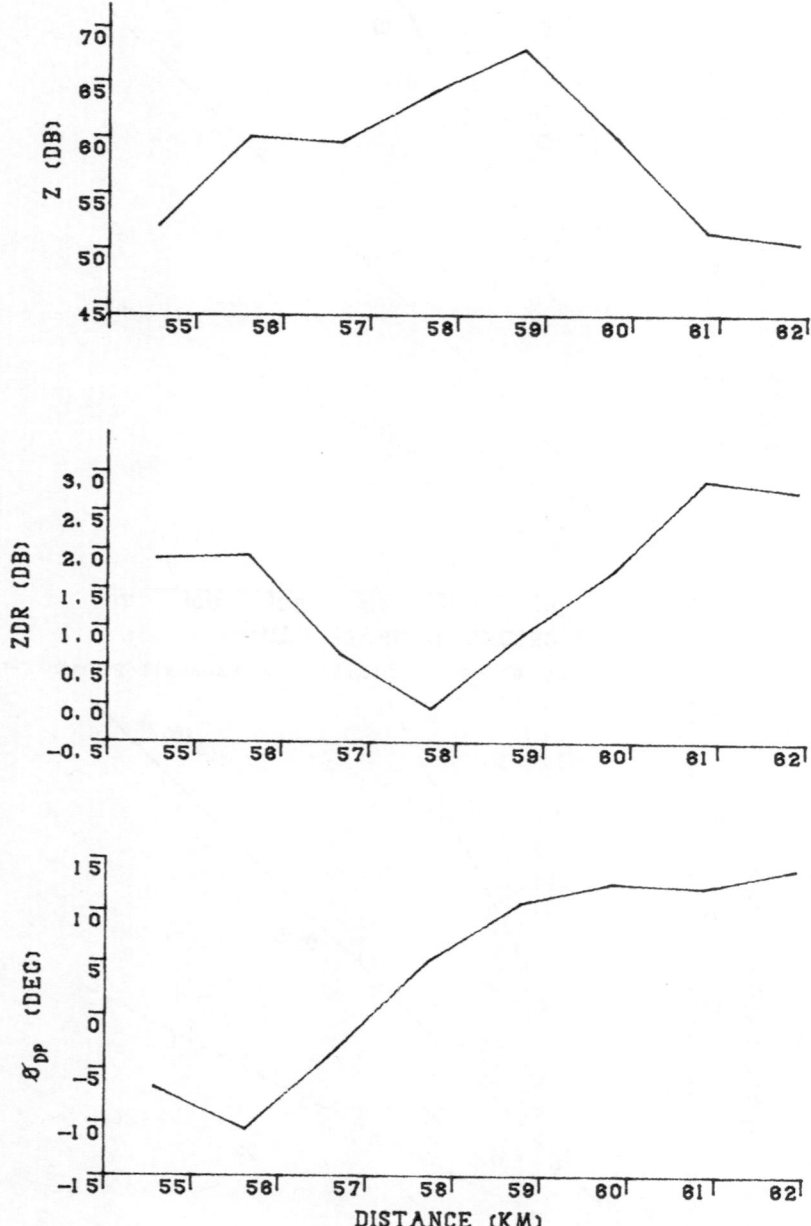

Fig. 4. Radial plot of Z, ZDR and ϕ_{DP} on 11th July 1985. Time 2102; azimuth 338°; elevation 0.9°. The arrow marks the ground observation location.

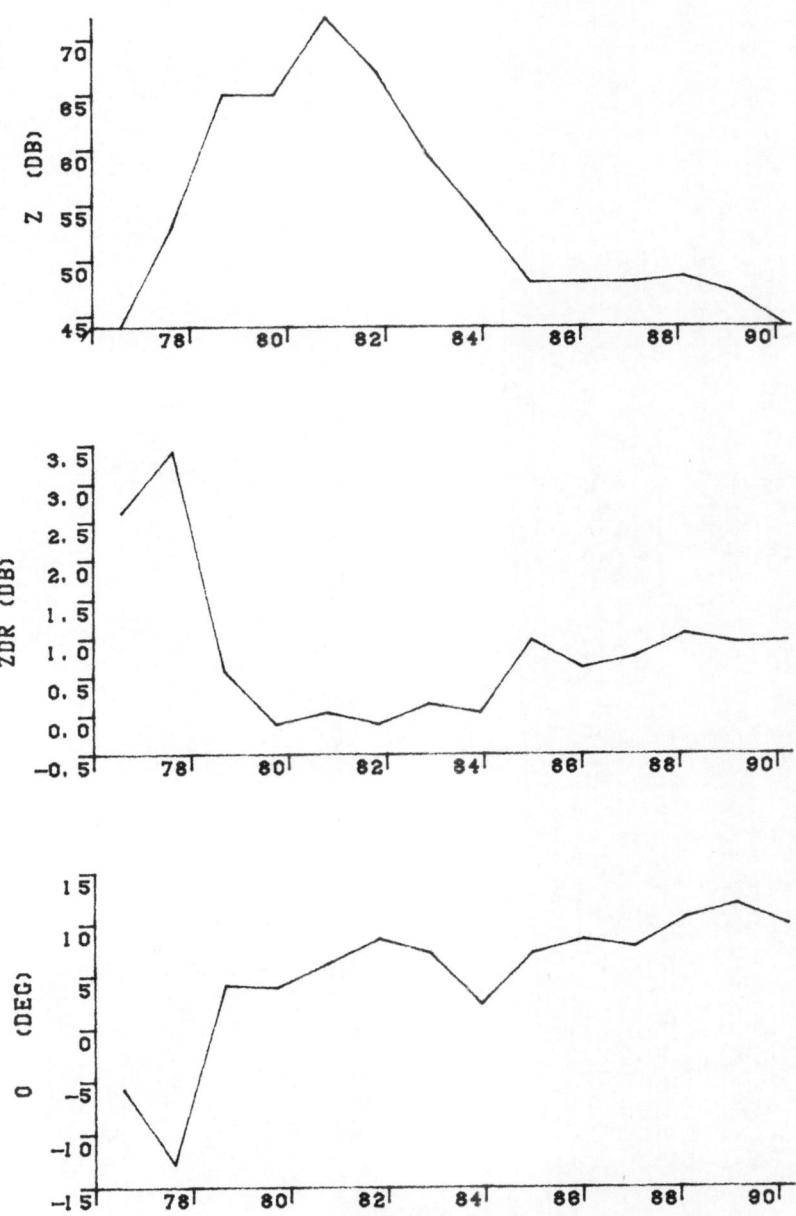

Fig. 5. Radial plot of Z, ZDR and ϕ_{DP} on 11th July 1985. Time 2035; azimuth 340°; elevation 0.9°. The arrow marks the ground observation location.

SESSION 4

Combining radar, satellite and
conventional meteorological data

Chairman : A. Van Gysegem

AN INTEGRATED APPROACH TO THE DISPLAY OF DOPPLER RADAR AND OTHER METEOROLOGICAL DATA

T. W. SCHLATTER
Program for Regional Observing and Forecasting Services
NOAA Forecast Systems Laboratory, Boulder, Colorado USA

Summary

Though a weather radar is one of the most important sources of information when hazardous weather threatens, the well-prepared forecaster
will want to consult other information also. For this purpose, effective merging of radar and other types of data in a common display is
vital. The choice of coordinate system should not be made lightly;
it depends upon many factors, including available computing resources,
volume of data to be remapped, size of area viewed, requirements for
position accuracy, and user preference. Once all data are in a common
projection, the way is open for many effective combinations of meteorological and hydrological data. Compositing radar data with complementary information from satellites, lightning detection systems,
drainage and basin maps, and geopolitical data has proven to be
valuable. Three compositing techniques meet most needs: a simple
overlay of graphical information upon an image, toggling between two
images, and fading out one image while another fades in. Forecasters
deal more easily with full-volume scan radar data if they can view
reflectivity and radial velocity data together and have the option of
slicing the data horizontally and vertically. They save precious
minutes in critical situations through use of a semi-automated system
for generating warning messages.

1. INTRODUCTION

One of the primary goals of the Program for Regional Observing
and Forecasting Services (PROFS) has been to develop meteorological
workstations capable of integrating all data relevant to nowcasting and
the short-term forecast. By this endeavor, PROFS is assisting the U.S.
National Weather Service in the design of a new communications and display
system for the 1990s. PROFS' experience with these workstations, especially in the display of radar data, forms the basis for this paper.

Aside from algorithms that extrapolate the movement of radar echoes,
cloud patterns, and other meteorological features, there is very little
numerical guidance, especially from dynamical prediction models, pertain-
ing to the first few hours of the forecast. This places a premium on
effective displays of current information. Indeed, images and graphical
data subjectively interpreted by forecasters are the foundation of
nowcasting and very short-range forecasting.

When hazardous weather threatens, perhaps no data source is studied
more intently or consulted more frequently than radar data. Nonetheless,
no single data source--not even Doppler radar--provides all the information
a forecaster needs. Thus, it is desirable to integrate information from
different sources on a single display and to provide easy access.

The three most important aspects of effective displays are animation,
use of color, and the ability to combine different sources of information

(compositing). Radar reflectivity and radial velocity images, high-resolution satellite images, lightning strike data, mesoscale surface observations, and information on landforms and drainages are highly complementary,
yet the strategies for bringing them together on a common display are non-
trivial. Compositing is therefore the focus of the next section.
Section 3 describes a number of radar products that have been found useful
at PROFS. An interactive program for generating severe weather warnings is
illustrated at the end. Section 4 summarizes the major points
of this paper.

2.   COMPOSITING RADAR AND OTHER KINDS OF DATA
2.1 Coordinate System Natural to the Data Source
      The first step in compositing is to become aware of the coordinate
system natural to each data source. Scanning radars use the spherical
coordinate system (range, azimuth, elevation angle), with origin at the
radar. Conversion from radar coordinates to latitude/longitude coordinates involves spherical trigonometry. Satellites use the line/element
system, in which the line traces the scan pattern of the radiometer.
Geosynchronous satellites scan in east-west lines. Polar orbiting
satellites scan in a direction roughly perpendicular to the orbital track.
An "element" is simply the location of a scan spot along a line. The
conversion from line/element to latitude/longitude coordinates is
complicated, involving spherical trigonometry, orbital geometry, and the
attitude (pointing direction) of the satellite. Most low-volume data
(e.g., surface and upper air reports, wind profiles, aircraft reports,
and geopolitical data) are associated with latitude/longitude coordinates.

2.2 Choosing a Coordinate System for Display
      The second step is to choose a target projection for display.
The decision is based upon many factors:
   • most desired property of the projection
   • volume of data to be remapped and frequency of remapping
   • location of sufficient computing resources
   • compromises on position accuracy
   • user preferences
Each of these factors will be considered in subsequent paragraphs.
      There are three mutually exclusive varieties of map projections.
Equidistant projections preserve distance relationships between specified
points. (It is impossible to construct a flat map which preserves distance relationships between all points on the earth.) Conformal projections preserve shapes, thus ensuring that the scale at a given point is
the same in all directions. Equivalent projections preserve areas. Any
one of these three properties is obtained at the expense of the others.
Meteorologists almost universally prefer conformal projections. One practical result of this choice is that the directions of wind staffs line up
with wind directions on the earth's surface.
      The volume of data to be remapped and the frequency of remapping
can dictate the choice of coordinate system. PROFS faced the following
dilemma. Numerical model output (analyses, forecasts, and a great variety
of diagnostic information in the form of contours) was to be overlaid on
satellite images of the United States. The contours were received in
display-ready format on a polar stereographic projection, that is, as
instructions for drawing vectors on a screen; the digital data from which
the contours were originally generated were not available. The satellite
data were received directly through an antenna in line/element format.

PROFS had to decide among three alternatives. 1) Remap infrared and visible images to a polar stereographic projection every half hour. 2) Remap several hundred sets of contours to the satellite projection every half hour. (The projection is not constant, but changes slowly with time.) A quarter of a million pixels make up each satellite image, but only a few hundred points define a set of contours. 3) Remap both kinds of data to a third projection. PROFS chose the option that minimized computation time and provided a constant projection regardless of satellite position and attitude: remapping the satellite images to a polar stereographic projection.

Sufficient computer power for timely remapping of data may not be available locally. Satellite data covering a substantial fraction of a hemisphere can be acquired through a single antenna. Should the data processing and remapping be performed centrally or should each local office assume this responsibility? The latter option requires more local computing power but allows far greater flexibility in choice of area and projection. The decision is more straightforward in the case of radar data, which cover a much smaller area and are necessarily acquired locally. If remapped at all, radar data will normally be remapped locally. A notable exception is a mosaic of all radars in a network.

The required position accuracy may influence the choice of coordinate system. For example, suppose one requires that all radar data for ranges less than 240 km be located on a map with an accuracy of 0.1 km or better. Calculations for a radar near latitude 40°N show that, if the radar is precisely located on a Lambert conformal map and the proper scaling is applied to the range coordinate, the criterion for position accuracy is satisfied merely by overlaying the radar data on the map. This requires only a simple transformation from polar coordinates to rectangular screen coordinates. On a polar stereographic map of the same area, however, the calculated position errors are nearly ten times larger than on a Lambert conformal map because the scale factor varies more across the map. To satisfy the criterion, one would first have to remap the radar data into the polar stereographic projection before displaying them.

Forecaster preferences should not be ignored when choosing a map projection for the display of data. For example, there is resistance in the United States to any display of data from a single radar in which north is not straight up. If satellite images are remapped centrally onto a Lambert conformal projection, north will be straight up at only one longitude, and thus radar images overlaid on the satellite images will normally have to be rotated by a few degrees. A second example is the running argument between those accustomed to viewing clouds from the satellite perspective and those who favor remapping. The first group claims that, when pixels are dropped or added during remapping, the original data are corrupted. When data near the limb of the earth are remapped, edge views of clouds are smeared over large areas. The second group counters that remapped images contain all the detail of the original images, given adequate screen resolution, and that a shape-preserving projection is superior to one that allows major distortions. Group two is a growing majority.

2.3 The AWIPS Example

The U.S. National Weather Service plans to install an Advanced Weather Interactive Processing System (AWIPS) at its field offices during the 1990s. Planning for AWIPS is well underway, and it is instructive to consider the rationale behind the choices of coordinate systems. The choice differs according to the area viewed and the types of data most often displayed in the area. See Table I.

Table I.
Information on AWIPS map projections.

Viewing area	Data most often viewed	Map Projection
Northern Hemisphere	composite infrared and water vapor images (from two or more satellites); global model output	polar stereographic
National (includes most of N. America)	visible, infrared, and water vapor images; plotted upper air data; model output	polar stereographic
Regional (1500 km on a side)	visible, infrared, and water vapor images; plotted surface and upper air data; model output	Lambert conformal
Local (750 km on a side)	visible, infrared, and water vapor images; plotted surface data; radar mosaics	Lambert conformal
WFO* scale (460 km on a side)	visible and infrared images, plotted surface data, radar reflectivity/radial velocity at 1 km resolution, NEXRAD products, output from local algorithms	Lambert conformal
Sub-WFO scale (230 km on a side)	visible images, radar reflectivity/radial velocity at 1/2 km resolution, plotted mesoscale surface data, NEXRAD products, output from local algorithms	local stereographic
Metro scale (115 km on a side)	radar radial velocity at 1/4 km resolution, plotted mesoscale surface data, NEXRAD products	local stereographic

*WFO - Warning and Forecast Office. A circle 460 km in diameter encloses the area within which the typical WFO will issue severe weather warnings. This is the area nominally covered by a NEXRAD radar.

At the hemispheric and national scales, coverage of large areas is very important for portraying the long-wave positions and the traveling synoptic-scale waves. Latitude coverage extends from the north pole to the equator. The polar stereographic projection meets these requirements. Model output from the National Meteorological Center is already on this projection.

The regional, local, and WFO maps cover sizable areas of the mid-latitudes, but minimizing scale variation is nonetheless important. The Lambert conformal projection is excellent for this purpose. The National

Weather Service plans to remap satellite data centrally onto two overlapping Lambert projections, one for the eastern half of the U.S. and one
for the western half. This will reduce the tilt of meridians from the
vertical along the Atlantic and Pacific coasts and thus satisfy forecaster
preference. Local offices will extract sections of the remapped image
according to their needs and overlay other data.

The sub-WFO and metro scales are designed primarily for viewing
high-resolution radar data. The curvature of the earth is no longer
important in such small areas and thus variation of scale can be kept
very small. A local stereographic projection with the point of tangency
at the radar site is ideal for this purpose. It can satisfy forecasters'
demands that north be vertical on the map. Moreover, properly scaled radar
data can be overlaid without remapping. On the sub-WFO scale, satellite
data will have to be remapped a second time, but, because the second
remapping never changes, it can be accomplished very quickly with a simple
look-up table.

2.4 The Mechanics of Compositing Data

Each pixel (addressable location) on the screen can independently
display any color. A bit pattern at each pixel determines the color. The
correspondence between color and the bit pattern is called a color table.
Compositing data is no more complicated than modifying color tables in the
appropriate way.

As a first example, consider a radar reflectivity image, to which
we assign 64 colors corresponding to 64 discrete values of reflectivity.
We represent these values in binary with the numbers 000000 through 111111.
We call this a six-bit image because the largest value of reflectivity can
be represented by six bits. When we design the image, we specify a
sequence of colors corresponding to the values from 000000 through 111111;
we will call this sequence a palette. We need six bit planes for this purpose, but suppose that the display device has eight bit planes available.
If we use the six lower order (rightmost) bits for the color image, the
two higher order bits remain for overlays.

Suppose we want to place two graphical overlays on the radar image:
a plot of mesonet data in cyan, governed by the leading bit, and the
county outlines in red, governed by the second bit in an 8-bit pattern.
The pattern 10XXXXXX means that a portion of the data plot occupies the
pixel; it should be colored cyan. The X represents a 0 or 1. The pattern 01XXXXXX means that a county boundary passes through the pixel; it
should be colored red. If the data plot and a county boundary both
occupy the same pixel (11XXXXXX), we can arbitrarily decide to color the
pixel cyan. Eight bits can represent the decimal numbers from 0 through
255. Following is the color table for the image and its two overlays:

255 ← 192	191 ← 128	127 ← 64	63 ← 0
cyan	cyan	red	palette

To "turn off" the image or its overlays independently, we can alter
the color table as follows:

	255 ← 192	191 ← 128	127 ← 64	63 ← 0
NO PLOTS	red	palette	red	palette
NO COUNTIES	cyan	cyan	palette	palette
NEITHER OVERLAY	palette	palette	palette	palette
NO IMAGE	cyan	cyan	red	black

It is important to remember that a change in color table does not alter the bit pattern stored at each pixel; it changes only the correspondence between the bit pattern and color. In the example just given, (eight bit-planes), the color table consists of only 256 numbers, which can be changed in less than a millisecond. (The PROFS displays have 12 bit-planes, which support an eight-bit image with four overlays. The bookkeeping is more complicated than in the above example, but the mechanics are the same.)

In the second example, we will combine two images. We begin with the simplest case of alternating views of two images, for example, radar reflectivity and radial velocity. We assume, as before, that each image is six bits "deep," that is, each image can use up to 64 colors. Each image has its own color table. To keep the discussion simple, we will extract the two high-order bits from each image: XX in the first image and YY in the second image. X and Y can take the values 0 or 1. This has the effect of reducing the number of possible colors to four in each image. Next, we construct a new four-bit image, XXYY. The object is to devise color tables so that only one of the original images shows at a time. If CX1, CX2, CX3, and CX4 represent the colors corresponding to XX = 00, 01, 10, and 11, respectively, and similarly for CY1 through CY4, then the scheme in Table II gives the desired result.

Table II
Color tables for alternate viewing (toggling) of two four-color images.

Image X on Image Y off	4-bit pattern XXYY	Image Y on Image X off
CX1	0000	CY1
CX1	0001	CY2
CX1	0010	CY3
CX1	0011	CY4
CX2	0100	CY1
CX2	0101	CY2
CX2	0110	CY3
CX2	0111	CY4
CX3	1000	CY1
CX3	1001	CY2
CX3	1010	CY3
CX3	1011	CY4
CX4	1100	CY1
CX4	1110	CY2
CX4	1110	CY3
CX4	1111	CY4

Meteorologists at PROFS frequently toggle between a four-bit radar reflectivity image and the corresponding four-bit radial velocity image; this procedure helps them to relate the wind field to the precipitation distribution within storms.

A simple generalization of this scheme allows one to mix the two images in any desired proportion. Any color on the screen is a known mixture of colors from the red, green, and blue guns inside the display device. Suppose we want a mixture of 60% of image X and 40% of image Y. If the bit pattern at a given pixel is 0110, we want a color that corresponds to 60% of CX2 and 40% of CY3. In the PROFS color monitors, each color is represented by 24 bits, consisting of three 8-bit segments corresponding to the intensities in each color gun. Eight bits for each color allow 256 intensity levels for each color. Let Rx, Gx, and Bx represent the intensity levels for the red, green, and blue guns for the color CX2. Define Ry, Gy, and By similarly for the color CY3. The desired color, corresponding to XXYY = 0110, is

$$Cxy = RxyGxyBxy, \text{ where } \begin{cases} Rxy = 0.6 \, Rx + 0.4 \, Ry \\ Gxy = 0.6 \, Gx + 0.4 \, Gy \\ Bxy = 0.6 \, Bx + 0.4 \, By \end{cases}$$

When the new colors are determined for each entry in the table and displayed on a screen, most observers agree that they see a little more of Image X than Image Y, although subjective interpretations differ slightly. Color tables can be calculated so rapidly that it is possible to bring one image into prominence while fading the other out. This can be accomplished in several ways, for example, by sliding a cursor along a bar. Only Image X is visible at the leftmost position, and only Image Y is visible at the rightmost position. These combined images can still have graphical overlays, as described above in the first example.

3. SAMPLE PRODUCTS AND APPLICATIONS AT PROFS INCORPORATING RADAR DATA

3.1 Products and Applications Based Exclusively on Radar Data

PROFS generates a broad range of radar products. Most familiar is the PPI (Plan-Position Indicator), which displays reflectivities or radial velocities as a function of range and azimuth. The elevation angle is fixed. The CAPPI is a Constant-Altitude PPI, a range-azimuth display at constant height. The CAPPI is obtained from the conical scans above or below the specified height, whichever is closer. This often results in annular discontinuities, which are distracting to some. The discontinuities can be virtually eliminated through interpolation, but at the expense of program efficiency.

A plate stack is available upon request. In Fig. 1a, the meteorologist centers a box on a storm to be examined in detail. From PPIs at different elevation angles, the system then generates a set of 30 x 30 km images in perspective view, one on top of the other (Fig. 1b). The plate stack, described by R. C. Lipschutz and his colleagues [1], is useful for studying storm structure. The rectangular grid of dotted reference lines enables the forecaster to infer storm tilt. At each level, the range has been projected onto a horizontal surface. In the example shown, the high-reflectivity core tilts toward the northeast. The range-azimuth diagram at left shows the location of the storm on a PPI. The diagram is essential for proper interpretation of the corresponding stack of radial velocity images (not shown).

Full-volume scans permit a generalization of the RHI (Range-Height Indicator) display, namely, a vertical cross section of reflectivity not constrained to lie along a radial, but arbitrarily oriented. Figure 2, also from Reference [1], shows a high-reflectivity overhang in midtroposphere and a weak-echo region at bottom center. The same information conveyed by the contours in Fig. 2 can also be presented as a color image.

Fig. 1 (next page) (a) A selected radar echo in a low-level PPI for which a plate stack is desired. (b) A plate stack of radar reflectivities. Elevation angles are given at left, altitudes of the center points at right. Note the range and azimuth of the selected echo, just above the range-azimuth diagram. The color bar at bottom gives the reflectivity scale in decibels db(Z).

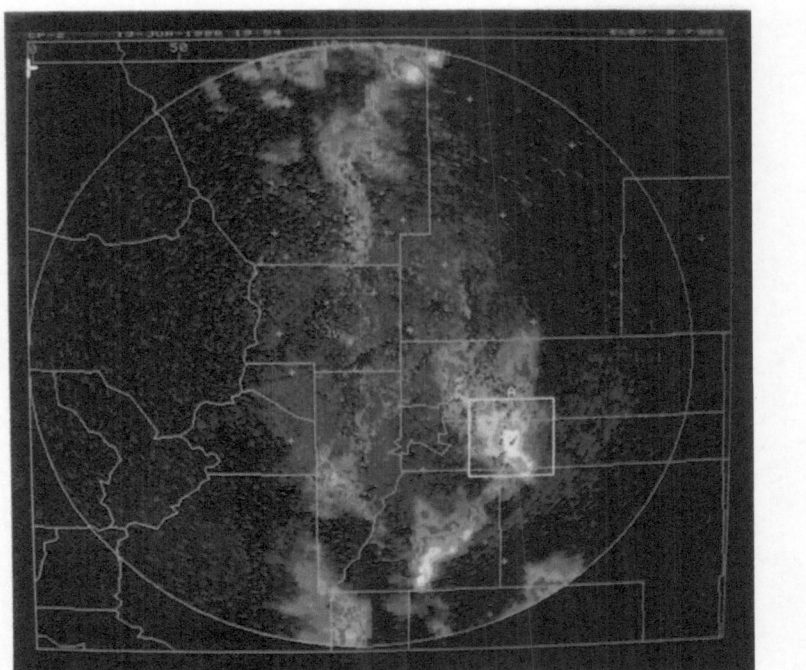

(a)

Figure 1

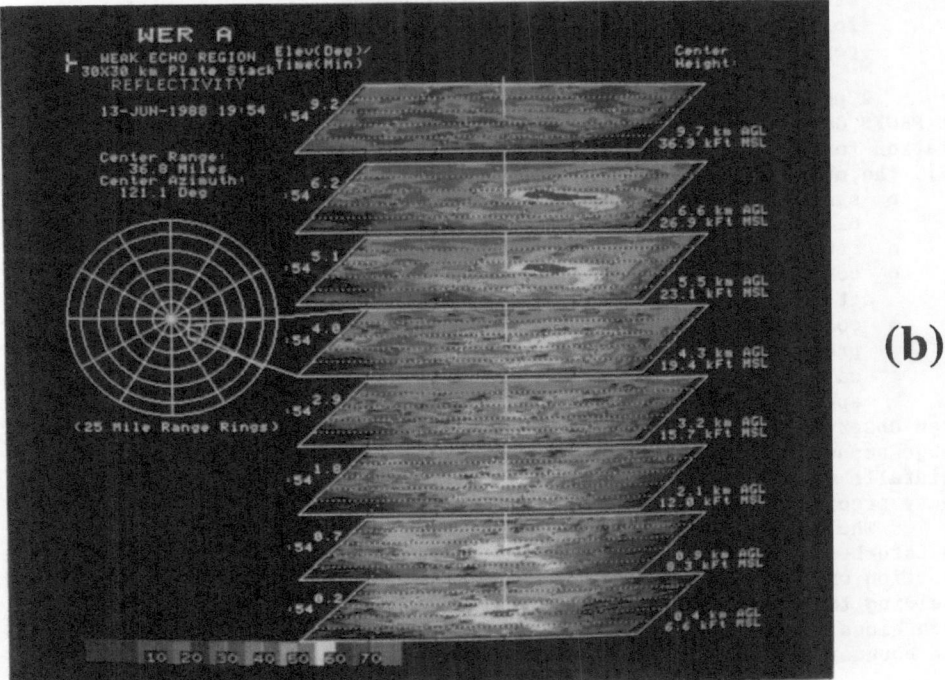

(b)

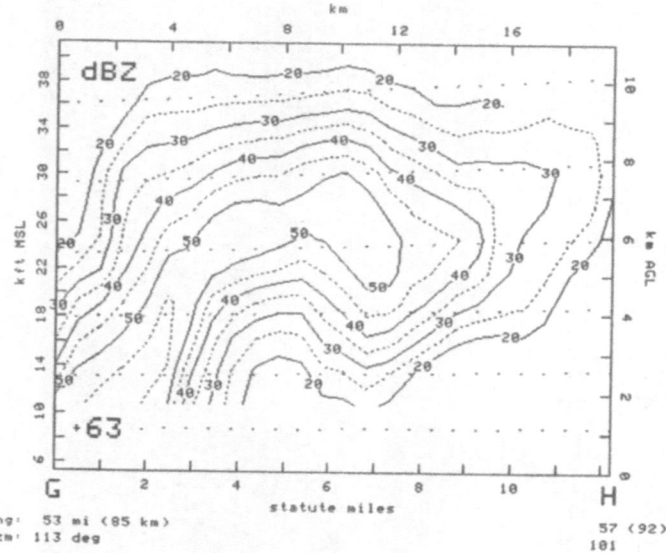

Fig. 2   Contoured reflectivities in vertical cross section for a Colorado
         storm, 2120 UTC, 25 June 1985 (extracted from reference [2]).   The
         forecaster chooses the end-points of the section prior to its
         generation.

     A precipitation accumulation algorithm was first tested in real time
at PROFS during summer 1988.   It produces estimates of accumulated precipitation for one and three hours and for an entire event.   As described in
[2], the main processing steps are:
   • synthesizing low-level reflectivity data to reduce non-meteor-
     ological echoes
   • removing point targets from the data
   • correcting for range, beam occultation by terrain, and beam
     attenuation by precipitation
   • converting reflectivity values Z to rainfall rates R using a
     predefined Z-R relationship
   • calibrating the Z-R relationship based upon raingauge data and
     applying a multiplicative bias to the rainfall estimate.
When observed rainfall exceeded 2.5 mm, estimated rainfall agreed with
gauge amounts to within a factor of two more than 70% of the time.   Light
rainfalls were severely overestimated.   Hail caused some overestimates of
heavy precipitation.
     The VAD (Velocity-Azimuth Diagram) estimates the horizontal wind in
undisturbed conditions.   Radial velocity at a fixed range is graphed as a
function of azimuth.   When the wind field is uniform, a sine wave results,
yielding the horizontal wind vector at the elevation of the range ring.
Such winds, computed at several elevation angles, define a wind profile in
the boundary layer.

Differential reflectivity, ZDR, depends upon a ratio of reflectivities. The numerator is the reflectivity as measured by horizontally polarized pulses, ZH; the denominator is the reflectivity as measured by vertically polarized pulses, ZV.

$$ZDR = 10 \log (ZH/ZV) \propto (a/b)^6,$$

where a and b are the diameters of hydrometeors in the horizontal and vertical directions, respectively. Differential reflectivity is used to distinguish between hail and heavy rain. Because of frictional forces exerted by the air on raindrops, large raindrops are flattened as they fall; their horizontal diameter exceeds their vertical diameter. In contrast, hailstones, though irregularly shaped, tend to tumble rapidly as they fall. Since they have no preferred orientation, the radar views them as spherical. ZDR for large raindrops is positive whereas ZDR for hail is close to zero. Large concentrations of big drops may be as reflective as hailstones, but ZDR reliably differentiates between the two. Figure 3, taken from [3], is the basis for rain-hail discrimination. PROFS has experimented with ZDR data taken from the 10-cm CP-2 Doppler radar, operated by the National Center for Atmospheric Research. Unfortunately, NEXRAD radars will not measure differential reflectivity.

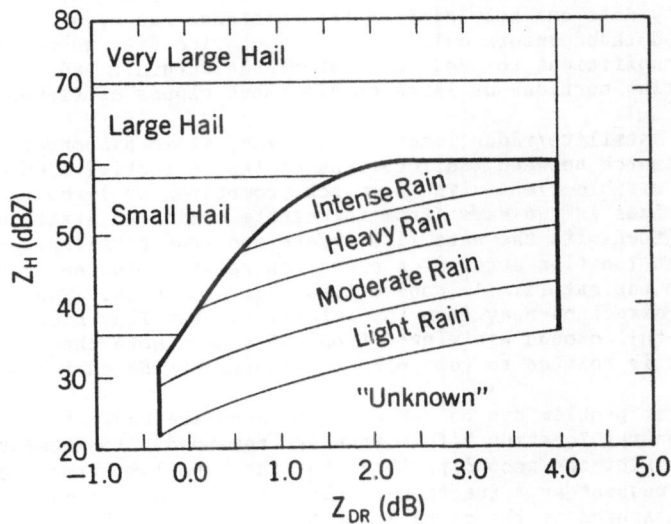

Fig. 3 A precipitation type/intensity model from reference [3]. The abscissa is differential reflectivity. The ordinate is reflectivity in the horizontally polarized beam.

3.2 Useful Combinations of Radar and Other Types of Data

Map backgrounds demonstrate the spatial relationships between radar echoes and the terrain, population centers, political boundaries, landmarks, and drainage basins. All backgrounds described here are available at the WFO scale and smaller scales.

Four backgrounds are generated with vector graphics: county and state boundaries, county names and station identifiers, a drainage map with rivers and streams, and a map of basin boundaries. Forecasters use the first two backgrounds whenever they issue severe weather warnings. They use the latter two backgrounds to assess the flash flood potential on specific streams when rainfall is excessive.

Two backgrounds are images. One is a grey-scale image of Colorado topography, in which lower ground is a darker shade of grey and higher ground is a lighter shade. The other image is a high-resolution, black-and-white road map, created by digitizing a real road map with an optical scanner. To transform the digitized road map into the target map projection, a technique called "inverse free sampling" [4] is used. Well-known to remote sensing experts, especially aerial photographers, this least-squares technique "warps" the source map into the target projection by determining a mapping formula from control points whose locations on both maps are known precisely. The road map is a valuable source of landmark information.

Several combinations of radar and other meteorological information are especially useful. The radar/lightning-strike product has helped to generate interest in the possible connection between lightning and the life cycle of thunderstorms. Surface mesonet data add a second dimension to radial velocity data and help confirm the existence and movement of convergence zones and thunderstorm outflows. They provide data where clear-air tracers are insufficient to produce a detectable return. Radar data show the precipitating portions of large cumulonimbus clouds as viewed by satellite.

On the satellite/radar image combination, it is disconcerting to discover a mismatch between the locations of the reflectivity echo and the corresponding visible cloud. The mismatch, sometimes as large as tens of kilometers, arises in two ways: from inaccurate orbit and attitude information transmitted with the satellite images and from parallax. Parallax is the apparent location error of a cloud top relative to the ground, resulting from the satellite's non-vertical viewing angle. The error is always in the direction away from the satellite. Parallax becomes serious only when tall clouds are viewed from the side. Note that neither of these problems is related to remapping procedures or the choice of map projection.

The first problem can be corrected by accurate navigation of the images at a central location before they are remapped. No attempt is made at PROFS to correct the second problem, but the parallax error could be computed upon request as a function of cloud height and the satellite zenith angle measured at the cloud location.

By examining radar and other kinds of data in combination, forecasters can make an informed judgment on the potential for severe weather. Upon deciding that a severe storm is imminent, forecasters normally display the latest PPI image, center a box on the echo, and adjust its size (Fig. 4). The system then solicits information on the type of warning and its duration. Using a network of locator points, each associated with part of a county, the system decides which points are inside the warning box and thereby determines which counties or parts of counties are at risk. It lists cities within the warning area. It describes the warning, gives

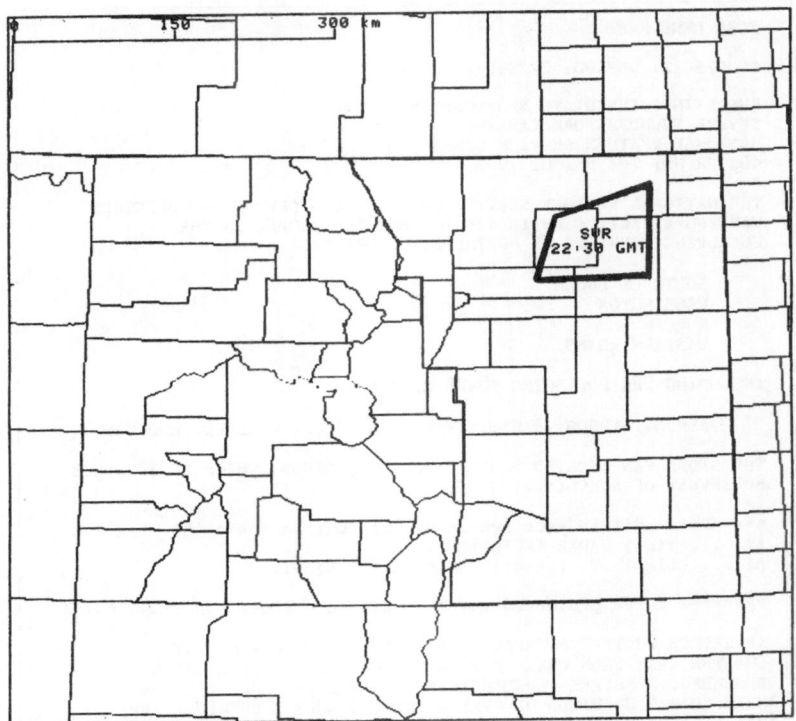

Fig. 4 A sample severe weather warning box, drawn by the forecaster
        and including parts of several counties in northeast Colorado.
        Colorado is the rectangular state occupying most of the map.

valid times (by default, a warning is effective upon issuance), and com-
poses a call to action. Before the warning is disseminated, the forecaster
edits the preformatted message by deleting unnecessary sections and adding
optional comments. A sample unedited message appears in Fig. 5.
        The semi-automated issuance of warnings will mark a major improve-
ment in warning procedures in the United States. Currently, local radar
displays are separate from displays of satellite data and both are separate
from the AFOS (Automation of Field Operations and Services) terminals,
which forecasters use to edit and transmit warnings. To generate a severe
weather warning, the forecaster must look at the radar image, estimate
which counties and parts of counties are at risk, mentally draw a box (or
actually draw one with a grease pencil), and then move to the AFOS console
to compose the message.

```
ZCZC DENwrkadm

COC075-121-087-001-022230-

BULLETIN...IMMEDIATE BROADCAST REQUESTED
SEVERE THUNDERSTORM WARNING
NATIONAL WEATHER SERVICE DENVER CO
403 PM MDT TUE MAY 02 1989

THE NATIONAL WEATHER SERVICE HAS ISSUED A SEVERE THUNDERSTORM
WARNING EFFECTIVE UNTIL 430 PM  MDT FOR PEOPLE IN THE
FOLLOWING COUNTIES OF NORTHEAST COLORADO:

        SOUTHERN LOGAN
        WASHINGTON
        MORGAN
        EASTERN ADAMS

INCLUDING THE FOLLOWING TOWNS/CITIES:

        AKRON..ATWOOD..BRUSH..FT MORGAN..OTIS..WILLARD..AND WOODROW

THE STORM WAS LOCATED 5 MILES SOUTH OF BRUSH..WHICH IS 37 MILES
SOUTHWEST OF STERLING.

********* edit in your source of information *********
AT ....(time) RADAR INDICATED ....
AT ....(time) ....(event) WAS REPORTED BY ....

******** enter projected movement of storm (location/time) *****

IF SEVERE WEATHER APPROACHES YOUR AREA..GO TO A STURDY
SHELTER AWAY FROM CREEK BEDS AND OTHER FLOOD PRONE AREAS.
REMEMBER..A SEVERE THUNDERSTORM CAN PRODUCE LARGE DAMAGING
HAIL..DAMAGING WINDS OF OVER 60 MPH..DEADLY LIGHTNING..AND
HEAVY RAINS THAT CAN QUICKLY FLOOD A LOW LYING AREA.

******** if intense lightning is occurring ********
INTENSE LIGHTNING IS REPORTED WITH THIS STORM.  IF OUTDOORS..
STAY AWAY FROM HIGH OBJECTS.  MOVE INDOORS IF POSSIBLE.  WHEN
INDOORS..STAY AWAY FROM WINDOWS AND DOORS AND AVOID USING
TELEPHONES AND ELECTRICAL EQUIPMENT.

******** if a tornado watch is still in effect ********
THESE STORMS CAN PRODUCE A TORNADO WITH LITTLE OR NO ADVANCE
WARNING..SO BE ON THE LOOKOUT AND BE PREPARED TO MOVE TO A PLACE
OF SAFETY IF ONE IS SPOTTED.

******** if a watch of any kind is still in effect ********
A TORNADO/SEVERE THUNDERSTORM WATCH REMAINS IN EFFECT FOR ....

TO REPORT SEVERE WEATHER HAVE THE NEAREST LAW ENFORCEMENT
AGENCY RELAY YOUR REPORT TO THE NATIONAL WEATHER SERVICE OFFICE
IN DENVER..OR CALL YOUR REPORT TO THE NATIONAL WEATHER SERVICE
AT 361-0663.  DO NOT CALL THIS NUMBER FOR INFORMATION.

********* enter your name on the next line *******
....(name)

LAT..LON  4044 10392 4071 10286 3988 10290 3987 10413
```

Fig. 5 A sample unedited severe weather warning message.

4. CONCLUSIONS

The major points made in this paper are:
The choice of a common coordinate system should not be taken lightly;
it depends upon many factors:

- which property of the projection is most desired--equidistance, conformality, or area-equivalence
- the volume of data to be remapped and frequency of remapping
- the division of labor between central and local processors
- requirements for position accuracy
- user preferences

Being able to view radar data along with other relevant information is
vital. Independent image and overlay control facilitates the analyst's
interpretation of the information.

Appropriate backgrounds help convey the spatial relationship between
precipitating clouds and political boundaries, population centers, familiar
landmarks (natural and man-made), and drainage basins.

To better analyze individual storms or complexes of storms, forecasters profit by slicing radar data from full-volume scans in a variety
of ways. The process of issuing severe weather warnings has been greatly
streamlined. Moreover, when forecasters view radar data superposed on
the appropriate maps, the area to be warned can be precisely defined.

ACKNOWLEDGMENTS

I thank Peter Amstein for his lucid explanation of image combination.
Joe Wakefield taught me how images and graphical overlays are combined.
Paul Schultz provided information on severe weather warnings and Figs. 4
and 5. I am grateful to Robert Lipschutz for his discussions of radar
products and for the loan of the first three figures.

REFERENCES

[1] LIPSCHUTZ, R.C., E.N. RASMUSSEN, J.K. SMITH, J.F. PRATTE, and C.R. WINDSOR (1989). PROFS' 1988 real-time Doppler products subsystem. Preprints, 24th Conference on Radar Meteorology, 27-31 March, Talla- hassee, Florida, American Meteorological Society, Boston, 211-215.

[2] RASMUSSEN, E.N., J.K. SMITH, J.F. PRATTE, and R.C. LIPSCHUTZ (1989). Real-time precipitation accumulation estimation using the NCAR CP-2 Doppler radar. Preprints, 24th Conference on Radar Meteorology, 27-31 March, Tallahassee, Florida, American Meteorological Society, Boston, 236-239.

[3] LIPSCHUTZ, R.C., J.F. PRATTE, AND J.R. SMART (1986). An operational ZDR-based precipitation type/intensity product. Preprints, 23rd Conference on Radar Meteorology and Conference on Cloud Physics, 22-26 September, Snowmass, Colorado, American Meteorological Society, Boston, JP91-JP94.

[4] ROSENFELD, A., and A.C. KAK (1982). Digital Picture Processing, 2nd edition, Vol. 2, Academic Press, New York, 22-36.

NEAR-REAL-TIME PRECIPITATION ANALYSIS OVER EUROPE

D M GODDARD and B J CONWAY
Meteorological Office, Bracknell, Berks, UK.

Summary

The UK Meteorological Office prepares hourly analyses of the
extent and type of precipitation over a large part of Europe and the
NE Atlantic. The analyses, which are available within about two
hours of datum time, are primarily for use with a numerical
dispersion and deposition model to predict the distribution of
contamination in the event of a nuclear accident like that at
Chernobyl in 1986.
    Quality-controlled radar data from the Meteorological Office's
FRONTIERS system and radar data from other western European countries
under the COST-73 project provide the most detailed information, but
are supplemented by conventional surface observations, Meteosat
imagery and products from NWP models. Analyses are generated at a
resolution of about 5km within the COST-73 area and at the lower
resolution of the Meteorological Office's "fine mesh" regional NWP
model beyond this, covering in all an area from 30°W to 40°E and 35°N
to 70°N. It is planned to extend the area of high-resolution
coverage as data from other national radar networks become available.
    Combining the different observations and products into a
coherent analysis is difficult because of their diverse
characteristics. At present, where more than one type is available
at a single location, they are selected according to an empirical
preference order. Work continues on better ways to combine the data
types, on using the observations to check and correct the NWP model
predictions, and on the derivation of confidence values.

1 INTRODUCTION
    On April 26th 1986, an accident occurred in Reactor 4 of the
V.I.Lenin Nuclear Power Station at Chernobyl in the Ukraine (USSR). This
resulted in the world's most serious accidental release of radioactive
materials into the atmosphere to date. Much of the released material was
carried aloft over great distances, resulting in contamination over much
of Europe (1).
    The Chernobyl plume crossed the UK during 2-4 May 1986, causing
significant contamination in the north and west of the country, mostly
over high ground and where rain had fallen. In the aftermath of this
event, plans were made in the UK to ensure that rapid assessment of the
distribution and amount of surface contamination could be made in any
future incident. To this end, a network of monitoring stations was
established thoughout the UK and work was started on the Nuclear Accident
Modelling Exercise ("NAME") - the development of a numerical dispersion
and deposition model (Figure 1). The purpose of the model is to provide
initial estimates of surface contamination to aid the organization of
emergency action and of detailed confirmatory measurements.

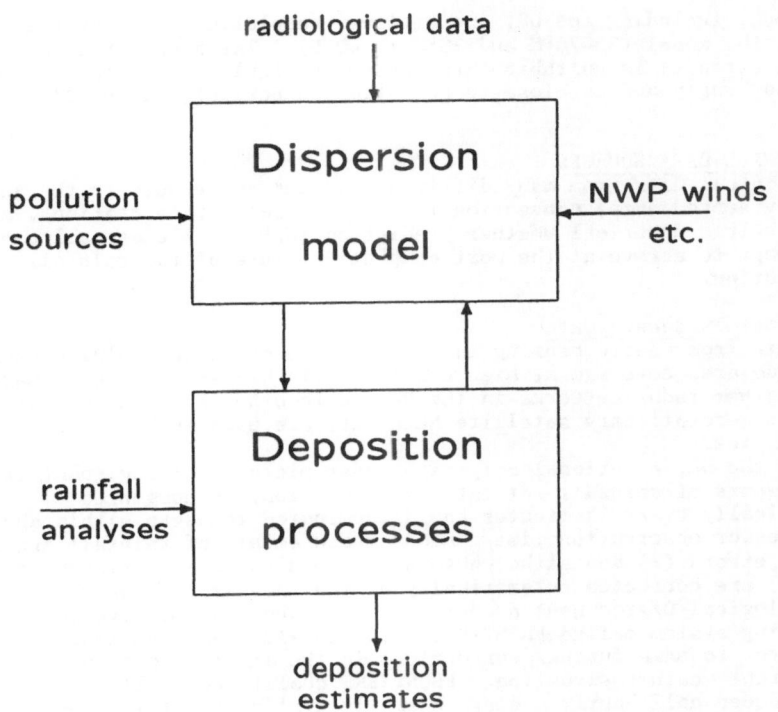

radiological data

Dispersion

pollution   NWP winds

sources   **model**   etc.

Deposition

rainfall

analyses   **processes**

deposition
estimates

Figure 1. Simplified diagram of the NAME dispersion/deposition model,
showing the main inputs and outputs.

1 THE IMPORTANCE OF RAINFALL

Airborne pollutants (nuclear or chemical) can reach the ground by
either dry or wet deposition processes, and both of these will be
represented in the model. Dry deposition is the direct transfer of
airborne material as it comes into contact with the surface by
sedimentation, absorption or impaction. Sedimentation is particularly
important in the immediate vicinity of the event, where larger particles
are preferentially removed by gravitational settling, but further afield
wet deposition and other aspects of dry deposition predominate. Wet
deposition is the uptake of material by precipitation which subsequently
falls to the surface. The particles can either be absorbed by water
droplets while they are growing in cloud or be swept out by precipitation
as it falls. Deep convective clouds are particularly effective at
reducing the amount of airborne material because they entrain large
volumes of contaminated air, and a heavy rain-storm can halve the local
mass of airborne pollutant in less than thirty minutes (1).

Because of the importance of rainfall in producing high local
concentrations of surface contamination, and the need to model the wet
deposition process, means are being developed to provide the dispersion
and deposition model with frequent, up-to-date analyses of rainfall
distribution. These analyses will include details of the type (frontal or
convective) and intensity of rainfall at a resolution of about 5km over

NW Europe, including the UK, and at lower resolution over the remaining area of the model (35–70°N and 30°W to 40°E). The higher resolution area will be extended as suitable data become available. The analyses will be provided hourly and as close to real-time as possible (at present within 2 hours).

3 RAINFALL DATA SOURCES

Rainfall data from many different sources contribute to the analyses. Remotely sensed data, conventional reports from surface stations, and products from numerical weather prediction (NWP) models are all used, in an attempt to arrive at the most complete picture of the rainfall distribution.

3.1 Remotely Sensed Data

Data from remote sensing instruments are of unique value because they give wide-area coverage at high resolution in near-real-time. Image data from weather radar networks in the UK and in other European countries, and from the geostationary satellite Meteosat, are used as basic components in the analyses.

In the UK, a national composite radar picture, giving quantitative measurements of rainfall rate at 5km resolution, is assembled automatically every 15 minutes and disseminated to users within about 6 minutes of observation time. Radar measurements of rainfall are subject to many errors (2) and although some of these (such as permanent ground clutter) are corrected automatically at the radar-site, the UK Meteorological Office uses a specially developed interactive image processing system called FRONTIERS (3,4) to allow an experienced forecaster to make further corrections in the light of his knowledge of the current weather situation. FRONTIERS quality-controlled composites are produced half-hourly and are normally available within about 25 minutes of observation time. It is these corrected analyses that are used in the NAME rainfall analyses.

FRONTIERS products cover only England, Wales and Ireland at present, though there are plans to extend the coverage, at least to include Scotland, in the next few years. To provide sufficiently wide area analyses for the NAME dispersion and deposition model, the FRONTIERS data are merged with other European radar observations shared under the COST-73 agreement.

The European radar data are assembled hourly at Bracknell to form the COST-73 composite, which is merged with Meteosat infrared cloud images (processed to show cloud colder than –15 C), and disseminated to the COST-73 participants within one hour of observation time. At the moment, the COST-73 composite uses only the site-corrected data from the UK, so these are replaced by FRONTIERS quality-controlled data in the composite for the NAME project.

The radar networks do not yet give complete coverage of the COST-73 area, though this situation continues to improve as further radars are added. Meteosat images provide information in the gaps between radars, and outside the range of the radar networks, but they do not indicate rainfall directly. Various methods have been developed to derive rainfall rates from satellite cloud data, with varying levels of success (5). The techniques are often specific to a location or not suitable for short period estimation. A method developed by Lovejoy and Austin (6), which is used in FRONTIERS to infer the presence or absence of rain beyond radar range, correlates the satellite data with radar measurements in areas where the two are both available and thus determines those cloud characteristics associated with rain. Correlation of the satellite data with other types of observation may also be attempted, though with extra difficulty if the observational characteristics are markedly different.

3.2 Conventional Observations

Conventional observations (mainly synoptic reports from manned surface stations, though automatic weather stations are assuming increasing importance) are useful in providing "ground truth" – reliable reports of whether rain is reaching the surface. The number of reports available depends greatly on the time of day and at best the coverage is sparse. These observations are subject to local effects and so great care must be taken in any attempt to apply them to neighbouring points in order to get an areal picture of the situation.

Weather reports from synoptic stations can be used to infer numerical rainfall rates using empirical methods (7). They can also provide information about rainfall type by distinguishing between showers and continuous rain and by reporting cloud types. This information is required to enable the wet deposition processes to be modelled correctly. Reports from European surface stations are usually available for examination within an hour of observation time.

A more accurate source of rainfall intensity comes from telemetered or hourly measured raingauges. Such gauges are fewer in number than the ordinary observing stations and at the time of writing only UK reports are routinely available for use in the Meteorological Office. Again, gauge measurements apply to point locations and care is needed in trying to infer areal values or to compare them with, for example, radar data.

3.3 Numerical Model Products

Products from NWP models complement the observational data by providing an indication of likely rainfall out to the boundaries of the dispersion/deposition model, and beyond the range of direct rainfall observations, for example over the sea. (It is important to estimate the deposition in sea areas if only to calculate how much pollutant has been removed from the aerial plume).

NWP models produce both analyses and forecasts. However, the model analyses are not available quickly enough to be used other than for investigations after the event, so it is the forecast products that will be used in the near-real-time rainfall estimates for the NAME model. Extra care must obviously be exercised in using these products, verifying the predictions against observational data wherever possible.

Two of the UK Meteorological Office's NWP models will be used in this way: the "fine mesh" model (8) which covers the full area of the dispersion/deposition model, and the higher resolution mesoscale model (9).

The fine-mesh model currently has a grid spacing of 0.75 degrees in latitude and 0.9375 degrees in longitude (approximately 75km over the UK) and has 15 levels in the vertical. It is run twice daily and outputs forecast fields at three-hourly intervals up to 36 hours ahead. This resolution is significantly coarser than that of the radar, but in remote areas it is all that is available. However the model will shortly be upgraded to 40km resolution over the UK and have 20 levels in the vertical.

The mesoscale model is currently undergoing operational trials. It is about to be extended to cover an area of about 2000 x 1800 km, centred on the UK, with a grid spacing of 15km and with 32 levels in the vertical. The intention is to run the model at 6-hourly intervals, each run producing forecast fields at hourly intervals up to 18 hours ahead.

Both models produce predictions of rainfall rates and cloud amounts. They also distinguish between rainfall types (dynamical and convective) and calculate the rate of each separately. The dynamical rainfall rate

(DRR) is averaged over the model grid-square while the convective rainfall rate (CRR) is the value for an average convective cell in the grid-square. In this form they are not directly comparable, but by multiplying the convective rainfall rate by the fraction of the model grid-square predicted to contain convective cloud (CCF), a convective rainfall rate averaged over the square is produced. Hence the total predicted rainfall rate = DRR + CRR x CCF, a quantity that can be compared with observational data.

4 COMBINING THE DATA

Combining the different data types to form a consistent analysis is not easy. Differences in spatial and temporal resolution, map projection, areal coverage, representation of the measured quantity, and reliability all make comparison difficult.

To give a feel for the problem of how to combine the data, a simplistic initial approach was taken, based on the use of an empirical preference order. The COST-73 composite's polar stereographic projection was chosen as the common projection on to which all the data would be mapped. A test case was sought with a rain-band stretching through areas covered by several data sources, and for which a complete set of data had been captured. The situation at 15.00 GMT on 25th August 1987 satisfied these requirements. Each data source was given a priority related to its accuracy and resolution. The data sets used for this composite are listed below with highest priority first.

1. FRONTIERS modified radar composite
2. COST-73 composite of European radar data (without satellite data)
3. Mesoscale model NWP forecast
4. Fine-mesh model NWP forecast

Other data sources were not incorporated into the first experimental composites.

At a pixel where a particular data source is reporting a non-zero rainfall intensity, then that intensity is used in the composite regardless of the intensities reported by lower priority sources for the same point. Under the present scheme, rain reported by a low priority source is believed even where higher priority data sources report none. The total area of rainfall therefore tends to be overestimated and it is planned to alter this rule to take account of the reliabilities of different sources. For example, the effective range of the radars varies with the atmospheric conditions (eg., with the depth of rain). This range is calculated by FRONTIERS for the UK network for each case and defines the so-called usable data boundary (UDB). Within the UDB the FRONTIERS radar data are considered reliable, therefore other reports of rain within the UDB (COST-73 and NWP) can normally be ignored when FRONTIERS indicates no rain.

This simple system of priorities was designed as a temporary expedient to get a pilot rainfall analysis scheme up and running as quickly as possible, so as to facilitate development work on other aspects of the NAME project. There is no attempt so far to combine different sources of data at a point and by so doing to arrive at the best estimate of rainfall amount at that point; the preferred data source is selected and any others are ignored. What is also missing at present is the ability to recognize errors (eg., timing errors in NWP forecasts), and then either to reject or correct individual data sources, by reference to other observations.

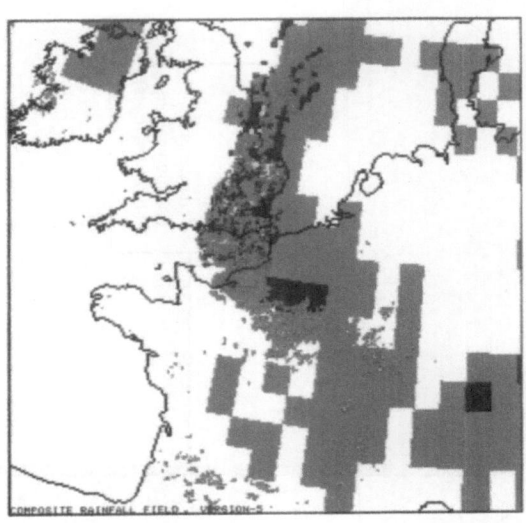

Figure 2. Composite rainfall field for 15.00GMT on 25 August 1987.
Colours indicate rainfall rates (mm/h):   blue = <1, mauve = 1-2,
pink = 2-4, red = 4-8, yellow = 8-16, green = 16-32 and brown = >32.

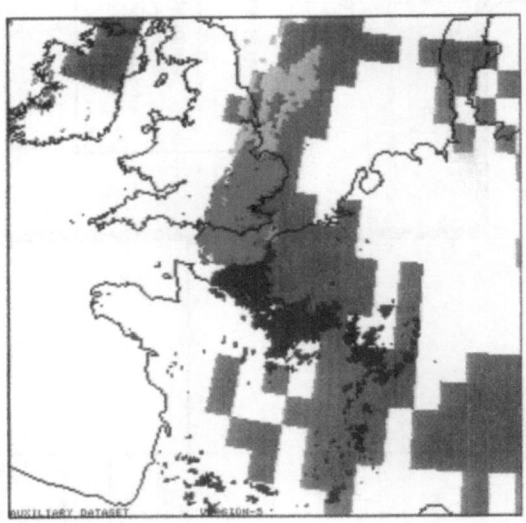

Figure 3. Map showing the data sources selected to form the rainfall
composite of Figure 2.   Blue = FRONTIERS, mauve = COST-73 radar,
green = mesoscale NWP model, brown = fine mesh NWP model.

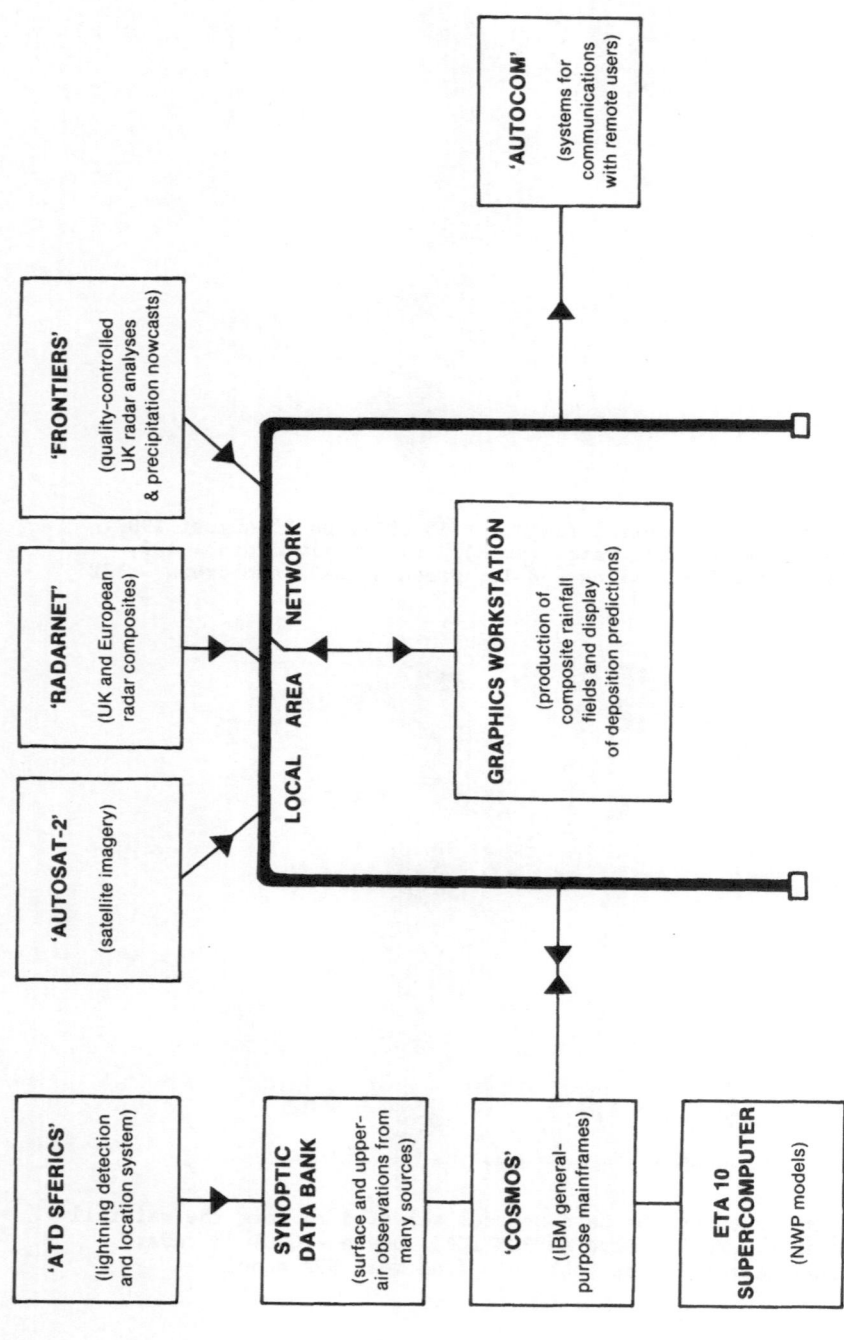

Figure 4. The interconnection of the NAME graphics workstation and other computers at Bracknell.

Nevertheless, the scheme probably makes reasonable use of the various data in situations where the individual data sources are well-behaved and in agreement. Figure 2 shows the resultant composite image within the COST-73 area for 15.00 GMT on 25 August 1987, and Figure 3 shows which data sources were used for which pixels. The results are generally encouraging in that the various data sources show reasonable agreement and combine to form a largely plausible picture, though the coarseness of the fine mesh predictions is unrealistic and gives an impression of too much rain.

5 IMPLEMENTATION

The rainfall analysis will be assembled operationally on a dedicated graphics workstation which will have the ability to display component images and final products. This arrangement offers the possibility of allowing intervention by a forecaster to modify the images before they are released, but there are considerable advantages in a completely automatic system and this is the present aim.

The data used by the analysis scheme reside on several different machines, connected by a local area network (Figure 4). The rainfall analyses are held on the IBM mainframe where the NAME model will run and where they are stored in a rolling archive of length 10 days, in case (as with Chernobyl) news of an incident is delayed and retrospective modelling of the event is needed. The rainfall archive will be continuously maintained and always available, but the NAME model will only be run for trials or if an incident occurs.

6 PROGRESS AND PLANS

The rolling archive of rainfall analyses has been established using historical data but by summer 1989 it will be routinely updated with current data.

The analyses will be based initially on the simple hierarchical selection scheme described earlier but, once the system is working routinely and reliably, priority will be given to investigating ways of combining the various data types in a less arbitrary way and obtaining, from the comparisons, a confidence factor which can be passed to the NAME model. One such possibility being examined is that of correlating the Meteosat infrared images with the model rainfall predictions in an attempt to refine and add detail to the latter.

The full NAME system is scheduled to to be complete and in operation by early 1991.

7 CONCLUSION

Hourly rainfall analyses at a resolution of about 5km within the COST-73 area and at lower resolution over a much wider area are being prepared routinely at the UK Meteorological Office.

The primary sources of data are the radar networks in the COST-73 project, but these are being augmented by observations from Meteosat and from conventional weather stations, as well as by forecasts from the Meteorological Office's fine mesh and mesoscale NWP models. Work is in progress on the problems of combining these very different data to produce the most reliable analyses.

The spur for this work was the accident at Chernobyl in 1986, and the need to be able to predict the distribution of surface contamination, known to be strongly dependent on the pattern of rainfall, in any future incident. However, once established, the composite rainfall analyses will be readily available on the Meteorological Office's central computer and are likely to find other applications.

8 REFERENCES

(1) SMITH, F.B. and CLARK, M.J. (1989). The transport and deposition of airborne debris from the Chernobyl nuclear power plant accident with special emphasis on the consequences to the UK. Met. Office Sci. Paper No 42, HMSO.

(2) AUSTIN, P.M. (1987). Relation between measured radar reflectivity and surface rainfall. Monthly Weather Rev., 115, 1053–1070.

(3) BROWNING, K.A. (1979). The FRONTIERS plan: a strategy for using radar and satellite imagery for very-short-range precipitation forecasting. Meteorol. Mag., 108, 161–184

(4) CONWAY, B.J. and BROWNING, K.A. (1988). Weather forecasting by interactive analysis of radar and satellite imagery. Phil. Trans. R. Soc. Lond., A, 324, 299–315.

(5) TSONIS, A.A. (1988). Single thresholding and rain area delineation from satellite imagery. J. Appl. Meteor., 27, 1302–1306.

(6) LOVEJOY, S. and AUSTIN, G.L. (1979). The delineation of rain areas from visible and IR satellite data for GATE and mid-latitudes. Atmos-Ocean, 17, 77–92.

(7) ANDERSSON, E., GUSTAFSSON, N., MEULLER, L. and OMSTEDT, G. (1986). Development of meso-scale analysis schemes for nowcasting and very short-range forecasting. Norrköping, Swedish Meteorological and Hydrological Institute, SMHI PROMIS-Rapporter, 1, 26–27.

(8) BELL, R.S. and DICKINSON, A. (1987). The Meteorological Office operational numerical weather prediction system. Met. Office Sci. Paper No 41, HMSO.

(9) GOLDING, B.W. (1987). Short range forecasting over the United Kingdom using a mesoscale forecasting system, In: Matsuno, T (ed), Short and Medium-range Numerical Weather Prediction, 563–572, Met. Soc. Jap.

ANALYSES OF A MESOSCALE HEAVY RAIN SYSTEM IN THE AREA OF BEIJING-TIANJIN-TANGSHAN

Xu Zi-Xiu and Wang Peng-Yun
Academy of Meteorological Science. State Meteorological
Administration. Beijing. Peoples Republic of China

Summary
       Data taken from a three conventional weather radar network as
well as densely spaced rain gauges, rawinsondes and synoptic maps
are used to analyse the process of mesoscale heavy rain systems in
Beijing-Tianjin-Tangshan area during July 29-30th, 1975. Analyses
show three types of mesoscale systems developed within 24 hours and
caused heavy rainfall of more than 200 mm in a large domain and 496
mm in some local districts. They are : meso-ß scale complex cells,
rainbands and meso-α scale cyclone. The complex cell with horizontal
scale of 50X50 Km appeared first in the early morning on July 29th,
new small convective cells merged in it continually. They produced
heavy rainfall of more than 200 mm in three hours in Tangshan area.
Their formation and development is associated with a boundary layer
jet and coastal lifting. Then six warmsector rainbands appeared and
aligned perpendicularly to the surface cold front. They have length
of about 40 - 150 Km, width ten to tens kilometers and consist of
many convective cells. The bands propagate northward with a pattern
of "younger bands develop and the older dissipate". Heavy rain or
local rainstorm occurred on each band within short period. Since the
bands are associated with the disturbance of east airflow and their
orientation is perpendicular to the vertical wind shear vector, it
is similar with the disturbance produced by shearing instability.
The model of shearing instability developed by Lalas and Einaudi
(1976) shows the wavelength of 19-43 Km and horizontal phase speed
of 20 Km/hr are in reasonable agreement with 20 - 50 Km and 30 Km/hr
of observations, respectively. When meso-ß rainbands entered in
Beijing area, three bands merged into cold front and leads to the
development of mesoscale cyclone with horizontal scale of 300X300 Km
and heavy rain. The theory of CISK is considered for its formation
and development.
In addition, the effects of orcgraphic uplift to the rainfall
enhancement are also considered.

I.   Introduction

       During 29-30 July 1975, there was a process of storm rainfall in
North China. It produced heavy rainfall in Beijing-Tianjin-Tangshan area
and southeast part of Hebei province.We mainly use the data of 3-cm and 5-
cm weather radars and a densely spaced network of raingauges (stations are
spaced by 20 Km on the average, but for some local area less than 10 Km),
hourly North China surface maps, rawinsonde data launched from Beijing and
Xintai and other synoptic maps to investigate the mesoscale systems in
this process.
       The analyses show that involved in this process there are three
types of mesoscale systems which are often observed by weather radar in
Beijing in the past years. Thus analyses of this process is very helpful

for us to understand the activities of mesoscale systems in storm rainfall. From the radar observations, these mesoscale systems are : meso- ß - scale complex cell, meso- ß - scale rainbands and meso- α - scale vortex. In this paper we present the main results of this case study : the structure of the mesoscale systems, their movement, propagation as well as their possible thermodynamical mechanisms.

II. Precipitation distribution and synoptic environment

This form produced total rainfall of more than 200 mm in the area of Tangshan, Hebei province, with maximum of 541 mm (496 mm in 24 hours) at southeast of Tangshan. In Beijing-Tianjin area and southeast Hebei province the rainfall was 100-150 mm. The precipitation mainly fell in the period of 0800 Beijing Time (here after called BT) 29 July - 0800 BT 30 July and ended at 2000 BT 30 July. Fig. 1 shows the distribution of 24-h rainfall from 0800 BT 29 - 0800 BT 30 July 1975. Several mesoscale systems embedded in a large area of precipitation, which will be discussed later in section III.

Before precipitating in North China, a cold front associated with upper level trough was moving eastward. In the east, a typhoon was moving northward along the south edge of stationary subtropical high pressure. The local heavy rain in Beijing-Tianjin- Tangshan area started at 0800 BT on 29 July, while the westerly jet on 200 hPa had moved southward to Huhehaote, Fengning and Haerbin and southwest jet on midlevel located south part of Hebei. At low level, the southeast wind form Bohai sea became strong. During 0800 BT 29 - 0800 BT 30 July the major rainfall area was between the location of the upper- and mid-level jet. Also the mesoscale systems developed in this area.

Fig. 2 shows a time-height cross section of θ_{se}, humidity, temperature and wind from 28-31 July 1975 at Beijing. From this figure we can see : In the warm sector there is a warm and wet airflow from southeast at low level (with average specific humidity 15 g/Kg). It brings abundant water vapor to Beijing-Tianjin-Tangshan area; Above the wet and warm airflow there is an airflow coming from southwest; Before the cold front arriving Beijing-Tianjin area, at 0800-1400 BT on 29 July, a shallow cold airflow from northeast under the warm and wet airflow entered Beijing-Tianjin area. All these arrangements are favourable for the establishing or releasing of the potential instability.

III Mesoscale systems and their possible mechanisms

During 0800 BT 29 July to 0800 BT 30 July the 3-cm and 5-cm radars in Beijing and 3-cm radar in Shijiazhuang observed one meso-ß scale complex cell, six meso-ß scale rainbands in front of the cold front, one meso-α vortex on the cold frontal line and several meso-ß scale rainbands associated with the vortex.

1. Meso-ß scale complex cell

The radar echos pointed by double-line arrow in Fig. 3a is a complex cell observed by 3-cm and 5-cm radar in Beijing in the morning on 29 July. It is comprised of several small and medium convective cells. Its horizontal scale grows from about 30 Km to 90 Km. and its height arrives up to 12 Km at 0615 BT (Fig.3b). The complex cell landed from north bank of the Bohai sea at 0740 BT (Fig.3c). The maximum 3-h rainfall from 0800-1100 BT generated by this complex cell is up to 232 mm which contributed to 50% of the rainfall pointed by three-line arrow in Fig.1. The complex cell moved and propagated continually,i.e. new cells form and develop in front of the main body of the complex cell and merge into it continually.

This process can be clearly in Fig.4.

Shown in Fig.4, at 0615 BT, the area of the complex cell is about 60X30 Km. In front of the main body of the complex cell, there are several new small convective cells. At 0740 BT these new small cells are merged into the main body and make it enlarged to 80X40 Km. After that time the area of the complex cell keeps almost unvariable till 0937 BT. Then several new convective cells appear again. At 1105 BT the horizontal area of the complex cell reached up to 90X90 Km. Among this process the strong echos of 30dB(z) enlarged continually and heavy rain was produced. The complex cell maintained for about 4 hours, then dissipated after 1200 BT.

Associated with the complex cell, the surface circulation is a pair of meso-high and meso-low pressure systems with scale of tens to one hundred Km. The radar echos of the complex cell located at side of meso-high pressure but the new small cells located at side of meso-low pressure. This feature is favourable to the new small cells' merging into the main body, driven by updrafts.

Since the complex cell was developed in the period of 0400-1200 BT 29 July. We use 0800 BT rawinsonde data at Beijing to analyses the thermodynamical features. This sounding is different from the others (0200 and 1400 BT soundings) in two aspects : 1) There is a jet like wind profile on the level of 600 m height (shown by the arrow in Fig.5). Shun (1978) pointed out that the development of heavy rain is closely related to jet near surface (1); 2) θ_{se} decrease with height by about $10°$ from surface to 900hPa. Richardson number (with considering the release of latent heat) of -0.25--056 shows the levels near the surface are convective unstable. Thus the planetary boundary layer instability may play important role for the formation and development of the complex cell. Furthermore, the enlargement of the complex cell due to the continually merging of new small cells is also favourable to its development. According to Levine (1959) the vertical acceleration in convective clouds depends on buoyancy, shape drag and entrainment between cloud and its surroundings (2):

$$\frac{dW}{dt} = \frac{\theta}{\theta_s} g - \frac{3}{8} \frac{3}{4} (- K + C_d) \frac{W^2}{V}$$

where W is the vertical speed at the mass centre of cloud, Cd the drag coefficient, K entrainment coefficient and r the radius of cloud.

Thus the influence of entrainment and shape drag deduced with the cloud's enlargement, which is favourable to the development of convective cloud further. In addition, the lifting of bank may not be neglected when the complex cell was landing.

2. Meso-ß scale rainbands

Six meso-ß scale rainbands are observed in this process. They are associated with mesoscale shear line in the warm and wet eastern airflow (See the surface wind in Fig.6a) and moved from southwest to northeast. Each band comprises of several large or small convective cells. The length of the band is 40-120 Km, width 20-40 Km and space 20-50 Km. They are nearly parallel each other and perpendicular to the front. The strong cells often generate heavy rain which contributes to the five rainbands (with rainfall of greater than 100 mm) in south part of Hebei province shown by small arrows in Fig.1.

In disparity of the complex cell the bands propagate non-continually. "Younger bands develop and older dissipate" is their main feature (Fig.6).

For example, in Fig.6a at 1105 BT we can see some small convective cells with scale of several kilometres were developing in front of the band labeled by A and they also arranged into a bandlike shown by the dashed line B. At 1248 BT these small cells had developed to stronger cells with scale of 20-40 Km. Fig. 6b shows the RHI pattern along the azimuth of 185°. From this figure we can see at 1103 BT the convective cell on band A had developed to 15 Km height while a new cell was developing in the middle level (4-7 Km) on band B about 50 Km away from band A. Two hours later the new cell developed to 14 Km height and horizontal scale of 40 Km. The total rainfall from 1100 - 1300 BT shows the structure of band-like rainfall (Fig. 6c).

Since the bands' orientation is perpendicular to the cold front hence to the vertical wind shear vector, similar with the perpendicular type of warm sector rainbands discussed by Xiu and Wang (1989), it is possible that the bands are generated by the shear instability (3).

In the model of shear instability developed by Lalas and Einaudi (1976), horizontal wind velocities in the shear zone are represented by a hyperbolic tangent profile given by :

$$V (z)=Vo + \delta V \; tanh \; (\; (Z-Zo)/h \;)$$

where V is the horizontal wind speed along the direction of the shear vector, Z the altitude above ground, Zo the altitude of the inflection point in the vertical profile of V, Vo the value of V at Zo. δV is one half the increase in V across the shear layer and h one-half the width of the shear layer (4). The profile is fitted to the component of horizontal wind perpendicular to the bands obtained from the rawinsondes launched from Beijing at 1300 BT 29 July 1975. The best fit parameters to the sounding are Vo=5.2 m/s, δV=1.9 m/s, Zo=1.65 Km and h=0.7 Km, respectively. According to Lalas and Einaudi (1976), the first three modes are most likely to be realised. The first mode disregard the lower boundary and the other two are due to resonance effects from the lower boundary. In our case, the wind profile are such (Yi=-Zo/h=-2.4) that the mode I and mode III are of importance. The wavelength and horizontal phase speed predicted by the model are shown in table 1.

Table 1. Comparison of model predicted parameters
of wave with observations

mode	wavelength (Km)	horizontal phase speed (Km/h)
I	7-10	18
III	19-43	20
observation	20-50	30

So the mode III is of consistent with the observations, in which the resonance effects from the lower boundary is considered.

Thus it is possible that the rainbands are produced by the shearing instability.

3. **Meso-α scale vortex on cold front**

After 1400 BT 29 July, a meso-α scale vortex was formed on the cold front in Beijing-Tianjin-Tanshan area which was responsible to the heavy rain of this area shown in Fig. 1 in addition the contribution of the complex cell and meso-ß scale rainbands discussed above.

Fig. 7 shows the radar echos of the formation and development of the vortex. Following the northward moving of three meso-ß scale rainbands in the warmsector and the shallow cold airflow entering the Beijing-Tianjin-Tangsan area from northeast, a cumulonimburs (Cb) group associated with the rainbands is developed due to the confluence of flows from northwest, northeast and east. Then this Cb group was nearing the precipitation echos on the front and combined with them by 1400-1500 BT and mesoscale vortex echo formed. Associated with this system a mesoscale system cyclone with horizontal scale 300X300 Km and cold and warm front circulation field on surface can be analysed (Fig.7a).

Major precipitations associated with this vortex is concentrated on the side of the warm front and in warm sector.

On the warm front, the precipitation is mixed type i.e. convective cells embedded in the uniform rainfall. These convective cells sometime arranged to band-like parallel to the warm front. The 5 meso-ß scale rainbands along the south slope of Yianshan mountain shown by double-line arrows in Fig. 1 are mainly produced by these warmfrontal rainbands. They were strengthened further by the orographic uplift.

The updraught Wo caused by the upslope of flow can be calculated by:

$$W_o = -\rho_o \, g \, V.\nabla h$$

where h is the terrain height. Using the south slope terrain data of Yianshan mountain and the southeast wind of speed 5 m/s, Wo is estimated about 0.22-0.30 m/s.

Then the orographic uplifting induced rainfall is calculated. The southeast wind maintained for 4-5 hours in the period of 0800 BT 29 July to 0800 BT 30 July and the estimated rain enhancement is about 28-34 mm.

In the warm sector, several meso-ß scale convective rainbands are observed by radar. They are parallel to the cold front, tens to one hundred Km long (shown by solid lines in Fig.7a b). The larger cells also produced heavy rain which are responsible to three southwest to northeast heavy rainfall band shown by small dashed arrows in Fig.1.

During 1500-1700 BT the vortex developed to its mature stage (Fig.7b), the echos both in warm sector and along warm front show clearly spiral structure. Accordingly the cyclonic circulation was strengthened. The south wind near the centre of cyclone enhanced from meters per second to 10-12 m/s. Fig.7d shows radar RHI reflectivity patterns along 360°, 115° and 185°. The mixed type of warm frontal precipitation in the cross sections of 360° and 115° and several strong convective cells in 185° can be seen.

From 1900-2200 BT 29 July rainbands continually formed near the cold front and spun towards warm front (Fig.7c). This is the main reason why the warmfront precipitation can be maintained. As long as the generation of the warm sector rainbands stopped the precipitation echos on warm front will weaken or even dissipate. The vortex maintained about 18 hours in Beijing-Tianjin-Tanshan area. After 0800 BT 30 July it moved northeast and weakened.

From the genesis and development of the mesoscale cyclone discussed above we can see clearly the interaction between Cb group and the cyclonic circulation. Before the cyclone genesis three currents confluenced, which induced a Cb group's formation and development and combined with the cold precipitation. From Fig.2 we can see that a deep wet layer (RH _100%) associated with a warm ridge appeared under 400 hPa from 1400 BT 29 to 0200 BT 30 July when the cyclone was developing. This is responsible for the convection development and release of latent heat. The warming by the latent heat caused the decreasing of surface pressure. Thus cyclonic circulation strengthened further and Cb cloud developed more. This process is similar to that associated with the release of the second kind of conditional instability (CISK).

IV. Conclusions
1.   The storm rainfall process includes several types of mesoscale systems : meso-ß scale complex cell produced local heavy rainfall; meso-ß scale rainbands produced local heavy storm rain and meso-α scale vortex continually produced heavy rainfall in a large area.
2.   The formation and development of these mesoscale systems are very complex. They may relate to the interaction of different scales and energy sources. The discussions in this paper are only primarily. However, the merging of various scale systems may be an important factor. Various scale systems developed at different time reflects the different characteristics of thermodynamics of atmosphere. They are associated with planetary boundary instability, shearing instability and CISK, respectively.
3.   Orographic influence on the development of mesoscale convective systems is another important factor, especially on the upwind slope there the precipitation enhancement is clear.

REFERENCES

(1) Shun Shu-Qing (1978). The relationship between heavy
    rainfall and boundary layer jet. Collecting papers of
    conference on heavy rainfall. (in Chinese). 40-46.
(2) Levine J. (1959). Spherical vortex theory of bubble-like
    motion in cumulus clouds. J. Meteor. 652-653.
(3) Xu Zixiu and Wang Pengyun (1939). Mesoscale rainbands
    ahead of cold fronts in northern China and their
    dynamical mechanism analyses. ACTA Meteor. Sinica. N°.3.
(4) Lalas D.P. and F. Einaudi 1976). On the characteristics
    of gravity waves generated by atmospheric shear layers.
    J. Atmos. Sci. 1248-1259.

<u>Figures caption</u> for "Analyses of a mesoscale heavy rain
                    system in the area of Beijing-Tianjin-
                    Tangshan" By Xu Zi-Xiu and Wang Peng-Yun,
                    China

Fig.1 Distribution of 24-h accumulated rainfall from
      0800 BT 29 July. Mountains and plain area is
      separated by the dotted-dashed line.

Fig.2 Time-height cross section of θse (in K), humidity,
      temperature and winds derived from rawinsondes
      launched every six hours from Beijing on 28-30
      July 1975. Thin solid lines - θse, dotted-dashed
      lines - RH, dotted lines - T, shaded area -
      regions of potential instability.

Fig.3 Radar reflectivity factor patterns of meso-β scale
      complex cell in the warm sector on 29 July 1975.
      (a) PPI; (b) RHI at 0615 BT. azimuth - 135°;
      (c) total rainfall from 0800-1100 BT produced by
      the complex cell landing. Double-line shows the
      landing direction.

Fig.4 Serial PPI radar reflectivity factor patterns
      shown the development of the complex cell from
      0600-1200 BT 29 July 1975. + represents the
      position of radar in Beijing at each time.

Fig.5 Wind profile near surface. Data taken from 0800 BT
      on 29 July 1975 sounding from Beijing.

Fig.6 (a) Composite map of PPI scopes of reflectivity
      taken at Beijing and Shijiazhuang on 29 July
      1975 and surface wind. Range markers are 100
      Km. The radar echos are contoured at 0 db with
      shaded areas representing 30 db; (b) RHI scope
      at 1300 BT and 1303 BT. azimuth - 185°; (c)
      Total rainfall (mm) from 1100-1300 BT on 29
      July 1975.

Fig.7 PPI scopes shown the genesis (a). mature stage (b)
      and maintaining stage (c) of mesoscale vortex on
      the cold front on 29 July 1975;(d) RHI scopes,
      azimuth-360°, 185° and 115°.

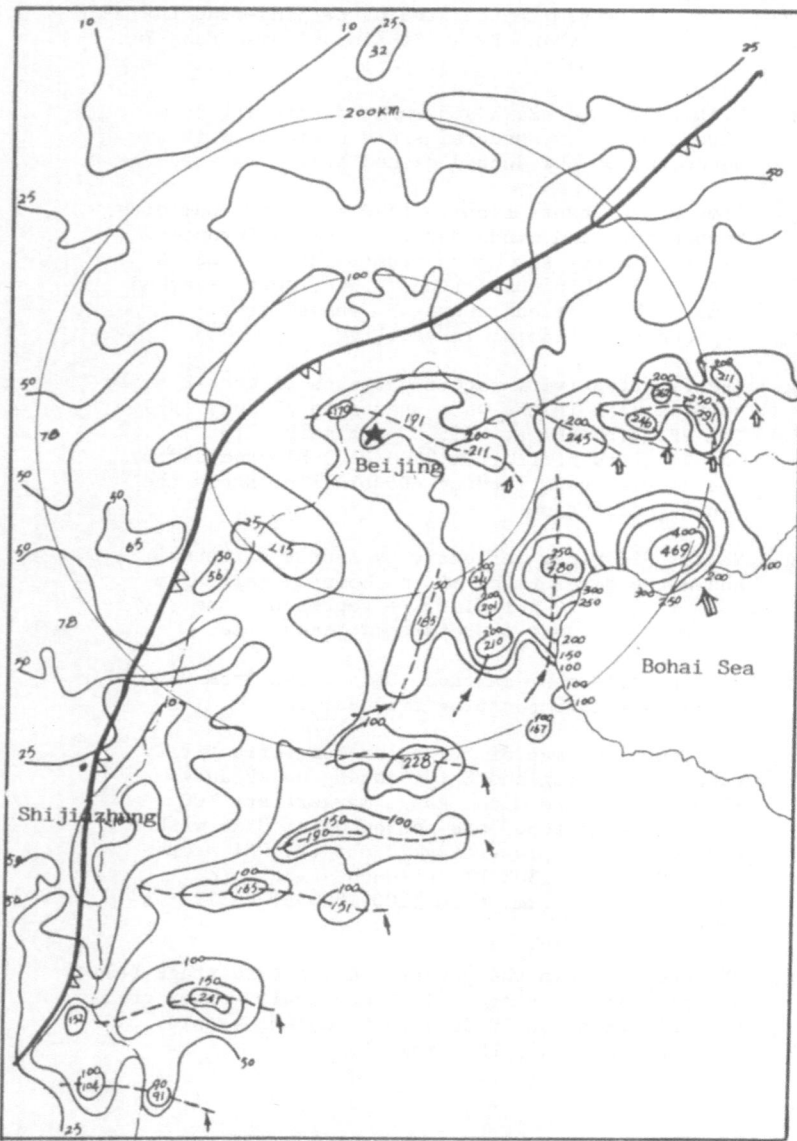

Figure 1

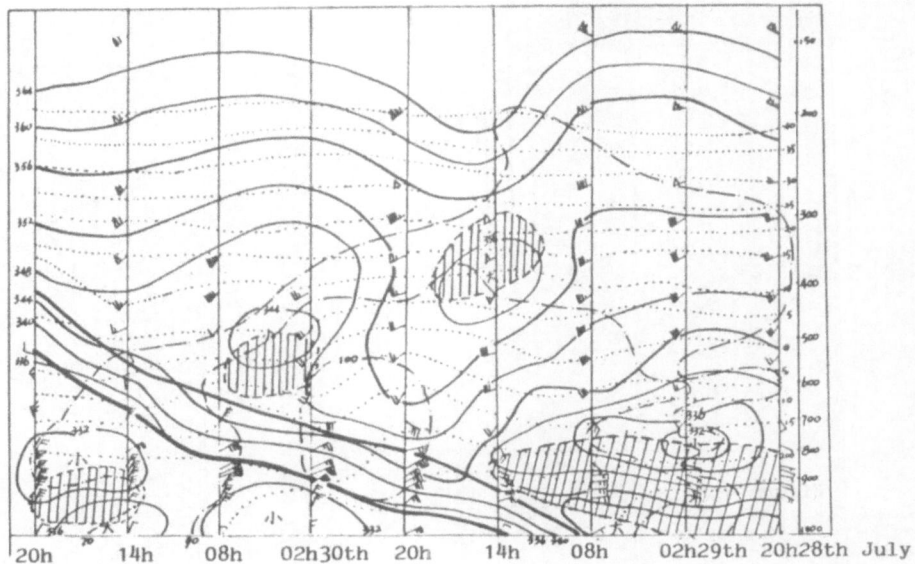

Figure 2

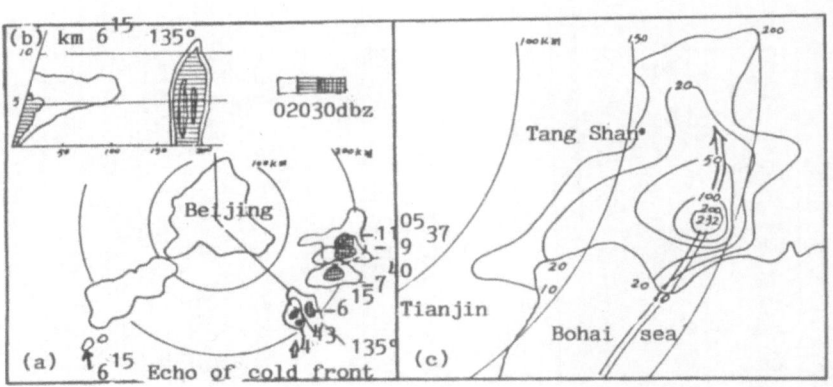

Figure 3 a. b. c.

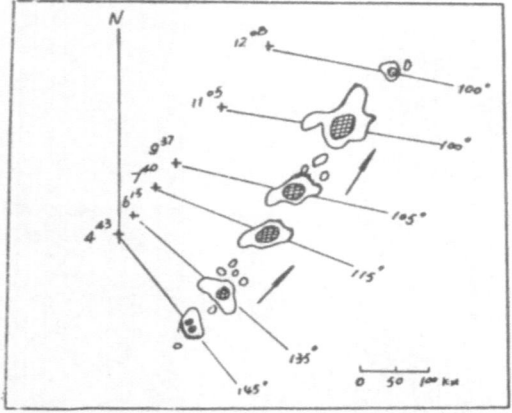

Figure 4

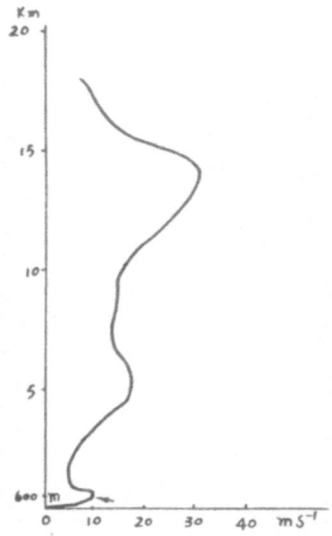

Figure 5

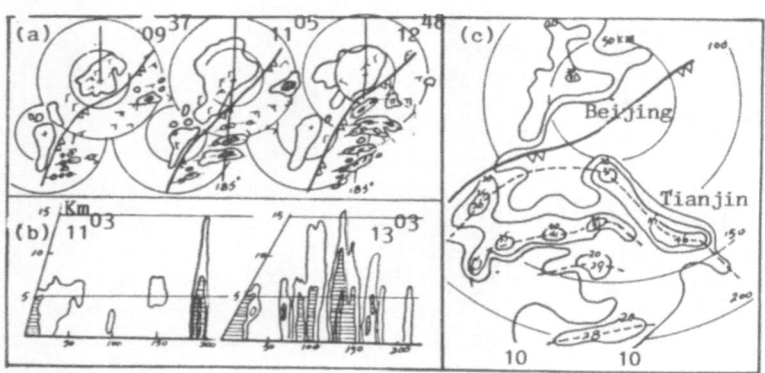

Figure 6

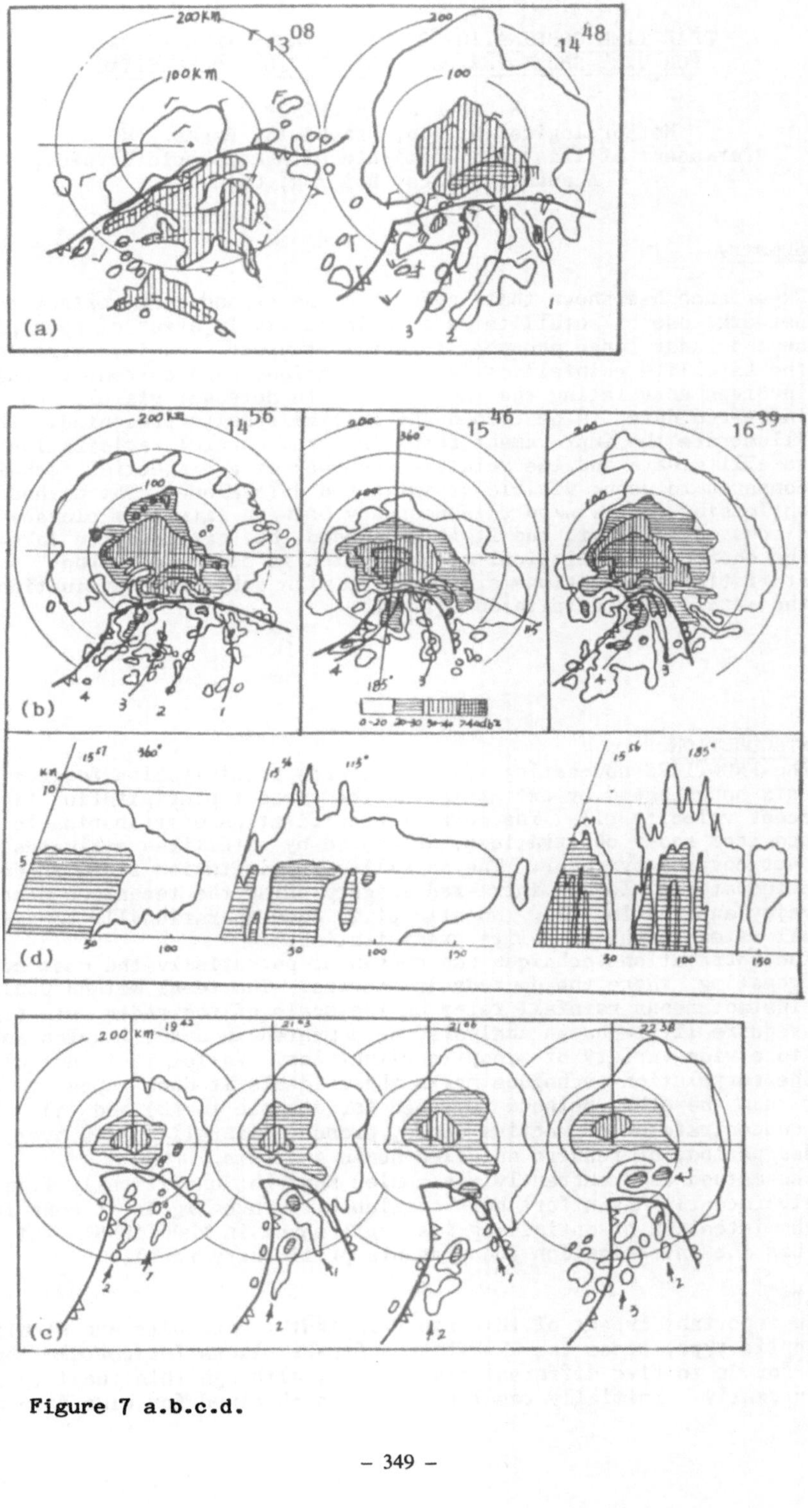

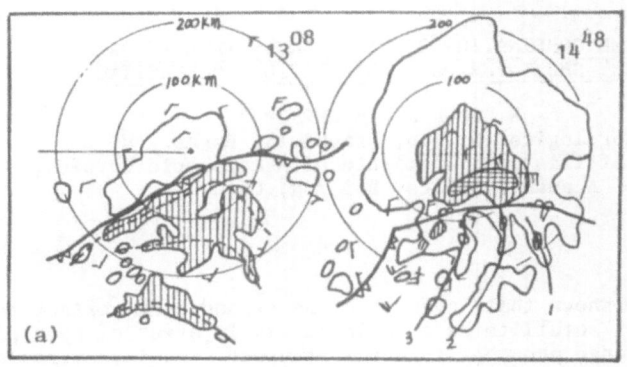

Figure 7 a.b.c.d.

REAL-TIME COMBINATION OF RADAR AND SATELLITE DATA
FOR VERY-SHORT-PERIOD PRECIPITATION FORECASTING

R BROWN and M CHENG*
Meteorological Office, Bracknell, Berks., UK
*Permanent affiliation: Institute of Atmospheric Physics,
Academia Sinica, Beijing, China

Summary

Experience has shown that, even with the expanded UK weather radar
network, use of satellite data to infer likely areas of precipitation
beyond radar range enhances forecast accuracy. An investigation of
the satellite rainfall estimation technique used currently, which
involves correlating the radar data with Meteosat visible and/or
infra-red data, is described and initial results presented. These
illustrate the improvement resulting from careful registration of the
satellite data and the relative accuracy of a bispectral technique
compared to using visible or infra-red data alone. The method of
optimising the rain/no rain boundary between satellite classes is
discussed in detail and it is concluded that care must be taken in
the choice of statistical measure which is maximised. The
statistical evaluations are compared with subjective evaluations of
the satellite precipitation fields.

1. INTRODUCTION

The FRONTIERS nowcasting system produces precipitation forecasts for
up to six hours ahead by extrapolating the current precipitation field at
its recent velocity (1). The current precipitation distribution is
obtained from radar observations, augmented by satellites estimates in
areas not covered by radar. The satellite precipitation field is derived
from Meteosat visible and infra-red imagery using the technique described
by Lovejoy and Austin, (2). No attempt is made operationally to assign
rainfall rates to the satellite precipitation.

The correlation technique was chosen as potentially the most suitable
for nowcasting, where the demands are severe. The ideal method would
yield instantaneous rainfall rates on the scale of the radar data (eg
5km), require little human analysis, be computed in a few minutes and
apply to a wide variety of synoptic situations. Whilst it is not claimed
that the correlation technique meets these ideals it does appear to come
closer than the other methods reviewed for example in (3) and (4). These
often concentrate on convective cloud, produce rainfall totals over an
extended period, or require detailed human analysis.

The method used currently was copied somewhat uncritically from (2)
and only recently has effort been available to investigate it more fully,
with the intention of optimising its performance in FRONTIERS. This paper
describes the investigation and presents preliminary results.

2. DATA

An important aspect of this study is that the results are stratified
by synoptic type, since the FRONTIERS software allows for correlation
tables for up to five different cloud types, although this facility is not
used currently. Initially results have been obtained for nine frontal

cases and nine cases of widespread convection. Because the main aim of
the work is to optimise the basic technique, frontal cases were chosen
where visual examination revealed at least a moderate correlation between
the cloud and precipitation patterns. Cases with a large area of
non-precipitating cirrus ahead of the (warm) front will be investigated
later. The convective cases contained precipitation on a variety of
scales from one or two pixels upwards but excluded mesoscale convective
systems and convection embedded within fronts. Although the convection
was generally fairly disorganised, on some occasions bands of convection
were present, generally triggered by topographic features. A typical
example of a convective case is shown in Figure 1a.

The radar data were obtained from the UK network plus Shannon in the
Republic of Ireland. The radars are operated in the PPI mode with a beam
elevation of 0.5° and have a maximum range of 210km. The minimum rainfall
rate detected is 1/32 mmh^{-1} within 50km of the radar but decreases to
about 0.2mmh^{-1} at maximum range. The quality-controlled composite radar
field produced by the FRONTIERS system was used for this study. Thus most
echoes not arising from precipitation had been removed manually, as was
any corrupt radar data. Because the radars often miss precipitation at
maximum range, the radar data was only used out to the FRONTIERS Usable
Data Boundary (UDB) for shallow layer cloud, which is shown in Figure 1a.
The UDB is a semi-theoretical estimate of radar range and allows for beam
occultation. The shallow layer cloud UDB is the most restrictive
category. Sometimes only part of the composite picture was used to avoid
a mixture of cloud types.

The satellite data comprised half-hourly visible and infra-red
imagery from the Meteosat geostationary satellite. The images were
registered manually using any coastline features discernible. Normally
the same registration offset, of typically three pixels, sufficed for one
case. After registration the images were reprojected from space view onto
the UK National Grid, used also for the radar data, with a resolution of
5km. The infra-red data were converted to temperature using the
transmitted calibration data and the visible data were normalised to zero
zenith angle.

3. THE CORRELATION PROCEDURE

Following (2), the basic procedure involved examining every pixel
within the UDB, assigning it a visible and infra-red class and noting
whether the coincident radar pixel indicated precipitation or not (ie was
wet or dry). For use in the bispectral correlation procedure as used in
(2), the visible and infra-red data were reduced to sixteen equi-spaced
classes each. Two tables were constructed, one containing the number of
dry pixels in each class and the other the number of wet pixels. From
these the final table was produced, containing the percentage of wet
pixels in each class, which is taken to be the probability of
precipitation in that class. Examples of such tables are found in Figure
1 of (2).

Although the case studies only covered periods when visible imagery
was available, FRONTIERS also uses two 1-dimensional tables built from the
visible and infra-red imagery independently. This is necessary because at
night of course only the infra-red imagery is available. Therefore the
performance of the 1-D tables has also been investigated. Although the
current 1-D tables use 16 classes, it has been felt for some time that
more classes would delineate the rain/no-rain boundary more precisely.
Preliminary experiments with some of the frontal cases using from 16 to 64
classes indicated that using more than 32 classes produced only a small
improvement. Therefore the 1-D table results quoted in this paper refer
to the use of 32 classes.

4. OPTIMISING THE RAIN NO-RAIN BOUNDARY

Having obtained the percentage of wet pixels in each satellite class it is necessary to choose a critical percentage (Pc) such that classes associated with a higher percentage of wet pixels are designated as raining whilst the rest are classified as dry, this assignation being used operationally beyond the radar area. The first step in determining Pc is to order the classes in decreasing percentage of wet pixels. The class exhibiting the highest percentage is then assigned as rain and the number of pixels of this class in the radar area counted up. The second highest percentage is then treated similarly. This continues until some statistic measuring the success of the rainfall diagnosis, by comparison with the radar data, is maximised. The statistics investigated for choosing Pc, and also used to evaluate the results, are derived from the following contingency table.

Precipitation diagnosed

	yes	no
Precipitation observed — yes	A	B
Precipitation observed — no	C	D

In (2) Pc was chosen by minimising the fraction of errors (f) defined by :-

$$f = (C+B)/N \qquad \text{where } N = A+B+C+D$$

with the additional constraint - "that the scheme classify approximately the same number of points into the rain category as there were radar classified rain points ". We have not used f but have investigated the maximisation of two standard statistics, the Critical Success Index (CSI) and the Hansen and Kuiper Skill Score (HK), defined by:-

$$CSI = A/(A+B+C)$$

$$HK = A/(A+B) - C/(C+D)$$

Maximisation of HK was used in an initial and rather unsuccessful version of the method operationally, before the current method was introduced.

The results have been compared with the method used currently in FRONTIERS, minimising the difference in the observed and diagnosed number of wet pixels, as in (2). This is referred to as the 'minimum percent' method

For completeness, two statistics used to evaluate the results by comparison with the radar data are also defined, these are the Probability of Detection (POD) and False Alarm Rate (FAR).

$$POD = A/(A+B) \qquad FAR = C/(A+C)$$

5. SENSITIVITY TO REGISTRATION

Although the misregistration was typically 3 to 4 pixels, registration appeared to improve even the frontal cases. The difference has been quantified by averaging several statistics for the registered and unregistered cases and the results are shown in Table I. The following points should be noted about the statistics presented in this and subsequent sections. They have been calculated by comparing the satellite predictions and radar observations within the UDB. No attempt has been made so far to evaluate the satellite fields outside the radar area. Although correlations were performed at half-hourly intervals, only fields two or three hours apart were used to calculate the averages, to achieve some degree of statistical independence. Standard deviations are also shown within parentheses. In Table 1, Max HK and Max CSI refer to the values obtained by maximising these statistics to define Pc, whilst the POD and FAR refer to the minimum percent method.

Table I COMPARISON OF REGISTERED AND UNREGISTERED RESULTS

Frontal Cases (30 times)

		Max CSI	Max HK	POD	FAR
2-D	Registered	0.56(0.08)	0.65(0.14)	0.69(0.09)	0.32(0.07)
Vis+IR	Unregistered	0.52(0.08)	0.61(0.14)	0.66(0.09)	0.34(0.07)
1-D	Registered	0.49(0.11)	0.57(0.15)	0.62(0.13)	0.38(0.12)
Visible	Unregistered	0.46(0.11)	0.52(0.14)	0.57(0.14)	0.42(0.12)

Convective Cases (33 times)

		Max CSI	Max HK	POD	FAR
2-D	Registered	0.27(0.08)	0.44(0.08)	0.40(0.10)	0.60(0.10)
Vis+IR	Unregistered	0.22(0.09)	0.33(0.10)	0.31(0.13)	0.68(0.11)
1-D	Registered	0.24(0.08)	0.40(0.08)	0.36(0.12)	0.64(0.10)
Visible	Unregistered	0.17(0.07)	0.23(0.09)	0.23(0.11)	0.76(0.10)

Although the differences are small for the frontal cases, they consistently show that the registered fields score better on every statistic. This is borne out by visual inspection which generally shows a discernible improvement after registration. Not surprisingly, the improvement is greatest for the convective cases, especially for the 1-D visible tables. This is probably because the visible imagery is particularly good at picking out the small convective cells. The infra-red convective tables (results not shown) show the least improvement.

6. THE OPTIMUM CORRELATION TABLES
    Casual visual inspection has always suggested that the 2-D tables tended to give the best results, followed by the visible and this is confirmed by Table II (which repeats some statistics from Table I for convenience).
    As pointed out in (2), the 2-D method is superior because infra-red alone cannot remove non-precipitating cold thin cirrus and visible alone cannot remove non-precipitating thick boundary layer cloud. Our experience suggests that Table II is rather flattering to the infra-red tables because cases with widespread cirrus have been excluded so far. For the convective cases there is a larger difference between infra-red and visible than for fronts. The reason is probably that the convection is better defined in the visible imagery.

Table II <u>AVERAGE SCORES FOR EACH CORRELATION TABLE</u>

Frontal Cases

	Max CSI	Max HK	POD	FAR
2-D	0.56(0.08)	0.65(0.14)	0.69(0.09)	0.32(0.07)
Visible	0.49(0.11)	0.57(0.15)	0.62(0.13)	0.38(0.12)
Infra-red	0.47(0.08)	0.55(0.17)	0.60(0.08)	0.41(0.08)

Convective Cases

	Max CSI	Max HK	POD	FAR
2-D	0.27(0.08)	0.44(0.08)	0.40(0.10)	0.60(0.10)
Visible	0.24(0.08)	0.40(0.08)	0.36(0.12)	0.64(0.10)
Infra-red	0.21(0.08)	0.34(0.08)	0.29(0.11)	0.70(0.09)

7.   CHOICE OF STATISTIC TO DETERMINE THE CRITICAL PERCENTAGE

The value of Pc and hence the position of the rain/no-rain boundary varied with the statistic maximised, emphasising the importance of choosing the right one. Table III shows the average difference between the actual and diagnosed area of precipitation as a percentage of the actual area (P), negative values indicating the satellite diagnosed area was larger. Only the 2-D Tables are considered. The average Pc value is also shown. It can be seen that maximising the CSI and HK produces too large an area of satellite precipitation, especially using HK and especially for convection.

Table III <u>PERFORMANCE OF THE OPTIMISATION STATISTICS</u>

	Frontal Cases		Convective Cases	
	P (%)	Pc (%)	P (%)	Pc (%)
Max HK	-58(38)	32(1)	-225(14)	14(5)
Max CSI	-31(18)	40(6)	-63(45)	22(5)
Min Perc.	-1(8)	52(9)	1(13)	28(6)

Because the satellite precipitation is not perfectly correlated with the observed field it is not obvious that forcing equal areas yields the most useful result for the forecaster. Therefore three forecasters have independently studied the satellite fields produced by each method for the times used to produce the averages and assessed the best and worst at each time by comparison with radar data and satellite imagery, but not using ground truth outside the radar area. Table IV shows the aggregate marks in percentage terms for the 2-D Tables.

Table IV <u>SUBJECTIVE EVALUATION OF THE 2-D FIELDS</u>

Frontal Cases	Min Perc.	Max CSI	Max HK
Best (%)	46	39	15
Worst (%)	18	24	58

Convective Cases	Min Perc.	Max CSI	Max HK
Best (%)	67	33	0
Worst (%)	2	5	93

It is clear from Table IV that maximising the HK skill score generally produces unacceptable results. This method was adopted in FRONTIERS initially because the Pc which maximises HK can be predicted theoretically, without ordering the satellite classes. Minimising P

- 354 -

clearly scores best for convection and is marginally better than maximising the CSI for fronts. The average value of the fraction of errors (f) as used in (2) was calculated and was a minimum for the minimum percent method. However it was only slightly different for the other two methods, which agrees with the suggestion in (2) that it has an ill-defined minimum.

An example of the field produced by each method is shown in Figure 1 for the convective case of 30 March 1988 at 1600. A showery north-westerly airflow covered the the UK and Ireland, associated with a low in the southern North Sea. Figure 1a shows the composite radar picture from which the satellite precipitation fields in Figures 1b - 1d were derived and with which they should be compared. The area of precipitation diagnosed by the 2-D method increases going from the minimum percent method (Figure 1b), the maximum CSI method (Figure 1c), to the maximum HK method (Figure 1d) of defining Pc. The latter is obviously a gross overestimate. The forecasters voted unanimously for the maximum CSI field at this time which diagnoses the convective lines over Wales and western England extremely well. Because of the depth of the convection, some radar echoes were detected beyond the shallow layer UDB, especially over SE Ireland. These were not used in the correlation procedure but are well diagnosed in Figures 1b, 1c.

Figure 1c shows too much precipitation in the south eastern part of the radar area. This was derived from cloud associated with a decaying cold front wrapped around the low pressure centre. A band of precipitation associated with this was also diagnosed at the eastern edge of the box. This illustrates that a correlation table derived for one synoptic type may not work for another.

8. CONCLUSIONS AND DISCUSSION

The preliminary results presented here show the importance of careful registration of the satellite imagery, even for fronts. The order of merit, 2-D correlation, visible alone, infra-red alone, agrees with the results in (2). It was also found in (5) and (6) that the visible imagery was better than the infra-red at diagnosing precipitation. The choice of statistic used to optimise the rain/no rain boundary was found to be important, with equalising the diagnosed and actual area, as used in (2) and (7), the most successful, especially for convection.

The scores in Table II are somewhat lower than those found in the references quoted here. Reference (5) quotes a POD of 0.55 for the mid-latitude data in (2), which is similar to the average of our 2-D frontal and convective results. However our scores for convective cases are not as high as those in (7). Using the same 2-D correlation method for convective cases they obtained CSI=0.37, POD=0.54, FAR=0.46. Using a 2-D table with fixed class boundaries calibrated by radar, reference (5) obtained POD=0.7, FAR=0.4 for non-convective cases and POD=0.6, FAR=0.3 for convective cases. By applying fixed visible and infra-red thresholds to the same data as used in (5), for visible alone scores of POD=0.62, FAR=0.38 were obtained in (6) and for infra-red alone, POD=0.6, FAR=0.4.

We tentatively suggest two reasons for the lower scores found here. First, the correlation was performed over a much larger area. All the other studies used data from a single radar, normally to a range of less than 200km. It is suggested in (5) that the skill of the correlation method decreases with range from the training radar. Unless this was entirely due to encountering cloud of a different synoptic type to that within the radar area, it suggests some decrease of skill may be expected with a larger area. Probably more significant is that (2) and (7) used radar data in the form of CAPPIs at a height of 3km, whilst in (5) and (6), where the highest scores were achieved , radar echo top maps were

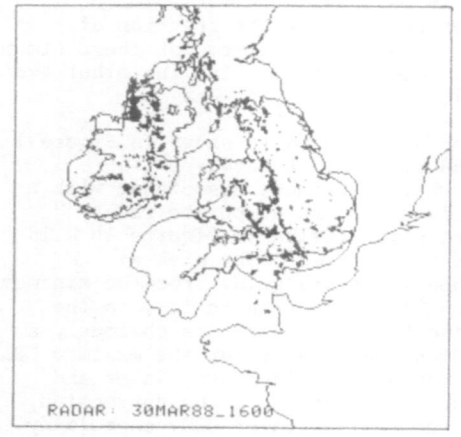

1a Radar field

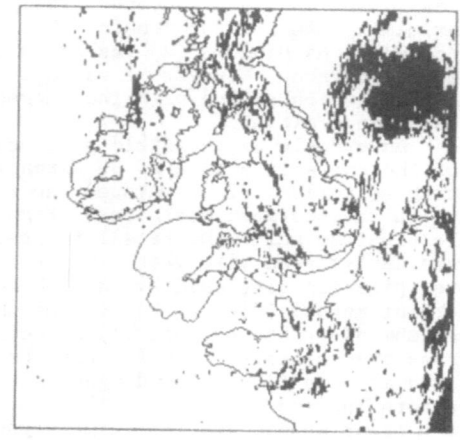

1b Satellite - Min Percent

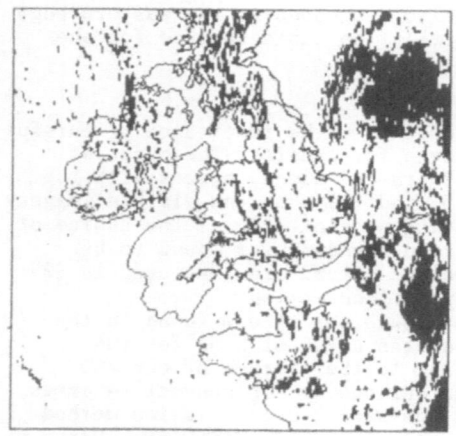

1c Satellite - Max CSI

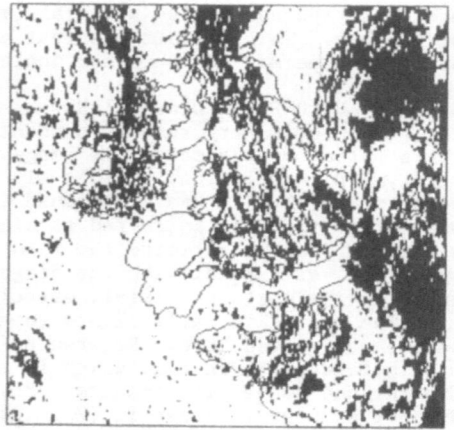

1d Satellite - Max HK

Figure 1 Comparison of the radar observed and satellite diagnosed precipitation fields for the convective case of 30 March 1988 at 1600 GMT. The radar data is shown in 1a, together with the shallow layer UDB within which the correlation was performed. The 2-D satellite fields are shown in 1b to 1d, with Pc obtained by the method annotated.

used. In the current study the radar data varied in height from close to the surface to about 3km. It seems highly plausible that the closer to cloud top the radar echoes are located, the higher their spatial correlation with cloud-top features.

It is pointed out in (7) that the low scores produced by a pixel-by-pixel evaluation (on a scale of 4km) masked the fact that when the satellite was in error it often diagnosed precipitation adjacent to its true location, which can still be valuable for short-range forecasts. This was demonstrated in (7) by evaluating over 3X3 pixel box. We have not attempted that here. However as an indication of the severity of the scoring , note that for the field in Figure 1c, CSI=0.29, POD=0.5 and FAR=0.6.

REFERENCES

(1) BROWN, R. (1987). The use of imagery in the FRONTIERS precipitation nowcasting system. Proc. Workshop on Satellite and Radar Imagery Interpretation, Reading, 20-24 July 1987, 459-472.

(2) LOVEJOY, S. and AUSTIN, G.L. (1979). The delineation of rain areas from visible and IR satellite data for GATE and mid-latitudes. Atmos-Ocean, 17, 77-92.

(3) BARRET, E.C and MARTIN, D.W. (1981). The use of satellite data for rainfall monitoring. Academic Press, 340pp.

(4) MANIKIAM, B. (1986). Rainfall estimation from satellite data - A review. Vayu Mandal, 10-14.

(5) TSONIS, A.A. and ISAAC, G.A. (1985). On a new approach for instantaneous rain area delineation in the midlatitudes using GOES data. J. Climate Appl. Meteor., 24, 1208-1218.

(6) TSONIS, A.A. (1988). Single thresholding and rain area delineation from satellite imagery. J. Appl. Meteor., 27, 1302-1306.

(7) CHERNA, E., BELLON, A., AUSTIN. G.L. and KILAMBI, A. (1985). An objective technique for the delineation and extrapolation of thunderstorms from GOES satellite data. J. Geophys. Res., 90, 6203-6210.

A PROFESSIONAL METEOROLOGICAL INFORMATION PROCESSING SYSTEM

J. Steranka and J.C. Chen, Ph.D.
General Sciences Corporation
Laurel, Maryland, U.S.A

Summary

The METPRO system offers the meteorological community an integrated, comprehensive system to collect and process meteorological data, and to generate and display graphical weather products from multiple space- and Earth-based remote observing platforms. This professional information processing system meets the challenging requirements of the diverse interests of the weather community from realtime operations to research. The METPRO system features strength, flexibility, and user friendliness in a single package and is operated on state-of-the-art and off-the-shelf hardware with industry standard software protocol.

1. INTRODUCTION

The METPRO system is a professional meteorological information processing system designed to operate on off-the-shelf hardware. The modular design of the METPRO system allows for efficient customization to meet the distinct needs of the user and provides for expansion or modification as requirements or systems change.

The METPRO system was developed to meet the international market demand for realtime meteorological satellite operations with the expanded capability to acquire, process, and analyze meteorological data from multiple sources. The METPRO system meets this demand through the automated realtime data ingest and processing from the NOAA/TIROS, GOES, and GMS satellites; radar; and conventional surface and upper air weather observations. The system derives meteorological parameters from the raw data sets and generates finished meteorological products.

The METPRO system evolved through the redesign, modification, and enhancement of METPAK (1) and the development of new applications software. METPAK is a meteorological data processing and analysis software package developed by GSC under contracts with the National Aeronautics and Space Administration at Goddard Space Flight Center and enhanced over a 12 year period to support meteorological research activity in the Severe Storms Research Program. The METPAK system is in use on the Atmospheric and Oceanographic Information Processing System workstation and has aided in the data analysis of over 200 research papers.

2. DESIGN

The METPRO system is designed for application in the broad range of activity from realtime operations to research. The system utilizes a mix of automated and interactive functions with full operator control.

a. Functional Design

The METPRO system provides automated functions for realtime data collection and processing, and interactive functions for parameter derivation and product development. The METPRO system provides specific support to:

    o   Realtime data collection, processing, and product development

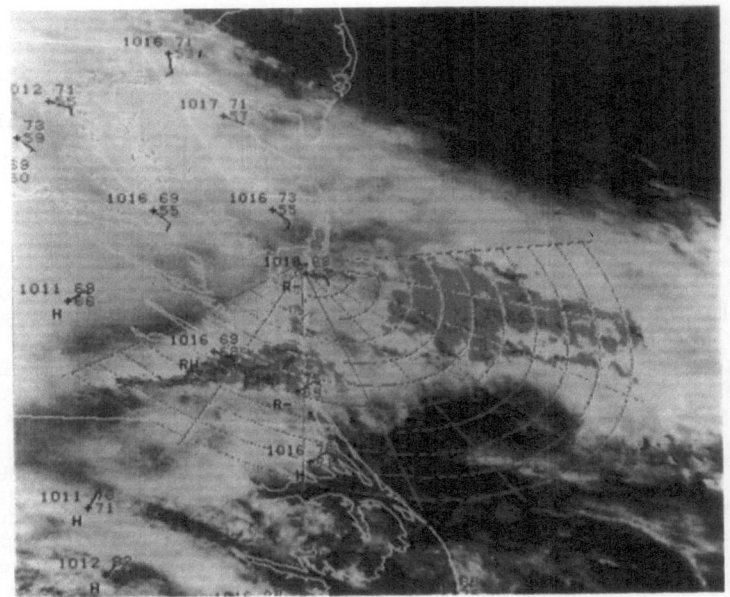

Figure 1. Satellite picture of clouds overlayed with radar
rainfall rate and surface weather reports.

o   Post realtime meteorological and oceanographic parameter
    derivation and product development
o   Operational forecasting and nowcasting
o   Weather briefings and summaries
o   Research

METPRO realtime data collection is designed to receive the High
Resolution Picture Transmission (HRPT) of the TIROS, the Stretched-
Visible and Infrared Spin Scan Radiometer (S-VISSR) transmissions of the
GMS and GOES, the volume scan data from radar, and the conventional
weather broadcast from the FAA Family of Services 604 line.   Data
collected in realtime is extracted separated, sectorized, calibrated, and
formatted into data sets for use in parameter derivation and product
development.   Satellite pictures are typically displayed and output to
hardcopy devices in realtime.

The derivation of meteorological and oceanographic parameters is
made with METPRO applications software after data sets are generated by
the realtime processing functions.   The METPRO system provides functions
to derive the following:

o   Cloud motion wind vectors from satellite imagery
o   Cloud top height and temperature from satellite imagery
o   Sea and land surface temperature from AVHRR measurements
o   Vertical temperature and moisture profiles from TOVS
    measurements
o   Geopotential heights, temperature, dew point, and wind from
    vertical profiles
o   Rainfall rate, dBz, velocity, variance, and covariance fields
    from radar

o   Divergence, vorticity, and streamline fields from cloud
        motion winds, vertical profile data, and rawinsonde reports
o   Precipitable water fields from vertical profile data and
        rawinsonde reports
o   Convective levels and stability indices from vertical profile
        data and rawinsonde reports
o   Vegetation index
o   Gridded data sets
o   Stereographic (3-D) analysis of satellite imagery

The METPRO system provides functions to perform objective analysis
of data sets and to generate map products with geographical gridding and
geopolitical boundaries. The map products consist of plotted values at
station locations or gridpoints and contours of selected values. Map
products are generated at user selected projections and scales.
Remapping functions allow any data set to be mapped to the projection of
other data sets. Maps and images prepared to a common scale are
displayed as overlays using color enhancements to clarify salient
features.

The METPRO system provides functions for the development of
satellite image, radar, and map products used for operational forecasting
and for nowcasting. Mapping to common scales allow quick interpretation
and association of weather events observed by different sensors (Figure
1). Sequential series of products may be looped in a time series to show
the movement, development, and decay of weather events. The map and
image products at multiple atmospheric levels and multiple times provide
a comprehensive four-dimensional portrayal of the atmosphere.

The meteorological products developed with METPRO are applied in
weather briefings to describe the distribution of weather events for
planning purposes in operations and aviation. Weather summaries are
generated to describe a history of weather events and to develop
climatology data bases.

The METPRO system provides research with an immensely powerful
software tool through its handling of data collected from multiple
platforms. Archived data may be ingested from magnetic tapes or other
data sources and formatted for use with the METPRO system. Products
developed for research may be displayed on video monitors and output on
hardcopy devices.

b.   Structural Design

The METPRO system is structured with major software packages which
function independently. A BASIC FUNCTION package (xxxPAK) provides data
base management and performs data set manipulation common to all of the
major packages. A REALTIME package (REALPAK) performs satellite data
acquisition and processing. A RADAR package (RADPAK) collects and
processes radar data, and generates derived products. A SATELLITE
package (SATPAK) performs image navigation, registration, and cloud
tracking. A STEREOGRAPHIC package (STEREOPAK) is provided to develop
true stereographic imagery from the observations of two different
satellites. Stereographic analysis provides highly reliable cloud
height. Sequential stereo imagery looping allows true altitude
determination of cloud motion winds. Psuedo stereographic images are
developed from visible and infrared images. A TIROS package (TIROPAK)
performs sea surface temperature, vertical temperature and moisture, and
vegetation index derivations; and develops map products of derived
parameters. An ANALYSIS package maps weather data sets, derives

meteorological parameters, performs objective analysis, develops gridded data sets, draws map backgrounds, plots values, and draws streamlines and contours.

The METPRO user is provided with descriptive menus to select and initiate functions. An on-line help allows the user to type HELP at any stage to obtain descriptive instructions. A direct control language capability allows the experienced user to bypass menus and directly initiate functions.

The METPRO system uses a Transportable Applications Executive (TAE)(2) as a user interface between operating system and program languages. The TAE interface requires users to learn only the TAE procedures to perform tasks and to let the operating system act transparently. The TAE interface allows flexibility through its independence to specific computers and devices.

3. HARDWARE

The METPRO system is designed to operate as a comprehensive data acquisition and processing system with major equipment components for realtime operations and with multiple workstations for product development, or on a single workstation dedicated to a specific need. The nature of the METPRO mission will prescribe the equipment need. A comprehensive system will require front-end equipment components (antenna, receiver, bit and frame synchronizer, computer interface) to acquire signals from observing platforms, ingest processor, host processor, system disks, input and output devices, and workstations. A forecast or research requirement may be filled with single or multiple workstations.

The configuration of a typical system is shown in Figure 2. The comprehensive system provides front-end systems for data acquisition from polar-orbiting satellites, geosynchronous satellites, radar, and conventional weather broadcasts. Ingest processor computers provide initial data collection before ingest to the host computer. Multiple image processing workstations, access terminals, and input/output devices provide for product development and display. An Ethernet links the system components.

The configuration of a workstation is shown in Figure 3. The workstation is made up of a host processor (or access to a host processor), an image processor with image and graphics display, color copier, color graphics camera, remote color monitor, remote alphanumeric and low resolution terminals, and line printer.

The METPRO system is designed to operate on the Digital Equipment Corporation (DEC) VAX series of computers linked to other system components with Ethernet. The METPRO system operates with state-of-the-art off-the-shelf front-end reception equipment, bit and frame synchronizers, video display devices, system disks, tape drives, image processors and hardcopy output devices.

4. SOFTWARE

The METPRO system is designed to operate with the DEC VAX processors using the Virtual Machine System (VMS) operating system. The METPRO programs interface with the system software of the host computer components, front-end reception components, preprocessors, synchronizers, image processing workstation, and display or output devices. DECnet protocol is used on the Ethernet. The METPRO system language is standard FORTRAN 77. TAE is the user interface system.

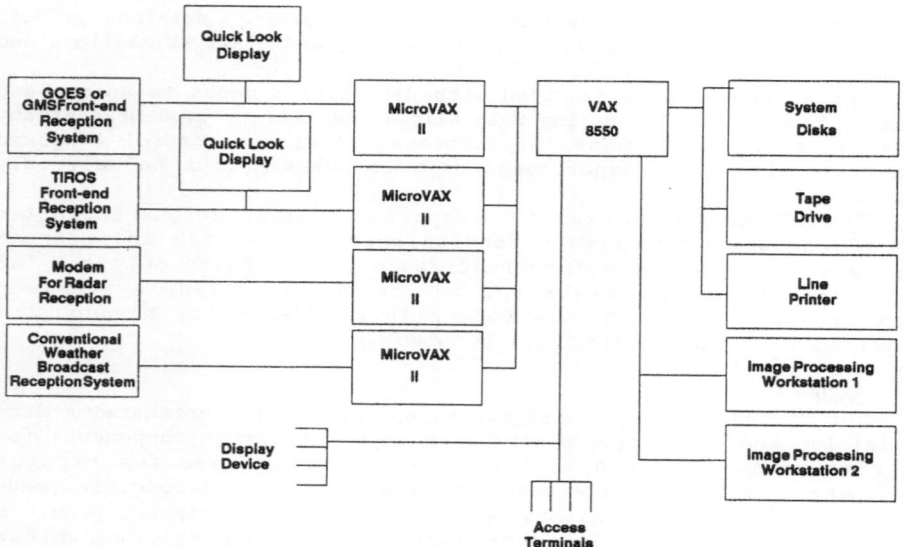

Figure 2. Configuration of comprehensive METPRO system for data acquisition and processing from multiple observing platforms. Parameter derivations and finished meteorological products are generated at the workstations.

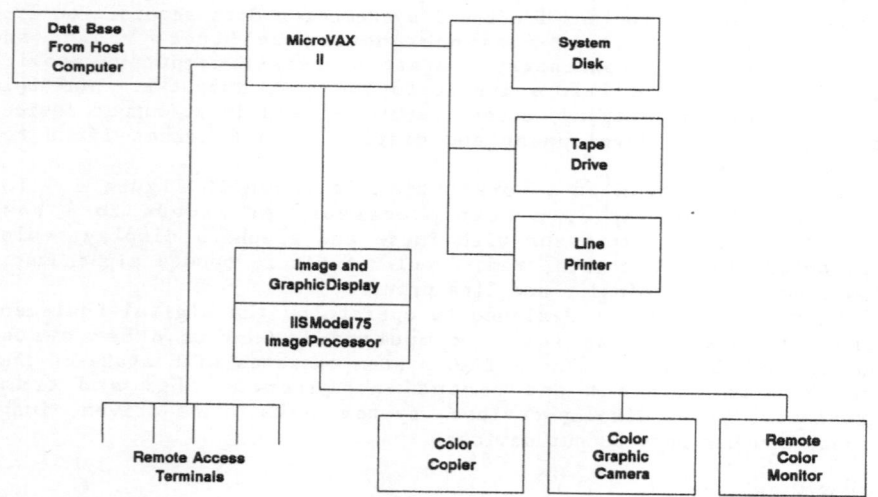

Figure 3. Example configuration of METPRO system workstation.

REFERENCES
(1)   HASLER, A.F. and M.L. des Jardins, (1987).   AOIPS/2:   An
       interactive system to process, analyze, and display meteorological
       data sets for nowcasting. Adv. Space Res. Vol. 7, No. 11, pp. (11)
       375-(11)388, 1987. Printed in Great Britain.

(2)   Developed by NASA for general use in the development of application
       software. For further information contact TAE Office, NASA/Goddard
       Space Flight Center.

AUTOMATING PRECIPITATION NOWCASTS BASED ON SATELLITE AND RADAR IMAGERY COMBINED WITH NUMERICAL MODEL PRODUCTS.

G SUTTON and B J CONWAY
Meteorological Office, Bracknell, Berkshire, UK

Summary

At the Meteorological Office, Bracknell, experienced forecasters use a specially developed display system (FRONTIERS) to interact with images from a network of ground-based radars and from Meteosat to generate, every half-hour, analyses and very-short-period forecasts of precipitation. As more radars are added to the network, the forecaster will have increasing difficulty keeping pace with the flow of real-time data. To reduce the burden, and to allow other improvements to the forecast, a new forecast stage is being developed to automate some of the interpretive and judgemental functions at present performed by the forecaster. A layered structure is envisaged, with the lowest level containing numerical algorithms for specific operations on images or other data, and upper level(s) of 'intelligence' to perform the decision-making functions using an expert system. The lower level algorithms will include correlation techniques and will also take advantage of products available from numerical models, such as wind fields, to represent the velocity field of the precipitation patterns and to indicate the likelihood of development or decay.

1. INTRODUCTION

Accurate and timely very-short-range (0-12h) precipitation forecasts are required by many users including the water industry, agriculture, transport, the construction industry, and for recreation and leisure activities.

Very-short-range forecasting and the nowcasting subclass (forecasts based upon extrapolation methods, usually for the period up to 2 or 3 hours ahead), require up-to-date observations of high spatial and temporal resolution. The UK Meteorological Office uses remotely sensed images from a ground-based radar network and from the European geostationary satellite, Meteosat, to observe the detailed structure and behaviour of precipitation fields. These data are used in combination as a routine part of the UK Meteorological Office's precipitation forecasting operation.

Two main methods are currently being used for very-short-range forecasting in the UK: linear extrapolation of the current weather, using a specially developed interactive workstation called FRONTIERS (1); and integration of the physical equations of the atmosphere, using the UK mesoscale numerical weather prediction (NWP) model (2). The geographical areas covered by FRONTIERS and the latest version of the mesoscale model are shown in Figures 1 and 2 respectively. Although both make short-period predictions for the UK, FRONTIERS and the mesoscale model have different strengths and so complement one another rather than compete. The first few hours of any detailed forecast are dominated by the current situation and recent trends, whereas for longer term forecasts these will be less important than nonlinear development. The mesoscale model predicts these longer term developments, which cannot appear in the FRONTIERS extrapolation forecasts, so that beyond a few hours ahead the numerical model will produce better guidance. However, the model cannot properly represent features with scales of less than three or four grid lengths, so the current situation and its immediate development are shown better by the FRONTIERS products. The point at which the crossover in skill occurs varies from case to case, but is normally in the range 2-6h.

A new forecast system is being developed to combine the strengths of the FRONTIERS system and the UK mesoscale model, to produce the best possible very-short-range forecast of the timing, intensity and duration of precipitation. This paper briefly describes the FRONTIERS system, the UK mesoscale model and the proposed new forecast system.

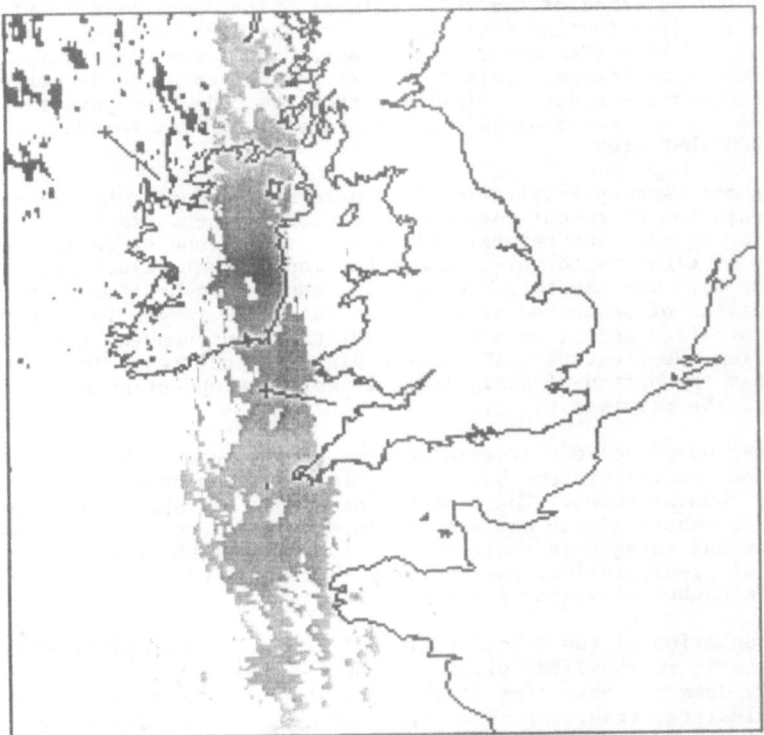

Figure 1. The geographical area covered by FRONTIERS. The combined radar/satellite-derived rainfall is shown, at a resolution of 5km, divided into two clusters (a N-S oriented front in light grey and a region of showers in dark grey). The centroid of each cluster is denoted by a cross, and each cluster has been assigned a velocity vector, pointing in the direction that the cluster will be moved.

2. FRONTIERS

FRONTIERS is a special-purpose interactive display system, which allows an experienced forecaster to examine, modify and combine radar and satellite images to produce a sequence of extrapolation forecasts of precipitation for the period 0-6h (1). The system operates on a half-hourly cycle. In each cycle the three major tasks described below are performed, each of which comprises a sequence of smaller steps.

- A quality-controlled precipitation analysis from the radar network is produced. The FRONTIERS operator aims to produce the best possible estimate of the intensity and areal extent of the rainfall, at a resolution of 5km, within the area covered by the radars (mainly the land area of the British Isles but excluding, at present, Scotland). He does this by using data from other sources (conventional observations, Meteosat cloud-images, etc.) and his understanding of the meteorological situation to help recognize and correct errors in the radar data.

- A larger-area precipitation analysis is derived using Meteosat to extend the coverage of the radar network. The operator attempts to deduce the distribution (but not the intensity) of rainfall over the rest of the FRONTIERS area, beyond the range of the radars, from Meteosat cloud-images. This satellite-derived rainfall is merged with the quantitative radar rainfall map from the radar analysis stage to provide as complete a picture as possible of current rainfall within the FRONTIERS area.

- Very-short-range precipitation forecasts are produced, by linear extrapolation of recent movement, covering the next six hours. The operator divides the rainfall field from the second stage into a number of clusters to which he assigns independent velocities (Figure 1). The clusters are defined, and the velocities measured, by examination of sequences of radar and satellite images to determine areas of differential movement. FRONTIERS then computes the forecast by moving the clusters with their assigned velocities. The operator also has the option of assigning a linear development/decay factor to each of the clusters.

The period of valid extrapolation may be six or more hours in the case of some frontal systems but may be limited to an hour or less for individual thunderstorms. The reasons for simple extrapolation breaking down after so short a time are the development and decay of rain and cloud formations, and changes in their velocities. FRONTIERS produces useful forecasts of precipitation, particularly in frontal cases, but the system does have a number of weaknesses and limitations:

- Representation of the velocity field. If he is to complete the three major tasks in FRONTIERS within the half-hour cycle, the operator usually does not have time in the forecast stage to define more than 3 or 4 clusters, resulting in a very coarse representation of the velocity field. Each cluster moves in a straight line at constant speed and without changing shape, so that nonlinear motion such as rotation cannot be well-represented within FRONTIERS.

Development and decay. The operator can subjectively assign a linear development/decay multiplier to the rainfall intensities in each of the defined forecast clusters, but this provides only a crude representation of development or decay; rainfall is never completely removed by this technique, nor can it be generated from nothing. The operator has the option of sketching in new areas of rainfall or deleting existing areas in each of the 6 forecast images, but often lacks the time to exploit this facility.

- Lack of time. One of the main causes of errors in FRONTIERS forecasts
  is the operator having insufficient time to use all the facilities
  provided. It is planned to add more radars to the UK network over the
  next few years, to increase the geographical area covered to include
  Scotland, and possibly to make use of data from some of the
  continental radars. This will significantly increase the amount of
  work that FRONTIERS has to do, especially in the radar analysis stage,
  and will leave the operator with even less time to spend on producing
  the forecast images.

Thus, to improve the quality of FRONTIERS forecasts we need to
introduce more realistic movement and development/decay, and to automate
the system as much as possible, to relieve pressure on the operator.

Figure 2. The domain and topography of the enhanced version of the UK
Meteorological Office Mesoscale Model. The grid points have a 15km
spacing and the contour interval is 50m.

3. THE UK MESOSCALE MODEL

Numerical models in current operational use in the UK give valuable guidance to forecasters on the broad scale atmospheric structure. A grid-spacing of about 150km is used for global predictions and half that for the regional model covering the North Atlantic and Europe. A mesoscale numerical forecast model with finer resolution is being developed at the UK Meteorological Office to provide forecasters with guidance on the short-term local variations of weather (2). This model has a 15km horizontal grid-spacing covering the British Isles (Figure 2), with 32 levels in the vertical spread between 1.25m and 14025m. It will be run every 6 hours, producing forecasts of many parameters, including precipitation, up to 18 hours ahead.

The mesoscale model produces a much finer representation of the velocity field than does FRONTIERS, with physically consistent velocity vectors for each of the grid points in its area (Figure 3). The model can also represent nonlinear motion such as rotation.

The treatment of development and decay within the model is also superior to that offered by FRONTIERS; for example, the model can indicate areas of likely new development which in FRONTIERS can only be added subjectively by the forecaster (perhaps with guidance from the model).

However, although the model can represent motion and development/decay better than FRONTIERS, it does have a number of limitations which prevent it producing good-quality precipitation forecasts for the first few hours:

- Spin-up. Although the model produces hourly forecasts for the period 0-18h ahead, the first few hours cannot be used because the model has a spin-up time of 2-3 hours. Longer-term forecasts from the previous model-run therefore have to be used instead. These forecasts depend on observations at least 6 hours old; they may also have timing or development errors which become evident when the model's predictions are checked against recent observations.

- Resolution. The resolution of the model is not as fine as that of FRONTIERS, since the model cannot represent features with scales less than about 3 or 4 grid lengths.

- Timing errors. The mesoscale model is closely tied to the regional model through its boundary conditions and therefore retains any timing errors in systems that are passed through the boundaries.

Therefore, although the model is capable of representing the motion and development/decay better than FRONTIERS, its forecasts for 2-3h ahead cannot be used, and it cannot represent features to the same resolution that FRONTIERS is capable of. The model may also have timing or development errors which may be apparent through consideration of recent observations.

Radar and satellite imagery often provide some of the earliest indications that numerical predictions may be coming, or failing to come, to fruition. Sometimes the imagery will reveal a feature predicted by the model, thereby giving confidence in the model output though perhaps indicating that the feature is predicted in the wrong place or at the wrong time. Provided the model predictions are broadly credible and the forecaster is able to interpret the imagery in terms of the variables predicted by the model, it may be possible for him to adjust the model predictions in space or time to fit the observed imagery as it unfolds (3,4).

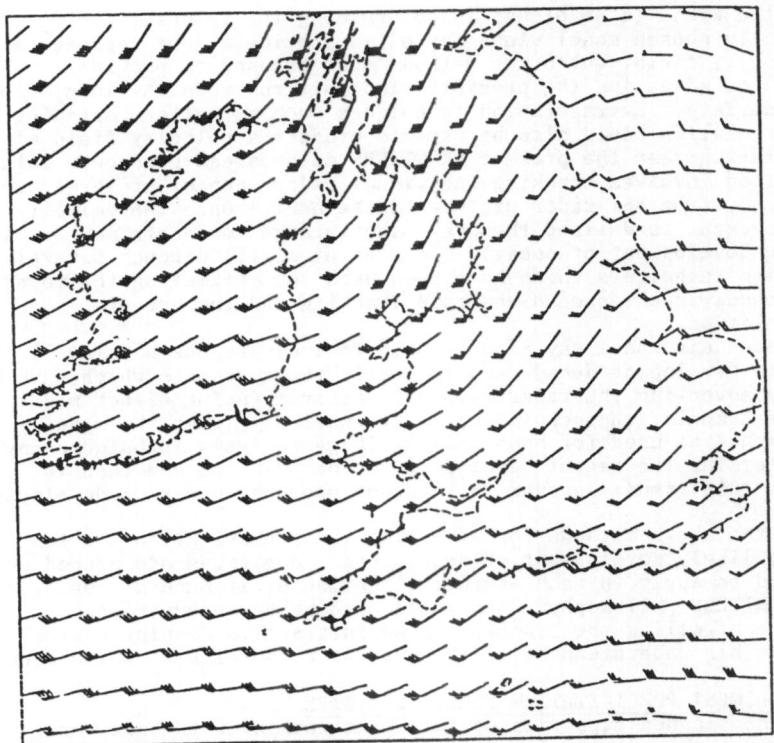

Figure 3. A typical mesoscale model wind field. The area shown is not the entire domain of the model and the wind vectors are not shown for every grid square.

4. COMBINING NWP OUTPUT AND EXTRAPOLATION TECHNIQUES

In spite of the prospect of using image data to verify and possibly to correct numerical model predictions, it is likely that in the first few hours of the forecast period the best predictions will remain those produced by translating and modifying high-resolution observations in near-real-time. What is needed is a way of combining the latest observations, as used by FRONTIERS, with the model's more realistic representation of the movement of the weather patterns, and of development and decay.

An obvious approach is to move the rainfall patterns in accordance with the physically consistent wind fields predicted by the mesoscale model. Unfortunately the height at which the echo motion is associated with the wind velocity will vary depending upon the meteorological situation. It is therefore necessary to obtain the depth and type of precipitation in order to assess the steering level for the radar echoes, and it may be necessary to use winds from different heights in different parts of the radar coverage, for example across a front. In some situations, for example orographically induced rainfall, the precipitation pattern may not be moving with the model winds at any level. In addition, there may be occasions when the errors in the model predictions are too large for the winds to be of use in determining the velocity field.

Correctly chosen model winds may often provide a good representation of the velocity field, but other methods must be used to provide velocities for advecting the precipitation patterns when the model winds are inappropriate. Extrapolation techniques such as pattern matching or pattern correlation could also be used to produce a velocity field of finer resolution than the present FRONTIERS methods can produce. This type of method involves tracking individual radar echoes, or large areas of radar echo, from one radar picture to the next. Once the velocity of the echo movement is defined then very-short-range forecasts may be made (neglecting development or decay). Collier et al (5) discuss several extrapolation techniques which have been used for estimating the movement of radar echoes, such as echo centroid tracking and the use of cross-correlation.

The mesoscale model may also be used to indicate how a rainfall field is likely to develop or decay, and to modify the forecasts which have been produced by advecting the radar rainfall patterns (using either model wind fields or extrapolation techniques). Alternatively, when the model products cannot be used for some reason, image analysis techniques may be used to determine the recent development or decay areas and then to extrapolate this trend; however this cannot show new areas of development or decay.

Given the above methods for advecting the precipitation and for determining likely development or decay areas, decisions are needed about which method to apply in each weather situation as it occurs. Under the present FRONTIERS philosophy, this would be done by an experienced forecaster controlling the process via an interactive display system, in the light of his understanding of the current meteorological situation.

5. AN INTELLIGENT PRECIPITATION FORECAST SYSTEM

With the planned expansion of the UK network over the next few years the FRONTIERS workload will increase considerably and the forecaster will have increasing difficulty keeping pace with the flow of real-time data. To reduce the burden, and to allow other improvements to the forecast, a new forecast system is being developed to automate as much of the forecast process as possible, including some of the interpretive and judgmental functions performed by the forecaster at present, or demanded by the new techniques proposed in section 4 above. The new forecast system will use products from the mesoscale model to improve the quality of the forecasts by introducing more realistic movement and development/decay, as discussed in the previous section, but under automatic control.

Work has now started on developing the new system for forecasting rain, using artificial intelligence techniques for the decision-making tasks inherent in combining the mesoscale model products with extrapolation techniques. A simplified diagram of the probable structure of this system is given in Figure 4.

'Intelligence' resides in the upper layers in the form of one or more expert systems (6) which have the tasks of recognizing the type of meteorological situation (by forming hypotheses and checking them against current data), deciding upon the details of constructing the forecast, and then of controlling the production of the forecast itself and of any accompanying explanation or justification.

Lower layers contain numerical algorithms for specific operations on the various types of current data. Broad questions posed in symbolic form at high levels (eg., 'Is the rainfall frontal or convective?') would, as part of the reasoning process, prompt a series of more specific questions or sub-goals. This would eventually lead to numerical algorithms being selectively accessed to obtain appropriate evidence by processing

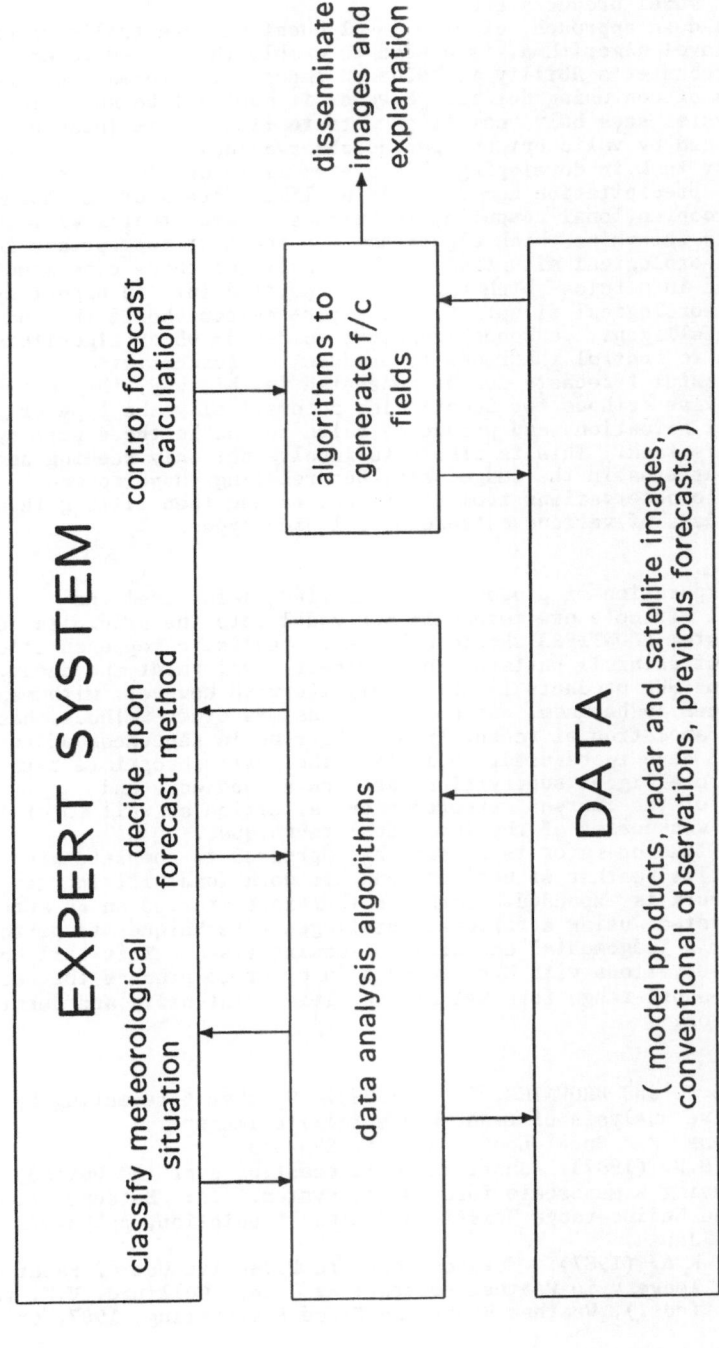

Figure 4. The probable structure of the proposed intelligent precipitation forecasting system

observations, model products etc.

This top-down approach, of high level questions eventually being fed down to low level algorithms, is needed to enable the system to emulate the human forecaster's ability to focus on important information embedded within a mass of confusing detail. However it must not be so rigid that the expert system sees only what it expects to find and is incapable of being influenced by valid but unexpected observations.

The first task in developing this system is to develop algorithms for advecting the precipitation and determining likely areas of development or decay, using conventional computing techniques. Case studies will then be carried out to determine which algorithms are the most appropriate in different meteorological situations. The results of these case studies will be formed into rules which can be incorporated into an expert system. Given the meteorological situation, the expert system should then be able to make an intelligent, reasoned judgement to decide which algorithms to use, and then to control their use to produce the forecasts.

If successful forecasts can be generated in this way, the next step will be to devise methods for identifying automatically the type of meteorological situation, and producing rules to enable the expert system to perform this task. This is likely to involve not only seeking and identifying patterns in the image data, but relating these to model products and to observations from other sources and then fitting them to conceptual models of various meteorological archetypes.

6. CONCLUSION

The incorporation of products such as wind fields from the Meteorological Office's new mesoscale NWP model into the precipitation nowcasting system FRONTIERS should allow more realistic representation of the movement of rainfall patterns and of their development and decay.

The use of NWP products is not straightforward however, with many detailed choices to be made, and on some occasions other methods, based purely on extrapolation of recent trends observed in sequences of rainfall analyses, will have to be used. Selecting the most appropriate methods will require intelligent supervision based on a knowledge and understanding of the current meteorological situation as well as of the strengths and weaknesses of the individual techniques.

The FRONTIERS operator is already hard-pressed to complete his tasks on time in active weather situations, and his work-load will increase as the radar network is expanded. Work has therefore started on an automatic forecasting system, using artificial intelligence techniques to perform the forecaster's judgemental and decision-making tasks. This will combine the latest observations with NWP products in order to produce the best possible very-short-range forecast of the timing, intensity and duration of precipitation.

REFERENCES
(1) CONWAY, B.J. and BROWNING, K.A. (1988). Weather forecasting by interactive analysis of radar and satellite imagery. Phil. Trans. R. Soc. Lond., A, 324, 299-315.
(2) GOLDING, B.W. (1987). Short range forecasting over the United Kingdom using a mesoscale forecasting system. In: Matsuno, T. (ed), Short- and Medium-range Numerical Weather Prediction, pp563-572, Met. Soc. Japan.
(3) BROWNING, K.A. (1987). Towards the More Effective Use of Radar and Satellite Imagery in Weather Forecasting. In: Collinge, V.K. and Kirby, C. (eds.), Weather Radar and Flood Forecasting, 1987, Ch.16, 239-269.

(4) SMITH, W.L. et al. (1988). The integration of meteorological
     satellite imagery and numerical dynamical forecast models.
     Phil. Trans. R. Soc. Lond., A, 324, 317-323.
(5) COLLIER, C.G., BROWN, R. and CONWAY, B.J. (1989). On the future of
     very-short-period rainfall forecasting using radar data. Preprint
     Vol., 5th Int. Conf. on Interactive Information and Processing
     Systems for Meteorology, Oceanography and Hydrology, 29 Jan - 3 Feb,
     1989, Anaheim, Cal., AMS, Boston.
(6) CONWAY, B.J. (1989). Expert Systems and Weather Forecasting,
     Meteorological Magazine, Vol.118, 23-30.

Dr G. VALENTINI
Director of Scientific and Technical Research
with third-countries (including COST Cooperation)
Commission of the European Communities
Speech of welcome at Château Ste-Anne, Brussels

Château Ste-Anne (6/09/89)
Speech of the Chairman C.G. COLLIER

Conference Dinner at
Château Ste-Anne (6/9/89)
(A table amongst 20 others)

Left to right :
Dr A. KLOSE
Head of Department (CEC)
responsible for COST Cooperation
Mrs and Mr R. SORANI

SESSION 5

Meteorological and other
applications of weather
radar data

Chairman : H.R.A. Wessels

THE IMPACT OF RADAR AND SATELLITE IMAGERY IN A MESOSCALE NWP SYSTEM

B.J. WRIGHT and B.W. GOLDING
U.K. Meteorological Office

Summary

The UKMO Mesoscale Model is used routinely to predict the weather over the British Isles on a 15 km grid. Synoptic scale development is largely controlled by the time-dependent boundary conditions. However the mesoscale evolution is sensitive to the initial conditions within the domain, and especially to the humidity distribution. The Interactive Mesoscale Initialisation (IMI) allows a forecaster to monitor the use of surface observations, satellite and radar imagery in the production of a set of key analyses of surface variables and the cloud distribution. Conceptual models are then used to make the remaining model fields consistent with these analyses. Thus, an inverse of the model's precipitation scheme is used to initialise cloud water content from the analysed cloud cover profile and the surface precipitation rate. The results are presented for two cases of interest, taken from an assessment programme which is currently underway: An anticyclonic stratocumulus forecast and a frontal precipitation forecast. It is illustrated how, in these cases, the use of satellite and radar data within the framework of the IMI produced a better set of initial fields which in turn resulted in improvements to the forecast.

1. INTRODUCTION

It has always been of great importance to obtain the best possible analysis from which to run a numerical weather prediction model, and this holds true for short range mesoscale forecasting. The problem of initialising the UK mesoscale model, with its 15 km resolution, is one of data sparsity. The surface observing network has at best a resolution of 50 km, with upper air observing stations being spaced more than 300 km apart. Over the sea areas the problem becomes an order of magnitude greater with only a few ships and oil rigs providing regular observations.

In an attempt to solve this problem, the UK Meteorological Office has developed the Interactive Mesoscale Initialisation (IMI), which provides an environment within which to make the best possible use of all the available data. A broad range of surface observations is incorporated, including cloud reports, visibility and snow depth, with satellite and radar imagery acting as additional data, providing much needed information over sea areas. The whole system is under the control of a human analyst, who, in addition to having control over the use of the data, is able to modify the analyses in order to incorporate his own knowledge of the situation. Conceptual models are used to relate the other model variables to the analysed quantities.

An assessment programme is currently underway, with subjective comparisons being made between the forecasts run from the current objective initialisation scheme and those run from the IMI. In particular, cases where the operational forecast was deficient in some way are being investigated.

2. THE MODEL

The UK Meteorological Office mesoscale model (1) uses a non-hydrostatic, compressible formulation of the primitive equations with a semi-implicit finite difference scheme allowing a forecast timestep of one minute. An Arakawa C grid of 63x79 points covers the British Isles, with a resolution of 15 km, giving good representation of orography (fig. 1). The vertical coordinate is height above orography, with 16 levels in the vertical, the lowest at 10 m and the highest at 12010 m, with the level spacing increasing linearly with height from 100 m to 1500 m.

The model has a detailed boundary layer mixing formulation with turbulent kinetic energy carried as a variable. Cloud water and humidity are predicted separately to give a good representation of cloud, and both grid scale and convective precipitation processes are modelled. The radiation scheme is also particularly detailed with respect to cloud.

A continuous three hour assimilation cycle is run. The observations are analysed using a combination of a three hour mesoscale model forecast and fields interpolated from the most recent forecast from the regional model, as a first guess. This "Hybrid" first guess takes the synoptic scale component of the upper air fields from the regional model, with the short wave component derived from the mesoscale model forecast. The cloud and surface fields are also taken from the mesoscale model forecast. The analyses are used to modify the other model variables to obtain a consistent set of model fields. The IMI can be used in the place of this objective initialisation.

Two 18 hour forecasts are run each day, from midnight and midday. Hourly predictions are obtained of pressure, wind, grid scale and convective precipitation, cloud cover and base, visibility, temperature and humidity.

Three hourly boundary conditions are taken from the regional model; a 15 level sigma coordinate model, with a resolution of about 75 km in the area of interest. Cloud water mixing ratio is not carried as a variable in the regional model, and so has to be diagnosed from the relative humidity. This and the interpolation required can cause undesireable effects close to the boundaries. Because the edges of the model domain are so close to the forecast area the forecast is generally driven by the boundary conditions in the latter stages, although it can develop topographically driven mesoscale features well. But in the first 6 to 9 hours, and longer in static situations, the initial conditions are very important.

3. THE INTERACTIVE MESOSCALE INITIALISATION

The Interactive Mesoscale Initialisation is a menu driven system, controlled by mouse input (and keyboard where required), which is operated by a human analyst on an interactive graphics workstation. It allows the analyst to monitor the use of surface observations, and satellite and radar imagery in the production of a set of key analyses of surface variables and the cloud distribution. The analyst is able to modify the data used in the analyses and the analyses themselves, so incorporating his own knowledge of the situation into the final result. Conceptual models are used to make remaining model fields consistent with these analyses.

The analyst has a choice of using the Hybrid (see chapter 2) or a set of fields interpolated from the most recent regional model forecast as a first guess, the latter only normally being selected if a serious timing

Figure 1.   Model domain and orography.   The gridpoints have a 15 km spacing and the contour interval is 50 m.

error is present in the forecast from the mesoscale model.   The first guess is used as a background to analyse mean sea level pressure, 10 m wind, surface precipitation rate, cloud cover, cloud top, cloud base, visibility, snow depth, screen temperature and dewpoint.   The analyst has the opportunity to use the radar image to add to or replace the analysed precipitation, within the radar coverage. The first guess cloud cover and cloud top fields may be calculated from the infrared Meteosat image, using the model temperatures for calibration. The cloud cover is generated by comparing the surface temperature in the first guess with the corresponding nine pixels within the Meteosat image (approximately 5 km resolution). The satellite image is then adjusted for surface radiation effects using the final cloud cover analysis, before being used to calculate the cloud top height.

Geostrophic corrections from the mean sea level pressure analysis are applied to the 10m winds before the observations are analysed. The pressure analysis is also used to calculate adjustments to the upper air pressures, winds and temperatures, which are applied with height dependent weighting, decreasing linearly to zero at 8 km. A one in a hundred slope, parallel to the average pressure gradient, is assumed for the pressure corrections, with the wind adjustments being geostrophic above 1 km and temperature corrections being calculated hydrostatically. Below 1 km a height dependent linear combination of the surface correction and a geostrophic correction is applied to the winds.

The low level relative humidity, cloud water mixing ratio, and boundary layer cloud condensation nucleus concentration are initialised using the analysed temperature, dewpoint and visibility. The cloud condensation nucleus concentration is used to diagnose visibility in the forecasts, but does not interact with other variables.

The three cloud analyses are used in conjunction with the first guess cloud distribution and additional cloud information from the surface observations to generate a three dimensional cloud analysis. The cloud and temperature structures may be examined in more detail by selecting up to twenty locations within the model area for which profiles of cloud cover and potential temperature are displayed. These profiles may be adjusted, with the temperature correction and the cloud cover values imposed over an area selected on the screen. The final cloud analysis is used to set the upper air relative humidity distribution and an improved first guess cloud water mixing ratio distribution. A single column version of the model precipitation scheme is used to iteratively adjust the cloud water mixing ratio profile to produce the analysed precipitation rate at each gridpoint.

The screen temperature analysis is used to modify the soil, surface and first level temperatures assuming either a logarithmic or a linear profile depending on the first guess temperature profile. Upper air temperatures are adjusted to be stable, using the modified temperature at the level below, assuming a linear combination of the dry adiabatic lapse rate and the saturated adiabatic lapse rate, dependent on the cloud fraction. The temperature immediately above cloud top is adjusted to be stable to an air parcel following a saturated adiabat from anywhere within the cloud deck. Where there is excess stability this is eroded in order to preserve the original pressure pattern, if necessary. The temperature stability adjustment is to try to limit excess convection in the early stages of the forecast. After the surface temperature has been initialised, it is possible for the analyst to modify the field either by hand or by copying in a sea surface temperature field generated using infrared Meteosat images over the period of a week.

The divergence profile is adjusted to give zero vertical velocity at the upper boundary. A single column version of the model precipitation scheme is used to calculate the height dependent precipitation rate using the analysed cloud water mixing ratio. Where precipitation is being produced, the vertical velocity is reset so as to lift enough saturated air to replace the precipitation falling out of the layer. Cloud top is ceinforced by a small negative vertical velocity. The vertical velocity, in

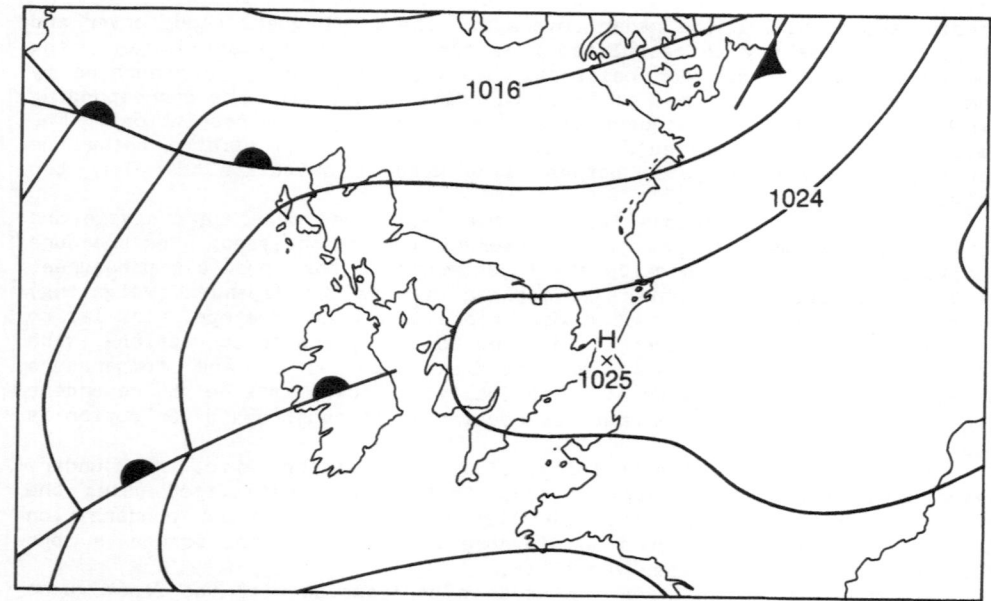

Figure 2.   Mean sea level pressure analysis with fronts for 6 GMT on 6th November 1988.

the lowest kilometre, is set equal to an exponentially weighted, height dependent mixture of the velocity recalculated from the horizontal winds and the initialised vertical velocity. A heavy smoothing is applied and values are limited to eliminate sudden changes and unreasonably high values. The divergence structure is adjusted again to be in balance with this final vertical velocity.

The analyst can apply his own corrections in a broad variety of ways. Areas, lines and points to be altered can be selected on screen or from a range of values in any of the fields available. Within the area selected it is possible to set a value, apply a correction, multiply by a factor, copy values from another field, or smooth the current values with an input smoothing radius. Finally, if he is not happy with the result he has the option of restarting the analysis.

4.   THE OBJECTIVE INITIALISATION SCHEME

The main differences between the objective initialisation and the IMI are the use of radar and satellite imagery and interactive control given to the analyst. In the objective initialisation a similar set of key analyses is carried out and a more limited set of conceptual models is used to make the other model variables consistent. But overall a less thorough knowledge of the situation is incorporated into the initial fields.

The Hybrid (see chapter 2) is taken as a first guess, and thus is used as a background for the analyses of mean sea level pressure, wind, precipitation rate, cloud cover, visibility, snow depth, screen temperature and dewpoint.

The soil, surface and first level temperature are adjusted to be consistent with the analysed screen temperature, preserving the initial model lapse rate. The temperature, relative humidity and winds within the

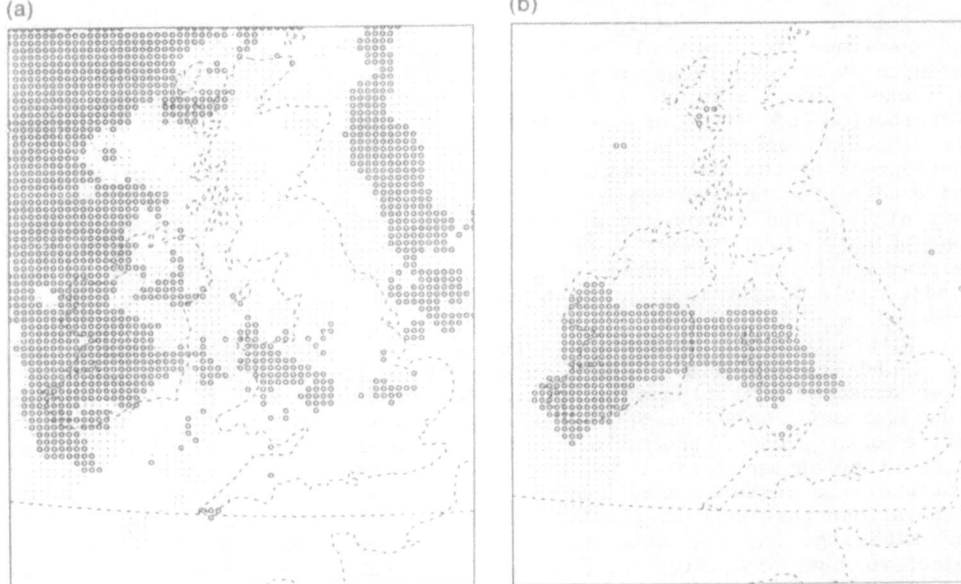

(a) (b)

Figure 3. Model analyses of low cloud cover for 12 GMT on 5th November
1988. (· > 4 oktas, 0 > 7.5 oktas).
(a) IMI analysis. (b) Objective analysis.

boundary layer (whose depth is diagnosed from the first guess temperature
profile) are adjusted to be consistent with the analysis values.
Super-adiabatic lapse rates are removed at all levels. No geostrophic
adjustment is made to the winds, but the upper level pressures are
recalculated hydrostatically, starting from the analysed mean sea level
pressure.
    The first level relative humidity and cloud water mixing ratio are
diagnosed from the analysed visibility. The surface observations of cloud
are used in conjunction with the cloud cover analysis to generate a three
dimensional cloud analysis, which in turn is used to initialise relative
humidity and to obtain a first guess cloud water mixing ratio distribution.
A single column version of the model precipitation scheme is used
interatively to adjust the cloud water mixing ratio until it produces the
analysed precipitation rate.
    The divergence profile is recalculated to give zero vertical velocity
at the upper boundary, but no account is taken of precipitation within the
vertical velocity initialisation.

5. AN ANTICYCLONIC STRATOCUMULUS CASE
    The use of satellite imagery and manual intervention within the IMI
should lead to a better cloud analysis, and thus a better cloud forecast.
This is especially true for static situations involving persistent layer
cloud, such as anticyclonic stratocumulus, where the boundary conditions
become less crucial. The first case investigated here, is such a cloud
forecast.

On the evening of November 5th 1988, under the influence of anticyclonic conditions (fig. 2), fog formed over much of southern England and remained for several days in parts. The thickness of the fog experienced in many places may have been a consequence of the large amount of smoke ejected into the air by bonfires on that evening, but the actual distribution of the fog was very much dependent on the low cloud distribution, as fig. 4 shows. A good forecast of the movement and development of the stratocumulus cloud present in the high cell circulation was crucial for the prediction of the onset and the distribution of the fog that night. The midday run of the mesoscale model did not produce a very good 18 hour cloud forecast and failed to develop the observed fog, instead keeping visibilities in excess of 20 km. The forecast run from the IMI had a better cloud distribution which was reflected in the lower visibilities produced.

Within the IMI, the Meteosat image was used in both the cloud cover and the cloud top analyses. After incorporating the observations, the cloud cover appeared slightly deficient to the north of East Anglia, so 7 oktas cloud was set where the satellite image was colder than 0tC. With very little data present to influence the analysed cloud bases over the sea areas, they appeared to be too high over the North Sea and to the west of Scotland. So in these areas, the bases were adjusted to agree with the few observations that were available. The resultant cloud analysis produced by the IMI (fig. 3a) has much more low cloud over the sea areas than the objective analysis (fig. 3b), which illustrates the importance of the satellite image over data sparse areas. Otherwise the two analyses are generally similar over much of the British Isles, except over Wales and the Irish Sea, where the cloud is more broken in the IMI analysis. There is also less high cloud in the IMI analysis and cloud tops over Ireland are higher.

The 18 hour forecast run from the IMI has a cloud sheet covering Wales and just beginning to extend into England, but with the majority of southern England cloud free (fig. 5a), which compares favourably with the observed low cloud distribution at 6 GMT (fig. 4). By contrast, the forecast run from the objective initialisation has much of southern England under a veil of low cloud, with the only significant breaks being on the south coast and in Humberside (fig. 5b), this being a very poor reflection of reality. So the incorporation of satellite imagery has had a marked impact on the 18 hour low cloud forecast.

The forecast visibilities were not as impressive, with the IMI forecast failing to develop the observed fog. But its visibilities were of the order of 4 km over most of southern England, which is a marked improvement on the objective forecast which kept visibilities at over 20 km. The improvement in the visibility forecast is a result of the improved cloud forecast and the higher aerosol concentrations initialised in the visibility analysis. Possible reasons for the failure to forecast the fog are the non-representation of the increase in aerosol concentration which occured during the evening as a result of the bonfires across the country, and the omission of any interaction between the cloud condensation nucleus concentration (which acts only as a tracer in the forecast) and other model variables.

The use of the satellite imagery within the IMI was reflected in the overall improvement in the forecast. The failure to produce the observed fog is disappointing, but the cloud forecast in itself would have been good guidance to the forecaster trying to predict the fog distribution on the morning of the 6th.

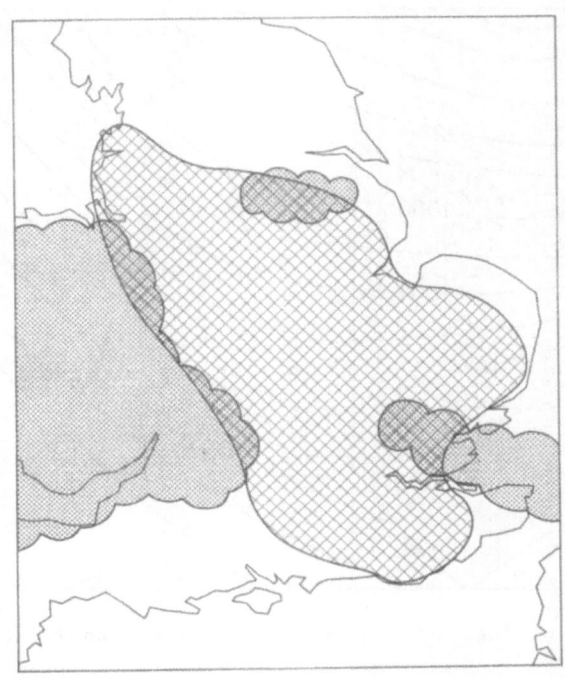

Figure 4. Observed low cloud and fog distributions for 6 GMT on
6th November 1988. (Stippled area denotes greater than 5 oktas low cloud
cover, cross-hatched area denotes less than 1000 m visibility).

(a)   (b)

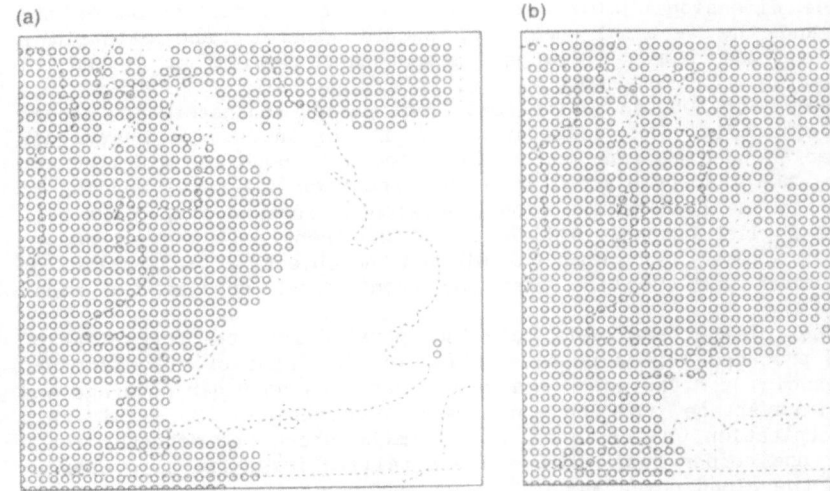

Figure 5. 18 hour forecast low cloud cover for 6 GMT on 6th November 1988.
(· > 4 oktas, 0 > 7.5 oktas).
(a) IMI forecast. (b) Objective forecast.

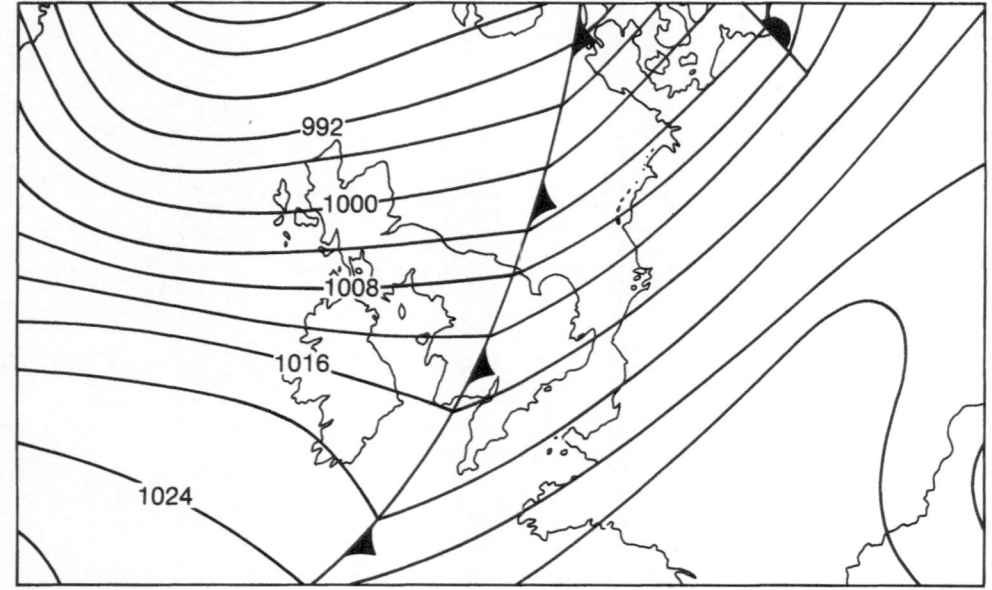

Figure 6. Mean sea level pressure analysis with fronts for 12 GMT on
15th February 1989.

6. A FRONTAL PRECIPITATION CASE

Radar imagery is currently the only available source of data that
captures the mesoscale detail present within frontal rainbands. Thus the
incorporation of these data within the IMI should lead to a better
precipitation analysis, which when used in conjunction with the satellite
derived, three dimensional cloud analysis, in the initialisation of the
cloud water and the vertical velocity, should generate and maintain a
realistic frontal structure. This case investigates the impact of the IMI
in the forecasting of a cold front.

On February 15th 1989 an active cold front passed southeastwards across
the British Isles (fig. 6). Associated with it was an area of heavy rain
and a well marked band of line convection (fig. 8). Both the midnight and
midday runs of the mesoscale model moved the front on too fast, tending to
lose the rain to the rear of the front, developing instead a spurious area
of rain in the English Channel, well ahead of the front. It was decided to
run the mesoscale model, both with the IMI and the objective initialisation
from 6 GMT, which was the time when the front first entered the radar
domain.

Within the IMI the radar data were incorporated into the precipitation
analysis, and produced a good representation of the front over Ireland and
northern England. Over the North sea where there was no radar coverage and
observations were sparse, the front was badly represented. So within that
area the precipitation was replaced by 0.4 mm/hr where the Meteosat image
showed cloud tops colder than -20°C. The satellite image was also used in
generation of the cloud cover and cloud top analyses. The front was badly
represented in the pressure pattern, so a trough was inserted before the
observations were analysed, and was maintained in the analysis. Some
spurious fog over northern France was removed. The resultant precipitation
analysis produced by the IMI (fig. 7a) had a more continuous band of

Figure 7. Model analyses of precipitation for 6 GMT on 15th February 1989.
(Rain: - · > 0.05 mm/hr, 0 > 0.1 mm/hr, ● > 0.5 mm/hr,
Snow (rainfall equivalent): - x   0.05 mm/hr, ✱ > 0.5 mm/hr).
(a) IMI analysis. (b) Objective analysis.

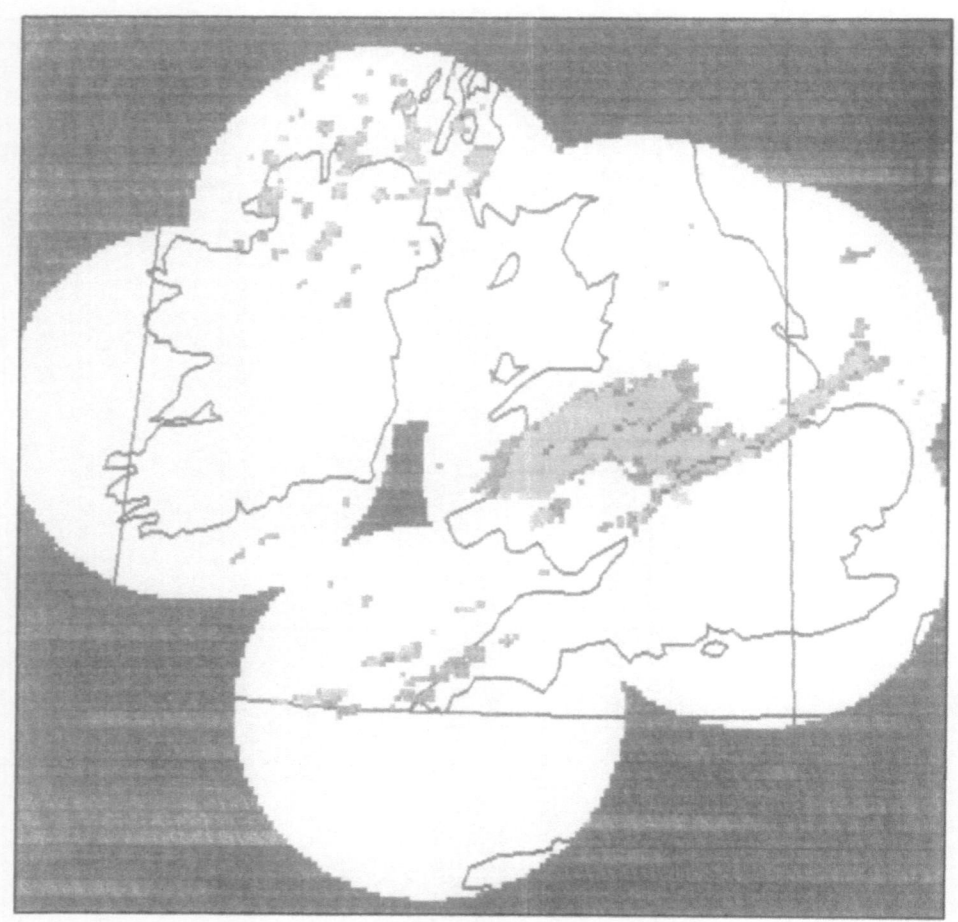

Figure 8. Composite radar image for 12 GMT on 15th February 1989.

precipitation than the objective analysis (fig. 7b), with the main area of rain further north over Ireland, leaving southern Ireland and Wales dry at 6 GMT.

The radar image for 12 GMT (fig. 8) shows a large area of moderate rain over Wales, with a narrow band of line convection stretching from the Wash through to the Bristol Channel. There are some smaller areas of rain apparent over Cornwall, and there is an area of light rain over southwest and central southern England which does not show up on the radar.

The 6 hour forecast run from the objective initialisation (fig. 9b) has a band of rain which is roughly coincidental with the line convection, but far too wide, with too much rain over East Anglia. It has the main area of rain too far south, over southwest England, where there was only light rain in reality, and much of Wales is completely dry. It has also developed a large area of precipitation over the English Channel, giving southern England and the north coast of France some rain . This area of rain was not supported by observations or radar and appears to be totally spurious.

(a)

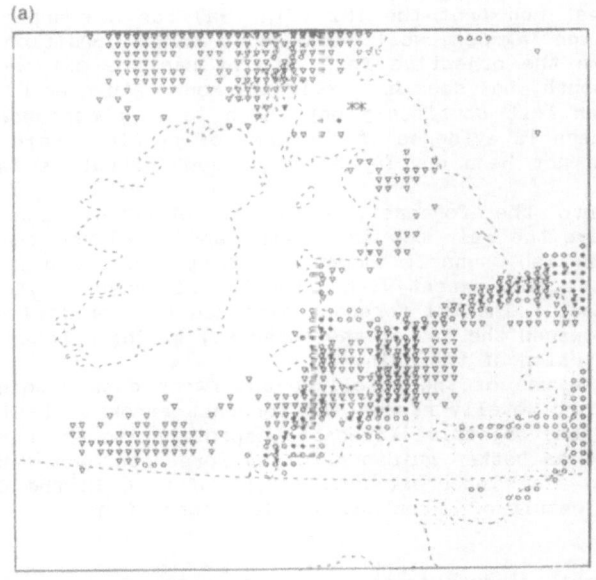

(b)

Figure 9. 6 hour forecast precipitation for 12 GMT on 15th February 1989.
(Rain:- • > 0.05 mm/hr, o > 0.1 mm/hr, ● > 0.5 mm/hr,
Snow (rain equivalent):- x   0.05 mm/hr, ✳ > 0.5 mm/hr,
Showers (local rate): ▽ > 0.4 mm/hr, ▼ > 2 mm/hr)
(a) IMI forecast. (b) Objective forecast.

The forecast run from the IMI (fig. 9a) has a sharper band of quite heavy rain which agrees well with the actual position of the line convection. Like the objective forecast, it has the main area of moderate rain too far south, but does have slightly more rain over Wales. Southeast England has been left completely dry which is in accordance with reality, and although there is evidence of the area of spurious rain in the English Channel, it has not been developed to the same extent as in the objective forecast.

Further into the forecast, both the objective and the IMI runs continued to have the rain too far south, and developed the spurious rain area over the English Channel. However, the spurious rain was not developed so soon or to the same extent within the IMI forecast as it was within the objective forecast. The IMI forecast also had the convective band slightly further north towards the end of the forecast, giving a better indication of the real distribution of the rain.

Although the use of the satellite and radar data within the framework of the IMI did not totally remove the forecast errors, it did provide some improvements on the objective forecast, especially in the early stages, and would have acted as better guidance to the forecaster attempting to predict the rain areas. The incorrect development of rain in the English Channel may have been a result of erroneous boundary conditions.

7. CONCLUSION

The two cases investigated here illustrate that the use of an interactive mesoscale initialisation scheme incorporating satellite and radar imagery has a positive impact on model forecasts. Furthermore, in static situations improvements to the predicted cloud distribution can last throughout the forecast period. The impact is generally not so long lived in frontal precipitation forecasts, which are normally associated with mobile situations, with significant improvements tending to last only for the first 6 to 9 hours. This is probably due to the influence of the boundary conditions.

Future improvements are likely to come from increasing the model domain, the use of more conceptual models and the utilisation of other sources of data as they become available.

REFERENCES

(1)   GOLDING, B.W.   (1987).   Short-range forecasting over the United Kingdom using a mesoscale forecasting system. Short- and Medium-Range Numerical Weather Prediction. Meteorological Society of Japan, Tokyo.

PROMIS 600; AN OPERATIONAL SYSTEM FOR VERY SHORT RANGE WEATHER FORECASTING IN SWEDEN

S. NILSSON and J.O. BRUNSBERG
The Swedish Meteorological and Hydrological Institute

Summary

PROMIS 600 is an operational system for especially very short range weather forecasting. The workstation, which has been set up in Norrköping, is a big step in the realization of PROMIS 90; the Swedish Weather Service System of the 90's. PROMIS 600 will test several aspects of very short range forecasting, as the observation system, mesoscale analysis and forecasting methods, workstation and methods of dissemination of new forecast products. During spring 1989 a realtime test of its operational workstation has been conducted and PROMIS 600 will during the autumn be incorporated in the regular weather service. The achievments of the technical systems of PROMIS 600 have been gained as a joint project between Swedish industry and the Swedish Meteorological and Hydrological Institute. The main contractor of the PROMIS 600 system, including the Doppler radars and the PROSAT system is Ericsson Radar Electronics AB.

1. INTRODUCTION

The Swedish Meteorological and Hydrological Institute (SMHI) is developing a system called PROMIS, which stands for a PRogramme for an Operational Meteorological Information System. In the initial study of PROMIS 90, (Bodin et al, Ref 1), an increasing demand was identified from various sectors of the society for very short range and detailed forecasts, which should be effectively disseminated and tailored to the customers needs (Liljas, Ref 2).

In order to accomplish the very short range part of PROMIS 90, which covers the complete weather service, including short range (1-2 days) and medium range (3-10 days) weather forecasting, a pilot station for very short range forecasting, PROMIS 600, has been built up in Norrköping at the headquarters of SMHI. On its colour graphic monitors meteorological data from different sources can be displayed on five spatial scales ranging from an area covering Europe to the local scale, which covers a small portion of southeast Sweden. Forecasters may display the various data on common map projections (on the smaller scales lakes and roads are included) and animate and zoom any product at will.

During spring 1989, PROMIS 600 was subjectively evaluated by forecasters in an quasi-operational environment. The primary forecasting area is located within an area around Norrköping with a radius of approximately 200 km. (Observational data with high resolution are available within a radius of about 120 km). During the test-period the workstation was used to issue very short range forecasts of temperature, wind, cloudiness and probability forecasts of precipitation. The forecasters agreed in that the ability to monitor mesoscale phenomena in detail was improved by using the PROMIS 600 system. In the extension this also has important impacts on the nowcasting and very short range forecasts.

Already today, the volume of the incoming meteorological data is very large (the PROMIS 600 system is producing around 3000 images each day) and it will probably grow substantially during the 90's. The diverse data will place heavy demands on computers and storage devices, but also on the forecasters. In order to handle the enormous data stream the forecasters have to determine the "meteorological problem of the day" to be able to eliminate some classes of products. But even after an elimination procedure there will still be problems with the selection and integration of meteorological data. A successfull mixture of various data will undoubtedly be a challenge for the next years to come. Another great problem to face is also the meteorological understanding of the mesoscale processes and the scale-interaction, a point which was stressed by the forecasters during the test-period.

2. MAJOR COMPONENTS OF PROMIS 600

Nowcasting and very short range forecasting demand an observation system based on remote sensing techniques (satellites, Doppler radars etc) and a dense network of automatic stations with a high frequency of updating. The short time scale also puts heavy demands on rapid communication, processing and dissemination of various forecast products. The design of the PROMIS 600 technical systems reflects these requirements and in Figure 1 an overview of the technical system is given.

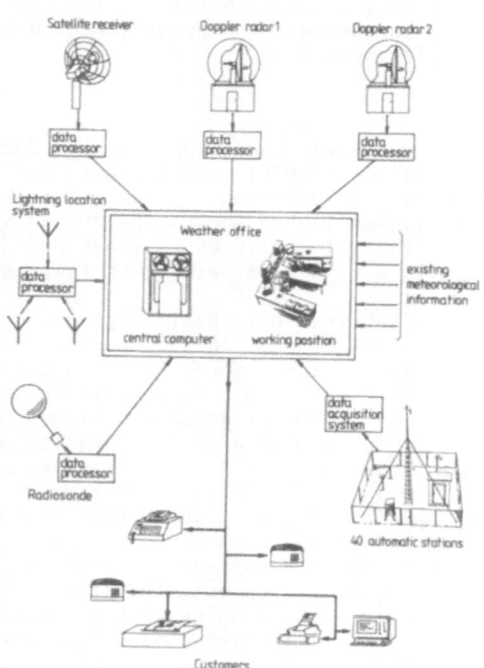

Figure 1. The PROMIS 600 technical system - an overview.

The observational data are displayed to the forecaster in alphanumeric or graphical form. The forecaster has also access to various mesoscale analysis and forecasts. The research and development in this area has been performed in cooperation with the universities and the Swedish military weather service.

The dissemination system is also of great importance. When we have produced sufficiently accurate, site or area specific forecasts, we must be able to deliver this information to the customers who need it to make correctly and timely decisions. This is especially important for the emergency services. Therefore, the development of customer-tailored dissemination techniques and formats has had a high priority in the PROMIS 600 project.

3. THE OBSERVATION SYSTEM

The observation system forms the base for PROMIS 600. In Figure 1 its principal components are shown together with the workstation and the dissemination system.

The data sources are:
- automatic station network
- Doppler weather radars
- satellite data subsystem, "PROSAT"
- lightning location network

Meteorological data from these sources together with existing conventional meteorological information are stored in a central computer. Processed data are available to the forecaster at the workstation.

3.1 The automatic station network

A network of 40 automatic stations is estabilished in the PROMIS 600 area (see figure 2). All stations are equipped with sensors for wind, temperature and humidity. In addition, a number of stations measure precipitation, surface pressure and radiation and some stations report cloudbase and visibility.

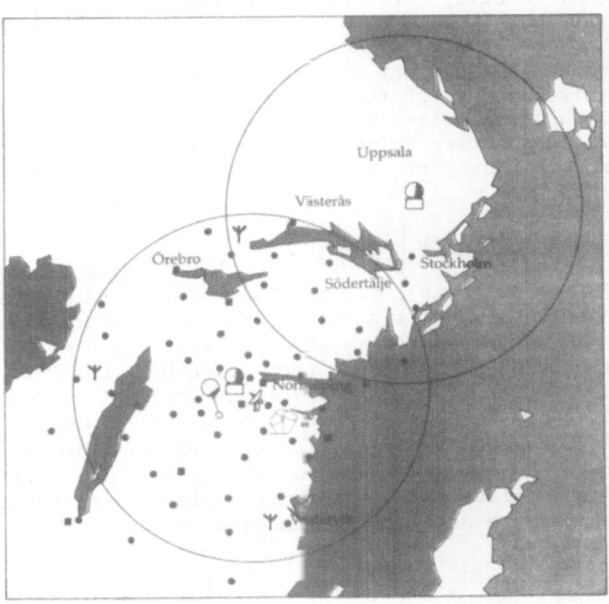

Figure 2. The PROMIS 600 area and the location of the data sources.

Data from the sensors are collected and preprocessed locally by an Automatic Data Acquisition Terminal (ADAT) and transmitted via telephone to the acquisition computer ADAC (Automatic Data Acquisition Central). The stations deliver data every 15 minutes.

The observational data are displayed in a conventional, "synoptic form" or as time-series for each parameter or a combination of parameters.

3.2 The doppler weather radars

Two 5 cm, 0.85° 3-dB beamwith, dual pulse repetition frequency (dual PRF) scanning Doppler radars are so far included in the observation system. The radars are localized in Norrköping and at Arlanda Airport north of Stockholm. Recently an additional Doppler radar has been installed on the west coast near Gothenburg and soon a digital radar will be set up on the island of Gotland. These radars will also be incorporated in the PROMIS 600 observation system and in the Nordic Radar network (NORDRAD).

The radars can operate in Doppler or amplitude (non-Doppler) mode. Both modes are used during each data collection cycle and the mode selection is controlled by a computer. The amplitude mode is for the largest measuring range (radius 240 km), while the Doppler mode is measuring only within a radius of 120 km around the radar.

The radars make three dimensional scans and they have the capability of reducing ground clutter echoes by blocking the Doppler spectral channels near zero radial velocity and utilize a 32-pulse Fast Fourier Transform (FFT) processing technique to provide radial velocities for each PRF. Even the amplitude mode is relatively free from ground clutter echoes, due to a developed algorithm for suppressing ground echoes.

The radars display images of the following basic parameters:
- reflectivity at several CAPPI levels and maximum values
- precipitation intensity and accumulated precipitation
- radial velocity, PPI and CAPPI levels
- turbulence
- echoe top heights

Vertical cross-sections may also be obtained.

The composite image of the two radars is at present only displaying maximum values of reflectivity, but will in a near future also display pseudo-CAPPI reflectivities with a fixed geometric boundary between the radars. Other methods of merging data from radars will also be investigated.

3.3 The satellite data system

The satellite data subsystem, PROSAT (PROsessing system for meteorolgical SATellite data), was installed in the early part of 1988 and consists of an antenna/receiving system and an advanced processing system. An overview of the PROSAT system is given in Figure 3.

The system handles both high resolution NOAA and Meteosat data. The products from PROSAT are images on different scales from 1000x1000 km (1km resolution) to 4000x4000 km (8km resolution), all in polar stereographic projection. Colour composite images (for Meteosat (VIS in Red, VIS in Green and IR in Blue) in the RGB-image and for NOAA (Ch1 in Red, Ch2 in Green and Ch4 in Blue), cloud and precipitation classifications, cloud top and surface temperatures are some of the standard products.

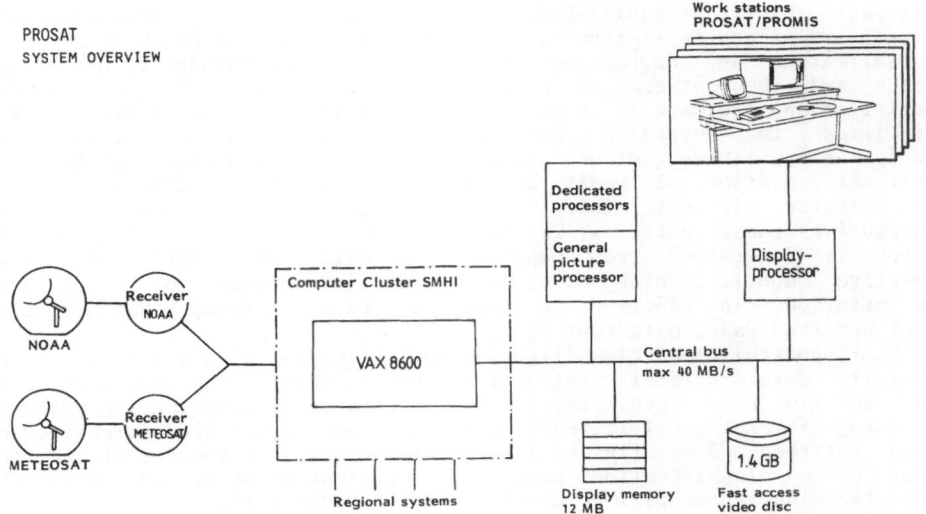

PROSAT
SYSTEM OVERVIEW

Figure 3. The PROSAT system - an overview.

3.4 The lightning location system

A lightning location network, LPATS (the Lightning Position And Tracking System) is since last year covering the whole of Sweden. The system utilizes Time-Of-Arrival (TOA) technique and the information from six optimally located detection stations are received in realtime and compiled by a central analyzer in Norrköping.

The detected Cloud-to-Ground (CG) strokes (mean CG strike location error is less then 5 km in PROMIS 600 area), its polarity and time are displayed to the forecaster at the workstation.

4. ANALYSIS AND FORECAST METHODS (METEOROLOGICAL MODELS)

The meteorologist at the workstation needs effective tools for analysis and forecasting of the mesoscale weather systems. Due to operational demands, e g short computing times, general three-dimensional mesoscale forecast models cannot be used. Instead we have tried a different approach to mesoscale forecasting. Simple methods for interpretation of the new observational information and for analysis of small scale features have been developed together with a number of specialized forecasting models, which still are able to catch the most essential atmospheric processes. The forecasting method also involves the use of subjective judgements from the meteorologists.

In the following a broad survey is given of some of the analysis and forecast methods that have been developed for PROMIS 600. The operational tests will reveal the meteorological merits and drawbacks of the various software moduls.
- Velocity Azimute Display, VAD and Uniform Wind Techniques derived from Doppler radar data: The VAD-technique permits the realtime conversion of radial wind into vertical profiles of wind speed and direction, divergence and deformation. Further analysis of these profiles can also give vertical velocity and thermal advection. The Uniform Wind Technique is designed to compute horizontal wind vectors and it shows the horizontal winds at different heights in a conventional form and is therefore much easier to

interpret than the radial-Doppler winds. (Ref 3) The application of the methods are of course limited to occasions when radar echoes exist.

- Estimation and nowcasting of precipitation and thunder by use of the radars and the network of automatic stations: Estimates of accumulated precipitation are made for a specified time interval, where radar data are correlated to automatic station data. For nowcasting purposes of precipitation, three methods have been developed. One method is based on pure extrapolation of radar echoes or a precipitation field, where the extrapolation is determined by a cross-correlation analysis of two consecutive radar patterns. The second method is analogous, but the wind vector is determined from Doppler wind data. The third method is an advective model, which gives a probability nowcast of accumulated precipitation. In addition, a forecast index for thunder has also been developed from radar data (Ref 4).

- Multispectral cloud classification and estimation of precipitation from satellite data: A first version of a cloud classification scheme (Ref 5), has recently been developed. The scheme uses box classification methods, including meteorological filtering processes and texture classifications of cloud entities. Presently a large data base is created for storing the input of a classification scheme for estimation of precipitation. The classification scheme will be correlated with radar data.

- The mesobeta and mesogamma analysis system: An analysis scheme for the mesobeta scale (grid distance 22 km) and the mesogamma scale (grid distance 5 km) has been developed (Ref 6). The analysis system is based on "optimum interpolation" and in the lowest tropospheric layer non-isotropic correlation functions have been developed for analysis of different parameters. The anlysis is three-dimensional and a multivariate scheme has been developed for analysis of vertical profiles.

- The Small Area Model, SAM: This model is based on a concept of a model developed by Danard (Ref 7). The model simulates the horizontal surface wind field from the mesoscale analysis of sea surface pressure, temperature, vertical stability, and the topography and roughness of the earth surface. The model also gives a crude estimate of the divergence and convergence fields.

- The Vorticity Advection Model, VAM: This model gives a very short range forecast of sea surface pressure. The model is based on vorticity advection that is evaluated from the mesobeta scale sea surface pressure analysis. The vorticity is advected by a flow determined from the synoptic scale model, the Swedish Limited Area Model (LAM). The utilization of all other physical effects from LAM, except vorticity advection will modify the result further, resulting in a model version called VAM-B (Ref 8). Verifications of VAM-B show that the model generally gives a satisfactory forecast up to 6 to 9 hours.

- The Air Mass Transformation model, AMT: The AMT-model is used to predict variables in the planetary boundary layer, like temperature and humidity and to give the possibilities of fog/stratus formation/dissolvement at some arbitrary point in space and time. AMT is an one-dimensional boundary layer model which is advected along a trajectory and influenced by both upper and lower boundary conditions (Ref 9).

5. THE WORKSTATION AND OPERATIONS

The very short range forecasts will partially depend on how well the data can be displayed in an optimal, integrated form. For this reason the display system is very important and considerable time has been devoted to the development of the graphical display software. In the design of the workstation great care has also been taken to ensure good enviromental and ergonometrical conditions.

The workstation of PROMIS 600 consists of an alpanumerical terminal for control, menu handling and editing, and a colour graphic monitor with a "mouse" and a graphic tablet for presentation, animation and combination of "pixel images" (radar and satellite data) and "graphic images" (lines, symbols, etc) as different plots, analysis and forecast products.

Forecasters can view the various products on five different spatial scales: the European-scale, the Scandinavian-scale, the southern Sweden-scale, the regional scale - East (the eastern part of southern Sweden), and the local scale - the PROMIS 600 area. All scales have common map projections and on the smallest scales map backgrounds of big lakes and roads are included as options. A topography background will also soon be implemented. The distribution of the various products on different scales are schematically illustrated in Table I.

Table I.

	RADAR Norr- köping	RADAR Compo- site	NOAA	METEO- SAT	LLS	OBS	ANA- LYSIS Sur- face	ANA- LYSIS Height	FORE- CAST Sur- face	FORE- CAST Height
P600	x		x				x	x	x	x
EAST	x	x	x		x	x	x	x	x	
SSWE			x	x	x	x	x	x	x	x
SCAN			x	x		x	x	x	x	x
EURO				x					x	

The display system is easy to handle by a forecaster and allows him to concentrate solely on meteorological problems. A key to this is the flexibility of the menu selection system. In PROMIS 600 the mostly used products are defined as "standard products" and they can be reached through the predefined Direct-Function-Keys (DFK) at the keyboard. As DFK are also some other important image manipulation functions defined, as animation and overlays. Using the mouse the zoom factor, and the forward and backward looping at desired speeds of the selected images can be performed. Other products, including the predefined standard products and order products, e.g. parameters from the AMT-model and time-series may easily be reached through the menu selection tree.

The system is very flexible regarding the application of different overlays. On a satellite or radar image it is possible to add up to four different graphic overlays, which could include e g various analysis and forecasts, wind observations and extrapolations of selected cloud/precipitation entities. It is also very easy to switch the selected overlays on or off.

With a draw-function it is possible to perform interactive image manipulations. Conventional fronts and symbols as well as text may be added to an image and the product then be specially designed for various customers. This function will be further .developed in a production-distribution system, which will be used as a complement system to the PROMIS 600 workstation.

The production-distribution system is also installed at the other four regional weather offices in Sweden, where analysis and forecast products, satellite and radar information can be received on the system. In the future much of the interactive man-computer forecast production will be managed on this system.

6. FORECAST PRODUCTS AND CUSTOMERS

A great number of weather sensitive customers in the Norrköping-region are served with the PROMIS system. Great efforts have been made to find methods to produce products with both high quality and high efficiency. At the PROMIS 600 workstation only very short range forecasts are produced, but short range and medium range forecasts produced at other workstations are distributed as well.

6.1 Education

Before the operational start of PROMIS 600 an intensive training and education programme for the forecaster staff was achieved. The courses were focused on the theory of meso-scale weather phenomena, analysis and forecasting technique, interpretation of radar and satellite images and on practical handling of the equipment at the workstations.

The educational programme has so far extended over 4 weeks and has been achived for the particular PROMIS staff. The education regarding the meso-scale meteorology will continue during the next years.

6.2 Methods

Forecasts are produced for a lot of meteorological parameters e.g. wind, temperature, humidity, cloudiness, precipitation, thunder and probability of strong winds. For most of the customers precipitation is the most important parameter. New forecasting models together with appropriate radar and satellite information have resulted in great improvements in this field. In order to give the customers a better base for decision-making the categorical precipitation forecasts have been replaced by probability forecasts. To obtain continuity and high quality in the forecasts, special forecast maps for the basic parameters are prepared at fixed times. Furthermore a method using forecast matrices has been initiated during the test-period. A forecast matrix will be valid for a small area and include different parameters at given forecast times. It will be updated both with results from computer models and with subjective forecast data entered into the computer by the operational meteorologist. Tailored products to customers consisting of forecast texts, tables and images will then be possible to distribute to the customers. The procedure is schematically illustrated in Fig 4.

Experiments have also been made by an automatic telephone answering system. From a directory of phrases containing about 1000 predefined words and sentences a message can be put together. The messages can be created either from a text generated from the forecast matrices or from a text formulated by the meteorologist and entered into the computer.

So far the experiments have been very successful and this technique will be operational in the winter season 89/90

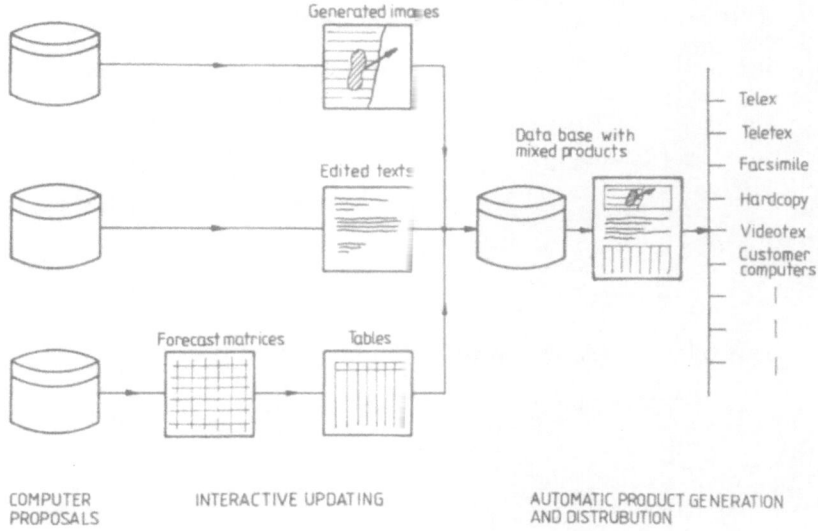

Figure 4. A schematic illustration of the new production technique.

6.3 Customers

During the test-period forecasts have been produced to the following customers: Energy and construction companies, local radio stations, local telephone weather services, the Swedish Rail Road Authority and the Swedish National Road Administration. During the summer season special forecasts for farmers and companies which are sensitive to thunder are produced.

To optimize the economy and efficiency in winter maintenance of the roads, SMHI has established an extensive cooperation with the Swedish National Road Administration. In recent year a pilot project has started to develop the weather service for the winter maintenance staff in the provinces within the PROMIS area.

The road surveyors need good and detailed weather forecasts which improves and facilitates proper decisions, especially in situations where preventive activities are needed, like salt spreading on roads, organization of personnel and so on. Both current weather data and forecasts from the SMHI data system are directly transferred to a personal computer at the road surveyors' particular office.

This meant that the availability of detailed weather information was dramatically improved for the road surveyors engaged in this project. Observations from road stations and from PROMIS automatic weather stations, as well as forecasts of road temperature and precipitation were sent to the personal computer. The data were displayed in both alfanumeric and graphical form. (Figure 5).

A great wish from various customers is also to receive radar and satellite images into their personal computers and preoperational tests to transfer this type of data have been made. This technique will be operational at the end of 1989.

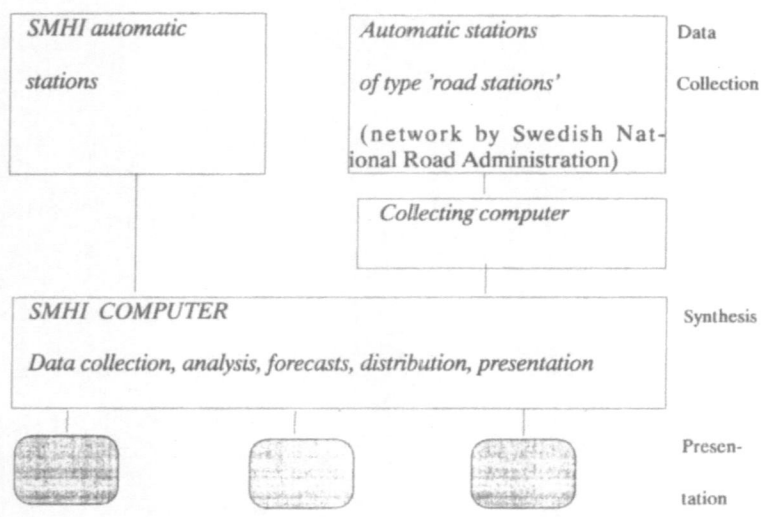

Figure 5. Main component of the subsystem for "road products".

REFERENCES

(1) BODIN, S., LILJAS, E. and MOEN, L. (1979). The future weather service at the Swedish Meteorological and Hydrological Institute – PROMIS 90. (In Swedish).

(2) LILJAS, E. (1984). Benifits resulting from tailored very-short-range forecasts in Sweden. Proc Nowcasting-II Symp, Norrköping, 3-7 Sep 1984, ESA SP-208, 503-507.

(3) PERSSON, P.O.G. and ANDERSSON, T. (1987). A real-time system for automatic single-Doppler wind field analysis. Proc Mesoscale Analysis Forecasting, Vancouver 17-19 Aug 1987, ESA SP-282.

(4) ANDERSSON, T., ANDERSSON, M., JACOBSSON, C. and NILSSON, S. (1989). Radar indicies for thunderstorms in southern Sweden, Geophysica, to be published.

(5) LILJAS, E. (1988). Effective weather monitoring with Meteosat second generation ideas stemming from experience of NOAA satellites. Proc 7th Meteosat Scientific Users Meeting, Madrid, 27-30 Sep 1988. EUMETSAT.

(6) ANDERSSON, E., GUSTAFSSON, N., MEULLER, L. and OMSTEDT, G. (1986). Development of meso-scale analysis schemes for nowcasting and very short-range forecasting. PROMIS reports Nr 1, SMHI.

(7) DANARD, M. (1977). A simple model for mesoscale effects of topography on surface winds. Mon. Weather Rev., 105, 572-581.

(8) GOLLVIK, S., OLSSON, E., PETERSON, E., ROBERTSON, L. and SALLIN, I. (1986). Vorticity Advection Model. PROMIS Reports, Nr 4, SMHI.

(9) GOLLVIK, S. and OMSTEDT, G. (1988). An Air-Mass-Transformation Model for short range weather forecasting, a bulk model approach. PROMIS Reports, Nr 8, SMHI.

RADAR FOR ELECTRICAL STUDIES
AND STORM WARNINGS

P. S. Ray
Department of Meteorology
Supercomputer Computations Research Institute
Florida State University
Tallahassee, Florida 32306 USA

Summary

Cyclonic storms are vital to many aspects of the earth system. The ubiquitous thunderstorms restore a significant amount of charge to the ionosphere, maintaining the earth–atmosphere electrical balance. The difference in tropical rainfall is literally the difference between desert and rain forest, between the least and the most productive land, and most of growing season rains on all continents are associated with the weather that accompanies cyclones. The role of tropical convection in redistributing heat and its effects in the general circulation and climate are also being revealed.

This report is a survey of some of the significant weather that is detectable by Doppler radar, specifically operational Doppler radar. Each phenomenon is described and its social impact assessed. Radar detection and structure is described, and, finally, the prospect for automated algorithms is assessed.

1. RADAR DETECTION OF STORMS

Radar detection is in two modes, clear air and storm. Clear air detection can provide wind sounding up to a height of the boundary layer in most locations. There is still debate about what fraction of the return is due to point scattering (from insects) and what fraction from fluctuations in the refractive index. Regardless, processing techniques include the VAD and the VVP. The value of such studies is the better prediction of impending storms.

2. CONVECTIVE INITIATION

Early Doppler radar studies produced methods for estimating wind speed and direction and identifying divergence and deformation fields (1,2,3). The Velocity Azimuth Display (VAD) scan, introduced in 1961, provided a simple method for estimation of wind speed and direction when suitably reflective elements are present. A key assumption in this technique is the existence of a linear variation of the wind in a horizontal plane (3). This assumption does not hold for strong cyclonic storms, such as hurricanes and typhoons, which possess non-linear effects. From VAD type scans it is possible to map the airflow. Also, gust fronts and other small scale phenomena which generate convergence lines (and may initiate convection) have been observed using Doppler radar (4). The VAD algorithm, as well as an existing gust front identification algorithm, may useful in assessing these phenomena and their potential for squall line generation.

3. RAINFALL

Several factors determine the quantity of precipitation produced by storms. Some of these are:

- **Rainfall rate.** A function of the vertical distribution of winds and moisture.
- **Foreward motion of the cyclone.** Obviously, a slow–moving or stationary storm will produce more accumulated rainfall than a fast–moving system (if rainfall rates are the same in both cases).
- **Topography.** If the airflow is lifted over an orographical feature such as a mountain, an enhancement of rainfall amount can occur.
- **Environmental conditions surrounding the cyclone.** An analogous situation to orographic rainfall enhancement occurs when airflow is lifted at a pre-existing frontal boundary.

A rainfall algorithm should be one of the easiest to develop. The most frequently used method of deducing rainfall rate from radar data is to empirically relate reflectivity (Z) and rainfall rate (R) using an assumed drop size distribution, although other methods have been proposed (5, 6, 7). Z/R relations do have their problems, as drop size distributions are not only dependent upon location, but vary from storm to storm. Great care must be taken in selecting a Z/R relation, as different regions in a storm are composed of precipitation brought about by different processes (i.e., the convective precipitation region and the stratiform precipitation region), and may have different drop size distributions. Inadequate beam filling results in the underestimation of rainfall rates in cyclones (signal losses of 3–9 dB at 230 km for the WSR–57). The greatest success is found when this relationship is made adaptable and calibrated in real time with raingage data. This is because

- different storms exhibit different Z/R relationships;
- the appropriate Z/R relationship varies within storms; and
- different geographic locations are characterized by different Z/R relationships.

Automated raingages may, or may not, be available. In the absence of gauges and the mechanism to poll them, a Z/R relationship will have to come from climatology and would be a site adaptable parameter. Ground–clutter filters that are based upon cancelling echo return at zero or near zero frequency shift may not work well for slow moving or stationary storm systems. Calibration of rainfall rates with site–measured precipitation may be difficult in high winds, since the wind drives rain horizontally, and picks up rain that has already fallen to the ground. When winds are in excess of 50 kt (25 m s^{-1}), it has been estimated that as much as 50% of the rainfall misses the rain gauge.

Additionally, flash floods can accompany heavy downpours. The selection of the proper Z/R relation is important. This makes hydrology algorithms even more dependent upon the location of the radar. Storm tracking algorithms may be useful in predicting the movement of the convection, thus aiding short-term flood forecasters. On a display, topographical contours might be overlayed, or drainage basins might be overlayed. It is not clear over what interval rainfall amounts should be integrated, but certainly a flash flood depends upon rainfall rate, length of rainfall, areal extent of rainfall, and character of the drainage basin. Other variables such as soil saturation, etc., probably contribute to the flash flood potential, but probably are not essential to making a successful short term flash flood forecast.

For a radar, the challenge is to

- effectively remove the contaminating effects of ground clutter in the mountainous terrain;
- to relate the measured returned power to rain–fall rate; and
- to relate the rainfall rate to the potential for flash floods.

3.3 TORNADOES

Based on this information, beam filtering has been found to be a problem since it results in the underestimation of perturbation amplitudes on the scale of the beam dimensions. At ranges ≤ 350 km, the ratio of the diameter of the $1°$ beam to the radius of the maximum sustained winds is, at most, 0.2, resulting in a 5% underestimate in peak velocities, which is a tolerable margin of error for these applications (8,9).

3.4 STORM TRACKING

Radar is probably the best tool for the close observation and analysis storms. Geostationary satellites can provide a great deal of information about movement, especially over the oceans, where ships and planes are infrequent and data is scarce. However, the satellite image is of the top of the storm; because of this the main information provided is that of the storm's outflow. Indeed, this is the reason it is possible to evaluate a storm's intensity based on its satellite image, because a good outflow mechanism is essential to the survival of a storm. Some of the factors that must be considered in the long–range viewing of storms include:

- compromises in range and velocity aliasing;
- beam filtering;
- beam filling;
- azimuthal resolution;
- beam height versus storm height.

Forecasters must try to pinpoint a region of coastline where they believe a storm will strike. Monitoring a storm's progress by radar is the best way to observe changes in motion that may lead to a need to revise current warnings. Recently, probability forecasts have been introduced. In such a scheme, threatened locations are assigned a probability rating, expressed as a percentage. This idea has been greeted enthusiastically by the media and civil defense officials.

Storm tracking algorithms already in existence can be used in tracking the line elements (and thus the squall line), and may aid in forecasting the movement of the squall line. Currently, the NEXRAD system for forecasting storm movement is performed using a set of three separate algorithms which locate and track cells, identify new cells, and forecast cell movement. These algorithms may need to be modified for a squall line situation, in which many cells form and decay within several hours.

4. STORM INTENSITY

Squall lines generally form ahead of surface fronts. Squall lines are associated with strong, gusty winds. Doppler observations of tropical and mid-latitude squall lines clearly show gust fronts (10, 11, 12). Also, radar signatures such as those mentioned above (especially bow echoes) have been correlated with severe weather events in the United States. Algorithms do exist which can identify gust fronts.

The term cyclone is derived from Latin meaning "coils of a snake", a reference to the spiralling nature of some storms as described above. Most of the globally occurring severe weather is produced by baroclinic or extratropical storms. Such systems generally form at boundaries between warm and cold air masses (fronts), and draw their energy from the temperature difference. Indeed, these extratropical storms can become quite powerful and occasionally produce winds in excess of hurricane force.

An essential element of cyclone structure is the convective activity (heavy showers and thunderstorms). The importance of the storm's convective structure cannot be overstated. The amount of organized convective activity present in a cyclone can be taken as a measure of the storm's strength.

The growth process is similar for all cyclones. Given favorable conditions, central pressure will continue to fall, and the winds will increase. The growth process can be extremely rapid. Hurricanes can rapidly strength (e.g., Allen, July–August, 1980, underwent three periods of rapid strengthening; during one of these, a rapid pressure fall of 4 mb hr^{-1} was observed over a 12 hr period, indicative of explosive intensification). Winter–time cyclones along the U. S. east coast can deepen at the rate of 60 mb in 24 hours, as did the Queen Elizabeth II day storm on 9–10 September, 1978 (13).

Mathematically, the mean cyclone wind profile may be approximated by the combined Rankine vortex, an idealized non–divergent rotating circulation in which the area inside the RMW is considered to be in solid body rotation, i.e., u/r is a constant. Here we define u as the tangential or rotational wind velocity and r as the radial distance from the center to the point of observation. Outward from the RMW, the tangential wind velocity diminishes with distance from the center of the hurricane at the rate:

$$ur^x = constant \tag{1}$$

where, in a Rankine vortex, $x \equiv 1$, although modelling and observations suggest a value of $0.4 \geq x \leq 0.8$ may be more accurate for the cyclone (14). A completely precise value for x does not exist for hurricanes; however, $x = 0.5$ appears to be a good average (14). Detection of weak or developing cyclone centers could pose problems if the storm is poorly–organized. In situations such as this, the storm's center position must be extrapolated based on the position of the squall bands, which curve in toward the center.

A great advantage the Doppler radar networks will have over the current class of weather radars is their ability to provide information about the windfield of a large cyclonic storm. Analysis of windfields would be much easier with dual or multiple Doppler radars. However, techniques will have to be developed to examine the storms using single–Doppler radar (16).

Donaldson and Harris developed a method for recovery of the four first–order derivatives of the cyclonic wind field (17, 18). Diffluence is the tendency for a stream of air to fan out; crosswind shear is the horizontal change in speed of the wind normal to the flow, while the downwind shear is the downstream change in velocity of the horizontal wind.

Using these wind field derivatives we can express the kinematic properties of the wind field essential to the problem of cyclone wind observation, namely:

- divergence;
- stretching deformation;
- shearing deformation;
- vorticity.

The first three of these properties are easily calculated from the Doppler VAD patterns. Vorticity, however, cannot be determined directly from measurements using a single Doppler radar. By expressing the other three kinematic properties in terms of the Fourier coefficients and assuming that curvature is inversely proportional to distance from the center of the hurricane, we can estimate the wind speed (to within a few percent) with respect to the radar scanning circle (3, 16).

To measure the strength of the cyclone, Donaldson and Harris developed an approach that may prove useful in the detection and determining the proximity and/or intensity of a cyclone (17). The index, referred to as the Cyclonic Intensity Indicator (CII), may be expressed mathematically as:

$$CII = \sin(\delta) \left(\frac{V_o}{r} \right) \tag{2}$$

where δ is the angle of deviation of the maximum and minimum Doppler velocities from diametric opposition, and provides an estimate of the combined effects of curvature and crosswind shear. V_o is an estimate of the wind speed at the radar site and r is the radius of the VAD circle. The wind speed at the radar site is estimated by empirically combining the mean wind as determined from Fourier analysis and the gradient of Doppler velocities in the VAD circle. The δ angles, however, are extremely difficult to measure objectively. The CII is also very dependent upon the curvature, which always increases as the storm approaches. Thus, even if the storm strength is steady or even decreasing, the CII may indicate strengthening if the storm is approaching (Donaldson and Harris, 1984). Therefore, the CII may be a good indicator of the approach of a storm, but it is not a good indicator of the true storm strength.

A modified form of this index has been developed where the contribution of the curvature term has been removed. This function is called the Storm Strength Indicator (SSI) and is given by:

$$SSI = \frac{2b_2}{r} - \frac{b_1}{R} \tag{4}$$

where b_2 and b_1 are the moduli of the complex Fourier coefficients of the first and second harmonics, respectively, and R is the distance of the radar to the center of wind–field curvature. Studies indicate that the SII only increases if the storm is strengthening.

Another intensity indicator, introduced by Donaldson as the potential–vortex fit, also shows promise (18).

$$PVF = \frac{b_2 R}{r \left[b_1' - V_0 \sin(\alpha_0) \right]} \tag{5}$$

It is defined as the ratio of the shearing deformation to twice the curvature, and is essentially a measure of how well the sampled flow field fits the theoretical profile of the Rankine combined vortex. Depending upon how well the measured flow fits the rankine vortex, estimates of the maximum winds can be made, even if they are outside the range of observation. Studies of this parameter are continuing.

5. STORM SURGE

In many parts of the world, population density, limited arable land, or topography, force large concentrations of people into flood–prone coastal areas. Cyclones are feared worldwide because of their characteristic high winds and torrential rains. Actually, it is the storm surge that is the greatest killer in these storms. A few of the more catastrophic examples of the impact of the storm surge include:

- Bombay, India (1882): 100,000 deaths;
- Galveston,Texas (1900): 6,000 died when a hurricane- storm surge inundated Galveston Island;
- Bangladesh (1970): 300,000 deaths attributed to storm surge at the head of the Bay of Bengal.

The storm surge is a product of the circulation and transport of water due to wind stress and changes in atmospheric pressure created in the vicinity of a storm. In mid–latitude storm systems, high waves and tides are caused by winds blowing across an open stretch of water for long periods (defined as the fetch).

To summarize, the storm surge is the result of the interaction of several variables:

- **Storm Intensity** – the height of the storm surge will always increase with increasing storm intensity;
- **Forward Motion** – the maximum storm surge height occurs when the storm is moving nearly perpendicular to the coast at a speed of about 34 kt (17 m s^{-1});
- **Coastline Irregularities** – Bays and inlets serve as funnels and increase the height of the storm surge. Superelevation of the water level in bays may exceed that on the open coast by 50% or more for slow-moving storms;
- **Storm Size** – a storm with a RMW of about 27 n. mi. (as in a large storm) will produce the largest storm surge, all other factors being the same;
- **Shoaling Factor** – Studies indicate this component has the maximum effect when the ocean bottom slopes up slowly to the beach from about 60 n. mi. out;
- **Astronomical Tide** – The storm surge is superimposed on the astronomical tide. This factor is important in a real-time assessment of storm surge threat, particularly in regions with large tidal ranges.

For cyclonic storms, storm surge is created by convergence of water toward the storm center. In deep ocean areas, this convergence at the surface is compensated by divergence at lower levels, resulting in no net sea level rise (19). But as the system moves into shallower water, this convergence of sea water results in a superelevation of the ocean surface. It should be noted though that the wind stress–induced component is about twice as large as the pressure–induced component of the storm surge (19). When the waves reach shallow water near a coastline they grow to characteristic size.

Another factor in the size of the storm surge is the angle of the storm's approach to the coast. The greatest storm surge occurs when the storm is moving rapidly, at an angle nearly perpendicular to the coast (19).

Of the six factors governing the cyclone storm surge, Doppler radar will be able to give information on the storm's intensity, size and forward motion. Astronomical tide charts are available for any location and time. Therefore, three parameters can be measured directly or inferred from Doppler radar, and the remaining three are site adaptable parameters.

Although radar can be an excellent tool, it is notable that wind velocity data from the Doppler system is only available to a radius of a few hundred km of the radar site. By this time, a storm moving at an average foreward speed (~12 kt) is less than 12 hours from landfall. Since most land-based radar sets are situated on coastal regions (as is the case in the U.S.) rather than on islands offshore, warning that a severe storm surge will strike an area is limited to this time.

An alternative is direct input of radar wind data into a more sophisticated dynamical storm surge model. Dynamic models such as these calculate storm surges on the open coast by solving the basic hydrodynamic equations of motion (20). These models require, as an initial condition, the direct input of the wind field of the storm. Previously,the maximum wind speed was known, but for the sake of real–time calculations, the wind field was usually estimated.

6. ELECTRICAL STUDIES

Other than in situ data, our best estimates of the microphysical structure of storms come from analysis of the synthesized Doppler wind fields. Using techniques such as outlined in Ziegler, it is possible to examine the co-evolving wind and water field (21). All the various contributions to the rain process can, in principle, be quantified for a more complete understanding of the role of microphysics in storms. It is believed possible to quantify the dominant mode of precipitation formation in the convective cells.

The retrieval of microphysical fields requires a Doppler radar synthesized wind field with little error; indeed, retrieved anomalies often reveal errors in the synthesized winds. Microphysical retrieval was first done by Rutledge and Hobbs for mesoscale rainbands and later extended to tropical squall lines by Rutledge (22, 23, 24). More recently, Rutledge and Houze have used the model to diagnose middle-latitude squall lines (25). Ziegler retrieved microphysical fields in three dimensions for a convective cell (21). Hauser et al., reformulated the problem allowing the microphysical and thermodynamical variables fields to be jointly derived, consistent with the whole set of governing equations and the wind fields (26). Every opportunity to check the results with independent measurements has confirmed its accuracy.

Ziegler outlines a method similar to that of Rutledge and Hobbs for retrieving the temperature and water substance distributions within storms, given an accurate storm wind field (such as might come from Doppler radar analysis) and a proximity sounding (21, 22, 23). The time-dependent, three-dimensional cloud model incorporates transport processes, as well as uses a microphysical parameterization which includes stochastic warm cloud coalescence effects and wet and variable density dry hail growth.

The gamma function is used to represent cloud and rain distributions, while an inverse exponential function represents the graupel/hail and snow distributions.

The gamma distribution used in Ziegler is of the form

$$N_v(v) = N(\nu + 1)^{\nu+1}(v/v_o)^\nu \exp\left[-(\nu + 1)v/v_o\right)/\Gamma(\nu + 1)v_o] \qquad (6)$$

where N is the total concentration (cm^{-3}), v_o the mean volume (cm^{-3}), v the volume, ν the shape parameter, and $N(v)$ the concentration per unit volume interval (21). The value of ν is fixed (typically in the range 0 to 3), while N and v_o vary.

The inverse exponential function is of the form

$$N_D(D) = N_o e^{-\Psi D} \tag{7}$$

with N_o the intercept parameter (concentration of particles per unit size range at zero size), Ψ the shape parameter, and D, the particle diameter.

An alternative formulation is the finite difference method of Berry and Reinhardt in which the particle distribution is represented by a set of discrete size categories and the appropriate microphysical processes applied to each category (27). For kinematic retrieval, this would be excessively computationally burdensome. When compared with Berry and Reinhardt's highly resolved finite difference coalescence model, Ziegler found very good agreement using the parameterized model in the Lagrangian frame (28, 21).

Variations of potential temperature, water vapor mixing ratio, cloud droplet number concentration and mixing ratio, rain drop number concentration and mixing ratio, snow mixing ratio, cloud ice mixing ratio, and graupel/hail concentration and mixing ratio are diagnosed from the kinematic field according to a system of continuity equations. The conservation equations for heat and water substances can be represented as follows. For potential temperature (θ)

$$\frac{\partial \theta}{\partial t} = -u^o \frac{\partial \theta}{\partial x} - v^o \frac{\partial \theta}{\partial y} - w^o \frac{\partial \theta}{\partial z} + \nabla_3 * (K_d \nabla \theta) + S_\theta \quad . \tag{8}$$

For nonprecipitating water substance (q_v, N_c, q_c, q_i)

$$\frac{\partial N}{\partial t} = -u^o \frac{\partial N}{\partial x} - v^o \frac{\partial N}{\partial y} - w^o \frac{\partial N}{\partial z} + \nabla_3 * (K_d \nabla N) + S_N$$
$$+ N w^o \frac{\partial \ln \rho}{\partial z} \tag{9}$$
$$\frac{\partial q}{\partial t} = -u^o \frac{\partial q}{\partial x} - v^o \frac{\partial q}{\partial y} - w^o \frac{\partial q}{\partial z} + \nabla_3 * (K_d \nabla q) + S_q \quad .$$

For precipitating water substance (q_r, N_r, q_h, q_s, N_h):

$$\frac{\partial N}{\partial t} = -u^o \frac{\partial N}{\partial x} - v^o \frac{\partial N}{\partial y} - w^o \frac{\partial N}{\partial z} + \nabla_3 * (K_d \nabla N) + S_N$$
$$+ N w^o \frac{\partial \ln \rho}{\partial z} - \frac{\partial (N V_t)}{\partial z} \tag{10}$$
$$\frac{\partial q}{\partial t} = -u^o \frac{\partial q}{\partial x} - v^o \frac{\partial q}{\partial y} - w^o \frac{\partial q}{\partial z} + \nabla_3 * (K_d \nabla q) + S_q$$
$$- \frac{1}{\bar\rho} \frac{\partial}{\partial z}(\rho V_q)$$

where S is a microphysical source or sink rate, V_t is the precipitation fall speed, ρ the air density, and the rest of the symbols retain their usual meaning. The airflow components u^o, v^o, and w^o are supplied from the synthesis of multiple Doppler radar wind fields. Liquid clouds form in supersaturated air, while other water phases and ice types form from the ensuing interactions. Further sensitivity studies were conducted in Ziegler (29).

ACKNOWLEDGMENTS

Partial support for the preparations was provided by NSF Grants ATM 8604143 and ATM 8619957 to Florida State University, and by the Department of Energy Contract No. DE-Fco5-85ER250000 to the Florida State University Supercomputer Computations Research Institute. This research was supported in part by the Florida State University through time granted on its Cyber 205, ETA[10] supercomputers.

REFERENCES

(1) PROBERT-JONES, J. R. (1960). Meteorological use of pulsed Doppler radar. *Nature*, **186**, 271-273.

(2) CATON, P. G. F. (1963). Wind measurement by Doppler radar. *Meteor. Mag.*, **92**, 213-222.

(3) BROWNING, K. A. and WEXLER, R. (1968). The determination of kinematic properties of a wind field using Doppler radar. *J. Appl. Meteor.*, **7**, 105-113.

(4) WILSON, J. W., and SCHREIBER, W. E. (1986). Initiation of convective storms at radar–observed boundary–layer convergence lines. *Mon. Wea. Rev.*, **114**, 2516-2535.

(5) BRANDES, E. A., and WILSON, J. W. (1986). Measuring storm rainfall by radar and rain gauge. *Thunderstorms: A Social, Scientific, and Technological Documentary, Vol. 3: Instruments and Techniques for Thunderstorm Observation and Analysis* (E. Kessler, ed.), 2nd edition, University of Oklahoma Press, Norman, OK.

(6) DOVIAK, R. J., and ZRNIC, D. (1984). *Doppler Radar and Weather Observations.* Academic Press, Orlando, FL., 480 pp.

(7) SELIGA, T. A., and BRINGI, V. N. (1976). Potential use of radar differential reflectivity measurements at orthogonal polarizations for measuring precipitation. *J. Appl. Meteor.*, **15**, 69-76.

(8) BROWN R. A., and LEMON, L. R. (1976). Single Doppler radar vortex recognition: Part II - tornadic vortex signatures *Preprints, 17th Conf. on Radar Meteorology*, Seattle, Amer. Meteor. Soc. 104–109.

(9) HARRIS, F. I., HAMANN, D. J., and DONALDSON, R. J. Jr. (1989). Hurricane monitoring with Doppler radar: a simulation. *Preprints, 24th Conf. on Radar Meteorology*, Tallahassee, Amer. Meteor. Soc. 203–206.

(10) ROUX, F., TESTUD, J., PAYEN, M., and PINTY, B. (1984). Pressure and temperature perturbation fields retrieved from dual–Doppler radar data: an application to the observation of a West African squall line. *J. Atmos. Sci.*, **41**, 3104-3121.

(11) ROUX, F. (1988). The West African squall line observed on 23 June 1981 during COPT 81: kinematics and thermodynamics of the convective region. *J. Atmos. Sci.*, **45**, 406-426.

(12) VASILOFF, S. V., and BLUESTEIN, H. B. (1988). Analysis of the Oklahoma segment of the 10–11 June 1985 severe squall line: maturity to decay. *Preprints, 15th Conf. Severe Local Storms*, Baltimore, MD., Amer. Meteor. Soc., Boston, 288-291.

(13) LAWRENCE, M. B. and PELISSIER, J. M. (1981). Atlantic hurricane season of 1980. *Mon. Wea. Rev.*, **109**, 1567-1582.

(14) RIEHL, H. (1979). *Climate and Weather in the Tropics.* Academic Press, London, 611 pp.

(15) DONALDSON, R. J., Jr., and HARRIS, F. I. (1984). Detection of wind field curvature and wind speed gradients by a single Doppler radar. *Preprints, 22nd Conf. on Radar Meteorology*, Zurich, Amer. Meteor.

(16) RUGGERIO, F. H., and DONALDSON, Jr., R. J. (1987). Wind Field Derivatives: A new tool for analysis of hurricanes by a single Doppler radar. *17th Conference on Hurricanes and Tropical Meteorology*, 178-181.

(17) DONALDSON R. J., Jr., and HARRIS, F. I. (1987). Estimation by Doppler radar of curvature, diffluence, and shear in cyclonic flow. *J. of Atmos. and Oceanic Tech.*, (Dec. 1988).

(18) DONALDSON, R. J. (1989). Potential - vortex fit. *24th Conf. Radar Meteorlogy*, Tallahasse, FL., Amer. Meteor. Soc., 186–189.

(19) ANTHES, R. A. (1982). Tropical cyclones: Their evolution, structure and effects. *Meteor. Monogr.*, **19**, No. 41., Amer. Meteor. Soc., Boston, 208 pp.

(20) SIMPSON, R. H., and RIEHL, H. (1981). *The Hurricane and Its Impact*, Louisiana State University Press, Baton Rouge, La., U.S. and London, U.K., 398 pp.

(21) ZIEGLER, C. L. (1985). Retrieval of thermal and microphysical variables in observed convective storms. Part I: Model development and preliminary testing. *J. Atmos. Sci.*, **42**, 1487–1509.

(22) RUTLEDGE, S. A., and HOBBS, P. V. (1983). The mesoscale and microscale structure and organization of clouds and precipitation in midlatitude cyclones. VIII: A model for the "seeder-feeder" process in warm-frontal rain bands. *J. Atmos. Sci.*, **41**, 2949–2972.

(23) RUTLEDGE, S. A. (1984). The mesoscale and microscale structure and organization of clouds and precipitation in midlatitude cyclones. XII: A diagnostic modeling study of precipitation development in narrow cold-frontal rain-bands. *J. Atmos. Sci.*, **41**, 2949–2972.

(24) RUTLEDGE, S. A. (1986). A diagnostic modeling study of the stratiform region associated with a tropical squall line. *J. Atmos. Sci.*, **43**, 1356–1370.

(25) RUTLEDGE, S. A., and HOUZE, R. (1987). A diagnostic modeling study of the trailing stratiform region of a midlatitude squall line. *J. Atmos. Sci.*, **44**, 2640–2656.

(26) HAUSER, D., ROUX, F., and AMAYENC, P. (1987). Comparison of two methods for the retrieval of thermodynamic and microphysical variables from Doppler radar measurements: Application to the case of a tropical squall line. *J. Atmos. Sci.*, **45**, – 1285–1303.

(27) BERRY, E. X, and REINHARDT, R. L. (1974a). An analysis of cloud drop growth by collection: Part I: Double distributions. *J. Atoms. Sci.*, **31**, 1814–1824.

(28) BERRY, E. X, and REINHARDT, R. L. (1974b). An analysis of cloud drop growth by collection: Part II: Single initial distributions. *J. Atoms. Sci.*, **31**, 1825–1831.

(29) ZIEGLER, C. L. (1987). Retrieval of thermal and microphysical variables in observed convective storms. Part II: Sensitivity of cloud model output to variation of the microphysical parameterization. *J. Atmos. Sci.*, **45**, 1072–1090.

THE OPERATIONAL USE OF A HIGH SPEED INTERACTIVE
RADAR DATA PROCESSING SYSTEM - RDPS
PART II : APPLICATION IN REAL TIME SEVERE THUNDERSTORM
DETECTION AND FORECASTING.

H.-P. BIRON

Centre Météorologique du Québec
Environnement Canada
St-Laurent, Québec
Canada   H4M 2N8

G.L. AUSTIN and A. KILAMBI

McGill Radar Weather Observatory
Ste-Anne-de-Bellevue, Québec
Canada   H9X 1C0

Summary

A prototype Radar Data Processing System has been in use at the
Centre Météorologique du Québec (CMQ) during the summer of 1988.
This system allows real-time reception and processing of full volumetric data from the McGill University radar at 5 minutes intervals
for use by operational meteorologists, thus providing them with a
powerful tool to analyse and display radar data in multiple ways and
to produce short-term forecasts. The capabilities of the system
allow an in-depth determination of thunderstorm structure and evolution, resulting in a better diagnosis of thunderstorm severity in an
operational setting. Users are able to study horizontal and vertical cross-sections (as well as many other products) through a
thunderstorm in real time in order to determine whether or not typical features associated with severe thunderstorms are present. The
system also allows fast animation of radar echoes, enabling the easy
recognition of typical severe storm behaviour such as storm propagation to the right of the mean propagation direction and new cell
growth in a multicellular complex.

1. INTRODUCTION
    The McGill Radar Weather Observatory developed, for the Canadian
Atmospheric Environment Service, a Radar Data Processsing System (RDPS)
designed to provide operational meteorologists with a powerful, user-
friendly tool to detect and forecast all types of precipitation and
particularly severe thunderstorms. The system is described in Austin et
al. (1), and the prototype is currently in use at the Centre Météorologique du Québec in Montréal.
    This paper will briefly describe how this system is used in weather
forecast operations.

2. THE DATA
    Datasets come from the McGill University radar (10 cm wavelength,
0.8 degree beamwidth, 10 m diameter antenna) and consists of a volume

scan of 24 elevation angles up to a range of 240 km, and up to 480 km
for the lowest elevation angle.

A new volume of data is transmitted every five minutes to the
weather centre and stored on disc. The Product Processing System (PPS)
runs on a MV-4000 Data General computer and is used to generate a wide
range of products that are displayed on a high resolution colour monitor.

3. OPERATIONAL USE OF RDPS

Radar data are incorporated into the everyday forecast programs at
CMQ. With RDPS, meteorologists have access to numerous radar outputs
every 5 minutes; the horizontal products have a spatial resolution of 1
x 1 km up to 120 km from the radar, and 2 x 2 km from 120 to 240 km.

For more than 12 years CAPPI's (Constant Altitude PPI's) have been
considered the basic operational products used for radar display; Echo
Tops (showing the highest tops of precipitation over each cartesian
point), accumulation maps and forecasts [SHARP algorithm, Bellon et al.
(2)] are other products used on a routine basis.

The system has been designed to automatically generate specific
products after reception of each volume scan; the schedule of such
products can be set up by users depending upon the season of the year,
weather types, etc. Table 1 shows a typical summer operational schedule
of automatically generated products.

TABLE 1. Operational schedule for RDPS automatic product generation	
every 5 minutes:	1.5 or 3 km CAPPI 5 km CAPPI 7 km CAPPI Echo Tops - intensity threshold at 18 dBZ Echo Tops - intensity threshold at 40 dBZ
every half-hour:	Forecast CAPPI Point forecast for 16 locations Long Range (to 480 km)
every hour:	Accumulation Forecast accumulation (for the next 2 hours) TIMMAXR - location of highest reflectivities over each cartesian point for the past "x" hours.

In addition, RDPS can generate other products requested interactively by users such as vertical cross-sections, CAPPI's at different
heights, point forecasts, etc. Finally the display system allows fast
animation of radar echoes, zooming, panning, look-up table selection and
modification, geographic overlay selection, statistics over a given area
of precipitation, etc.

4. SEVERE WEATHER DETECTION

RDPS is an excellent tool for "ordinary" precipitation events but its potential is fully revealed during severe convective situations.

Several studies have shown that severe thunderstorms have a distinctive structure; in particular, the reflectivity distribution in the cell at all levels is an important characteristic to identify a severe storm; the overhanging of mid-level reflectivity core associated with a weak echo region underneath is a good example of a distribution generally associated with severe weather. Access to volumetric radar data becomes essential to weather forecast operations since it allows meteorologists to display and study data in such a way as to easily identify and pinpoint specific patterns related to severe weather.

The capability of RDPS to generate several CAPPI's every five minutes proved to be extremely useful to CMQ meteorologists. Looking at the precipitation at a constant level (CAPPI's) is a much better way than looking at PPI's, where the height of echoes increases with range. Moreover, as shown in recent studies, it is at mid-level that organization of reflectivities will first exhibit a potential for severe weather. A CAPPI at 7 km is thus an easy and accurate way to detect reflectivity cores, precursors of severe thunderstorms and this, over the entire radar coverage area (240 km radius for the McGill radar).

At CMQ, a thunderstorm having reflectivities of 45 dBZ or more at 7 km is considered as potentially severe. Several studies have shown that high reflectivity cores develop at 7 km at least 30 minutes and very often one hour before they reach the ground and produce severe weather. As an example, fig. 1 and 2 show the 1.5 km and 7 km CAPPI's for August 3, 1988 at 2110 UTC. The area indicated by the arrow on the 1.5 km CAPPI (fig. 1) shows no evidence of any significant echo pattern. However on the 7 km CAPPI (fig. 2) one can already see a well defined circular cell with intensities of 100 mm/h or 55 dBZ. 40 to 50 minutes later, this storm produced 2-4 cm hail and winds in excess of 100 km/h, causing heavy damage over the Montreal area.

This one criterion alone gives very few false alarms and has a high probability of detection of severe weather over southwestern Quebec (it might be different elsewhere). However, one must consider the whole structure of the precipitation volume in order to discriminate between different types of severe events. Large hail will be associated most often with supercells showing an overhang with very high reflectivities; the vertical cross-section of a hail storm that occurred over Montreal on May 29, 1987 is a good example of a supercell (fig. 3); heavy rain however will show as a vertical column with the highest intensities located mostly in the lower part of the storm. "Solid" walls of reflectivities at mid-levels, having very often a crescent or arc shape or even spiral shape will most certainly be precursors of strong winds (maybe a tornado ?) on the ground.

RDPS is designed specifically to give the forecasters the possibility to do such an in-depth study of storms and thereafter issue timely and accurate warnings for the threatened regions.

With RDPS forecasters can also request short term (1 to 5 hours) forecasts of CAPPIs; these forecasts give a quick view as to where precipitation will be located. For severe thunderstorms which deviate from the mean flow, the system allows the meteorologist to input a new displacement vector and get a new forecast for these storms. At the same time, point forecasts can be generated for specific locations such as airports or cities; these forecasts will give the time of occurrence of precipitation and the total possible accumulation based on intensities and the assigned Z-R relationship. These products help the forecaster in issuing heavy rain or flash flood warnings.

Animation of precipitation echoes allows forecasters to constantly monitor the propagation and the evolution of severe thunderstorms cells.

5. CONCLUSION

A Radar Data Processing System was developed by the McGill Radar Weather Observatory and the prototype is currently used operationally at the Centre Météorologique du Québec. Operational use proved that RDPS is a powerful and user-friendly tool allowing meteorologists to perform an in-depth study of precipitation volumes and particularly of severe thunderstorms. As our knowledge of severe thunderstorms increases, access to volumetric radar data becomes essential to operational meteorologists. RDPS is a superb example of what can be done to use these data to their maximum benefit.

REFERENCES

(1) AUSTIN, G. et al., 1986: RAPID II: An operational, highspeed interactive analysis and display system for intensity radar data processing, Preprints, 23rd Conference on Radar Meteorology, Snowmass, Colo., JP79-JP82

(2) BELLON, A. and G.L. AUSTIN, 1978: The evaluation of two years of real-time operation of a short-term precipitation forecasting procedure (SHARP). J. Appl. Met., 17, 1778-1787.

(3) BIRON, H.-P., 1988: Radar Data Processing System: On the use of full-scan radar data in severe weather forecasting. 22nd Canadian Meteorological and Oceanographic Society Congress, Hamilton, Ont. Canada.

(4) BIRON, H.-P., 1986: Operational use of meteorological radars in identifying summer severe weather at Quebec Weather Center, First AES/CMOS Workshop on Operational Meteorology, Winnipeg, Man., Canada.

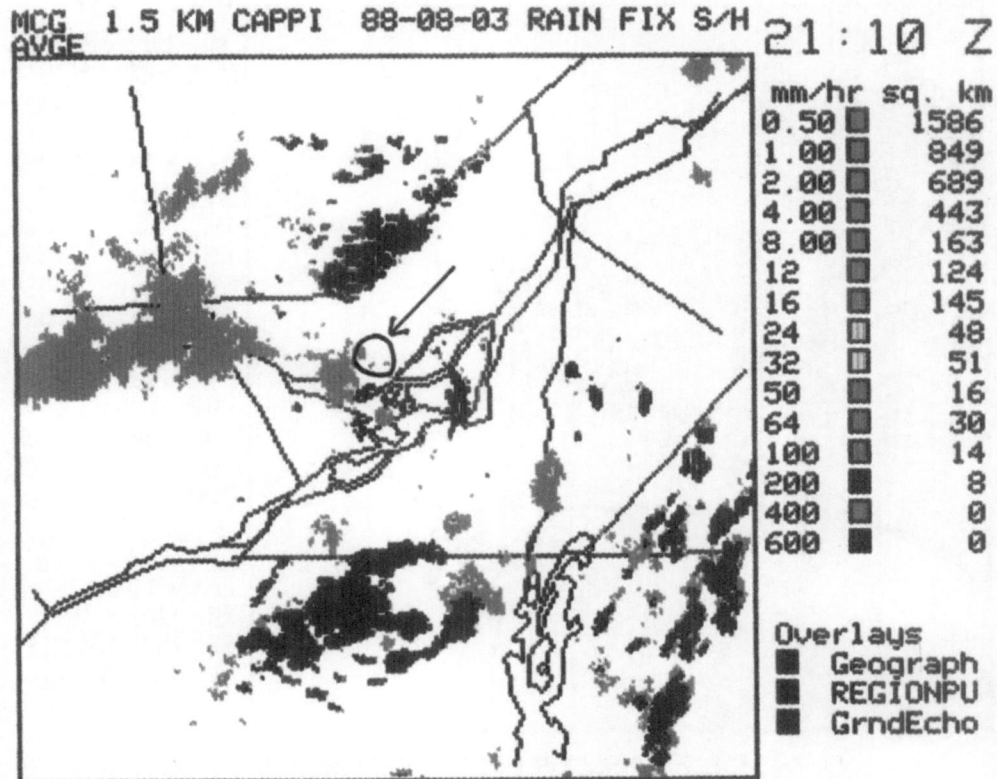

fig. 1.   1.5 KM CAPPI from the McGill radar at 2110 UTC August 3, 1988.
Arrow indicates the area of interest.

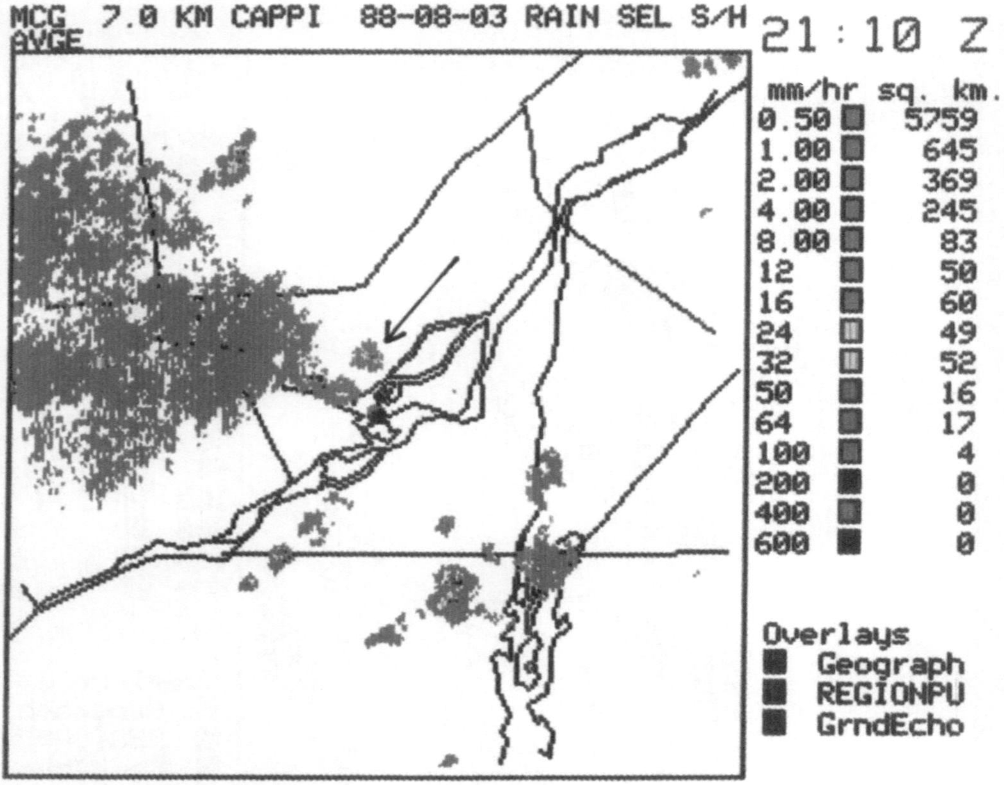

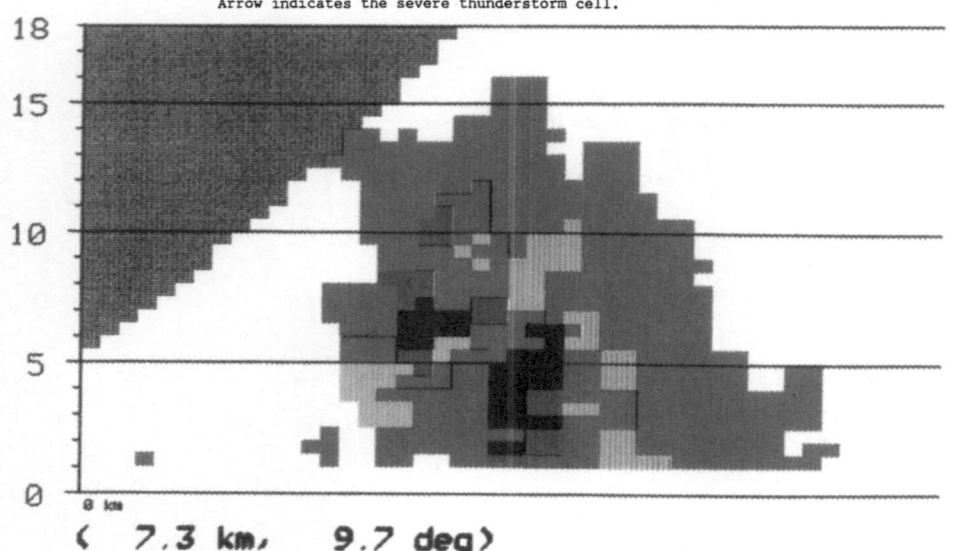

MCG 7.0 KM CAPPI 88-08-03 RAIN SEL S/H
AVGE
21:10 Z
mm/hr sq. km.
0.50 5759
1.00 645
2.00 369
4.00 245
8.00 83
12 50
16 60
24 49
32 52
50 16
64 17
100 4
200 0
400 0
600 0
Overlays
Geograph
REGIONPU
GrndEcho

fig. 2. 7 KM CAPPI from the McGill radar at 2110 UTC August 3, 1988. Arrow indicates the severe thunderstorm cell.

18
15
10
5
0
0 km
(7.3 km, 9.7 deg)

fig. 3. Vertical cross-section of a hailstorm showing overhanging of strong mid-level reflectivities and a weak echo region (Montreal hailstorm, May 29 1987, 2308 UTC).

DETERMINATION OF CLOUD FIELD EVOLUTION BY THE METHOD
OF CORRESPONDENCE BETWEEN RADAR REFLECTIVITY ISOSURFACES

D. PODHORSKÝ and Ľ. VLČÁK

Slovak Hydrometeorological Institute

Summary

On the basis of specific limited parts of the cloud
field radioecho, its horizontal and vertical evolution
is computed by the method of correspondence between
limiting isosurfaces. The changes assessed of the radar
reflectivity values in a four-dimensional space (x,y,z,t)
determine the cloud evolution tendency and the associated
phenomena. The computed radar reflectivity values
determination vector enables to dissect a cloud field
into separate parts which are of dominant significance
for the classification of weather phenomena as well as
of cloud types, including their prediction for up to
two hours.
Input digital data are provided by the Automated Radar
Meteorological System (ARMS) for the MRL − 5 two-wave
radar. The software and computation of cloud field
evolution is caried out by IBM PC-AT computer.

Radar measurement automation is of importance
particularly when it is based on a comprehensive assessment
of the individual methods applied for manual measurements in
the different countries up to the present. At the same time
attention should be paid to the fact that through a single
instance of automated scanning of radar characteristics in
a three-dimensional matrix those data should be registered
and transformed into the unified cartographic system of the
meteorological radar network which are necessary for the
evaluation of the output information to be provided for the
individual users, i.e. for the hydrological forecast service,
weather forecasts, aviation meteorology and the air traffic
control for nowcasting and very short-range weather
forecasts, for the radar remote sensing climatology as well
as for the modification of the phenomena associated with
cloudiness.
    The Czechoslovak Automated Radar Meteorological System
(ARMS) provides synchronous measuremnts on two wave-lenghts
of an MRL-5 radar (USSR), namely 3.2 and 10.0 cm respectively

(1,2,3). On the basis of the operator—computer relationship, the radar software enables to change the measurement mode in an interactive manner prior to each measurement within the scope of manual control of the aerial and the simultaneous complete automation of return signals at the one end, right up to the comprehensive data processing automation on both channels at the other. The control programme furnishes the possibility to operatively change the aerial elevation, to choose the pulse length, to set the limit values, the average number of pulses,etc. Despite the fact that, technologically, the MRL—5 belongs to 1970's, the ARMS philosophy — based on the works by J. Podlešák, M. Ružička and D. Podhorský — even nowadays enables theoretico—experimental studies of the different radar characteristics of the cloud field and of other associated phenomena.

The methodology of manual weather radar measurements in Czechoslovakia proceeds from the Soviet school established by E. M. Salman (1965 — 1975) at the A. I. Voyeykov Central Geophysical Observatory, Leningrad. After being modified in accordance with the physico—geographical and synoptic conditions of Central Europe, it allows assessments to be carried out in one—hour intervals (under favourable synoptic conditions also in 30—minute intervals) for each discrete square of the radar—scanned area (30 x 30 km for 300 km range; 15 x 15 km for 150 km range). The information thus obtained concerns the top level echo, the cloud system type and its respective accompanying phenomenon, the radar reflectivity at two levels selected (for example, at the levels representing spontaneous crystallization of cloud particles in stratiform clouds : -15°C and at the level of the -25°C isotherm which characterises the evolution stage of convective cloudiness). Further, radar reflectivity values are estimated at the height of one km, i.e. precipitation intensity derived from the $Z \sim I$ relation, the development tendency of the echo field as well as its direction and propagation velocity. More than fifteen years' continuous observations in Czechoslovakia by means of radars, evaluation of case studies and the climatological processing activities have provided us with a sufficient ground for the selection of a certain set of feature vectors, and having used these as a basis, expert aproach has then been applied to select the standard measures for the classification of the individual classes of clouds and phenomena associated with them.

The set of feature vectors valid for each radar volume column comprises the following items :
a) top level echo
b) temperature at top level echo
c) maximum radar reflectivity level
d) temperature at the maximum radar reflectivity level
e) depth of supercooled cloud section, i.e. the difference between top level echo and 0°C isoterm
f) cloudless layer thickness above the echo
g) parameter determining the possible existence of another echo level
h) radar echo vertical profile, representing the respective

cloud and phenomenon class
i) maximum radar echo value,representing the respective class
j) base level echo
k) curve indicating the vertical movement of echoes, which represent the selected radar reflectivity values for the individual classes
l) vertical movement value of radar echo volume, possessing maximum radar reflectivity.

These twelve feature vectors are,in principle, the basis of the dynamic (four-dimensional) approach to the radar echo field classification. The number of feature vectors is not limited and it can experimentally be complemented by either Doppler or pseudo-Doppler radar data and by the features used in the METEOTREND system, i.e. the features typical of advective-convective tendencies (4,5,6,7,8) calculated from geostationary satellite-based primary data, etc.

The calculation for the individual cloud classes and phenomena classification is based on a modified "method of the minimum distance from the standard measures".

At present, experimental verification is under way of both the set of feature vectors and standard measures for the following classes :
1. hailstorm
2. thunderstorm
3. Cumulonimbus with a shower
4. Cu con
5. Nimbostratus, Altostratus
6. Altostratus, Cirrostratus
7. Cirrostratus
8. ground

The algorithm used in the first stage of our research into the volume movement calculation, which represents the selected radar reflectivity values, proved exceptionally succesful for the METEOSAT geostationary satellite IR-channel isotherm correspondence in the software applied to nowcasting based on the top level echo temperature data.

Finally, it should be pointed out that the experimental knowledge acquired through the verification of the algorithm in question requires certain corrections to be made of our theoretical assumptions with regard to the combination of the satellite and radar observation - based features in a unified algorithm suitable for the classification procedure. This fact complicates the demand for machine time, i.e. for both the classification and the METEOTREND computations in real time, in the sense of the nowcasting definition. The possible solution lies in the utilization of associative - parallel processors and expert systems.

Estimation of Cloud Movement and Development by Means of a 4 - Dimensional Method of Correspondence between Isosurfaces of Radar Reflectivity Values.

Let $r(i,j,k,t)$ be the value of radar reflectivity at point (i,j,k) and in time t. Since reflectivity is recorded in digital form and measurement is carried out at a certain time only, a discrete form has to be taken into account

$$\left\{ I_{\vartheta_t} \right\}_{t=0}^{T} \quad , \quad I_{\vartheta_t} = \left\| r_{ijk}^{\vartheta_t} \right\|_{i=1, j=1, k=1}^{n \quad m \quad l} \quad , \quad R_1 \leq r_{ijk} \leq R_2 \quad [1]$$

where n, m, l represent the dimensions of the specified space, $\vartheta_T < \ldots < \vartheta_1 < \vartheta_0$ represents the succession of measurement times considered $r_{ijk}^{\vartheta_t} = r(i,j,k,\vartheta_t)$, R_1 is the minimum and R_2 the maximum reflectivity value ($R_1 \leq r_{ijk} \leq R_2$).

Based on the succession in question (1) and estimation of cloud movement and development represented by reflectivity value is intended, i.e. the succession

$$\left\{ X_{\vartheta_t}, Y_{\vartheta_t}, Z_{\vartheta_t} \right\}_{t=0}^{T} \qquad\qquad\qquad [2]$$

$$X_{\vartheta_t} = \left\| x_{ijk}^{\vartheta_t} \right\|_{i=1, j=1, k=1}^{n \quad m \quad l} \qquad Y_{\vartheta_t} = \left\| y_{ijk}^{\vartheta_t} \right\|_{i=1, j=1, k=1}^{n \quad m \quad l}$$

$$Z_{\vartheta_t} = \left\| z_{ijk}^{\vartheta_t} \right\|_{i=1, j=1, k=1}^{n \quad m \quad l} \qquad\qquad\qquad [3]$$

The vector ($x_{ijk}^{\vartheta_t}$, $y_{ijk}^{\vartheta_t}$, $z_{ijk}^{\vartheta_t}$) expresses the estimation of the movement velocity in time ϑ_t (x is the direction of coordinate i, y is the component in the direction of coordinate j, z is the component in the direction of coordinate k).

The motion field determination consists of determination of the individual velocity matrices X,Y,Z.

The algorithm of the cloudiness movement and development estimation, which is represented by radar reflectivity starts from the presumption that reflectivity is (relatively) invariant within the time interval between two measurements. More exactly reflectivity changes of the corresponding clouds at differents times are smaller than the reflectivity changes between different types of clouds in the same moment. This means that such a reflectivity value (and/or reflectivity value vector) can be found through which - in case that the threshold value is detected at different times - the images of the same clouds can be obtained at the points across which these were travelling at any particular moment.

Based on the cloud parts thus specified the cloud movement is then calculated by a method of correspondence of limit isosurfaces between these cloud sections. The changes of reflectivity values in the individual cloud parts are a basis on which their further development is predicted. The reflectivity values vector by means of which the clouds are divided in the individual parts is called the reflectivity determination vector.

When reflectivity vector d_s= 1, ... , $\varkappa$, is selected a threshold of a 3 - dimensional picture $I = \| r_{ijk} \|$ can be calculated

$$B^S = \left\| b_{ijk}^S \right\|_{i=1,j=1,k=1}^{n \quad m \quad l} \qquad b_{ijk} = \begin{cases} 1 \text{ if } r_{ijk} \overset{\leq}{=} d_S \\ 0 \text{ othervise} \end{cases} \qquad [4]$$

for s = 1, ... , $\varkappa$

In objects within the B^S horizon a continuous point set is implied, for which b_{ijk}^S = 1. This object is separated from the rest of the space enclosed by the isosurface d_S. Starting from above invariance supposition, the isosurface into subsequent images will not differ too much in shape , but in position only.

Let us now consider the binary images $B_{\vartheta_t}^S$ and $B_{\vartheta_{t-1}}^S$ in time ϑ_t and ϑ_{t-1}. $\vartheta_{t-1} < \vartheta_t$, formed by the isoplane d_S , containing the objects O_{t-1} and O_t (images of objects O).The movement velocity of the object O will be determined in two steps.

In the first step, the isosurface position differences in each direction and for each object point will be determined. Let us consider a point (i_0, j_0, k_0) of the object O_t, whose position is not further beyond the object boundary than h_{max} (a selected constant) along the coordinate i. The distance λ_t of this point from the closer object boundary O_t alongside axis i is determined.Then the distance λ_{t-1} of point (i_0, j_0, k_0) from the object boundary O_{t-1} alongside axis i is sought. The search is conducted up to the distance of $\lambda_t + P_{max}$ (P_{max} — a selected constant) in turns, both in the direction i and direction −i. Only a boundary of the same kind is regarded as a valid one O_{t-1} (if the front boundary O_t in direction i is looked for, then the front O_{t-1} is valid. Similarly the same goes for the rear boundaries). After this boundary is found

$$X_{i_0, j_0, k_0} = (\lambda_{t-1} - \lambda_t) / (\vartheta_t - \vartheta_{t-1}) \qquad [5]$$

is calculated. If the object boundary of the object O_{t-1} is not found within the distance $\lambda_t + P_{max}$, it is not possible to primarily determine X_{i_0, j_0, k_0} .

In step 2, the values are being averaged. Let us calculate the $\overline{X}$

$$\overline{X} = \frac{1}{\gamma_A} \sum_i \sum_j \sum_k X_{ijk} \quad , \qquad (i,j,k) \in A \qquad [6]$$

where A is a point set of O and γ_A represents the number of these points. In these points it was possible to determinate the component X_{ijk} primarily. Let us calculate values y and z analogously. Supposing that object O is solid that it is all moving simultaneously and its boundaries are symetrical, at least in three axes perpendicular to each other and parallel to axes i,j,k, then values x,y and z are identical to object velocity components of the object O in directions i,j,k. If the object does not satisfy the above conditions, it must be taken into consideration that $\overline{x}$, $\overline{y}$ and $\overline{z}$ as velocity estimates show method − dependent errors.

Under the condition mentioned above, all points move at the same velocity and, additionaly estimated velocity vectors $\overline{x}$, $\overline{y}$ and $\overline{z}$ may be assigned to those points of the object O in which the vectors could not be determined primarily. The algorithm, however,is to be applied to radar reflectivity representing the clouds and thus a fact has to be taken into account that the objects are not absolutely stable, they become deformed partly in time $\vartheta_t - \vartheta_{t-1}$. If the velocities of various object parts are to be the characteristic velocity features in the respective part of the object, local averaging is applied. Therefore, the velocity X_{i_0, j_0, k_0} at the point (i_0, j_0, k_0) alongside axes i is calculated :

$$X_{i_0,j_0,k_0} = \frac{1}{\gamma_{i_0 j_0 k_0}} \sum_{i=i_0-q}^{i_0+q} \sum_{j=j_0-q}^{j_0+q} \sum_{k=k_0-q}^{k_0+q} X_{ijk} \qquad [7]$$

where q represents the surrounding area size. The sum only comprises the points in which the velocity has been determined in the primary way, or those the velocity of which has been calculated by the algorithm as follows : ($\gamma_{i_0 j_0 k_0}$ is the number of such points in the individual surrounding area). Applying this local calculation to all points of the object (first, the procedure along each axis is ascending i=1, ...,n then descending (i=n, n-1, ...)), the velocity values will be obtained for all object points.

In this way, all images B^S , S=1,....,$\varkappa$,will be evaluated. The procedure is followed by merging the estimates (X^S, Y^S, Z^S) made according to individual isosurface reflectivities into a single estimate of the field of motion (X,Y,Z). The estimate X_{ijk}^S (and Y_{ijk}^S , or Z_{ijk}^S), which is calculated using an iso-surface with the greatest reflectivity is placed at the point (i,j,k).

After the estimated field of motion is determined, it is possible to calculate the field of reflectivity changes $\| \delta_{ij} \|$. For each point (i,j,k) of object O_{ϑ_t} at time ϑ_t , the supposed position of this point at the previous point in time is calculated

$$i_{t-1} = i - X_{ijk}^{\vartheta_t} (\vartheta_t - \vartheta_{t-1})$$

$$j_{t-1} = j - Y_{ijk}^{\vartheta_t} (\vartheta_t - \vartheta_{t-1})$$

$$k_{t-1} = k - Z_{ijk}^{\vartheta_t} (\vartheta_t - \vartheta_{t-1})$$

Thence the reflectivity change in the point (i,j,k) is obtained as the difference between reflectivities at points corresponding to each other

$$\delta_{ijk}^{\vartheta_t} = r_{ijk}^{\vartheta_t} - r_{i_{t-1},j_{t-1},k_{t-1}}^{\vartheta_{t-1}}$$

The estimated field of motion (X^t, Y^t, Z^t) can be made use of for the prediction of radar reflectivity, i.e. cloudiness. First, the extrapolation of velocity values (X^t, Y^t, Z^t) by means of a polynomial for the forecast time ϑ_p, ϑ_0, ϑ_p is carried out. Then, the supposed location of the point (i,j,k) in time ϑ_p and the prediction of reflectivity value is calculated.

$$i_p = i + X^{\vartheta_p}_{ijk} \; (\vartheta_p - \vartheta_0)$$

$$j_p = j + Y^{\vartheta_p}_{ijk} \; (\vartheta_p - \vartheta_0)$$

$$k_p = k + Z^{\vartheta_p}_{ijk} \; (\vartheta_p - \vartheta_0)$$

$$r^{\vartheta_p}_{i_p, j_p, k_p} = r^{\vartheta_0}_{i,j,k} + \delta^{\vartheta_p}_{i,j,k}(\vartheta_p - \vartheta_0)$$

where $\delta^{\vartheta_p}_{ijk}$ represents the reflecivity changes prognosis for the time point ϑ_p.

On the basis of the experimental evaluation of the algorithm in question a conclusion has been reached that the data provided by he algorithm on volume motion velocities, representing the selected radar reflectivity values, correspond neither with the theoretical nor with the experimental knowledge acquired about convective cloudiness in the process of intensive development. Following the principles of the physics of clouds and precipitation, a new algorithm has recently been worked out and is currently under verification; so far it has proved good in representing objects and their changes both in space and in time. A high frequency of automated radar measurements (by means of the ARMS) is creating a new expert approach to classification.

The reason for the failure of the algorithm — which is successfully used in continuous operation at the WMO Activity Centre for Very Short-Range Weather Forecasting at Malý Javorník near Bratislava, namely for computing the horizontal movement of cloud objects scanned by the METEOSAT — can be explained by the fact that the IR — channel primary data contain a much greater number of picture elements (pixels) than in the case of ARMS output data (where a single object, eg a Cumulonimbus with shower, can be represented by a single column sized 8x8 km and 7,500 m in height).

References

(1) NEMEC,J., PODHORSKÝ, D. (1987): ARMS and its function in hydrometeorological service. Radar Meteorology. Materials of the methodic centre of radar meteorology of socialist countries, Leningrad, Gidrometizdat ,3 - 6.

(2) NEMEC, J., SENCAKOVA, D., JUSKO, M., PODHORSKÝ, D., KALINA,L. (1987): Using of the automatized radiolocation meteorological system on the base of mini-computer SM 4-20 and its input data. Radar meteorology, Materials of the methodic centre of radar meteorology of socialist countries, Leningrad, Gidrometizdat, 10 - 14.

(3) NEMEC,J., PODHORSKÝ, D., LIETAVA, L. (1987) : Automatized radiolocation meteorological system and its function in the uniform meteorological radar network. Meteorological reports, vol. 40, N° 2, 33 - 37.

(4) PODHORSKÝ, D., VLCAK, L. (1987) : Advective - convective tendency calculation on the base of primary data from Meteosat satellite. Meteorological reports, vol. N° 2, 38 - 43.

(5) PODHORSKÝ, D., (1987) : Use of satellite imagery forecasting the evolution of mesoscale phenomena. Satellite and radar imagery interpretation. (Preprints for a Workshop held at Reding, England, 20 - 24 July 1987 : Eumetsat' 87) 415 - 436.

(6) PODHORSKÝ, D., VLCAK, L. (1987) : Advective - convective tendency of the cloud field on the base of satellite data. Proceedings of scientific works, Leningrad, LGMI, N° 96, 52 - 65.

(7) PODHORSKÝ, D., OLSINA, O., VLCAK, L. (1988) : Meteorological interpretation of satellite data for DIGISAT, WMO technical conference on instruments and methods of observations (TECO' 88), Leipzig, 207 -212.

(8) PODHORSKÝ, D. (1987) : Nowcasting for the national economy needs. Mesoscale analysis and forecasting. Proceedings of an International symposium, Vancouver, Canada, 7 - 12.

COMPARISON OF RADAR PRECIPITATION MEASUREMENTS
WITH A DENSE NETWORK OF DAILY RAINGAUGES -
APPLICATIONS TO RADAR NETWORKING

J.L. BROWNSCOMBE and B.D. HEMS
UK Meteorological Office

Summary

The UK Meteorological Office maintains a 5 km spatial resolution archive of hourly and daily rainfall totals derived from integrations of the radar images obtained at 5 minute intervals from each radar in the UK network. The archive is named PARAGON (1). Since the radars are mostly separated by more than the ~75 km generally accepted as the effective range for quantitative use of radar data, the daily rainfall totals derived from radar are 'adjusted' using the daily totals from a synoptic raingauge network of average spacing ~40 km. These 'adjusted' fields of daily rainfall are also archived. The archive also has the facility to 'adjust' the daily radar rainfall totals using data from the UK climatological network of daily raingauges (average spacing about 8 km). However in this study we use the climatological gauge values as the 'truth' with which to compare radar and 'adjusted' radar fields. Comparison maps can be produced for each radar of the network and areas of low sensitivity or noisy output identified. The effect of 'adjustment' on performance can be identified. Examples of these maps over a complete year are given and long term and seasonal variations of performance are also explored.

1. INTRODUCTION

The UK weather radar network (2) which forms a subset of the COST 73 network (3) currently uses data from seven radars in the UK and one in the Republic of Ireland to produce a network radar rainfall image of 5 km resolution every 15 minutes. In addition both 2 km (within 75 km of the radar) and 5 km resolution single site images are produced for immediate operational use by Water Authorities. The radar data are calibrated in real time using a small number of dedicated calibration gauges for each radar. This calibration scheme is applied to defined areas called domains (4). The lowest elevation radar beam is used where possible but is automatically infilled with data from higher beams where appropriate to avoid potential problems of ground clutter or obscuration. The 5 km resolution data recorded on tape every 5 minutes at each radar site form the basic data for the PARAGON radar archive. These data are integrated over time to produce 1 hour radar rainfall totals on a 5 km grid at distances out to 210 km from each radar.

Data from areas of known ground clutter are excluded and (for some radars) a scheme is implemented for automatic rejection of data subject to anomalous propagation errors (anaprop). Further integration to daily totals is followed by 'adjustment' of the daily totals using an adjustment field of gauge to radar ratios $(^G/_R)$ derived from values of $^G/_R$ determined at the synoptic gauges of average spacing ~40 km. The surface fitting techniques used are adapted from those given in (5) for raingauge data.

The resulting PARAGON data sets are:
 (i) Hourly unadjusted but real time calibrated data; R_H
 (ii) Daily unadjusted but real time calibrated data; R_D
 (iii) Daily adjusted real time calibrated data; R_A
In this study we compare R_D and R_A with the daily rainfall totals from the gauges of the climatological raingauge network (G).

A further 5 km resolution daily rainfall field (S) can be produced by interpolating the ratios of the synoptic gauge daily totals to annual average rainfall (AAR) using a 5 km gridded AAR data set. This represents the best rainfall field that could be obtained in real time in the absence of radar. The surface fitting procedures are similar to those used for adjusting the daily radar rainfall totals. Note that the adjusted radar values (R_A) could be considered as radar assisted interpolation of synoptic gauge daily rainfall fields as compared with the climatologically interpolated S.

We aim to demonstrate
 (i) The use of comparisons of R_D with G to show the extent of good radar performance from each network radar to allow better informed choice of inter-radar boundaries in the real time network output and the appropriate location of calibration gauges.
 (ii) The improvements of radar performance by adjustment using rapidly available synoptic gauge data.
 (iii) The advantages of adjusted radar daily totals (R_A) over climatologically interpolated synoptic gauges (S).
In addition, we use longer period comparisons of radar daily rainfall totals with synoptic gauge totals to identify any long term trends and seasonal variability of radar performance.

2. COMPARISON MAPS

For each radar in the network, comparison maps can be produced over any selected period of days. Selections can be by year, by season or (with suitable identifiers) by weather type. It is necessary to have sufficient comparisons at each gauge site to get a statistically representative sample. So far we have produced maps for individual years of data (1986 and 1987) for three radars of the network (Chenies, Clee Hill and Hameldon Hill). We can select limits on daily precipitation totals for radar and/or gauges and can display the results as contoured maps of:
 (i) Mean and standard deviation of log $^R/_G$ (the distribution of $^R/_G$ is approximately log normal).
 (ii) Median and interquartile range of $^R/_G$.
 (iii) Percentage of occasions that $^R/_G$ lies within N to $^1/_N$ where N can be chosen as required.
Maps displaying these performance contours can be produced for adjusted and unadjusted radar rainfall fields (R_D and R_A) and for the climatologically interpolated synoptic gauge field (S). It is not possible to present more than examples of the maps here but the general conclusions reached so far from a study of the log $^R/_G$ and percentage of occasion maps for Chenies, Clee and Hameldon can be summarised as follows:

(i)   The performance contours can be very non-circular and not centred on the radar. Although there are minor differences between years, contour shapes are broadly similar.   Separate contour maps of standard deviation and mean values allow the identification of areas of persistent clutter/anaprop problems.

(ii) If we use all daily radar rainfall amounts greater than 0.5 mm per day then increasing the lower limit of gauge rainfall totals from 0 to 2 mm per day increases substantially the area within 75 km of the radar where a high (80) percentage of daily radar rainfall totals lie within a range of $^1/_{1.6}$ to 1.6 of the climate gauge daily rainfall totals. However, it does not greatly extend the area of 68% agreement to longer ranges.

(iii) Adjustment of radar using synoptic gauges improves the level of agreement substantially both close to the radar and particularly at longer range. The effect is primarily an improvement in the mean level of agreement but the standard deviation of log $^R/_G$ is also decreased from about 0.3 to 0.2 at longer ranges (ie the range of $^R/_G$ is reduced from about 2 to $^1/_2$ down to 1.6 to $^1/_{1.6}$).

(iv) In general, the area of good agreement between adjusted radar, R_A, and climate gauges, G, within 200 km of the radar is substantially greater than that of the climatologically interpolated synoptic gauges, S.

Using data from the Clee Hill radar in 1987 as an example we show in Table I the fractions of the areas within different range circles where agreement with climate gauge rainfall within the range 1.6 to $^1/_{1.6}$ is achieved on a given percentage of occasions. Data are presented for the daily precipitation fields derived from unadjusted radar, R_D, adjusted radar, R_A, and climatologically interpolated synoptic gauges, S. Results are shown for total range annuli and the eastern and western halves separately. This reveals much poorer performance in the west which is attributed to the effects of the more mountainous terrain. Both the radar and the synoptic gauge fields show this east/west performance bias but the ratio is not so great for the synoptic gauge field. The east/west ratios of the fractional area with percentage of agreement greater than 68% are 1.2 for S, 1.6 for R_A and over 4 for R_D within 75 km of the radar and 2.0, 2.3 and > 3.0 respectively at ranges from 75 to 125 km. These results cannot be generalised to other radars as the shape of each comparison field is different. However the three radars for which we have mapped comparisons all show significant azimuthal variability of performance. Examples of maps for Clee Hill in 1987 are shown in Figures 1 and 2 for unadjusted, R_D, and adjusted, R_A, radar rainfall fields. The significant improvements after adjustment are evident as is the better performance to the east of the radar.   Inspection of Table I reveals that the fields of adjusted radar, R_A and climatologically interpolated synoptic gauges, S, have similar areas with more than 68% of comparisons with the dense gauge daily rainfall values between $^1/_{1.6}$ and 1.6 times the gauge values. However, R_A has a much greater area than S with more than 80% of comparisons in the same range.

Before such maps can be used to decide priorities between radars in networks it is necessary to investigate more fully the seasonal and weather dependent variation of performance. This has not yet been done but in the next section we explore some features of the long term and seasonal changes of radar performance.

Table I
The fraction of the areas at different ranges from the Clee Hill Radar for
which a given percentage of field values lie between 1.6 and $^1/_{1.6}$ times
the climatological gauge value.

Field	Area	Percentage within 1.6 to $^1/_{1.6}$	Fraction of area within the given percentage limits		
			Range <75 km	Range 75 to 125 km	Range 100 to 125 km
Unadjusted calibrated radar R_D	Total	80-95	0.03	0.02	0
		68-80	0.35	0.22	0.08
		<68	0.62	0.76	0.92
	East	80-95	0.05	0.03	0
		68-80	0.57	0.43	0.16
		<68	0.38	0.54	0.84
	West	80-95	0	0	0
		68.80	0.15	0.02	0
		<68	0.85	0.98	100
Adjusted radar R_A	Total	80-95	0.21	0.33	0.33
		68-80	0.57	0.39	0.37
		<68	0.22	0.28	0.30
	East	80-95	0.38	0.56	0.60
		68-80	0.59	0.43	0.39
		<68	0.03	0.01	0.01
	West	80-95	0.04	0.09	0.07
		68-80	0.55	0.34	0.35
		<68	0.41	0.57	0.58
Synoptic gauges interpolated using climatology S	Total	80-95	0.07	0.06	0.06
		68-80	0.76	0.60	0.61
		<68	0.17	0.34	0.33
	East	80-95	0.10	0.11	0.10
		68-80	0.81	0.81	0.78
		<68	0.09	0.08	0.12
	West	80-95	0.05	0.01	0.02
		68-80	0.69	0.40	0.45
		<68	0.25	0.59	0.53

3. CHANGES OF RADAR PERFORMANCE WITH TIME

The radar return signal for a given rainfall rate depends not only on
rainfall rate and dropsize distribution but also on many other factors
which are functions of radar performance and calibration, beam elevation,
temperature and humidity profiles, distance from the radar, height of
rainfall generation region, phase of precipitation etc. The PARAGON
archive provides a source of radar rainfall data which can be used to
investigate the performance of the individual radars of the network as a
function of many of these parameters in both detailed case studies (for
example (6)) and in more statistical evaluations.

A three monthly assessment report is produced to record the
performance of radars measured against a small number of hourly and daily
comparison gauges. These are used to identify long term performance
trends. The mean and standard deviation of log $R_{D/G}$ are calculated for
each monthly period for each assessment gauge and plotted as a time

series. We have assembled data for a larger number of daily assessment gauges from 1984 to 1988 (from 1985 for Chenies) and assessed the results as a function of range from the radars. (The gauges centred on Clee Hill are shown in Figures 1 and 2.) Generally this results in samples of ~20-400 comparisons per month in each range annulus from the radar. The smallest samples are within 35 km of the Clee Hill radar. Taking into account the standard deviations of the monthly comparison samples and the number of comparisons we can estimate standard errors of estimate (s.e.e.) of individual monthly values to be mainly in the range 0.02 to 0.10 with further reductions for averaged yearly and seasonal values. Table II shows mean values of log $R_{D/G}$ as a function of range for the Chenies, Clee

Table II
Mean log $R_{d/G}$ as a function of radar, range, year and season

Radar	Year or Season	Range in km				
		<35	35-70	70-105	105-140	140-210
Clee Hill	1984	-0.06	0	-0.05	-0.12	-0.26
	1985	-0.02	-0.04	-0.18	-0.19	-0.30
	1986	+0.08	-0.02	-0.17	-0.20	-0.31
	1987	+0.03	+0.01	-0.12	-0.18	-0.28
	1988	-0.06	-0.07	-0.17	-0.25	-0.38
	Winter	+0.02	-0.03	-0.16	-0.22	-0.39
	Spring/Autumn	+0.01	-0.01	-0.14	-0.20	-0.31
	Summer	-0.03	-0.02	-0.13	-0.16	-0.22
	Overall mean	0	-0.02	-0.14	-0.19	-0.31
Hameldon Hill	1984	+0.02	+0.02	-0.10	-0.04	-0.17
	1985	-0.08	-0.10	-0.20	-0.13	-0.23
	1986	-0.14	-0.14	-0.31	-0.23	-0.27
	1987	-0.05	-0.15	-0.34	-0.27	-0.32
	1988	-0.08	-0.12	-0.26	-0.16	-0.24
	Winter	-0.10	-0.12	-0.33	-0.31	-0.39
	Spring/Autumn	-0.07	-0.10	-0.27	-0.18	-0.25
	Summer	-0.02	-0.09	-0.15	-0.03	-0.12
	Overall mean	-0.07	-0.10	-0.24	-0.17	-0.25
Chenies	1985	-0.08	-0.15	-0.15	-0.09	-0.31
	1986	-0.06	-0.12	-0.10	-0.04	-0.39
	1987	-0.11	-0.16	-0.08	-0.07	-0.41
	1988	-0.06	-0.16	-0.11	-0.10	-0.42
	Winter	-0.10	-0.19	-0.20	-0.22	-0.51
	Spring/Autumn	-0.06	-0.14	-0.11	-0.04	-0.36
	Summer	-0.07	-0.12	-0.03	-0.03	-0.29
	Overall mean	-0.08	-0.15	-0.11	-0.07	-0.38

Hill and Hameldon Hill radars averaged over individual years and over winters (N,D,J,F), summers (M,J,J,A) and spring/autumn (M,A,S,O) separately over the whole period. Remembering the non-circular range dependence shown in the comparison maps, we expect that better discrimination would be possible if gauges were more carefully chosen in relation to particular sectors. However several general features are evident in Table II. The most obvious is that the range dependence of $R_{D/G}$ is largest in winter and smallest in summer as would be expected from the different mean heights of precipitation generation processes and precipitation types in these seasons. The most regular decrease of sensitivity with range and smallest variation between winter and summer is for the Clee Hill radar which has a lower beam elevation of 0° compared with 0.5° for the other radars and to which no range correction is applied (other than the usual $1/_{r^2}$ dependence of signal, where r is the range from the radar). Hameldon Hill shows a response peak at 105-140 km range in summer and Chenies at 70-140 km. The variable effects are probably related to the use of a standard technique for range correction and (in the case of Chenies) may be affected by the lack of an anaprop correction system. However the non circular nature of range effects may interact with raingauge location to produce some anomalies.

The variations in performance from year to year are generally smaller than the seasonal effects but significant changes are evident - particularly for the Hameldon Hill radar at ranges from 70 to 105 km. A graph of 120 day running means shown in Figure 3 reveals a very significant decrease of sensitivity from 1984 to 1987 superimposed on the seasonal variation and a recovery in 1988. Yearly average sensitivity values taken from Table II are also shown for ranges of 0-35, 75-105 and 140 to 210 km in Figure 3.

4.   CONCLUSIONS

We have demonstrated the use of the PARAGON archived daily radar rainfall totals to display detailed performance assessments of individual radars in a network. Radar is most useful in providing information on rainfall distribution over shorter periods of 5 minutes to 1 hour. However accurate assessment of overall radar performance requires a dense network of raingauges for comparison. The shorter the measurement period, the denser the network that is required. In the UK we have a relatively dense daily gauge network (~8 km spacing) which forms the best available 'truth' for radar performance assessment. The resulting maps of radar performance can be of use in assessing optimum network configurations and can also be used to reveal areas in which it may be helpful to place additional calibration gauges. They also reveal the improvement which can be produced in the accuracy of daily rainfall totals by adjustment of the radar field using the daily rainfall totals from a less dense network of gauges.

Although we have displayed comparison maps based on data for one complete year we have shown that seasonal and other long term radar calibration factors also need to be considered.

This will not be possible without the maintenance of an accessible long term archive of radar data such as PARAGON.

The current study confirms the continuing need to improve basic radar rainfall data using both automatic objective and manual subjective techniques. It also demonstrates some of the problems involved in networking radars of differing performance characteristics and the need to monitor these characteristics - ideally in near real time.

RERERENCES
(1)   MAY, B.R., (1988). Progress in the development of PARAGON.
Meteorol. Mag. 117, 79-86.
(2)   COLLIER, C.G. and JAMES, P.K., (1986). On the development of an
integrated weather radar processing system. Preprint Vol. 23rd Conf. on
Radar Met. 22-26 September, Snowmass, Colorado. American Meteorological
Society, Boston, Mass.
(3)   COLLIER, C.G., FAIR, G.A. and NEWSOME, D.H., (1988). International
Weather Radar Networking in Europe. Bull. Amer. Meteor. Soc., 69, 16-21.
(4)   COLLIER, C.G., LARKE, P.R. and MAY, B.R., (1983). A weather radar
correction procedure for real-time estimation of surface rainfall. Quart.
J.R. Met. Soc., 109, 589-608.
(5)   SHEARMAN, R.J. and SALTER, P.M., (1975). An objective rainfall
interpolation and mapping technique. UGGI Association Internationale des
Sciences Hydrologiques, Bull. Sci. Hydrol., 20, 353-363.
(6)   HITCH, T.J. and HEMS, B.D., (1988). A comparison of radar and gauge
measurements of rainfall over Wales in October 1987, Meteorol. Mag., 117,
276-279

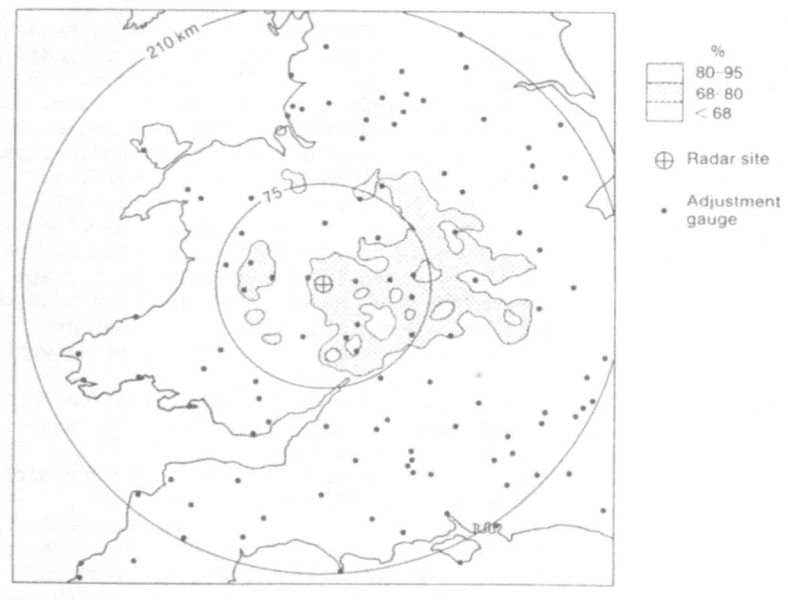

Figure 1
The percentage of daily rainfall totals derived from unadjusted radar,
R_D, which are between $1/1.6$ and 1.6 times the daily rainfall totals
measured by climatological raingauges, G.

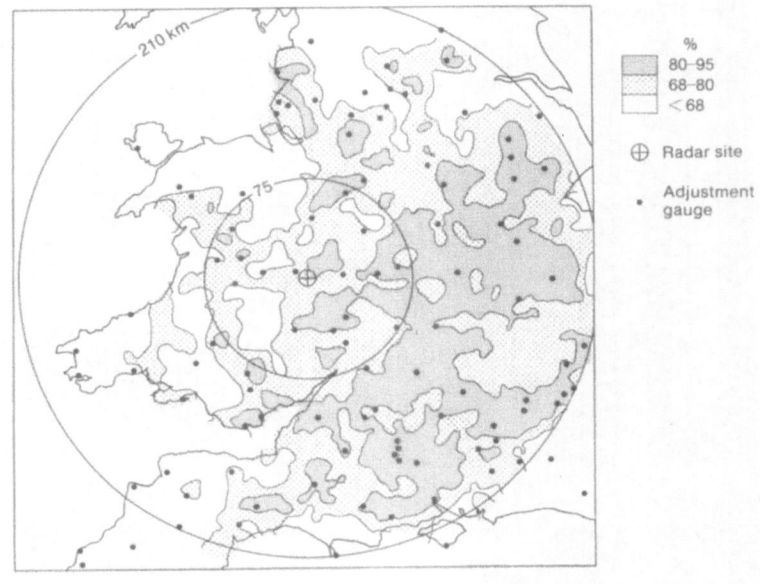

%
80–95
68–80
< 68

⊕ Radar site

• Adjustment gauge

Figure 2
The percentage of daily rainfall totals derived from adjusted radar, R_A, which are between $^1/_{1.6}$ times the daily rainfall total measured by climatological raingauges, G.

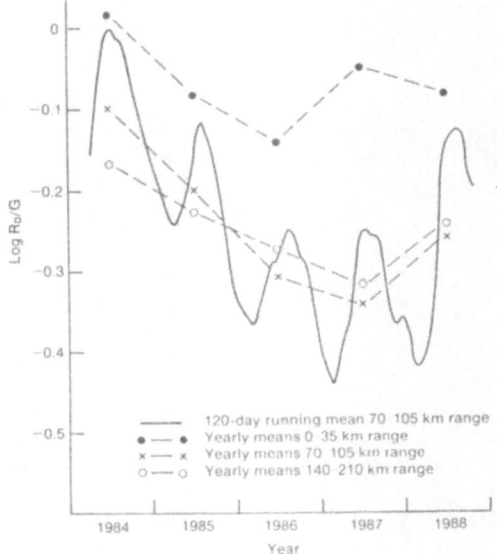

120-day running mean 70–105 km range
Yearly means 0–35 km range
Yearly means 70–105 km range
Yearly means 140–210 km range

Year

Figure 3
Seasonal and long term variations of the log $^R D/_G$ for the Hameldon Hill radar.

SESSION 6

Hydrological and other
applications of weather
radar data

Chairman : R. Sorani

Dr R. SORANI
Italian National Meteorological
Service. Vice-Chairman of
COST 73 project. Chairman of
Session 6 at the Seminar

Dr P. PODHORSKY
Slovensky Hydrometeorologicky
Ustav, Czechoslovakia

Keynote speaker at COST 73
Seminar.

PRACTICAL APPLICATIONS OF WEATHER RADAR DATA IN EUROPE

D. H. Newsome
CNS Scientific & Engineering Services

Summary

The practical application of data from weather radars with digital outputs began in Western Europe in the late 1960s. One of the first applications was hydrological and took place in the United Kingdom. Since then, the use of such weather radar data has become much more widespread and examples are given of uses to which they are put in some of the countries taking part in the COST 73 Project. The paper then forecasts some important possible future uses of weather radar data at all levels - locally, regionally, nationally and inter- nationally. It concludes there is a danger that all that has been achieved through the international co-operative efforts put into developments at these levels by the countries concerned could be lost if the work that has been started in the COST 72 and 73 Projects is not continued by some other body after the conclusion of COST 73 in 1991.

1. INTRODUCTION

It was in the late 1960s when the possibility of the practical appli- cation of weather radars with digital outputs for hydrological and meteo- rological purposes was first explored in western Eurpoean countries. One of the early research projects was the Dee Weather Radar Research Project in the United Kingdom. This Project was added to an existing water resources operational management research programme designed by the Water Resources Board (WRB) for the River Dee in North Wales. For the WRB research, the catchment had already been heavily instrumented with conven- tional hydrological and meteorological equipment. There was, for example, a network of some 70 raingauges recording on magnetic tapes and a network of streamflow gauging stations. There was also a distrometer to give an indication of the size of the raindrops. The Dee catchment thus provided a ready-made test bed for evaluating the utility of weather radar for a variety of purposes. After the installation of the radar at Llandegla, some of the raingauges were connected by dedicated lines to the radar control cabin and telemetered their data to the computer used in associa- tion with the radar so that any adjustments considered necessary could be made to the areal rainfall totals provided by the computer data.

The data from the radar were used as an input into a mathematical model of the River Dee whose management objectives were multiple and often conflicting. The Dee is a regulated river system with impounding reser- voirs in its headwaters which release water at times of low flows to sus- tain abstractions downstream. There is also a natural lake which has been modified for use in the regulation of floods. The catchment is largely agricultural, but there are towns, a large chemical factory and other industries. Treated industrial and sewage effluents are therefore dis- charged to the river which supports a large migratory fish population of

Atlantic Salmon (*Salmo salar*) and is heavily used for recreational and amenity purposes.

The impounding reservoirs have to be kept as full as possible to sustain the downstream abstractions, while the flood control lake must be kept with sufficient freeboard to contain run-off that might otherwise cause a flood. The river water is used for public supply; it is also used as an effluent carrier; industry and agriculture prefer relatively stable water levels in the river, whereas the salmon require floods called "freshets" to induce them to move upstream to spawn and to return again to the sea subsequently (1).

The processed radar data provided areal totals of rainfall which proved to be satisfactory for operational management purposes of the river system. Local meteorologists found the display of digital data useful for tracking the passage of weather systems and it gave them a better idea of the likely intensity of rainfall to be expected. This improved the accuracy of their forecasts, particularly for frontal systems, and enabled them to improve their prediction of the times of onset and cessation of precipitation at a given location.

Towards the end of the Dee Programme, the possibility of networking radars was explored and, in the mid-1970s, a semi-operational network was functional in the UK. At about the same time the Swiss Meteorological Service was introducing a service to its customers using its two radars at La Dole and Albis.

From simple beginnings such as these and, with increasing experience and gathering evidence that a network of radars would provide a valuable tool to meteorologists and hydrologists, a more ambitious plan was conceived; that of building an integrated weather radar network in Western Europe. This gave rise to the COST Research Project No. 72 (COST 72) and its successor, COST 73 (2). In the meantime, much progress had been made in merging data streams from different radars and integrating the result with data from METEOSAT to form a composite picture which, during the COST 72 Project, became known as the COST image. During the COST 73 programme, this image has been improved and, in a more ambitious pilot project than that started during COST 72, draws data from a larger number of radars in more countries .

Undoubtedly the impetus behind both the COST 72 and COST 73 projects was provided by the increasing use being made of weather radar data for practical applications locally, regionally, nationally and, latterly, internationally. Applications developed in some of the different countries taking part in the COST projects were quite varied, and examples of them have been included in the paper. They do not pretend to be exhaustive and the use described may not be the principal use of weather radar data in that country. By being selective, however, it is hoped that the diversity of uses will indicate, at least to some extent, the potential uses of single radars, regional, national and now, international networks.

2. USES MADE OF WEATHER RADAR DATA IN VARIOUS EUROPEAN COUNTRIES

2.1. Austria

Austria currently has a network of three weather radars situated at Schwechat (Vienna), Zirbitzkogel and Patscherkofel.

A picture resulting from the combination of the three data streams is transmitted to all the country's major airports every ten minutes for use in flight weather predictions. The image is in plan only, but there have been many requests to extend the image to include three-dimensional information (i.e. cross-sectional information on the weather systems). It should be noted that the airports nearest to each radar also receive a single-radar picture which does contain three-dimensional information.

The Central Meteorological Office in Vienna and its sub-office in Klagenfurt use the image for short-term forecasting and for the issue of severe weather warnings to the public.

Radar data are also used in the field of hail suppression, which is of major concern for the agricultural activities in Niederösterreich. The radar data are used to guide small aeroplanes which disperse silver iodide crystals in an effort to suppress potential hailstorms that can cause tremendous damage to the crops, which include grapes.

An interesting application is the use of the radars to provide data for research into the effects of precipitation on earth-satellite links. For this purpose, the output of the dual-polarised Hilmwarte radar was used and systematic scanning along several earth satellite links was carried out during the "rainy" season in 1988 (May-October). Good correlation was obtained between radar-derived and directly measured propagation and precipitation parameters, see Figure 1.

Future uses of weather radar data in Austria are expected to include:

* giving traffic warnings about treacherous road conditions and to help road authorities plan their road maintenance
* providing measurements of total rainfall for electricity companies ("Verbundgesellschaft")
* public presentation of severe weather events on the television network

2.2. Belgium

The use of weather radar data is in its infancy, but the data are already being used by hydrologists for predictions of river levels and flows (3). They are also used to determine the likely urban surface water run-off.

Of particular interest is the international River Meuse for which COST 73 weather radar data could have a part to play by providing real-time data to hydrologists in the countries through which the river flows. As yet this is an unused capability of the COST image, but one which is believed to have great potential utility, (see Section 3).

2.3. Denmark

The Danish Meteorological Institute (DMI) has operated a Doppler Radar situated at Kastrup airport, Copenhagen since 1986.

The principal use of the data is made by the weather service at the airport in daily operations. Short-term forecasts for the approaches to the airport are provided, together with a "weather watch" over Danish air space and, when necessary, pilots are guided round areas of heavy precipitation. For this purpose the weather radar images are updated every five minutes and the standard image contains:

i.   a horizontal pseudo CAPPI image
ii.   as i) but integrated vertical sections are included
iii. a horizontal CAPPI 1.5 km above the earth's surface
iv.   a horizontal maximum reflectivity image

The display system can show sixteen levels including the background and can store locally eight images which can be animated as required.

Of great interest is the service provided to farmers, (4). This was commenced experimentally in the summer of 1988 by the transmission of images to the Danish Agricultural EDP-Centre (LEC) which has the biggest privately-owned computer installation in Denmark. There are more than 20,000 registered users of this facility - mostly farmers and branches of industry related to agriculture.

For the experiment, twelve users were selected on the basis of their location. They had to be evenly distributed within the coverage of the radar. Of the chosen twelve, seven were farmers, four consultants and one was an agricultural machine pool.

The Danish Agricultural Advisory Centre, Department of Plant Production (LP-K), a research organisation, agreed to monitor and evaluate the experiment.

The end user was supplied with two programs, a menu-driven communication program to exchange data via dial-up PTN line to LEC and a display program to show the weather images. The latter program could unpack and display the packed images received by the communications program. Five images could be stored locally as a time series, together with a map of the area covered.

An evaluation questionnaire was devised by LP-K and given to each of the twelve users selected. At the end of the experimental period, the result of the evaluation was very clear.

The end users were usually capable of interpreting the precipitation images correctly and the planning of outdoor work had often been changed because of the radar image. In 80 per cent of the cases where a change had been made, it was judged to have been beneficial. There were, of course, some wrong interpretations of the image, some being due to meteorological phenomena such as inversions, but they were considered to be minor compared with the overall benefit of having the radar images.

As the experiment was judged to be a success, a full scale service was requested by the users and was commenced in May 1989. It comprised weather radar images and regional forecasts. Currently (June) about fifty users have subscribed to the service and the number is expected to continue to rise rapidly. The spatial resolution is 2 x 2 km and the reflectance resolution is 256 levels. These levels are reduced to three placed in the dbZ intervals:

1.   10 - 20 dbZ   2.   20 - 40 dbZ   3.   > 40 dbZ

Four colours are used:

0   background;   1.   shaded, level 1;   1.   solid, level 2;
            2.   solid, level 3;   3.   solid, map

The three precipitation levels represent light, medium and heavy precipitation. It has been established that three levels are about right; more than three appears to confuse the end users.

The end user pays an annual subscription of DKr 500 and a price of DKr 2 per image for this service. He must also, of course, invest in the necessary equipment which costs about DKr 10 000, which is not an exorbitant sum. Moreover, it can be used for other purposes and it is therefore generally regarded as a good investment.

Future plans include the installation of an additional radar on the north west coast of Denmark to improve the coverage of the country and its western approaches.

2.4. Federal Republic of Germany

Weather radar out-putting digital products are in their infancy but, apart from the uses made by the meteorological service itself, other organisations have started to experiment with and use weather radar data. The

Bavarian State Authority for example has started to use weather radar data in the operational management of its rivers and there is now a limited facility for obtaining weather radar products by accessing the radar computer via a PTN line and using a PC, (5).

2.5. France

The Direction de la Meteorologie Nationale has instituted the comprehensive METEOTEL service. Currently there are about 250 outstations receiving the service. The complete repertoire of products that are available comprise:

    i. Composite radar images every 15 minutes
    ii. Visible and infra-red satellite pictures of France every 30 minutes
    iii. Visible, infra-red & water vapour satellite pictures every three hours
    iv. Surface observations every hour
    v. Analysed forecast maps twice per day
    vi. Forecast maps three times per day

The information is transmitted to receiving stations within France every thirty minutes and to stations in Europe every three hours. Many combinations may be made of the available information. For example, it is possible to show four pictures of the same type on the colour screen simultaneously or, alternatively, a comparison of three types of picture - radar, visible and infra-red satellite pictures - and satellite data may be superimposed on the radar picture etc. Speeds and directions of moving radar echoes can be calculated, as can those of a cloud in a sequence of satellite pictures. All the other usual facilities such as replay sequences, zooming etc. are, of course available.

2.6. Italy

Again, in its infancy, weather radar data are used for flight assistance in a service operated jointly by the National Meteorological Service and the Azienda Autonoma per l'Assistenza al Volo ed al Traffico Aereo Generale (AAAVTAG) and in agro-meteorology (6, 7). One of the interesting applications is a joint venture with Yugoslavia in using weather radars from both countries for hail detection and suppression. Two radars, one in Italy (Fiuli-Venezia Giulia Region) and the other in Yugoslavia, (Republic of Slovenia) provide information to guide rockets into potential hail clouds where they will release chemical products to seed the clouds which, hopefully, will suppress the formation of hail.

2.7. The Netherlands

Real-time data are starting to be used to optimise pumping strategies in order best to utilise the storage capacity of urban sewerage systems, an idea which may well be of use in other countries which have low-lying urban areas.

The data are also being used for the planning of short-term activities which are weather sensitive, one example being road construction and maintenance.

2.8. Norway

At the time of writing, the radar data are not yet combined with other meteorological data but, nevertheless, the radar data are primarily

used by weather forecasters watching the development and movement of precipitation cells. The type of image most frequently used is one showing
the maximum intensity echo projected both horizontally and vertically.

An automatic picture distribution system has been implemented using
PTN lines. It is used for aiding road authorities at three sites to help
them to optimise the spreading of salt for de-icing roads. The meteorologist predicts the actual locations that will be affected by precipitation in a few hours time and the radar data now enable him to advise when
salt should be spread on the roads so that this can be accomplished ahead
of the precipitation. The image is up-dated every fifteen minutes.

2.9. Portugal

The digital data from the Lisbon radar are transmitted using TELERAD
to Direccao-General dos Recursos Naturais (DGRN) where they are used to
improve hydrological forecasting for the benefit of water resources
management and the reduction of flood damage, (8).

They are also used to advise civil aviation, both over-flying and, in
particular, aircraft approaching or leaving the airports.

2.10. Spain

At the present time, three out of the total network of fifteen radars
have been installed. Their principal initial purpose is to enable improved forecasts to be made of severe weather in coastal regions, but a
full range of services is planned when the network has been completed.

2.11. Sweden

Sweden uses weather radar for a number of services including flight
assistance; advective models for probability nowcasts and to provide an
index for summer thunderstorms.

Any system which will improve the accuracy of forecasts of snowfall
will almost certainly be cost-effective if, like in Sweden, snow clearance
costs almost SEKr 800 million each year. This sum is divided between the
counties using some complex rules and the county of Östergötland, which
contains Norrköping, receives about 1/20 of this sum i.e. SEKr 400M. The
Meteorological Service (SMHI) works closely with the road authorities and
produces a six-hour forecast of the type of precipitation and the expected
amount. This is divided into two-hour intervals and, for this purpose,
the radar is indispensable. The cost for a turn-out of the road clearance
personnel and equipment is SEKr 400,000, so a turn-out which proves to be
abortive is an expensive mistake.

Moreover, if the radar "sees" snow, it is possible to get the snow
clearing equipment out before the traffic begins to get stuck and thus it
is more likely that the traffic will be kept moving. Being able to disperse the equipment to the right places at the right time means that the
efficiency of the service is substantially improved.

Using PROMIS 600, (9) a new weather radar information system, the
Swedish Agricultural Industry, which may be considered to be the most
weather sensitive of all, is able to receive short-term forecasts containing information from a variety of sources including radar (see Figure
2). This system is undergoing a two-year experimental and evaluation
phase which will finish in the Spring of 1990. The results will be used
in designing the future observational network, taking into account meteorological economical and safety considerations.

2.12. Switzerland

Comparatively speaking, the Swiss Meteorological Institute (SMI), has
a long history of making weather radar products available. Apart from

three regional forecasting centres and two international airports which receive composited images every ten minutes over dedicated lines, the facility is available to a limited, but growing number, of other users. The main uses are, at the moment, mainly qualitative and the principal application is to give warnings of severe weather. These are issued to local authorities and to the police who have reposibility for the lakes.

One of the most recent applications is to road maintenance. Based on improved knowledge of the onset and cessation of snowfall through the availability of radar data, snow clearance and the de-icing of motorways can be optimised. Following local trials over two winters, the road authorities are now being equipped with dial-up display equipment with the interpretation of the images being given by SMI personnel.

Quantitative applications are still in a development phase, joint projects being undertaken between SMI, the Federal Hydrological Authority and research institutes to improve the forecasts of flows in major Swiss rivers.

2.13. United Kingdom

Data from the semi-operational network in the 1970s was used by the Meteorological Office and some water authorities as inputs to mathematical models of the river systems under their control. Radar data have proved to be particularly useful for river regulation and in the issue of flood warnings. There has been a significant improvement in the accuracy of these warnings in terms of the expected severity and the timing of the peak since radar data became available. There is now a well-developed system which operates routinely in most water authorities. A typical organisation chart is shown in Figure 3.

On the general weather forecasts transmitted by the television services to the public, pictures of the images derived from the radar network feature regularly, showing the distribution and, through different colours, some idea of the intensities of the precipitation.

Experimental use is being made of the data for managing the sewerage systems of urban areas with a view to optimising the flow to the sewage treatment works by utilising the capacity of the sewer network to avoid peak flows whenever possible.

The Meteorological Office is becoming more cost conscious and is looking for revenue-raising activities. Thus a comparatively new service is the "Open Road" service. The service, which provides *inter alia* 24 hour and 2-5 day forecasts; early morning summaries and site specific ice prediction information, is aimed primarily at local authorities for planning their road maintenance activities in summer and their de-icing and snow clearing programmes during the winter months. It is hoped however that, having proved its value to the end user, the service will spread to other customers such as road and rail transport organisations.

The service is available via the regional meteorological offices and costs about £315 per calendar month plus VAT at 15%.

3. FUTURE APPLICATIONS OF INTERNATIONAL WEATHER RADAR DATA

Many uses other than those outlined above can be envisaged for local, regional and national network weather radar data. Some activities in the paper industry, such as loading rolls of paper for transportation by ship or road are, for example, very sensitive to weather conditions and could benefit from the receipt of short-term forecasts and real-time precipitation information. The use of weather radar data in the management of sewerage systems to even-out the loads placed on the treatment plants and to minimise the deleterious effects on receiving waters is now being investigated. Their use as a real-time input to water quality models looks as though it would be helpful to operational personnel managing river systems that are used fairly intensively, but the use of near real-time "international" precipitation data has so far received little attention from meteorologists and hydrologists alike.

However, meteorologists are now finding that the regular reception of the COST image is a useful adjunct to their other sources of meteorological data, particularly, for instance, in flight assistance to pilots. Other possible meteorological applications of the COST image data include:

* the initialisation of meteorological meso-scale numerical models, by providing "ground truth" and hence improving their performance.

* international transport by road, rail and sea (in the North Sea, Atlantic Western Approaches and the Mediterranean Sea)

* air traffic control when there is a unified system throughout Europe.

* logging the areas of deposition of acid rain.

* logging the areas of wet deposition of nuclear fall-out (as in the case of the Chernobyl accident).

Within the current COST 73 programme some of these proposed applications are being investigated. However, perhaps because of the more stringent requirements of hydrologists compared with meteorologists, the potential of pan-European radar data in large-scale European hydrological projects is, as yet, untapped.

One possible application of these data might be as inputs to mathematical models of international rivers - of which there are several in Europe - see Figure 4. They may well also have a role to play in water quality models of major river systems and in the tracking and prediction, at points downstream, of the likely concentrations and, hence, the possible damage from pollution incidents such as the recent pollution of the River Rhine by chemicals spilt from a bank-side chemical factory in Switzerland - a pollution incident whose effects were felt along the entire river as far as the estuary.

It is suggested, therefore, that consideration should be given to initiating an international project to study the possible uses of the COST products for this type of application on an international river system which is currently experiencing problems, either of water resources management, quality of water in the rivers or, indeed, both. There are several river systems in Western Europe which would qualify under these criteria, one example being the Rhine and another, the Tagus. If co-operation could be achieved between East and West, perhaps the best example of all would be the Danube, the world's most international river.

4. CONCLUSIONS

Some possible uses of the COST image data have been described; undoubtedly there will be others and the potential for their use will become greater as the COST image is developed and improved.

There is a danger, however, that the momentum of the international co-operative work that has been successfully carried out so far under the COST 72 and COST 73 Projects will be lost, unless it continues to be energetically pursued under the aegis of an international organisation. Perhaps the World Meteorological Organization or World Health Organization would be appropriate bodies to supervise such a project, which could be carried out under contract, after the end of the COST 73 Project in 1991.

5. ACKNOWLEDGEMENTS

The Author gratefully acknowledges the contributions sent to him by the COST representatives of various countries, examples of whose use of radar data are given in the body of the text. Any opinions expressed, however, are those of the Author and are not necessarily held by those persons who made such helpful contributions.

The figures in the paper are reproduced by courtesy of Dr W. Randeu (Figure I); Mr Tage Andersson (Figure II) and North West Water, UK, (Figure III).

6. REFERENCES

1. Central Water Planning Unit, (1977). Dee weather radar and hydrological forecasting project. Report by the Steering Committee, Central Water Planning Unit, HMSO.
2. Collier C.G. (1989). COST 73: Progress towards a West European weather radar network. Paper 1.1, this seminar.
3. De Troch F.P. et al. (1989). On the usefulness of weather radar data in real-time hydrological forecasting in Belgium. Paper 6.3 this seminar.
4. Overgaard S. and Wienberg E. (1989). Distribution of weather radar images to agricultural end users. Paper 6.14 this seminar.
5. Riedl J. (1989). The weather radar network of the Deutscher Wetterdienst - state of the project and first results. Paper 1.9 this seminar
6. Sorani R. and Dietrich E. (1989). Overview of national and regional radar meteorological activities in Italy. Paper 1.12 this seminar
7. Salsi A. and Nanni S. (1989). Agricultural use of weather radar data in Emilia-Romagna, Italy. Paper 6.12 this seminar.
8. Dias B. et al. (1989). An operational system for display and analysis of hydrometeorological radar data. Paper 2.4 this seminar.
9. Nilsson S. and Brunsberg J-O. (1989). PROMIS 600: An operational system for very short range weather forecasting in Sweden. Paper 4.5 this seminar.

Figure I : <u>RADAR - RADIOMETER DIVERSITY DATA</u>

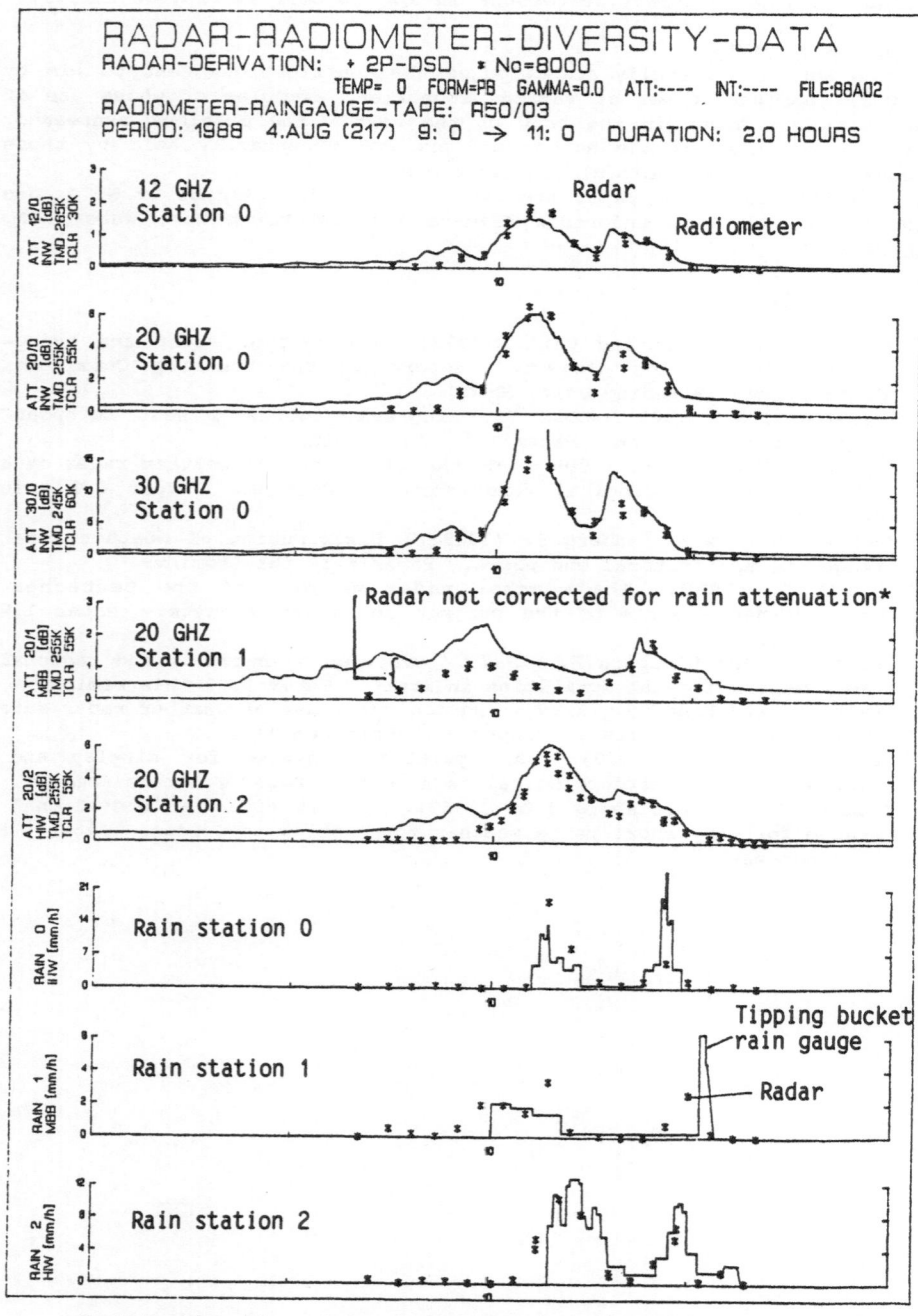

* Intentionally not corrected, to show the need for correction

Figure III : FLOW DIAGRAM FOR DATA AND

FORECAST INFORMATION

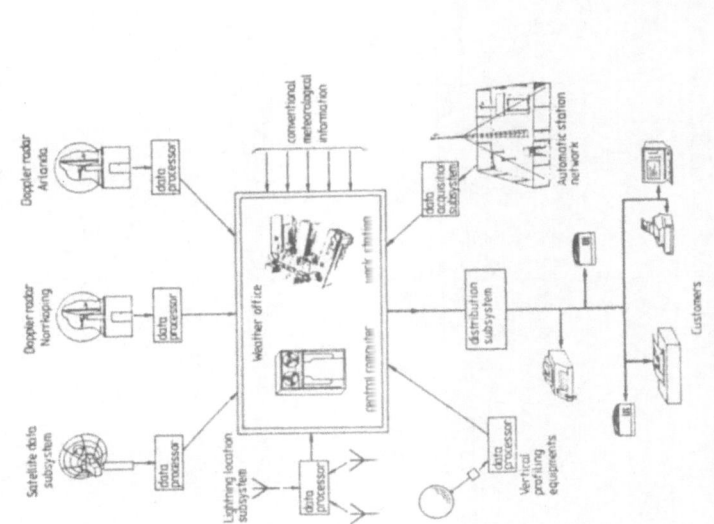

Figure II : PROMIS 600: SYSTEM ORGANISATION

Figure IV : <u>THE MAJOR RIVER SYSTEMS OF WESTERN EUROPE</u>

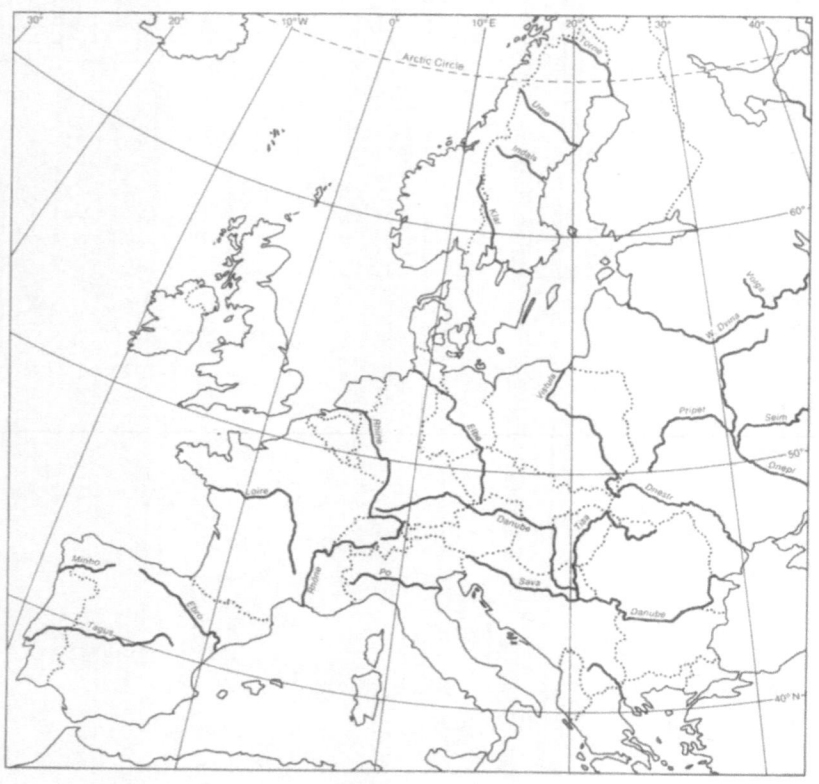

REAL-TIME FORECASTING: MODEL STRUCTURE AND DATA RESOLUTION

Ian D. Cluckie, Professor of Water Resources
Pao-shan Yu, Research Fellow
Kevin A. Tilford, Research Assistant

Department of Civil Engineering, University of Salford
Salford, M5 4WT, ENGLAND, U.K.

ABSTRACT

This paper introduces transfer function rainfall runoff models based on three different spatial structures: lumped, semi-distributed and grid-based fully distributed. All the models have been developed for use with radar derived estimated of rainfall, though raingauge data can also be used. The mathematical basis for each model structure is introduced and the underlying assumptions explained. The paper describes via case studies the application of each model structure to river catchments in the North-west and East of England - catchments which differ considerably in size and their physical characteristics. In addition to the spatial dimension, the intensity resolution of the rainfall data (a function of radar return signal quantisation) is considered as it pertains to flood forecasting with rainfall runoff models. A detailed analysis of high and low intensity resolution (eight and three-bit) radar rainfall data using spectral analysis techniques to study information content, and end-point use assessment of forecast quality shows the three-bit rainfall data to be sufficient for flood forecasting despite its lower intensity resolution. Exaplanations are provided by consideration of the nature of catchment processes, the information content of the rainfall data, and the mathematical characteristics of the rainfall runoff model. The implications of this for operational use of remotely sensed radar rainfall data are discussed.

1. INTRODUCTION

Flooding is one of the most destructive acts of nature, the annual cost of flood damage worldwide is estimated to be billions of dollars (1). The forecasting of flooding in advance enables suitable action to be taken in order to limit damage. In a cost-benefit study of flood warning Chatterton *et al* (2) showed that within the Severn-Trent Water Authority in the United Kingdom, substantial benefits would arise from the establishment of an accurate and reliable flood warning system. The development of short-term real-time flood forecasting over recent years reflects the importance of limiting flood damage for any given catchment. In addition to flash flooding, real-time forecasting has been developed for other applications including irrigation, power production, pollution control and on-line operation of multi-purpose reservoir systems. The versatility of the real-time forecasting model can be attributed to the fact that model structure can be chosen according to context, e.g. on the basis of: (i) forecast variables (ii) the purpose of the forecast and (iii) forecast period (1).

A number of constraints apply regarding the structure of a model for real-time flood forecasting. For the timely, and efficient dissemination of reliable flood warnings to the public, the model should be structurally simple and quick to execute. In addition it is desirable that the storage demands on the forecasting computer hardware system are small. The success of flood warning ultimately depends upon communication systems: both to transmit the hydrometric data collected in the field and to relay the flood warning to the risk area. Capital cost of equipment and the timing of transmission may also be important factors in the choice of communication system.

The decision regarding the choice of a suitable model for real-time forecasting of flow should take into account the following: (i) required forecast accuracy, (ii) the time available between forecast preparation and of flooding taking place, (iii) the manpower and computer resources required to operate the model (3), and (iv) the resolution of the rainfall data to be used as model input.

2. MATHEMATICAL MODELS

Within a large river basin, precipitation in the spatial domain may be highly heterogeneous and catchment response may vary according to position within the catchment. The spatial distribution of the rainfall field can be directly estimated by weather radar. Two approaches can be developed to utilise the information provided by the radar data:

(i) determine areally averaged rainfall at a basin or sub-basin scale, for input to a lumped or a semi-distributed model.

(ii) use the spatial rainfall data as a direct input to a fully distributed grid-based forecasting model (GBDM).

Lumped Transfer Function Rainfall-Runoff Model

A wide range of rainfall-runoff forecasting models with a lumped structure (i.e. using a single rainfall value taken to be representative of catchment rainfall as input) have been developed including conventional methods (e.g. the unit-hydrograph, S-curve, Clark method and linear cascade reservoir model (4), (5), (6) and (7)), conceptual models, non-linear storage models and transfer function models. These models have been reviewed by O'Connell and Clark (8) and Reed (9).

Transfer function models give a forecast based on recent and previously observed rainfall and flow. Essentially, the model is equivalent to a unit hydrograph though considerably more efficient parametrically. Harpin (10), Powell (11) and Owens (12) have demonstrated that both rainfall-runoff and flood wave transformation processes can be satisfactorily simulated by single input-single output (SISO) transfer function models with the structure shown below:

$$y_t - a_1 y_{t-1} - a_2 y_{t-2} - \cdots - a_p y_{t-p} = b_1 u_{t-1} + b_2 u_{t-2} + \ldots b_q u_{t-q}$$

(eq. 1)

Using the Z-transform eq.1 can be written as:

$$A(z)y_t = B(z)u_t$$

(eq. 2)

so rearranging,

$$y_t = \left[\frac{B(z)}{A(z)} \right] u_t$$

(eq. 3)

where,

$$A(z) = 1 - a_1 z^{-1} - a_2 z^{-2} - \cdots - a_p z^{-p}$$

(eq. 4)

$$B(z) = b_1 z^{-1} + b_2 z^{-2} + \ldots + b_q z^{-q}$$

(eq. 5)

In the case of a catchment where catchment response lags rainfall input, a pure time delay can conveniently be incorporated into the model. A transfer function model with order p+q (where p=number of a or flow parameters and q=number of b or rainfall parameters) when subjected to a unit impulse input at time t=1 has the following equivalent impulse response function:

$$y_t = \int H_l \, u_{t-l}$$

(eq. 6)

Before applying a transfer function model, the optimal model structure (i.e. p and q) is determined. Simultaneously, the parameters values (a1, a2, . . . ap, b1, b2, . . . , bq) are estimated. Harpin (10) and Cluckie and Ede (13) studied different parameter estimation approaches and found, though intrinsically biased, the recursive least squares estimator (RLS) to be adequate for the estimation of parameters for use in a real-time model. Owens (12) further refined a subjective identification method for determining the optimal structure of a transfer function rainfall-runoff model, based primarily on the model impulse response but also the model steady state gain, one-step-ahead (reconvolution) errors, and parameter redundancy.

The central properties of the model in operation are the use of past observed flow values to correct the model forecasts (feedback) and real-time updating. Three approaches to model updating are summarised by Reed (9); error prediction, state updating, and parameter updating. An alternative technique is employed in the transfer function model whereby forecast error (i.e. the difference between forecasted and observed flow) is used to update a model scaling factor, delta (Δ). This method is analogous to the variable proportional loss method of defining effective rainfall, the value of Δ essentially acting as a real-time rainfall correction factor and being applied on the rainfall terms as:

$$y_t = a_1 y_{t-1} + \ldots + a_p y_{t-p} + \Delta \left\{ b_1 u_{t-1} + \cdots + b_q u_{t-q} \right\}$$

(eq. 7)

and updated in real-time according to the following equation:

$$\Delta_t = \mu \, \Delta_{t-1} + (1 - \mu) \cdot \frac{y_t - \left\{ a_1 y_{t-1} + \ldots + a_p y_{t-p} \right\}}{\left\{ b_1 u_{t-1} + \ldots + b_q u_{t-q} \right\}}$$

(eq. 8)

Semi-Distributed Transfer Function Models

Due to their size, large river basins are commonly composed of several distinct subcatchments, over which precipitation and hence, subcatchment responses (volume, timing etc.) may vary considerably. Multiple-input single-output (MISO) transfer function models provide one means of addressing this problem. The main structure of the semi-distributed model is a cluster of several lumped models: each lumped rainfall-runoff transfer function model being first applied over the subcatchment to determine a forecasted output at the downstream end of each. A flow-flow transfer function model is then used to route flow to the main basin outlet. The structure of the multiple SISO and MISO models are shown in figures 1a and 1b.

The basic equation in the semi-distributed models describes the extension of the single lumped transfer function model to a system with m inputs.

$$A(z) y_t = \sum_{i=1}^{m} B_i(z) \cdot u_t$$

(eq. 9)

Model identification determines the parameters A(z) and Bi(z) in (eq. 9). Harpin (10), Snorrason et al (14) and Olason and Watt (15) have discussed the difficulties of model identification (due primarily to multi-collinearity of inputs), and proposed some methods to tackle it.

In the multiple SISO transfer function model, individual flow-flow transfer function models are identified for each tributary using the flow data at that tributary and catchment outlet, thus overcoming the problem of multi-collinearity during calibration stage. An area weight factor is applied on each as:

$$y_t = \sum_{i=1}^{m} \left\{ \frac{a_i}{a} \right\} \cdot \left[\frac{B_i(z)}{A_i(z)} \right] \cdot y_{i(t)} + N_t$$

(eq. 10)

where:

$$N_t = \left[\frac{B_n(z)}{A_n(z)} \right] \cdot Uu_t$$

(eq. 11)

and:

m	=	numbers of gauged subcatchments
a	=	the total area of the river basin
a_i	=	the area of each subcatchment
$B_i(z)/A_i(z)$	=	flow-flow transfer function model for ith subcatchment
N_t	=	noise model for ungauged area
Uu_t	=	rainfall at ungauged area
$B_n(z)/A_n(z)$	=	noise model

Identification of the model described in eq. 10 is divided into two parts: identification of the flow-flow transfer function model for each tributary (using the response function and RMSE [root mean square error] criteria), and the identification of a noise model (eq. 11) to predict up to six-hours ahead, runoff from the ungauged area using rainfall data over the area as input and a noise sequence as output (the noise being defined by the difference of actual flow and flow reconvoluted by the multiple SISO model).

$$N_t = y_t - \sum_{i=1}^{m} \left\{ \frac{a_i}{a} \right\} \cdot \left[\frac{B_i(z)}{A_i(z)} \right] \cdot y_i(t)$$

(eq. 12)

The second component of the semi-distributed model regards the MISO transfer function model structure to forecast total catchment outflow. The structure is shown in eq.13.

$$y_t = \sum_{i=1}^{m} \left[\frac{B_i(z)}{A(z)} \right] \cdot y_i(t) + N_t$$

(eq. 13)

As in lumped model response functions, each B_i/A is defined as the partial response of a unit pulse at time t=1 occurring at each source, and $\sum B_i/A$ defined as the total response of all sources. Since the partial and total impulse responses reflect the physical response to a unit input from each tributary, a MISO model needs to reflect a 'reasonable' hydrological response for each tributary. Hence, the shape of the partial impulse response of each input source and the total catchment response function is used as a guide for model identification. Owing to the high cross-correlation between flow data at the tributaries, it is very difficult to simultaneously obtain reasonable partial and total responses. One order differencing was found to be useful in reducing the degree of cross-correlation, and applied on the flow sequence before identification.

In this model, two updating techniques were investigated: error prediction (as in the multiple SISO flow-flow transfer function model), and the delta updating described earlier. Forecasting error is assumed to primarily come from the rainfall component of the model. The governing equation (with Δ updating based on eq. 8) is given by:

$$A(z)y_t = \Delta \cdot \left[\sum_{i=1}^{m} B_i(z) \cdot (y_t)_i \right] + N_t$$

(eq. 14)

Grid-Based Distributed Model (GBDM)

To further exploit the spatial information provided by the radar and simulate the hydrological processes in detail, a fully distributed grid-based rainfall-runoff model was investigated (16). The catchment is depicted as a gridded mesh in accordance with the grid used by the radar (i.e. 2km or 5 km squares) as shown in figure 2. The spatial rainfall distribution together with surface and subsurface catchment characteristics form the input to each grid square. Figure 3 shows the form of the input data. The hydrological processes of infiltration loss and rainfall-runoff transformation are simulated at each grid square, the runoff generated from each entering the channel and simulated by river flow transportation. These three major hydrological processes, abstraction loss, rainfall-runoff transformation and river flow transportation, are briefly described below.

(i) simulation of losses due to abstraction neglect evaporation and interception and consider only infiltration. These losses are simulated using the modified Horton equation (17), in which infiltration loss during intermittent rainfall sequence are taken into account.

(ii) a conceptual storage model using the effective rainfall determined from (i) is used for the simulation of the rainfall-runoff transformation. The basic equations are:

Dynamic equation:

$$S(t) = K.O(t)$$

(eq. 15)

Continuity equation:

$$\frac{dS}{dt} = R(t-L) - O(t)$$

(eq. 16)

where:
$S(t) =$ the storage in one grid
$R(t) =$ the net rainfall
$O(t) =$ the outflow discharge of the grid
$K =$ storage coefficient
$L =$ time lag. As the storage coefficient is equivalent to the lag time of a basin (18), the time in eq. 16 is estimated as $L=K/T_s$ where T_s is the time interval in the model.

The combination of conceptual linear channel and linear reservoir elements as described above is often referred to as the lag and route model and can be approximated by a transfer function model with an appropriate time lag (19):

$$O(t) = aO(t-1) + bR(t-L)$$

(eq. 17)

where a+b=1.

Eq. 17 is the major equation used at each grid to simulate overland flow. Before entering the river channel, the overland flow may be routed through several grids depending upon the path route. Hence, the whole rainfall-runoff transformation over the catchment is simulated by a cascade transfer function model.

(iii) When the runoff generated from previous process enters into the river channel, upstream to downstream river flow is simulated by a coefficient equation having the form of a single parameter transfer function model (eq. 18).

$$Q(t) = A.Q(t-1) + B.I(t) + B.I(t-1)$$

(eq. 18)

$Q(t)$ = outflow at river reach
$I(t)$ = inflow at river reach
A and B = parameters (A+2B=1)

State updating is applied in the GBDM to update the flow variable at each river reach using the observed flow at the outlet. The basic assumption is that error at the outlet flow $Q(t)$ in eq. 18 is attributable to the inflow $I(t)$ and that $Q(t-1)$ and $I(t-1)$ are correct because they have been updated at the previous timestep. When the observed flow $Q(t)$ is available, the estimated inflow $I(t)$ can be updated based on the difference between observed flow $Q(t)$, and estimated flow $Q(t)$ as:

$$I(t) = (1 + \beta).I(t)$$

(eq. 19)

where:

$$\beta = \frac{\left[Q(t) - Q(t)\right]}{B.I(t)}$$

(eq. 20)

and:

$I(t)$ = inflow at river reach after updating
$Q(t)$ = outflow at river reach after updating
$I(t)$ = inflow at river reach before updating
$Q(t)$ = outflow at river reach before updating

This procedure is repeated for all river reaches from the basin outlet 'upwards' (i.e. upstream).

3. CASE STUDIES

A small rural catchment in the Anglian Water Authority Region of Eastern England is used to demonstrate flood forecasting using a lumped transfer function rainfall runoff model. Willow Brook (see figure 4) is a tributary of the River Nene one of the major river systems draining Eastern England. The Brook is gauged by a standing wave flume at a station based at Fotheringhay, the catchment area to this point being about 90sq km. Subcatchment averaged rainfall data from the Chenies radar (London) were used as model input, the catchment some 85km from the radar.

Due to sparsity of data (the radar has only been on-line since 1985), a revolving calibration/ verification scheme was developed (described fully in (21)), enabling unbiased forecasting performance to be assessed. The optimal model structure was found to be 3,5,0 the model impulse response having a time to peak of 10 hours (average catchment response time of 12 hours), and representing runoff of 18%. Despite the distance of the catchment from the radar, the 'odd' shape of the catchment, and limited data for calibration; a parametrically efficient model performs well for flood forecasting. One, three, and six hour-ahead forecast RMSE's for three verificatiuon events are shown in Table 1.

The River Irwell catchment chosen for the semi-distributed and distributed model case studies is situated in the North-West Water Authority region of the U.K (see figure 4). The main basin consists of three sub- catchments gauged at Stubbins (headwaters of the River Irwell), Blackford Bridge (River Roch) and Farnworth (River Croal), the main catchment outlet being gauged at a station located on the site of the old Manchester Racecourse immediately upstream of the City of Salford flood risk zone. An ungauged subcatchment contributes about 20% of the total catchment area and flow volume. Total catchment area is about 551sq km.

TABLE 1: <u>Root Mean Square Error of One, Three and Six Hour-Ahead Forecasts for Willow Brook Model</u>

Event	To Event Peak			To Event End		
	1hour	3hours	6hours	1hour	3hours	6hours
06-06-85	0.033	0.152	0.385	0.028	0.118	0.296
08-10-87	0.055	0.190	0.341	0.034	0.122	0.232
14-10-87	0.055	0.239	0.578	0.042	0.186	0.443

Cluckie and Yu (20) compared three semi-distributed models with a lumped rainfall-runoff transfer function model applied over the whole catchment. One, three, and six hour-ahead forecast RMSE's and average improvement percentages are shown in Table 2. The improvement percentage, defined as the ratio of the difference of RMSE between the semi-distributed and lumped models to the RMSE from lumped model, was used to comparatively judge model performance. A positive improvement percentage (e.g. for the semi-distributed model) implies a reduction in forecast RMSE (e.g. relative to the lumped model) and hence, improved forecasting performance. The results show that consideration of spatial rainfall distribution on a subcatchment scale (via the the use of models with a semi-distributed structure; MISO-1 and MISO-2) improves forecast accuracy over a lumped model structure: average RMSE on the rising limb being reduced by 33% and 30% respectively.

TABLE 2: <u>Root Mean Square Error and Average Improvement Percentage for One, Three and Six Hour-Ahead Forecasts for Lumped and Semi-Distributed Models (from Cluckie and Yu, 1988).</u>

TO EVENT PEAK

Time ahead	31-10-86 (RMSE)			18-11-86 (RMSE)			09-11-86 (RMSE)			Ave. Imp. Per. (%)		
	1 hr.	3hrs.	6hrs.	1 hr.	3hrs.	6hrs.	1 hr.	3hrs.	6hrs.	1 hr	3 hrs	6 hrs
LUMP	5.0	20.2	43.2	5.9	18.8	23.2	9.2	49.8	105.1			
M-SISO	5.3	11.0	20.1	10.6	21.0	22.4	11.2	36.7	73.4	-36	15	29
MISO-1	3.1	10.4	17.4	6.7	12.6	15.5	6.8	26.7	59.8	17	43	45
MISO-2	3.9	13.0	19.4	5.1	10.8	14.8	9.2	27.1	61.4	12	42	44

TO EVENT END

time ahead	31-10-86(RMSE)			18-11-86 (RMSE)			09-11-86 (RMSE)			Ave. Imp. Per. (%)		
	1 hr.	3hrs.	6hrs.	1 hr.	3hrs.	6hrs.	1 hr.	3hrs.	6hrs.	1 hr	3 hrs	6 hrs
LUMP	3.6	14.3	31.2	4.6	13.4	17.0	5.6	29.5	62.5			
M-SISO	4.4	8.7	14.8	7.8	15.5	16.6	6.7	21.6	43.8	-37	17	28
MISO-1	2.7	9.2	14.0	5.0	9.5	11.8	4.2	16.7	36.4	14	36	43
MISO-2	3.2	10.8	15.9	4.0	8.4	11.6	5.9	16.9	37.5	6	35	40

M-SISO: Multiple SISO semi-distributed transfer function model
MISO-1: MISO semi-distributed transfer function model with a noise component
MISO-2: MISO semi-distributed transfer function model with (delta) updating

4. DATA RESOLUTION

The spatial and temporal resolutions of hydrometric data, and the models which use them have been the subject of thorough and continuing research and discussion. Intrinsically linked, easily overlooked though no less important, is the influence of intensity resolution on the quantitative representation of rainfall and the performance of flood forecasting models. The Meteorological Office in the U.K. uses two quantisation methods to convert the analogue radar return signals to a discrete form so that discrete weather radar data fall into two intensity resolutions: eight-bit data whereby rainfall intensity is represented across 208 intensity ranges (often referred to as 'quantitative' data), and three-bit data whereby the representation is across 8 intensity ranges ('qualitative' data).

Examination of a number of techniques for the assignment of actual rainfall intensities from within the qualitative ranges of the three-bit data (22) have shown that assigned values based upon the logarithmic mean of the slice ranges (see eq. 21) (i.e. geometric mean) produces three-bit rainfall values which closely replicate the eight-bit data. This is illustrated by figure 5, cumulative depths for about 500 hours of event rainfall over the River Roch subcatchment of the Irwell catchment; the cumulative rainfall depth being within 5% of the eight-bit cumulative rainfall depth. The comparison of cumulative rainfall depths was extended for gridded rainfall data for approximately 15 000 events, and the resultant enclosure scattergraph (figure 6) shows that providing the rainfall is significant (more than 5mm), the difference between cumulative rainfall depths derived from eight and three-bit rainfall data are generally less than 10% (23).

$$R_i = \frac{\log_{10}(i+1) - \log_{10}(i)}{2}$$

(eq. 21)

where:
Ri = three-bit rainfall intensity assigned from eight-bit data (mm/hr).
i = bounds of intensity ranges used to slice eight-bit data (mm/hr).

The mathematical properties of the rainfall data are examined in the frequency domain using spectral analysis techniques to determine a power spectra (i.e. a breakdown of signal variance with frequency). This provides a measure of the information content of the rainfall time series. The power spectrum (spectral density function) of a stochastic process is given by the continuous function:

$$f(\omega) = (2\pi)^{-1}\left[C_0 + 2 \sum_{i=1}^{\infty} C_i \cos \omega i \right]$$

(eq. 22)

where:
ω = frequency (radians) and may take any value in the region $[-\pi, \pi]$, though since $f(\omega)$ is symmetric about zero, all the information in the power spectrum is contained in the range $[0, \pi]$.

In practice, eq. 22 cannot be used to estimate spectra from sampled data, and an alternative method involving taking Fourier transforms of truncated sample autocovariance function and using a weighting procedure to smooth the sample spectrum is used (eq. 23).

$$f(\omega) = \frac{1}{2\pi}\left[w_0 C_0 + 2 \sum_{i=1}^{n} w_i C_i \cos \omega i \right]$$

(eq. 23)

where:
wi = set of weights: the lag window.
i = number of data.
m = window truncation point.

The power spectra of a series of event rainfall for the River Roch catchment (gauged at Blackford Bridge - an area of 183 sq km) is shown in figure 7. The spectra show that i. the major portion of the power is confined to the low frequency end of the spectra, and ii. despite degradation in intensity resolution, the spectra of the eight and three-bit rainfall sequences do not differ greatly. Figure 8 shows the impulse responses for lumped transfer function models calibrated for the River Roch subcatchment using the rainfall data on which the power spectra in figure 7 are based. The impulse responses for the models derived from eight and three-bit rainfall data are almost identical, a fact compounded on by the frequency responses of the respective models shown in figure 9. The frequency responses illustrate how the transfer function model processes predominantly low frequency data: indeed these (and most other rainfall runoff models) can be considered as signal filters, transforming a high frequency input (rainfall) to a low frequency output (flow).

Six hour ahead forecast hydrographs from the Blackford Bridge models and also for eight and three-bit Willow Brook models are shown in figures 10 and 11. The hydrographs illustrate negligible difference in forecast quality and in the context of operational flood forecasting, insignificant. Similar results have been observed from a large number of other verification events for catchments in the North-West Water Authority (24) and further preliminary work on catchments in Eastern England within Anglian Water Authority (21).

5. CONCLUSIONS AND RECOMMENDATIONS

The paper has introduced three differently structured rainfall runoff transfer function models. The structures differ primarily in terms of the spatial resolution of the rainfall input data: from a lumped structure which uses a single areally averaged rainfall data, a semi-distributed structure which considers spatial heterogeneities in rainfall at a sub-basin scale, and a fully distributed grid based distributed model which simulates in detail the rainfall runoff simulation process over 2km grid squares. The mathematical basis of each model structure is introduced in detail and the relative advantages and disadvantages of each discussed. The case studies for catchments of different physical characteristics in the North-West and East of England illustrate the operational modes of the models and forecasting performance. The rainfall input for all is determined from weather radars in the U.K network operated by the Meteorological Office.

Intensity resolution of the radar rainfall data is an important characteristic. The paper has shown that despite significant degradation in terms of the ranges used for quantisation of the analogue radar signal (and hence the intensity resolution of the data), the integrity of data is retained. This is illustrated both via comparison of cumulative rainfall depths for the eight and three-bit rainfall sequences, the information content of the respective signals, and the 'end-point use' assessment of flood forecasting using a lumped transfer function model. The results show that models calibrated from eight and three-bit rainfall do not differ significantly, and that forecasts produced from each differ insignificantly.

It may be that we are preoccupied with the 'accuracy' of the data we use (itself a contentious issue for rainfall data given the difficulty of defining 'ground-truth'). Consider the processing that implicitly exists within river catchments and the rainfall runoff process in general: physical processes (e.g. evapotranspiration, percolation, overland flow) introduce significant filtering transforming a high frequency process (rainfall) to a lower frequency one (runoff); for implementation of discrete models on digital computer systems the continuous hydrological processes are sampled entailing temporal filtering and the errors associated with it; data may be further (spatially) filtered if a lumped model is applied. The cumulative effect of this filtering modulates the high frequency rainfall signal to a low frequency one (flow). With such extensive filtering and processing within the natural system combined with the fact that most rainfall runoff models process predominantly low frequency information from the rainfall signal, it is perhaps not surprising that rainfall data with an intensity resolution much lower than we are used (or perhaps feel comfortable with) can be used quantitatively without compromising the accuracy of quantitative applications. The implications for the utilisation of weather radar/satellite rainfall data are significant, since it means that the use of a single product (e.g. the FRONTIERS product in the U.K. or even the European COST image potentially resulting in significant cost reductions for the Water Industry, is a viable one.

6. ACKNOWLEDGEMENTS

The authors would like to thank North-West Water Authority, Anglian Water Authority for their help and interest and the Meteorological Office for their continuing support and help regarding the reported work and the work of the Water Resources Research Group in general.

7. REFERENCES

(1)   Nemec, J., Hydrological Forecasting - Design and Operation of Hydrological Systems, 1986.
(2)   Chatterton, J.B., Pirt, B.A., and Wood, T.R., "The Benefits of Flood Forecasting", Jrnl of Inst. of Water and Env. Sci., Vol. 33, pp 237, 1979.
(3)   Noonan, G.A., "An Operational Flood Warning System", Weather Radar and Flood Forecasting ed. by V.K. Collinge and C. Kirby, 1986.
(4)   Chander, S., and Shanker, H. ,"Unit Hydrograph Based Forecast Model", Hyd. Sci. Jrnl, Vol. 31, pp 279, 1984.
(5)   Bobinski, E., and Mierkiewicz, M., "Recent Developments in Simple Adaptive Flow Forecasting Models in Poland", Hyd. Sci. Jrnl, Vol. 31, pp297, 1986.
(6)   Corradini, C., and Melone, F., "On the Structure of a Semi-Distributed Adaptive Model for Flood Forecasting", Hyd. Sci. Jrnl, Vol. 32, No. 2, pp 227, 1987.
(7)   Corradini ,C., ,Melone, F., and Uvertini , "A Semi-Distributed Model for Real-Time Flood Forecasting", Water Resource Bulletin, Vol. 22, No. 6, pp 1031, 1986.
(8)   O'Connell, P.E., and Clark, R.T., "Adaptive Hydrological Forecasting - A Review", Hyd. Sci. Bulletin 26(2), pp 179, 1981.
(9)   Reed, D.W., "A Review of British Flood Forecasting Practice", I.H. Report No. 90, pp42, 1984.
(10)   Harpin, R., "Real Time Flood Routing with Particular Emphasis on Linear Methods and Recursive Estimation Techniques", Ph.D Thesis, University of Birmingham, Department of Civil Engineering, 1982.
(11)   Powell, S.M., "River Basin Models For Operational Forecasting of Flow in Real-Time", Ph.D. Thesis, University of Birmingham, Department of Civil Engineering, 1985.
(12)   Owens, M.D., "Real-Time Flood Forecasting Using Weather Radar Data", Ph.D Thesis, University of Birmingham, Department of Civil Engineering, 1986.
(13)   Cluckie, I.D. and Ede, P.F., "End-Point Use a Criterion for Model Assessment", 7th IFAC/IFORS Symposium on Identification and System Parameter Estimation York, U.K., 1985.
(14)   Snorrason, A., Newbold, P., and Maxwell, W.H.C., "Multiple Input Transfer Function Noise Modelling of River Flow", Frontiers in Hydrology, ed. by Maxwell, W.H.C and Beard L.R, Water Resource Publication, Collins,1984.
(15)   Olason, T., and Watt, W.E., "Multivariate Transfer Function-Noise Model of River Flow for Hydropower Operation", Nordic Hydrology ,Vol. 17, pp185, 1986
(16)   Yu, P.S, "Real-Time Grid-Based Distributed Rainfall-Runoff Model for Flood Forecasting with Weather Radar", Ph.D Thesis, University of Birmingham, Department of Civil Engineering, 1989
(17)   Bauer, S.W., "A Modifed Horton Equation for Infiltration during Intermittent Rainfall", Hydrological Science Bulletin, Vol 19, pp 219-225.
(18)   Boyd, M.J., "A Storage-Routing Model Relating Drainage Basin Hydrology and Geomorphology", Water Resource Research 14(5), pp 921-928.
(19)   Young, P.C., "Real-Time River Flow Forecasting", International Postgraduate Course, July 26-Aug. 6, 1983, Wageningan, Netherlands.
(20)   Cluckie, I.D., and Yu, P.S., "Stochastic Models for Real-Time Riverflow Forecasting Utilising Radar Data", The 5th IAHR International Symposium on Stochastic Hydraulics, University of Birmingham, U.K. Aug. 1989.
(21)   Anglian Radar Information Project., "Transfer Function Models for Flood Forecasting in Anglian Water Authority", Report No. 3, 1989.
(22)   Anglian Radar Information Project., "An Evaluation of the Influence of Radar Rainfall Intensity Resolution for Real-Time Operational Flood Forecasting", Report No. 2, 1988.
(23)   Cluckie, I.D., Tilford, K.A., and Shepherd, G.W., "Radar Rainfall Quantisation and it's Influence on Rainfall Runoff Models", Proc. Int. Symp. on Hyd. App. of Weather Radar, University of Salford, U.K, 1989.
(24)   Tilford, K.A., "Real-Time Flood Forecasting using Low Intensity Resolution Radar Rainfall Data", M.Sc Thesis, University of Birmingham, Department of Civil Engineering, 1987.

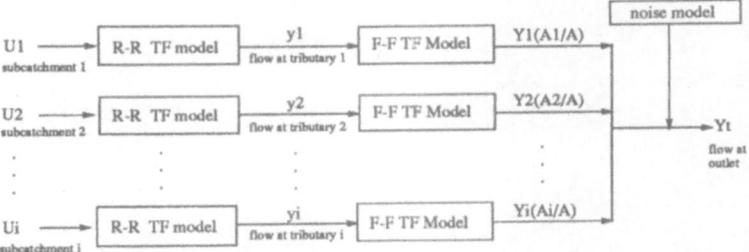

FIGURE 1A: The Structure of the Multiple SISO Semi-Distributed Transfer Function Model

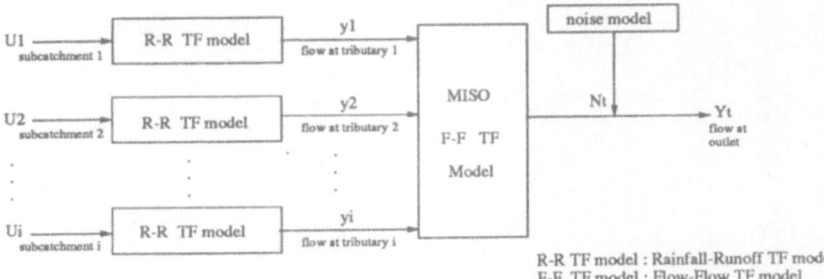

R-R TF model : Rainfall-Runoff TF model
F-F TF model : Flow-Flow TF model

FIGURE 1B: The Structure of the MISO Semi-Distributed Transfer Function Model

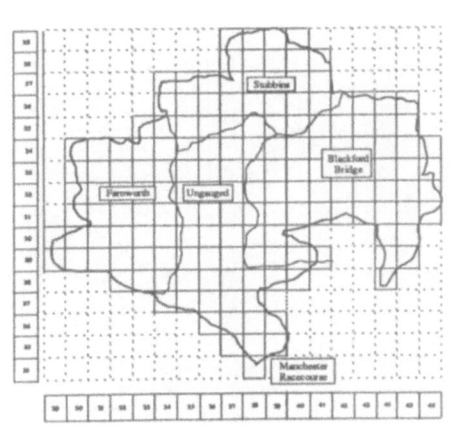

FIGURE 2: The Grid-Based Mesh for the Irwell Catchment

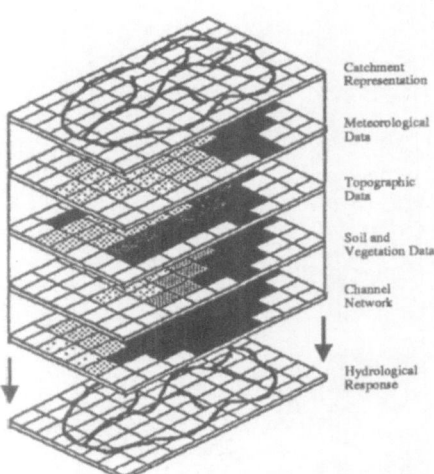

FIGURE 3: The Multi-Layered Simulation of the Grid-Based Distributed Model (GBDM)

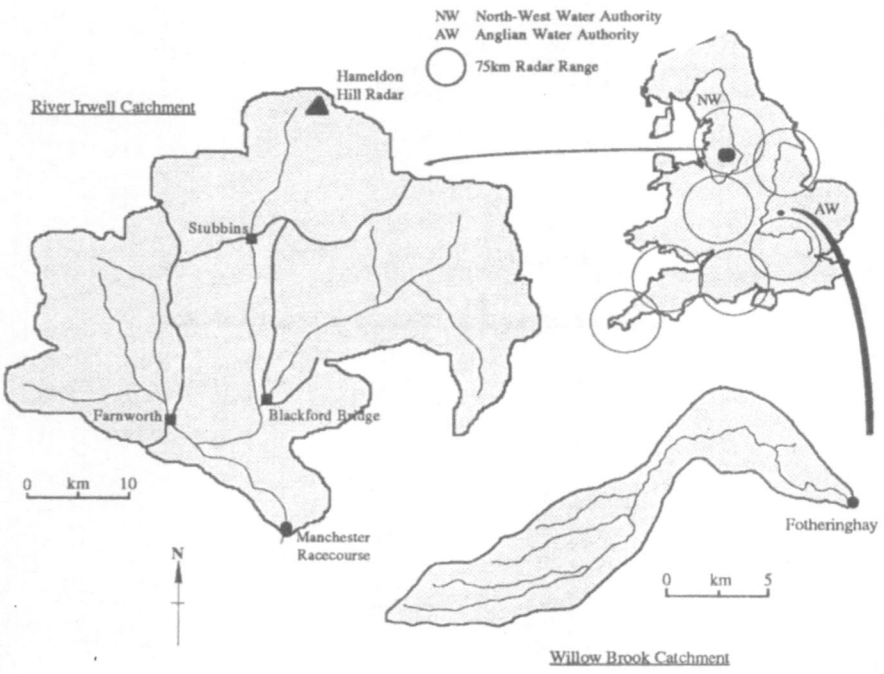

FIGURE 4: River Catchments Used In The Case Studies

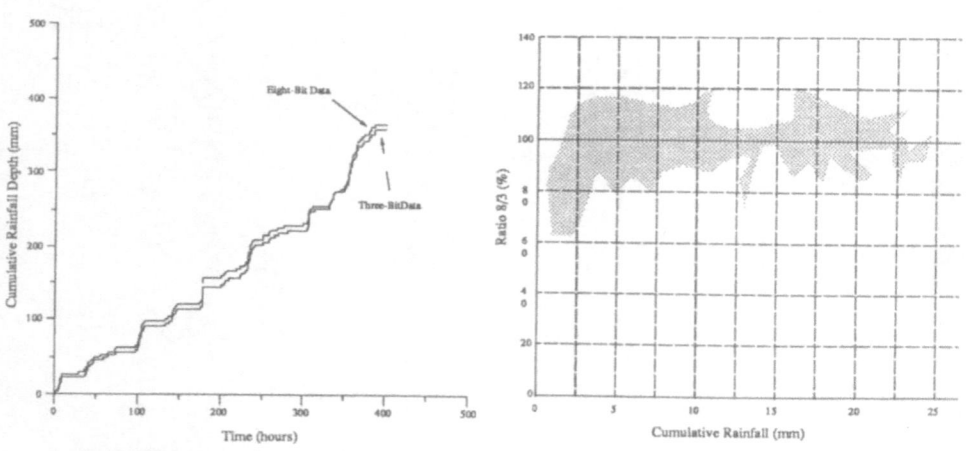

F IGURE 5: Cumulative Rainfall of Eight and
Three-Bit Rainfall Data

F IGURE 6: Enclosure Scattergraph: Conversion
Error as a Function of Total Rainfall

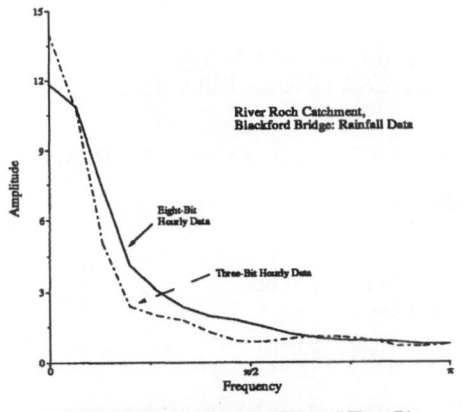

FIGURE 7: Power Spectra of Eight and Three-Bit Subcatchment Rainfall Data

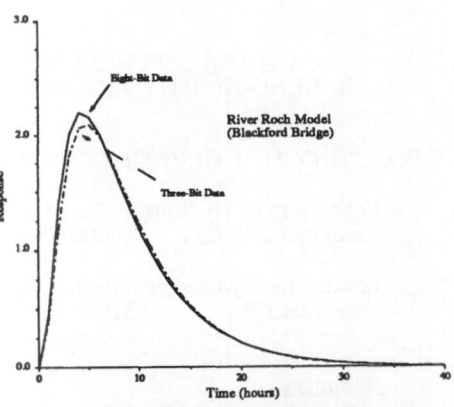

FIGURE 8: Impulse Responses of Lumped Transfer Function Rainfall Runoff Models

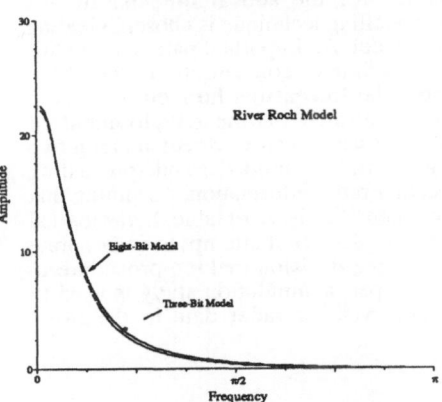

FIGURE 9: Frequency Responses of Lumped Transfer Function Rainfall Runoff Models

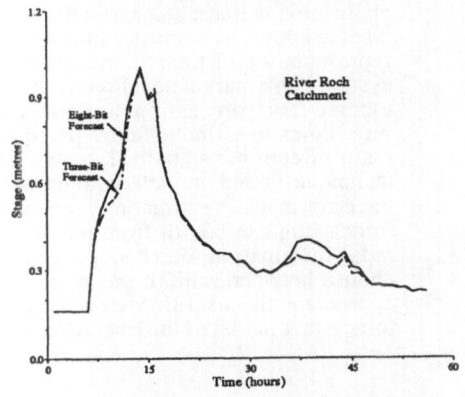

FIGURE 10: Example Flood Forecast Hydrographs for River Roch (Blackford Bridge) Model

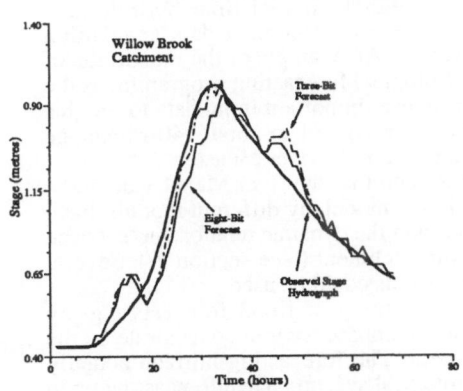

FIGURE 11: Example Flood Forecast Hydrographs for Willow Brook (Fotheringhay) Model

ON THE USEFULNESS OF WEATHER RADAR DATA
IN REAL-TIME HYDROLOGICAL FORECASTING IN BELGIUM

F.P. DE TROCH ; J. HEYNDERICKX ° ; P.A. TROCH and D. VAN ERDEGHEM

Laboratory of Hydrology, State University Ghent
Coupure Links 653   B9000 Ghent - Belgium

° Service for Hydrological Research, Ministry of Public Works
Wetstraat 155   B1040 Brussels - Belgium

Summary

The hydrologic part of the forecasting model of the river Meuse in Belgium describes the rainfall-runoff relationships for the subcatchments. In this application, a linear stochastic black-box modelling technique is chosen, yielding what is known as a transfer function noise model. An important parameter in this representation of the real system is the dead time or concentration time of the system. This parameter directly influences the forecasting horizon. Efforts to extend this forecasting horizon are concentrated on the development of procedures to estimate future precipitation. Research can be directed along two main directions: statistical analysis (e.g. rainfall generators) and forecasting techniques based on meteorologic and weather radar information. Assuming that the error made in estimating precipitation intensities is acceptable, hydrological forecasting can benifit from this information . As a first attempt to incorporate radar information, one can try to develop some decision making procedure to choose between rainfall scenarios. In this paper, a simulation study is used to appreciate the usefulness of incorporating weather radar data in the flood forecasting model of the river Meuse.

1. INTRODUCTION

In Belgium, the development of and research on real-time hydrological forecasting can be based on an extensive real-time hydrological data acquisition network, managed by the Ministry of Public Works. An example of the integration of this network within the development of an hydrological forecasting programme is the real-time flood forecasting model for the river Meuse. Important input data to run this flood forecasting model are hourly precipitation, recorded in about 40 raingauge stations, and hourly discharge of the major tributaries of the river Meuse.

The density of the monitoring network is such that the river Meuse catchment can be divided in subcatchments represented, in the model, by different moduli. Each modulus consists of a mathematical model describing the dynamic relationship between the hydrological variables in the considered subcatchment (see section 2). Several modelling techniques to develop such dynamic models could be used.

Many efforts are made to improve the accuracy of flood forecasts e.g. by developing adaptive forecasting models. In a second phase we will concentrate on the extension of the lead-time or forecasting horizon. The forecasting horizon actually depends on dynamic characteristics of the hydrological system. When forecasting up to several hours ahead, the accuracy of flood predictions is decreasing rapidly. For this reason we will try to appreciate the usefulness of weather radar to extend the forecasting horizon and to improve the accuracy of flood predictions.

2. REAL-TIME FLOOD FORECASTING FOR THE RIVER MEUSE, BELGIUM

The total basin area of the river Meuse is appr. 33 000 km^2. At Liège (nearby the Dutch-Belgian border) the basin area is appr. 20 000 km^2. The catchment can be divided in several subcatchments, mainly situated in the Ardennes (Fig.1). Flooding of the Meuse is caused by heavy rainfall in the Ardennes.

The real-time flood forecasting model, developed by the Laboratory of Hydrology of the State University Ghent in cooperation with the Service for Hydrological Research of the Belgian Ministry of Public Works, consists of two parts: hydrologic forecasting and hydraulic forecasting. The hydrologic part incorporates the rainfall-runoff relationship for the main tributaries of the river Meuse. The hydraulic part routes the contribution of these tributaries during floods using hydraulic flood routing based on the complete de Saint Venant equations. In this section we will concentrate on the hydrologic model building procedure.

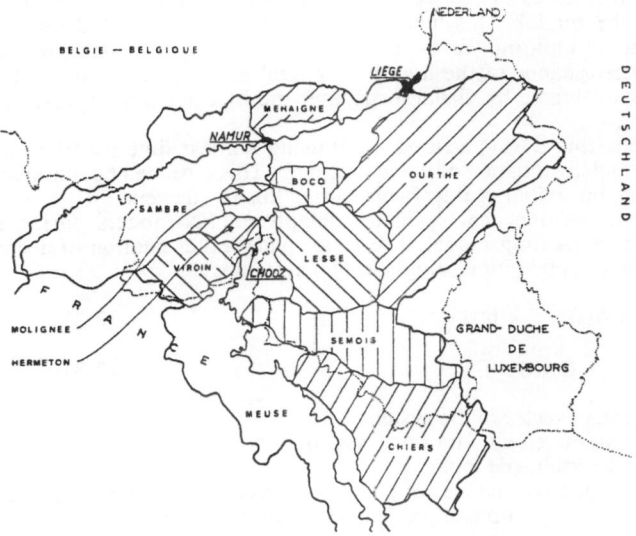

Fig. 1: Subcatchments of the river Meuse in the Ardennes.

Based on geomorphologic characteristics of the river Meuse and on considerations concerning the available hydrological information in real-time, a linear stochastic modelling technique was chosen (TROCH, SPRIET and DE TROCH (1988)). In a first approach time-invariant transfer function noise models are defined to represent the rainfall-runoff process for the subcatchments, leading to the following difference equation :

$$A(q^{-1}) y(t) = q^{-d} B(q^{-1}) u(t) + C(q^{-1}) e(t) \qquad (2.1)$$

where : $y(t)$: process output ; discharge (m^3/s) in outlet station
$u(t)$: process input ; catchment precipitation (mm/h)
$e(t)$: white noise sequence
q^{-1} : backward shift operator
$A(q^{-1}) = 1 + a_1 q^{-1} + ...+ a_{na} q^{-na}$
$B(q^{-1}) = b_0 + b_1 q^{-1} + ...+ b_{nb} q^{-nb}$
$C(q^{-1}) = 1 + c_1 q^{-1} + ...+ c_{nc} q^{-nc}$
d : dead time

System identification includes structure characterisation, parameter estimation and model validation.

First, the order of the polynomials $A(q^{-1})$, $B(q^{-1})$ and $C(q^{-1})$ must be chosen. The dead time d is found using the technique of correlation. Structure characterisation of the TFN-model is based on two different approaches. Namely, the heuristic BOX-JENKINS approach (BOX and JENKINS, 1970) and a more objective BIC search (SPRIET, 1985). For a detailed description of these structure characterisation methods used for river catchment modelling, we refer to SPRIET, TROCH and DE TROCH (1987).

The model parameters are estimated using the Instrumental Variable - Approximate Maximum Likelihood (IV-AML) algorithms. These schemes are fully discussed in YOUNG (1984). Model calibration is based on carefully selected historical flood events for the different tributaries of the river Meuse. To check for model validity, it is necessary to work with separate data sets for calibration and for verification of the model. This technique of split-data-set is used throughout the hydrological model building for the different subcatchments. The criterion used to judge for the performance of the identified model is forecasting power. Forecasted flood events are evaluated by visual inspection and by calculating statistics based on forecasting error.

Flow forecasting during a flood event using a linear time-invariant model can lead to over- or underestimation of the hydrograph. This error can be introduced due to different dynamic behaviour of the system during this flood event.

For this reason one can try to use adaptive TFN-models during real-time operation of the forecasting model. The state-space representation of a linear time-variant deterministic single-input, single-output system is :

$$\dot{\underline{x}}(t) = A(t)\,\underline{x}(t) + \underline{b}(t)\,u(t) \tag{2.2}$$

$$y(t) = \underline{c}^T(t)\,\underline{x}(t) \tag{2.3}$$

where: $\underline{x}(t)$: state vector of the system
$y(t)$: measured output of the system
$u(t)$: deterministic input
$A(t), b(t), c(t)$: matrices of proper dimension ; the index t indicates the time-dependent nature of the system

Equation (2.2) is called the system equation while eq. (2.3) is called the measurement equation. One can easily transform the state-space representation into a transfer function representation using Laplace transforms (or in the case of discrete systems using Z transforms).

The parameters in the transfer function can be estimated on-line using the following recursive algorithm (YOUNG, 1984) :

$$\underline{a}_k = \underline{a}_{k-1} - \underline{k}_k (\underline{z}_k^T \underline{a}_{k-1} - y_k) \tag{2.4}$$

$$\underline{k}_k = P_{k-1}\underline{x}_k (\delta_k + \underline{z}_k^T P_{k-1}\underline{x}_k)^{-1} \tag{2.5}$$

$$P_k = \frac{1}{\delta_k}[P_{k-1} - P_{k-1}\underline{x}_k (\delta_k + \underline{z}_k^T P_{k-1}\underline{x}_k)^{-1} \underline{z}_k^T P_{k-1}] \tag{2.6}$$

where :

$$a^T = [\, a_1, \ldots, a_{na}, b_0, b_1, \ldots, b_{nb}\,]$$

$$x_k^T = [\, -x_{k-1}, \ldots, -x_{k-na}, u_{k-d}, \ldots, u_{k-d-nb}\,]$$

: estimated instrumental variable vector

$$z_k^T = [\, -y_{k-1}, \ldots, -y_{k-na}, u_{k-d}, \ldots, u_{k-d-nb}\,]$$

P_k : $(na+nb+1) \times (na+nb+1)$ square symmetrical matrix

δ_k : forgetting factor ; $0 < \delta_k < 1$

    This on-line parameter estimation technique is used to update the rainfall-runoff model in real-time. We refer to DE TROCH, TROCH and VAN ERDEGHEM (1988) for a comparison in performance between time-invariant and time-variant modelling techniques.

3. EXTENDING THE FORECASTING HORIZON

    The hydrological models used so far are able to calculate future discharge at the outlet of a particular subcatchment up to d hours ahead with high accuracy (TROCH, SPRIET and DE TROCH, 1988). Due to the specific geographic characteristics of the catchment of the river Meuse it is of great importance to extend this forecasting horizon. Some important tributaries (namely Vesdre, Amblève and Ourthe) are situated near the city Liège. The dynamic behaviour of these subcatchments is such that heavy rainfall results in important increase of river flow in only a few hours. Therefore it is essential to concentrate on other hydrologic measuring and modelling techniques for the subcatchments in order to extend the forecasting time. The problem is illustrated in figure 2.

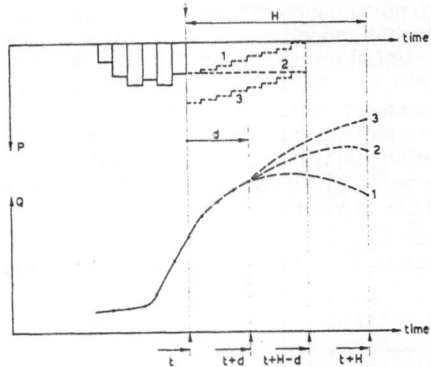

Fig. 2 : Extending the forecasting horizon using rainfall scenarios.

    Given the transfer function noise model (eq. 2.1) with dead time d, d-step-ahead forecasts of river flow Q can be based on measured precipitation P up to time t (time "now"). If the desired forecasting horizon is H, than future precipitation up to time t+H-d has to be known. In figure 2, calculated discharges up to time t+H given three different rainfall scenarios are indicated. The choise between these three (or more) rainfall scenarios can be based on on-line and off-line information.

Using historical storm events, statistical analysis of these storms can result in a real-time prediction model for rainfall distribution. This technique is used by CROLEY, ELI and CRYER (1978) and by CREUTIN and OBLED (1980).

Based on historical rainfall events the distribution functions of a few simple and independent variables, such as inter-arrival time (NSS), duration of a given storm segment (DURSS), etc... can be adjusted. During real-time operation of the flood forecasting system, it is tried to anticipate what is likely to occur during the coming hours, taking into account what has happened up to the current hour, leading to the use of a stochastic rainfall model in a conditional way.

Other rainfall generation models as described by BRAS and RODRIGUEZ-ITURBE (1976), MARIEN and VAN DE WIELE (1986), JAKUBOWSKI (1988) are designed for simulation purposes and are difficult to transform into a real-time operational model.

Recently, a lot of attention is paid to weather radar in combination with flood forecasting. CLUCKIE and OWENS (1987) investigate the performance of transfer function models using quantitative rainfall forecasts (up to 6 h ahead) generated by the FRONTIERS system (Forecasting Rain Optimized using New Techniques of Ineractively Enhanced Radar and Satellite). They conclude that the FRONTIERS data generally provide a helpful forecast and could be useful to the hydrologist, but, that further research is necessary. KLATT and SCHULTZ (1983) introduce a stochastic approach for forecasting rainfall based on weather radar information. With the aid of the probabilistic model, future rainfall is estimated according to its depth and duration and the chosen probabilility of non-exceedance. In order to serve as input into a rainfall-runoff model, also the time distribution of rainfall intensity within the duration is specified.

It is clear that weather radar offers the hydrologist tools to support future research in the domain of quantitative rainfall forecasting and that the hydrological community has accepted the challenge.

4. IMPORTANCE OF PRECIPITATION DATA

As mentioned in section 2, the hydrologic part of the flood forecasting model consists of rainfall-runoff models. The rainfall-runoff relationship is described by a transfer function, with precipitation data as the input variable. Before focussing on the effect of short-term rainfall modelling on flood forecasts using TFN-models, we want to check if real-time rainfall measurements in only a few stations can really improve flood forecasting. A check on the benefit of rainfall data in hydrological forecasting models can be based on a comparison between the forecasts of TFN-models and more simple autoregressive - moving average (ARMA) models. The last one does not take into account precipitation, but is based on measured discharges only. For both models the mean square error of the 4 hour ahead forecast was calculated for a verification set of 4 flood events of the river Vesdre.

Table I: Mean square error of the 4 hour ahead forecast for the river Vesdre (m^6/s^2).

	event number			
	1	2	3	4
ARMA-model	3,5	20,0	6,8	20,1
TFN-model	2,6	14,3	3,2	19,8

The results shown in Table I indicate that the forecasts of the TFN-model are more accurate than those obtained by the ARMA-model. Especially at the beginning of a flood event, the difference between the forecasts made by a TFN-model and an

ARMA-model is quite clear (figure 3). The ARMA-model does not take into account measured rainfall data and thus underestimates future discharges in the rising part of the hydrograph.

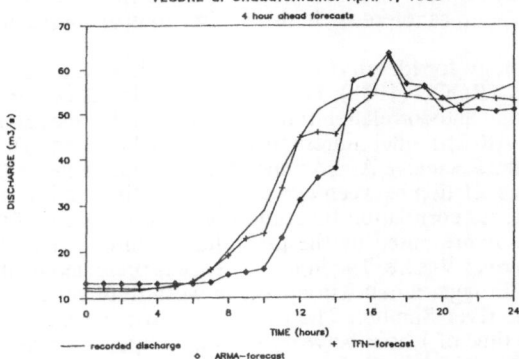

Fig. 3 : Difference between forecasts by a TFN-model and an ARMA-model.

As indicated in table I forecasts of discharge for the river Vesdre are made up to 4 hours ahead. This is equal to the dead time of the rainfall-runoff process. When forecasting up to the dead time, predictions can be completely made based on known rainfall data. If however, we want to generate forecasts up to time t + H with H > d (d: dead time), some kind of quantitative precipitation prediction is required.

5. RAINFALL SCENARIOS

Methods for generating or predicting rainfall are presented in section 3. Especially for real-time applications very simple assumptions of future precipitations or stochastic rainfall models are most suitable. They have to guarantee rainfall predictions only for a short time range.

In this study six rainfall scenarios are selected to investigate their influence on runoff predictions, when forecasting ahead of the dead time:
1. future precipitation = mean precipitation
2. future precipitation = instantaneous precipitation (at instant "now")
3. future precipitation = linear increasing (or decreasing) from the
   instantaneous to the mean precipitation
4. future precipitation is predicted by an autoregressive-moving average
   (ARMA) model for the rainfall time series

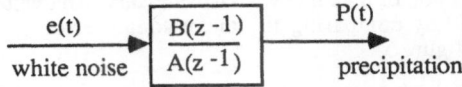

5. future precipitation is predicted by a transfer function noise (TFN) model
   for the rainfall time series. The input and output sequence are precipitation
   time series for different stations.

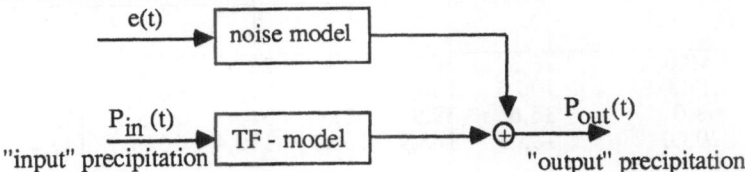

6. future precipitation is perfectly predicted.

Rainfall scenarios 1, 2 and 3 are very simple and crude assumptions of future precipitation. Scenarios 4 and 5 are stochastic rainfall models. The last scenario number 6 is only used as reference in this study and is of course not applicable in real-time.

A methodology for identifying stochastic models was worked out by TROCH, SPRIET and DE TROCH (1988). For the river Vesdre, the estimated autocorrelation function and partial autocorrelation function indicates that the rainfall sequence can be modelled by an AR(1)-model (autoregressive-model of order 1). The parameters are estimated using the Recursive Approximate Maximum Likelihood (RAML) method.

The dynamic relation between two precipitation time-series can be investigated by calculating the cross correlation function. For the case study of the river Vesdre the output sequence is presented by the precipitation time series of station Spa in the catchment of the river Vesdre. The input sequence is presented by the precipitation time series of station Ernage, ninety kilometers in western direction and belonging to the catchment of the river Sambre. The estimated cross correlation shown in figure 4 suggests a delay time of 1 hour between input an output. This can be explained by the movement of the showers in the eastern direction (in most cases). High values for the estimated cross correlation are obtained at lag 1 and 2, while the rest cannot be considered as significant. A TFN-model for the rainfall-rainfall relationship can be identified and represents rainfall scenario 5.

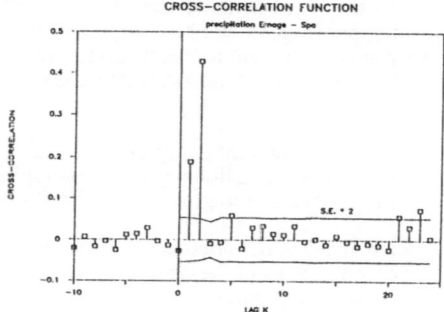

Fig. 4 : Cross correlation function between two precipitation time series.

Each of the six rainfall scenarios is used to make a runoff forecast for the river Vesdre of 12 hours ahead. Since we found that the dead time is equal to 4 hours, a particular rainfall prediction of 8 hours ahead is necessary. An evaluation of the rainfall scenarios can be made by comparing the mean square error of the 12 hour ahead forecasts (table II and figure 5).

Table II: Mean square error of the 12 hour ahead forecast for the river Vesdre (m^6/s^2).

event	ARMA-model	TFN rainfall-runoff model + rainfall scenario number					
		1	2	3	4	5	6
1	46,6	28,3	26,7	24,9	25,5	23,2	25,8
2	134,0	100,3	130,5	100,1	97,1	90,4	69,8
3	68,6	36,6	39,9	31,0	31,4	29,5	17,3
4	203,0	183,5	165,9	166,3	170,4	162,6	147,4

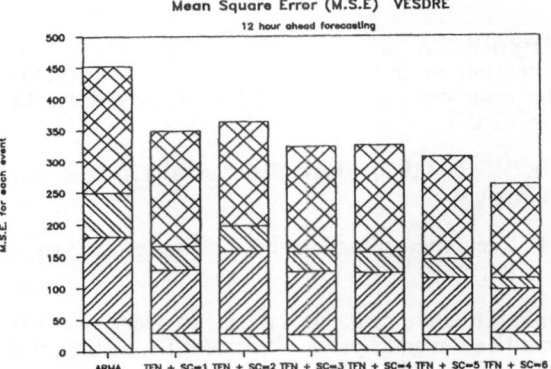

Fig. 5 : Mean square error of the 12 hour ahead forecast for the river Vesdre (m^6/s^2).

Except for the reference scenario 6, rainfall scenario 5 is observed as the most accurate among all five scenarios. The runoff forecasts with rainfall scenarios 3 and 4 are very similar and much better than forecasts with scenarios 1 and 2. The difference in accuracy between forecasts with different rainfall scenarios shows the importance of the knowledge of future precipitation in flood forecasting.

Weather radar may be considered as an important tool to improve rainfall forecasting. One can suppose that the accuracy of rainfall forecasts made by weather radar is at least comparable of those by rainfall scenario 5. The main advantage is that weather radar gives a spatial representation of rainfall patterns. This can lead to changes in the hydrological model building. Instead of the usual lumped models more distributed rainfall-runoff models will be applicated.

6. CONCLUSION

The hydrological part of the forecasting model of the river Meuse in Belgium consists of rainfall-runoff models. Linear transfer function noise (TFN) models to represent the real system of the different subcatchments are chosen. A parameter limiting the forecasting horizon is the dead time. To extend the forecasting horizon procedures have to develop to estimate future precipitation. Research in this direction can be based on weather radar information.

The usefulness of weather radar data in real-time hydrological data was investigated by a simulation study. Several rainfall scenarios, from very simple assumptions to more complicated stochastic rainfall models, were introduced. It was found that rainfall scenario 5, rainfall forecasting by a TFN-model, leads to the most accurate runoff predictions. Supposing that the accuracy of rainfall forecasts made by weather radar is comparable with that of rainfall scenario 5, weather radar may be considered as an important tool to improve the accuracy of flood forecasts.

ACKNOWLEDGEMENTS

We would like to express our sincere gratitude to the following organizations: the Belgian Ministry of Public Works for supporting this research and for supplying flow data and the Royal Meteorological Institute of Belgium for supplying historical precipitation data.

REFERENCES

TROCH, P.A. ; SPRIET, J.A. and DE TROCH, F.P. (1988). A methodology for real-time flood forecasting using stochastic rainfall-runoff modelling. In: Computer methods and Water resources : Computational Hydrology (Eds : D. Ouazar et al.), Rabat, 1988, pp. 243-255.

BOX, G. and JENKINS, G. (1970). Time Series Analysis, Forecasting and Control. Holden - Day, San Francisco.

YOUNG, P. (1984). Recursive estimation and Time series analysis. An introduction. Springer-Verlag, Berlin, 300 p.

SPRIET, J.A. (1985). Structure characterisation - an overview, in H. Barker and P. Young (Eds.), Identification and System Parameter Estimation, 7 th IFAC Symp., Pergamon Press, Oxford, pp. 749 - 756.

DE TROCH, F.P. ; TROCH, P.A. and VAN ERDEGHEM, D. (1988). Operational Flood Forecasting on the river Meuse, Belgium. NATO Advanced Study Institute on Recent Advances in the Modelling of Hydrologic Systems, Sintra, Portugal.

CROLEY, T.E. ; ELI, R.N. and CRYER, J.D. (1978). Ralston Creek hourly precipitation model, Wat. Resour. Res., n° 3, 485 - 490.

CREUTIN, J.D. and OBLED, Ch. (1980). Modelling spatial and temporal characteristics of rainfall as input to a flood forecasting model. Proc. of IAHS symposium on Hydrological Forecasting, Oxford.

BRAS, R.L. and RODRIGUEZ-ITURBE, I. (1976). Rainfall generation : a non-stationary time-varying multidimensional model. Wat. Resour. Res., vol. 12, n° 3, pp. 450 - 456.

JAKUBOWSKI, W. (1988). A daily rainfall occurence process. Stochastic Hydrol. Hydraul., n° 2 , pp. 1 - 16.

CLUCKIE, I.D. and OWENS, M.D. (1987). Real-time rainfall-runoff models and use of weather radar information. In : Weather radar and flood forecasting (Eds : Collinge, V.K. and Kirby, C), Chapter 12, Wiley & Sons, Chichester.

KLATT, P. and SCHULTZ, G.A. (1983). Flood forecasting on the basis of radar rainfall measurement and rainfall forecasting. Proc. of IAHS symposium on Remote sensing and remote data transmission, Hamburg.

A RADAR REFLECTIVITY-RUNOFF MODEL FOR USE IN FLOOD WARNING

L. R. TROVATI[1] and A. MATTOS[2]

[1]Faculdade de Engenharia de Ilha Solteira - UNESP.
15378 , Ilha Solteira - S. Paulo - Brasil.
[2]Escola de Engenharia de S. Carlos - USP.
13560 , S. Carlos - S. Paulo - Brasil.

Summary

A flood forecast linear model with a warning objective is proposed. This model connects directly the radar reflectivity data and hydrological variable runoff. The catchment is discretized in pixels (4 Km x 4 Km) with the same resolution of the CAPPI. Carefull discretization is made so that every grid catchment pixel corresponds precisely to CAPPI grid cell. The basin is assumed a linear system and also time invariant.
The forecast technique takes advantage of spatial and temporal resolutions obtained by the radar. The method uses only the measurements of the factor reflectivity distribution observed over the catchment area without using the reflectivity - rainfall rate transformation by the conventional Z - R relationships. The reflectivity values in each catchment pixel are translated to a gauging station by using a transfer function. This transfer function represents the travel time of the superficial water flowing through pixels in the drainage direction ending at the gauging station. The parameters used to compute the transfer function are concentration time and the physiographic catchment characteristics.
The reflectivity measurements in every pixel are stored in an arrival queue form taking into account their translation in time. The queue is updated with every new CAPPI by advancing positions of the reflectivity values. Reflectivity values stored in the queue may be examined at any moment to create a reflectivity histogram with the arrival time in the abscissa. Only intense rain events were considered for which this linear model may apply.
The Z-Q model calibration is made by the correlation of branch of the hydrograph (up to the peak) because the main purpose is flood warning. The model was tested in the catchment of the Jacaré-Guaçu River (São Paulo State - Brazil). The results showed that the Z-Q model shows a potential for operational forecast system in the catchments that have short lead time.

1. INTRODUCTION

Although radar systems have a great potential to flood forecasts, they are subject to the natural limitations of the remote sensing technique. As recently, Schultz (1) has pointed out that the main problem is the fact that the remote sensors never measure directly hydrological data. They measure only electromagnetic signals in certain spectral bands emitted or reflected by the analyzed environment. Usually, the majordifficulty is to convert the electromagnetical signals into hydrological informations.

In the case of the remote sensing of precipitation field by weather radars the conversion from the radar signal (Z) to rainfall rate in gauge (R) is, on majority of the operational systems, empirically obtained by a relation of the type $Z=aR^b$. However, the parameters a and b exhibit a large variability, Battan (2), and various error sources may appear when this relations are produced. The difference of the rain sampling by two instruments interferes on the comparison between their measurements, Wilson (3); Zawadzki (4). In short, the central question is that the conversion of the reflectivity factor Z to rainfall rate R is by no means invariant. A great deal of researches have been made to solve this question, such as: - recalibration of the Z-R relation in real time using telemetering raingauges as reference, N.W.W.R.P. (5); - correction of the Z-R relation with a radial distance, Calheiros (6). According to Reed (7), calibration techniques now available appear to be moderately successful.

Concisely, the present schemes of flood forecasting that use radar data suffer a transformation stage from the radar reflectivity data to rainfall rate (the Z-R relation which causes a degradation in the measurements) and this rainfall data are used as input on the hydrological simulation models.

This paper presents a technique to flood forecasting in real time for a warning purpose. The radar reflectivity factor is utilized directly as an input variable in a model whose structure is based purely in translation and incorporates the effects of the spatial distribution of rain and the storm movement over the catchment.

The main purpose of this paper is to connect directly the radar reflectivity factor Z with the hydrological variable runoff Q, so that the simple values of Z accumulated on time become sufficient to flood forecasting.

2. BASIN AND RADAR CHARACTERISTICS

The Jacaré-Guaçu basin, tributary of the Tietê River, is situated in the center of São Paulo State, Figure 1. The basin is geologically constituted by sandy soils, covered mostly by pasture and annual crops.

To test the model, two catchment areas defined by gauging stations were used: Ponte Preta and São José prefix RD6 and RD7 respectively, Figure 2.

A C-band radar located on Bauru city is at an average distance of around 100 Km from the catchments. The radar is equipped with a processing system to generate CAPPIs in a radius of 157,5 Km and in an average height of 3,5 Km. The spatial resolution of CAPPI data are pixels of 4 Km × 4 Km. Time resolution for one antenna cycle is of about 5 minutes.

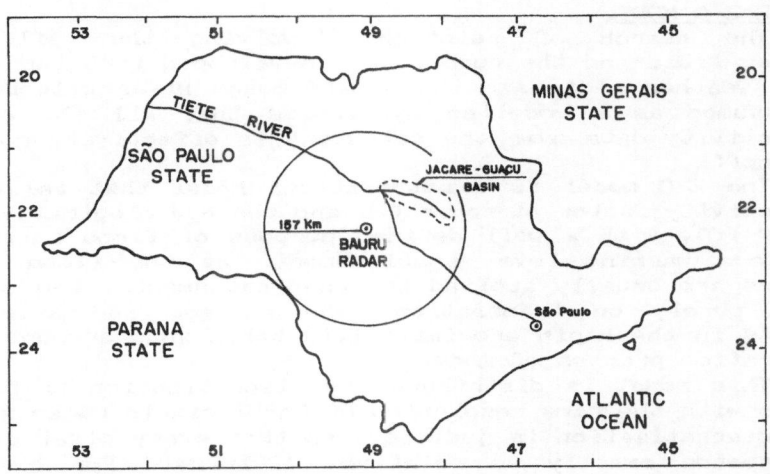

Figure 1. The Jacaré-Guaçu basin and the Bauru radar
localization in São Paulo State, Brazil.

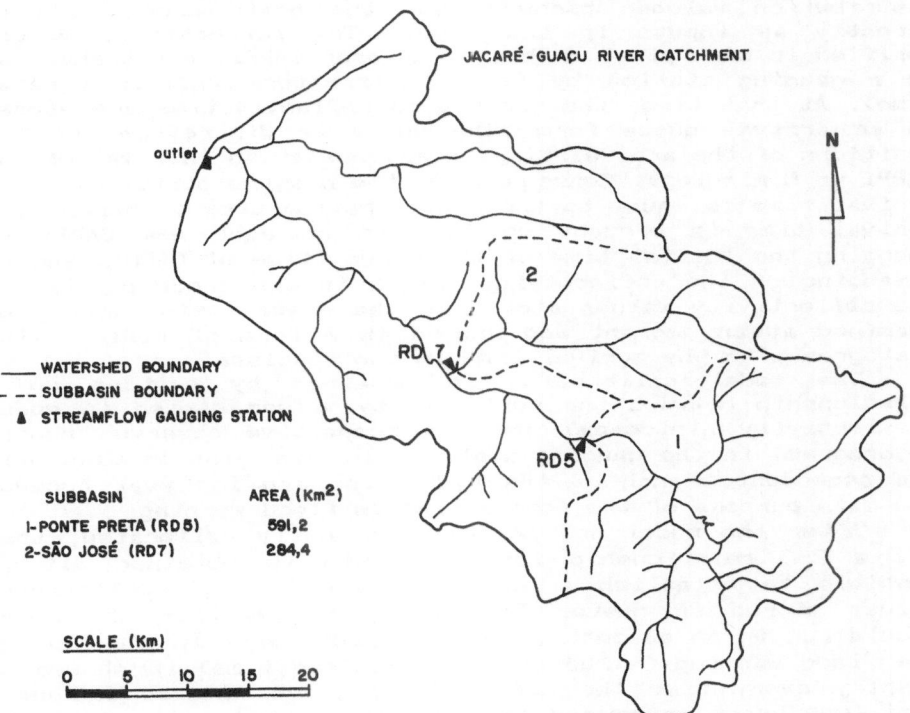

Figure 2. The Jacaré-Guaçu River catchment, subdivision
of two subbasins used on the model test.

3. THE Z-Q MODEL

The search of relations involving the reflectivity weather field and the runoff were developped from the data of Bauru weather radar and the Jacaré-Guaçu River catchment. It is assumed as a modeling hypothesis that all the observed reflectivity data over the catchment is effectively converted to runoff.

The Z-Q model is a mathematical model that relates the reflectivity factor of radar (Z) and the hydrological variable runoff (Q), with a well defined purpose of flood forecasting to flood warning level stablishment. As the flood warning systems are usually applied to urban catchments, the model is based purely on translation. The storage and attenuation effects in the basin are implicitly taken into account in the calibration process of model.

This model is distributed by discretization of basin in pixels with the same resolution of CAPPI pixels (4 Km x 4 Km). This discretization is judicious so that every pixel of basin corresponds exactly to relative CAPPI pixel. For every one pixel of basin the travel time until gauging station or outlet is computed.

Forecasting methods take advantage of spatial and temporal resolutions obtained by the radar. The reflectivity distribution values observed over the basin area are used directly as inputs in the model. The reflectivity values verified in each pixel of basin, at each CAPPI, are translated to a gauging station by using a transfer function (travel time). At that time, the translated reflectivities are stored in an arrival queue form. The queue is discretized on "n" positions of the arrival time, every one at an interval of one CAPPI (= 5 minutes). Consequently, the n-esime position of the arrival time is equal to the concentration time of basin. The arrival time is brought up to date on each new CAPPI by changing the initial time of the queue (time of CAPPI) and by advancing of the reflectivity values in the queue positions. The reflectivity values stored in the queue positions may be examined at any moment and showed in a form of reflectivity histogram with the arrival time in the abscissa.

The model calibration is obtained by setting up a relationship between the reflectivity values stored in queue (reflectivity histogram) and the respective observed runoff (hydrogram) in the gauging station. The correlation uses only the ascendent branch of the hydrograph and its peak because the main purpose of the forecasting is flood warning.

After the model has been statistically calibrated, that is, a Z-Q relationship for the basin is obtained, it is possible to establish a warning level to the reflectivity values stored in queue. In real time operation the model should run in an automatic and continuous way. Information of the flood warning (value of peak and lead time) is showed in display every time the reflectivity values stored in queue positions reach or exceed the relative amount settled to the warning.

4. THE TRAVEL TIME

The runoff hydrograph due to a given period of rainfall reflects all the combined physical characteristics of the basin. If a mean runoff hydrograph is obtained from some flood events, the attenuation and storage effects are taken into account implicitly in the time basin response. The assumption of linearity and time invariance is regarded to the catchment area.

The transference time of the superficial water between consecutive pixels of the basin on drainage direction is determined by declivity measurements on pixel vertices. To get declivity measurements we use a topographic map, which is overlaid on CAPPI grid pixels of the basin. Trovati (8), developed an algorithm to overlay and to calculate the drainage direction. This algorithm uses a digitalizer table to overlay the basin map over CAPPIs and uses the pixel vertices altitude to determine the drainage direction.

The transference time TF_{ij} between any pixel "i" and the subsequent "j" is computed by equation (1):

$$TF_{ij} = K \, d_{ij} / S_{ij} \qquad (1)$$

where K is the proportionality coefficient, d_{ij} is the distance between i and j catchment pixels and S_{ij} is the declivity of pixel i to pixel j.

The determination of constant K is based on time of concentration T_c of catchment, which is a calibration parameter. The time of concentration may be determined from a mean real hydrogram of some severe flood events observed in the catchment. In this context, similarity between time of concentration and time of peak flow is assumed.

The sequential addition on drainage direction order of the TF_{ij} of every pixel to the outlet of the catchment, results on a maximum travel time that can be equalized to the time of concentration by:

$$Max \sum TF_{ij} = K \, Max \sum (d_{ij} / S_{ij}) = T_c \qquad (2)$$

and with the rearrangement of equation (2), we obtain:

$$K = T_c / Max \sum (d_{ij} / S_{ij}) \qquad (3)$$

that when substituted in equation (1) permits to find the transference time of superficial water from pixel to pixel until the gauging station.

5. TRANSLATION OF REFLECTIVITY

The reflectivity translation is determinated from transference time of the superficial water of the pixels, whose summation in drainage direction results on travel time. Reflectivity values, Z (DBZ) = $[10 \log Z(mm^3 . m^{-3})]$, representative of the levels quantified on CAPPIs maps of those pixels affected by rainfall are translated to basin outlet. The time step of this translation is equal to about 5 minutes that is the average time between CAPPIs.

All inputs of the reflectivity observed over the basin are converted in runoff. With this assumption taken into consideration the total reflectivity translation observed by radar over the basin in each step of time is expressed by the following equation:

$$Z_j^\tau = \sum_{i=1}^{n} Z_i^{\tau - \Sigma TF_i} \qquad (4)$$

where,

Z_j^τ = the reflectivity stored on outlet of the basin "j" in time "τ";

$Z_i^{\tau - \Sigma TF_i}$ = the reflectivity of pixel "i" observed in moment "$\tau - \Sigma TF_i$" with a time delay "TFi" to come at the outlet of basin;

n = total number of pixels on basin.

The reflectivity values translated to the outlet of basin by the pixels travel time, are stored in an arrival queue form. This queue is constructed as a function of the arrival time of the reflectivity in the gauging station in order to create a reflectivity histogram. In the abscissa the arrival time at the subbasin "j", represents the CAPPI time increased of the travel time of each pixel until the basin outlet. This is equivalent to suppose that a spatial and temporal distribution of the reflectivities over all pixels was projected for the basin outlet.

6. CALIBRATION OF THE MODEL

Taking the historical series of the observed hydrograph at the gauging station of the subbasins 1 and 2, we may determine the time of concentration (Tc) for the two subbasins. Only the more significant floods were considered for the determination of that times. The Tc found for the subbasins were 15±3 hours and 11±4 hours, respectively. Using these values in equation (3), a K coefficient for each subbasin was computed and the travel time of pixels was obtained.

Figure 3 shows the travel times of each pixel until the outlet of each subbasin.

Six events of intense rain occurred in the period between January and July 1983 were utilized in the calibration of the model. To these events the hydrographs were prepared in order

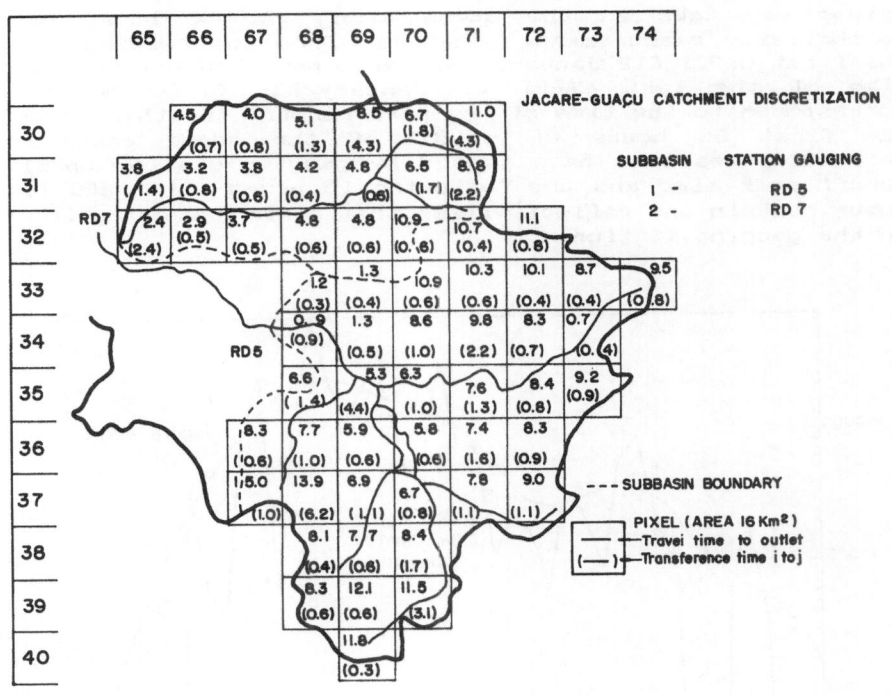

	65	66	67	68	69	70	71	72	73	74

JACARE-GUAÇU CATCHMENT DISCRETIZATION

	SUBBASIN	STATION GAUGING
1 -		RD 5
2 -		RD 7

30 | 4.5 (0.7) | 4.0 (0.6) | 5.1 (1.3) | 8.5 (4.3) | 6.7 (1.8) | 11.0 (4.3)

31 | 3.8 (1.4) | 3.2 (0.6) | 3.8 (0.6) | 4.2 (0.4) | 4.8 (0.6) | 6.5 (1.7) | 8.8 (2.2)

RD7

32 | 2.4 (2.4) | 2.9 (0.5) | 3.7 (0.5) | 4.8 (0.6) | 4.8 (0.6) | 10.9 (0.6) | 10.7 (0.4) | 11.1 (0.8)

33 | 1.2 (0.3) | 1.3 (0.4) | 10.9 (0.6) | 10.3 (0.6) | 10.1 (0.4) | 8.7 (0.4) | 9.5 (0.8)

34 | RD5 | 0.9 (0.9) | 1.3 (0.5) | 8.6 (1.0) | 9.8 (2.2) | 8.3 (0.7) | 0.7 (0.4)

35 | 6.6 (1.4) | 5.3 (4.4) | 6.3 (1.0) | 7.6 (1.3) | 8.4 (0.8) | 9.2 (0.9)

36 | 8.3 (0.6) | 7.7 (1.0) | 5.9 (0.6) | 5.8 (0.6) | 7.4 (1.6) | 8.3 (0.9)

37 | 15.0 (1.0) | 13.9 (6.2) | 6.9 (1.1) | 6.7 (0.8) | 7.8 (1.1) | 9.0 (1.1)

SUBBASIN BOUNDARY

38 | 8.1 (0.4) | 7.7 (0.6) | 8.4 (1.7)

PIXEL (AREA 16 Km²)
— Travel time to outlet
(—) Transference time i to j

39 | 8.3 (0.6) | 12.1 (0.6) | 11.5 (3.1)

40 | 11.8 (0.3)

Figure 3. Discretization of two subbasins with relation to CAPPI grid.

to let the temporal scale start on the instant: $T_p - 2T_c$ (runoff peak time minus twice the subbasin concentration time), and end up at the instant: $T_p + T_c$ (runoff peak time plus the concentration time). The initial time of the hydrographs (the runoff peak time minus twice the concentration time) was taken as a reference to start the arrival queue time, therefore, it corresponds to the reading time of the first CAPPI. In this time interval it is ensured the complete precipitation sampling that contributes for the ascendent branch of the hydrograph, at least. If the precipitation ceases before the time of the runoff peak, the arrival queue contains a complete sampling of the reflectivity values of that precipitation which produced the hydrograph.

Because runoff measurements have hourly registers, the reflectivity values stored in arrival queue positions were integrated in hour steps to make easy the Z-Q comparison.

7. THE Z-Q COMPARISON

A typical example of the Z-Q comparison for subbasin 1 (station RD5) is showed in Figure 4. The superior time scale (1 to 45) shows the 45 hourly positions that the reflectivity values may take in the queue along the 30 hours of the reflectivity measurements. That is, from the reading time of the first CAPPI (12/January/83; 19:00 hours) until the reading time of the last CAPPI (14/January/83; 01:00 hour) what corresponds to the time of the peak runoff. In other words, on the first 30 hours (1 - 30) of the queue contain the reflectivities that have passed by gauging station until the runoff peak time and the following 15 hours (31 - 45) of the queue contain the reflectivities that again are going to pass in the gauging station.

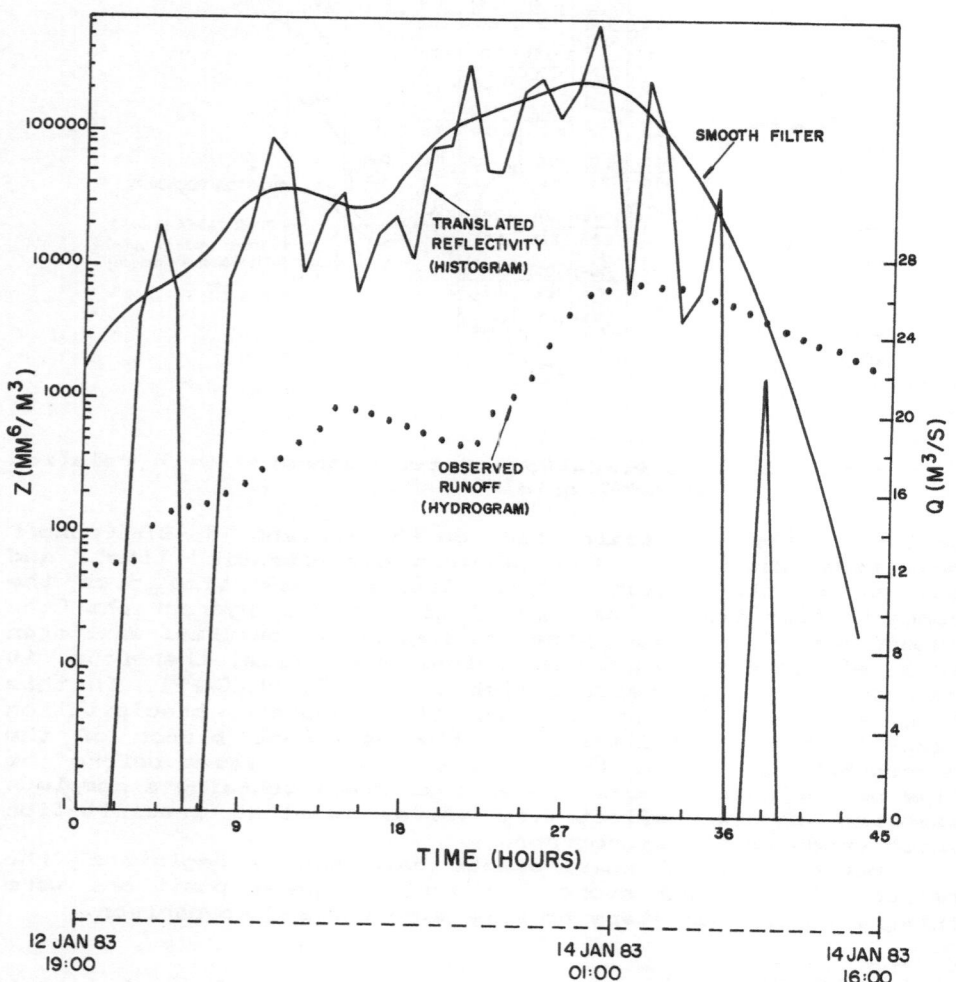

Figure 4. Example of the Z-Q comparison for the subbasin 1, station RD6.

Figure 5 shows an example of the Z-Q comparison for subbasin 2 on gauging station RD7. To this subbasin the time of concentration found was 11 hours therefore, the sampling reflectivities interval was 22 hours before the time of each runoff peak. This case presents the composition of the two consecutive queues, that is, it refers to two consecutive flood events with an intermittence of 8 hours in reflectivity values between them.

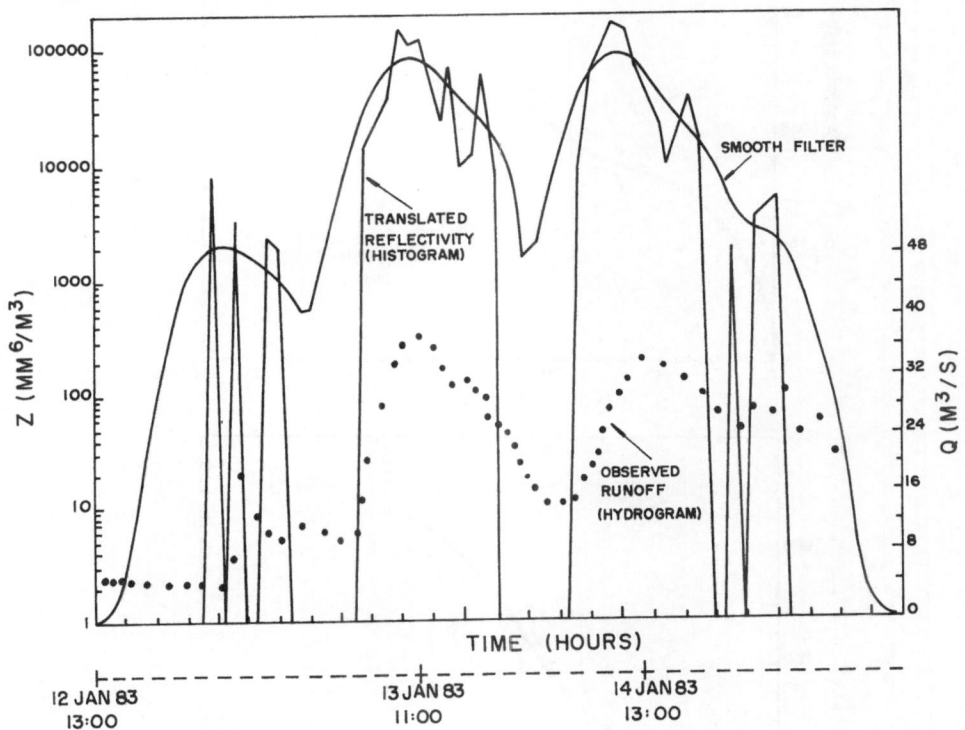

Figure 5. Example of the Z-Q comparison for the subbasin 2, station RD7.

8. THE Z-Q RELATIONSHIP

A preliminary analysis shows that the electromagnetic signal of radar (Z) translated according to the model is strongly correlated with the observed runoff (Q). Polynomial fitting between these two variables exhibited a determination coefficient around 0.90 for the six events independently tested.

Figure 6 shows the mean fit between Z and Q for the tested events in the subbasins 1 and 2, respectively. Note that for the Q variable the values of the accumulated runoff were used, expressing in this way the flow volume summation.

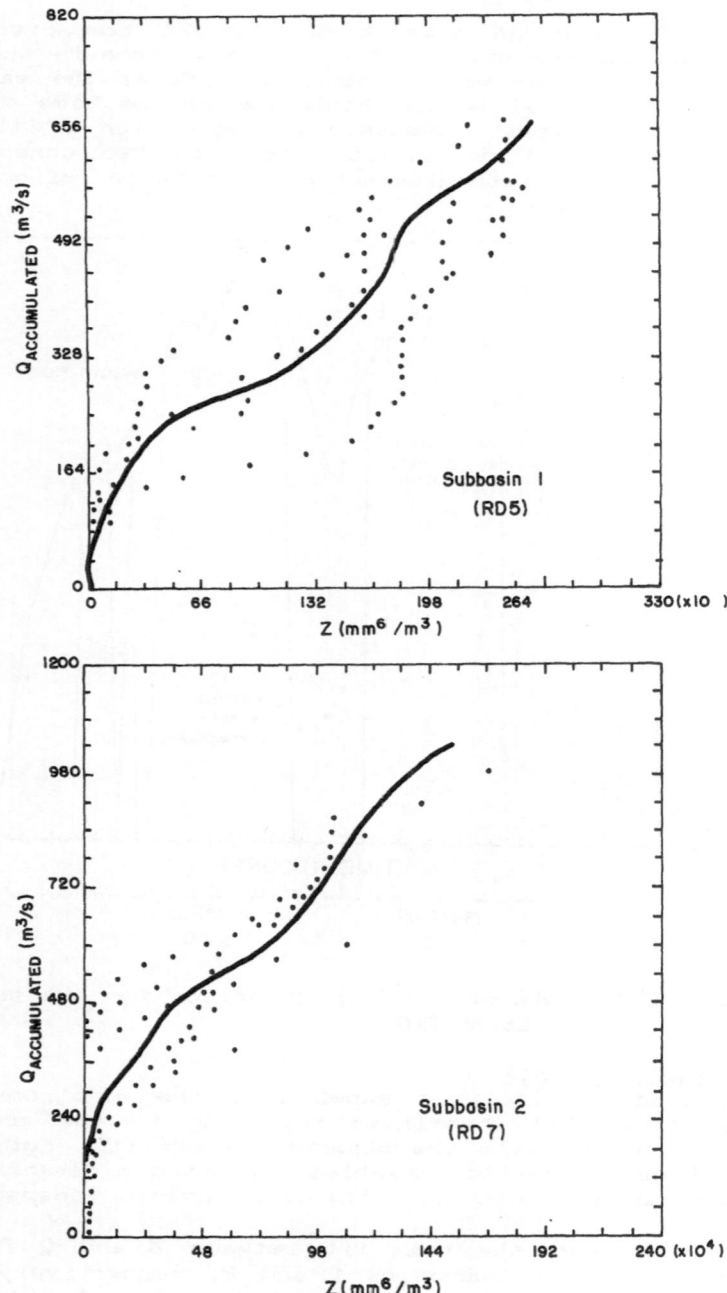

Figure 6. The average agreement of the Z-Q relation for the tested events.

9. CONCLUSIONS

The advantage of the described model is to suppress the radar rain gauge calibration step in view of its high variability and also due to the intrinsic difference between these two measurement systems. It appears more coherent to assume the similarity between the sampling of rainfall field with radar and the precipited volume over basin than over a raingauges network. Another advantage of the Z-Q model is its simplicity and the fact that the radar informations are obtained on real time and concentrated in a single local. Furthermore, after the model has been calibrated (the establishment of Z-Q relationship) a continuous management and a lead time optimization of the forecasting are possible. Results obtained from comparisons between reflectivity values of radar and observed runoff in the subbasins showed that the presented model may produce efficient responses to flood forecasting with warning purposes.

Although the model has been tested in rural catchments, which are subject to larger losses than in urban drainage systems, the model has shown good responses. It is observed the abstraction necessity of some reflectivity peaks which have isolated occurrence in the histogram origin and in general don't contribute effectively for the instantaneous runoff.

The model performance is conditioned to the determination of some parameters. The time of concentration is the parameter that needs carefull evaluation. With regard to the smooth filter used in the reflectivity histograms a more efficient filter than the moving average must be considered, because the reflectivity data set stored in the queue is not always continuous, sometimes it is discrete.

The Z-Q relations found for the two subbasins are preliminary since they are results of a simplified adjustment and presumably a no robust test series. Therefore, at present it is not possible to present a statistical analysis of the accuracy of the results.

At this stage we are considering the operationalization of the model in real time and its application in urban catchments to flood warning. Another perspective to this model is its use for flood reservoirs management at small hydroeletrics.

ACKNOWLEDGEMENTS

The authors wish to thank Dr. R.V. Calheiros of the Meteorological Research Institute - UNESP of Bauru by radar data and the School of Engineering of São Carlos - USP by basin data. We extend our thanks to Mr. P. Tepedino and Mrs. C. Teixeira by their support in data processing. Thanks also to FAPESP and CAPES for partial financial support for this research.

REFERENCES

(1) SCHULTZ, G.A. (1988) - Remote sensing in hydrology. Journal of Hydrology, 100; 239-265.

(2) BATTAN, L.J. (1973) - Radar observation of the atmosphere. University of Chicago Press, 324 p.

(3) WILSON, J.M. (1976) - Radar-rain gauge precipitation measurements: a summary. Proceedings, NATIONAL CONFERENCE ON HYDROMETEOROLOGY, 1st, Forth Worth, Texas, 72-75.

(4) ZAWADZKI, I.I. (1984) - Factor affecting the precision of radar measurements of rain. Proceedings, CONFERENCE ON RADAR METEOROLOGY, 22nd, American Meteorological Society, 251-256.

(5) N.W.W.R.P. (1985) - North West Water Radar Project; Meteorological Office; Water Research Center Department of Environment Ministry of Agriculture, Fisheries and Food. Report of the steering group. Consortium Report, September.

(6) CALHEIROS, R.V. (1982) - Resolução espacial das estimativas de precipitação com radar hidrometeorológico. (Tese de Doutorado - Escola de Engenharia de São Carlos / USP). 229 p., dezembro.

(7) REED, D.W. (1984) - A review of british flood forecasting practice. INSTITUTE OF HYDROLOGY, Report, nr. 90.

(8) TROVATI, L.R. (1988) - Modelo de comparação entre o fator de refletividade do radar e vazões para estabelecer alertas de cheias. (Tese de Doutorado - Escola de Engenharia de São Carlos/ USP), 150 p., julho.

THE IMPACT OF WEATHER RADAR ON ASPECTS OF OPERATIONAL MANAGEMENT

IN THE THAMES REGION

P.F. BORROWS and C.M. HAGGETT
National Rivers Authority, Thames Region

Summary

The operational use of Weather Radar for flood warning in the Thames Region has emphasised the need for reliable information that can be made available in an appropriate and useful format. Software has been developed to provide quick and simple access to radar data and to manipulate that data for the convenience and information of the operator. Recent development work has produced a technique for improving the calibration of the radar against local raingauges. The increased reliability of the radar data has implications for other weather dependant management activities and these are described.

1. PRELUDE

Sound operational and management decisions depend upon good quality information. Not only does the information need to be accurate, representative and timely, it needs also to be presented in a manner which allows easy, clear and consistent interpretation. These criteria apply equally to the functions associated with river catchment management. They apply also to the use of weather radar which is contributing increasingly to better informed decision making in a range of weather related disciplines.

During the past few years, development in the Thames Region of techniques to improve precision and reliability, coupled with a suite of computer driven programs for data display have generated increasing confidence in the operational use of weather radar. That is leading to an extension of applications in the fields of flood warning and flood forecasting, river control, water resources and pollution control.

2. BACKGROUND

Weather radar was first used by Thames Water in 1984 in the form of an experimental networked picture based on four sites in the United Kingdom. Since then, the network of radars has expanded across the British Isles and now comprises ten units including that at Chenies in Buckinghamshire to the North West of London. The London Weather Radar installation at Chenies became operational in 1984 and provides detailed local coverage over a large part of the Thames Water's catchment including the heavily urbanised area around London.

The Thames catchment extends over approximately 12000 sq.km from the Thames estuary in the East to the Cotswold Hills in the West. Thames Water presently provides the entire range of water related services to the 11.8 million customers in its area, including water supply, sewerage, sewage disposal and land drainage related activities. The intense development throughout the catchment, including residential, industrial and infrastructure development, imposes many operational problems on

Thames Water. One of these is the management of the various demands on the River Thames which serves as a major source for drinking water abstraction, as a receiving water for the discharge of treated sewage effluent, as a navigable waterway with a series of weirs and locks and as a major amenity and recreational asset for the populace.

Much development has taken place alongside the Thames and its tributaries, to the extent that the risk of flooding from non-tidal rivers has increased. A survey in London showed that over 9,500 properties were at risk from a flood with a 50 year return period and that in such an event, the costs of damage to residential properties alone could exceed £17M (1). Elsewhere in the Thames catchment, urbanisation since the major flood of 1947 has put many more residential and industrial properties at risk, over and above the thousands that were affected at that time. Furthermore, the social costs of disruption associated with flooding are only now beginning to quantified (2).

It is against this background that Thames Water has developed a catchment monitoring system that includes telemetered raingauges, flow gauges and river level gauges as well as weather radar to provide forewarning of flood events. Differing flooding problems are encountered in the catchment. To the West of London, the area is more rural, albeit with larger towns like Swindon, Oxford and Reading situated on or close to the River Thames. The Eastern area is fundamentally urban and includes London, with smaller but fast responding rivers.

The non-tidal flood warning systems which operated in the Thames Water area were largely based on those inherited from predecessor organisations and reflected the varying circumstances in the region. In 1987, rationalisation took place so that two systems now cover the Thames Region.

(i)   The Western area based at Reading includes the non-tidal Thames and its tributaries upstream of Teddington (the former Thames Conservancy area).

(ii)   The Eastern area based at Waltham Cross which includes all non-tidal tributaries in the Lee catchment and the London area.

Weather radar is also used at the Thames Barrier which monitors tidal conditions in the Thames estuary and controls the tidal flood defences of London.

In the Thames area, significant advances have been made both in terms of the integrity of radar derived rainfall information and in the presentation of the information. The benefits will extend beyond the original flood warning application.

3. WEATHER RADAR AND THE TELEMETRY SYSTEM

Weather radar information forms an integral part of the Thames environmental data aquisition system, based on control centres at Reading to the West of London, at the Thames Barrier, and Waltham Cross to the North of London.

At these three centres, radar data are received from the Chenies Site, north-west of London (Fig 1).

In addition, data from the composited national network are received at Reading from the Meteorological Office at Bracknell and displayed by a Software Science system using a Compaq micro-computer. The system includes a rolling archive of 5km picture data and the facility to store particular events, frame by frame, on floppy disk.

The single site high resolution radar data from Chenies are received by a Ferranti Argus 700G computer at 208 intensity levels, and stored on a rolling archive. The Chenies radar itself is calibrated using data from five telemetered raingauges, transmitted to it in real time from the Argus computer. This arrangement has been under review and will be dealt with later in this paper.

The weather radar data are presently used subjectively for operational flood warning, providing an indication only of developing weather conditions. Real time data from 31 telemetered raingauges and 40 river level recorders located throughout the Thames Valley area are used to complement the radar data. A programme to extend the network of river level gauges is in hand and this, coupled with planned improvements to weather radar data handling software, will provide a significantly improved monitoring ability in the early 1990's.

Coupled with this, flood forecasting models are being developed using raingauge, radar and river level gauge data. Until the existing Argus computer is replaced, this work is undertaken on an off-line Hewlett Packard micro-computer.

At the Thames Barrier, a VAX 11/750 mini computer receives radar data from Chenies via a PSTN private wire. The Control staff at the Barrier display the 5km Chenies picture which provides a view of prevailing weather conditions, supplementing river level data and synoptic information. This serves as a useful adjunct to the principal activity of monitoring tidal levels in Thames estuary.

The VAX computer at the Barrier is also connected to the Waltham Cross flood room which has continuous remote terminal access. Data from 30 raingauges and 50 river level gauges in the London and Lee area can be accessed. Under normal conditions, outside office hours, duty staff monitor the weather radar from home using a PSTN dial-up modem link. Both in the flood room and at home there is full access to the highly developed software which has been produced for the Eastern area covering London and the Lee Valley. This software displays concurrently both real-time and archived radar and outstation data and drives real-time flood forecasting models. The sophistication of the software allows a swift and reliable response to a developing flood by giving operational staff both a better overview of the situation and local detail.

Data display software is run on Tektronix 410X and 420X series terminals and incorporates the Tektronix Plot 10 International Graphics Library. Simple menu-driven commands assist the operator to produce single, multiple or 'movie' images either for 5km or 2km grid data, (fig 2). Subcatchment displays utilise graphical or tabular format and 2km grid and gauged estimates of rainfall can be compared visually. All displays can be routed to a colour plotter or line printer.

4. DATA TREATMENT AND PRESENTATION ON THE VAX SYSTEM

Radar data and gauge data, are captured and processed automatically. Displays are available immediately using simple, menu driven commands. Software is divided into integrated but independent units to provide system security in the event of failure of one unit. Error handling is accomplished by creating log files for each data transmission which are retained for examination and action in the event of error, but otherwise deleted. For security reasons, the software operates on two disks and can be categorised under the headings: capture, validation and sorting, processing, display and archiving.

(i)   **Capture** - RDCAPT is a permanently running detached process which monitors the Chenies link and captures all transmitted data.

(ii)   **Validation and Sorting** - RDSORT operates automatically after each transmission is completed. It validates and processes the various data types; 5km grid, 2km grid, subcatchment and calibration data. Further processing of the validated information proceeds either through other batch jobs or other programs.

(iii)   **Processing** - The batch jobs RDGRID5, RDGRID2 and LVGRID2 invoked by RDSORT are responsible for producing the 5km and 2km data displays, running at 5 and 15 minute intervals respectively. The 2km data can be displayed in a variety of ways ranging from the entire area covered by the 75km radius from the Chenies site, to individual subcatchments in the Eastern area. This latter facility is particularly valuable for tracking convective storms which cause most of the flooding problems in the urbanised area in and around London.

RDGAUGE is another batch job which relates the 2km grid square corresponding to the site of the 16 raingauges that are located in the principal subcatchments in London. Rainfall totals are updated at each 5 minute transmission and compared with the raingauge total for the same time period.

A similar program, LEESUBC enables real time comparisons to be made between rainfall estimates derived from telemetered raingauges in the Lee Valley area and the validated 2km data for the equivalent grid square. It also converts the intensities derived for the relevant 2km data into summations for 20 subcatchments in the Lee area. This progam contrasts with two others, RDSUBC and LVSUBC which produce data generated at the radar site to provide rainfall estimates for the 90 London and 100 other Thames region subcatchments respectively, for 15 minute, hourly and daily periods.

Five calibration raingauges are interrogated every 15 minutes by the Argus computer and the data transmitted to Chenies. At the site, a PDP 11/34 mini computer compares the ground based measurements with the radar estimate and adjusts the transmitted data in accordance with a predetermined algorithm. This process has proved to be a source of error. The factors used to calibrate the radar are transmitted to the VAX computer and selected and tabulated by the program RDCALIB.

(iv)   **Display** - the processing programmes pass the data into files for display at the command of the operator. Each of the programs described above is accompanied by a clear display which updates automatically.

(v)   **Archiving** - all the processing programs incorporate an archiving component which is continuous and transfers data into common format files.

5. RADAR CALIBRATION

A shortcoming of the Chenies weather radar is that it is calibrated using only 5 raingauges and that the synoptic type dependent domain procedure used introduces temporal and spatial discontinuties in the rainfall estimates supplied. The result is that in certain circumstances radar data loses accuracy leading to the generation of unreliable flood forecasts. To improve the situation Thames Water commissioned the Institute of Hydrology in 1987 to explore the possibility of developing a regional recalibration system to obtain a more accurate and reliable estimate of spatial rainfall variations using data from the Authority's network of 30 telemetered raingauges in the London and Lee Valley area. The adopted procedure has been implemented on the VAX computer at the Thames Barrier which processes radar and raingauge data in real-time.

The recalibration system described by Moore et al (3), first removes the effect of the domain calibration undertaken at Chenies and then uses rainfall totals over the last 15 minutes from the larger raingauge network to compute calibration factors. A multiquadric surface is then fitted to these factors which in turn is applied to the radar field to obtain the recalibrated product. The system became operational for the Eastern area on the 14th March 1989. Analysis of historical raingauge and radar data for 19 storm events indicated that the product was on average 25% more accurate than the at-site calibrated product. Displays of instantaneous radar estimates using recalibrated data are now available for operational use. In the near future 15 minute subcatchment totals will be generated from these data which may be used in catchment models to generate flood forecasts at points of interest.

6. WEATHER RADAR APPLICATIONS

Flood Warning
As indicated earlier in this paper, the first use of weather radar in the Thames Region was for flood warning. From a qualitative assessment using data in which there was not complete confidence, flood duty staff were able to discern relative rainfall concentrations and obtain information on the extent and movement of rainfall. This allowed some judgement to be exercised over the likelihood and severity of a flood but weather information obtained from the commercial service of the Meteorological Office had previously served almost as well.

The development of weather radar now offers a level of detail and interpretation that allows flood duty staff to build a complete picture of events in real time from a specific 2km grid square, through a particular subcatchment to the entire river basin.

With the advent of software to incorporate the radar data, flood warning has been transformed from something that was virtually a 'black art' into a discipline based on extensive data gathering, sophisticated processing and display and the use of modern mathematical modelling techniques. As a result, flood warnings have more credibility not only with those who receive them but, just as importantly, with those responsible for their issue. Some of the economic benefits may therefore be in the process of being delivered.

The accuracy of data obtained from Chenies now encourages the use of the precipitation information in real-time models. The spatial information represents a considerable advance over the use of a simple raingauge in a subcatchment, with the result that model calibration and model use for flood forecasting give better results, leading to more reliable and credible flood forecasts. The different conditions found in the Thames region favour the use of a variety of modelling techniques dependent on whether the catchment is urban or rural and the prevalence of particular geomorphological or geological features. (4)

A specialised development, under consideration at present, is the use of weather radar to predict surface water flooding as well as river flooding. There is no precedent for a water authority providing such a service, nor are there specific enabling powers. Generally heavy rainfall warnings are available from the Meteorological Office and Thames avails itself of this service which is particularly useful in respect of summer convective storms in the London area. However, the service is not locality specific. The applications software now in use on the VAX and operated from Waltham Cross, will allow individual 2km grid squares to be monitored and alarmed for any given rainfall rate or series of rates corresponding to the intensity and duration of storms known to cause surface water flooding problems. Inevitably, such information could only be transmitted to the civil authorities as the flooding was occurring, but it would enable traffic redirection and the emergency services to be targetted quickly and relevantly rather than wait for congestion to develop and emergency calls to be made. Because the costs of providing such a service are evident, whereas the benefits are intangible, for instance savings in journey times and social disruption, moves into this area will remain limited and experimental for the time being.

River Control

A considerable amount of effort is put into the collection and archiving of information relating to the behaviour of river catchments under a wide range of conditions. Thames Water maintains comprehensive records of river flows over many years, together with records of precipitation. In the Thames region data are available from some 350 gauges. In some cases however, comparison of stream flows, a function of rainfall run-off, are made with data from only a couple of raingauges in a catchment. The chances of either gauges adequately representing conditions over a catchment extending to perhaps 100 sq km are low. This is particularly true of convective storms whose active cells may be only 2km in diameter. Recalibrated radar data offer an opportunity to determine more accurately the relationship between rainfall patterns and the behaviour of a catchment as exemplified by river flows. This information is essential for assessing the effects on a catchment of development or river improvements or other works affecting the river channel. In this way, investment decisions can be optimised and properly focussed.

Water Resources

The same records collected for analysis of stream flows serve as a foundation for the assessment of water resources. Again, existing methods rely totally upon raingauge data which give a partial view only of precipitation patterns. Better water resources planning will be possible using radar data which can be tailored not only to suit particular subcatchments but also aquifers, whose boundaries may not be contiguous with those of catchments. In some catchments in the Thames Region, abstraction demands have adversely affected river flows leading to calls for corrective or remedial action. A better understanding of the relationship between aquifer behaviour and stream flows, should improve the chances of preventing a repetition of these problems in catchments presently unaffected and restoring affected catchments to a more satisfactory state, whilst at the same time allowing abstractions for drinking water, industrial or agricultural purposes.

River Pollution

A specialised application of weather radar data is made in the tidal Thames. During the summer months, natural river flows decline and water temperatures increase making the Tideway river vulnerable to pollution. The condition of the estuary is particularly important for migratory fish, such as salmon, and this is threatened in the summer by storm sewer overflows. These operate more frequently in the winter, but in long hot dry spells during the summer sewage residues become anaerobic and highly polluting. The high intensities of rainfall often associated with summer convective storms frequently results in the capacity of the combined sewers in London being exceeded. These overflow to storm sewers which then discharge their pollution load into the Tideway. To counter the problem, a novel scheme was devised involving the deployment of a barge equipped with the capacity to inject gaseous oxygen into the river to counter the pollution (5). The vessel, known as the 'Thames Bubbler' is expensive to operate and takes some hours to mobilise. It is important, therefore, to position it promptly and accurately to counter the effects of the depletion in dissolved oxygen.

Weather radar allows the conditions likely to cause problems to be identified fairly readily, thereby improving the effectiveness and cost efficiency of the operation which is of critical environmental importance to Thames Water and the subject of considerable publicity.

Sewer Safety

One application of weather radar worthy of note relates to sewer safety. In London, many of the Victorian sewers still serving the capital, are large enough for man entry and sewer gangs are constantly deployed removing accumulated debris or repairing the fabric of the sewer. It is vital that the men receive adequate early warning of the onset of rainfall in order that they may be evacuated before sewage levels rise. Presently, warnings from the London Weather Centre are utilised, but weather radar offers an opportunity for more detailed information leading to improved safety and the more effective use of manpower by deployment to sewers unaffected by a storm or to other maintenance work.

7. THE FUTURE

The Institute of Hydrology has also been asked to investigate whether a short term high resolution rainfall forecasting procedure could be developed using radar data to complement the Met Offices' FRONTIERS system. It is envisaged that 2km, 15 minute resolution forecasts would be of greater value particularly for the issue of accurate and timely flood warnings in small, fast responding urban catchments. To date, two procedures have been developed to identify the speed and direction of storm movement. Both techniques are based on maximising the correlation between two radar data fields for successive time frames through an appropriate displacement in space. One is based on a regional analysis of storm movement and the other on local analysis both of which are described in full by Moore et al (1989).

It is envisaged that the present study will be extended with the aim of producing an operational regional rainfall forecasting system to be implemented on the Barrier VAX computer. An evaluation of FRONTIERS forecasts will also be made in terms of accuracy in an absolute sense, and also relative to the regional procedure. The evaluation might, for example, indicate that the latter is to be preferred at shorter lead times and in situations where a product of higher resolution is required, but the sounder physical basis of FRONTIERS forecasts yields better longer-term predictions.

ACKNOWLEDGEMENTS

This paper was prepared with the agreement of Thames Water, NRA Region. It describes work to which George Merrick, Chris Richards and Joseph Jeyakumar contributed also. The views expressed are those of the authors and are not necessarily representative of Thames Water.

REFERENCES

1. HAGGETT, C.M., (1986). The use of weather radar for flood forecasting in London. Conference of River Engineers 1986, Cranfield 15-17 July, Ministry of Agriculture, Fisheries and Food, 11pp.

2. COLLINGE, V.K., (1989). Investment in weather radar by the UK Waters Industry. NERC Seminar - Weather Radar and the Water Industry, Opportunities for the 1990's. BHS Occasional Paper (in press).

3. MOORE, R.J., WATSON, B.C., JONES, D.A., BLACK, K.B., HAGGETT, C.M., CREES, M.A., RICHARDS, C.I., (1989). Towards an improved system for weather radar calibration and rainfall forecasting using raingauge data from a regional telemetry system. Surface Water Modelling - New Directions for Hydrologic Prediction. IAHS Third Scientific Assembly, Baltimore, USA, 10-19 May 1979, 9pp.

4. HAGGETT, C.M., MERRICK, G.F., RICHARDS, C.I., (1989). Quantitative use of radar for operational flood warning in the Thames area. International Symposium on Hydrological applications of Weather Radar, University of Salford, 14-17 August 1989 (to appear).

5. WOOD, L.B., BORROWS, P.F., and WHITELAND, M.R. (1979). Scheme for remedying the effects of storm sewage overflows to the tidal River Thames. IAWPR Workshop, Vienna.

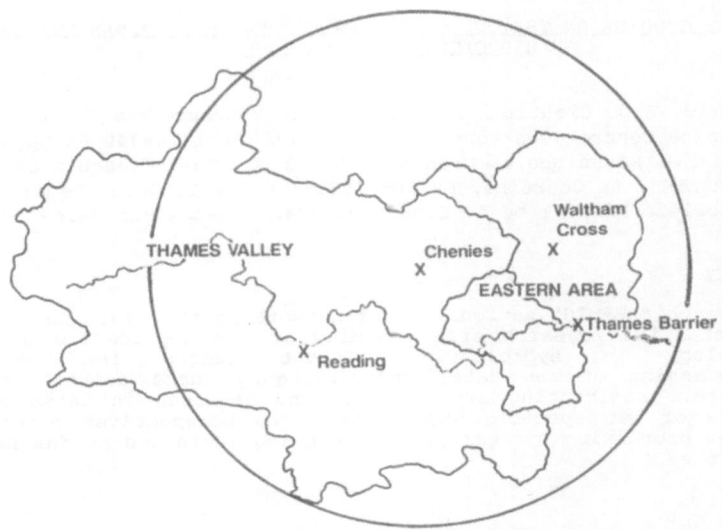

Figure 1 : Thames Catchment showing 75km radius of quantitative radar cover.

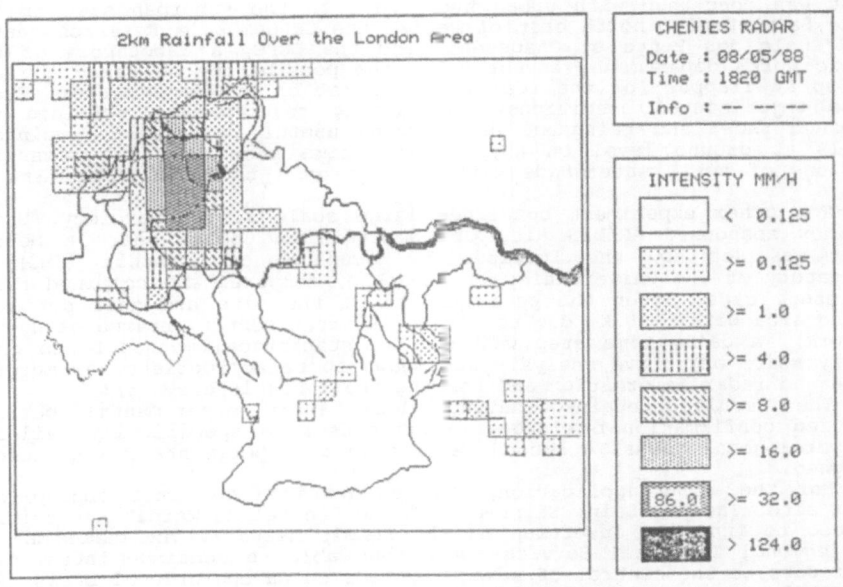

(c) Thames Water Rivers Division 1988

Figure 2.

FEASIBILITY STUDIES ON THE USE OF THE FRENCH "ARAMIS" RADAR NETWORK FOR HYDROLOGIC APPLICATIONS

H. Andrieu[1], J.D. Creutin[2], G. Delrieu[2], T. Denoeux[3] and G. Jacquet[4]
[1]Laboratoire Central des Ponts et Chaussées, BP 19,44340 Bouguenais
[2]Institut de Mécanique de Grenoble, BP 53 X, 38041 Grenoble Cédex
[3]CERGRENE, La Courtine, BP 105, 93194 Noisy le Grand Cédex
[4]Société RHEA, 1 bd A. Einstein, 77420 Champs sur Marne

Summary

The French "ARAMIS" national radar network is the starting point of a substantial research effort aiming at the application of radar technology in hydrology. Different stages including the preprocessing of raw data, radar/raingauge data combination and short-term forecasting are reviewed and assessed in terms of the results of two specific experiments. The perspectives offered by the new processing systems presently being installed on the network are evoked.

1. INTRODUCTION AND BACKGROUND

In France, hydrologists first became interested in weather radar back in the mid 60's (see (1)). However intensive investigations of the hydrologic capabilities of radar only became possible in the 80's with the development of the ARAMIS radar network by the National Weather Service (NWS) and with implementation of numerical processors.

A first experimental evaluation by the NWS ("Experience Hydromel" in (2)) showed that raw radar data could not be used as such for rainfall estimates. In 1982, two other experiments were initiated. The first was concerned with urban hydrology. In the suburban area of "La Seine Saint-Denis" north east of Paris, the CERGRENE, a Research center of l'Ecole des Ponts et Chaussées, and the LCPC, a laboratory of the public works department, investigated the possibility of using a radar set up at Trappes for the real time control of a sewage system. Under these experimental conditions, encouraging results were obtained and combined radar and raingauge data proved useful in assessing rainfall totals at ground level on a 15 minute time step and in forecasting outflows of small watersheds with areas from 1 to 20 km (see (3) and (4)).

The other experiment concerned large scale rural hydrology. Using the now abandoned ARAMIS site of Dammartin en Goële, where a Melodi radar was set up, the Institut de Mecanique de Grenoble (IMG), a laboratory of the Universities of Grenoble, compared and combined radar and gauge data. Given the poor quality of the site and data set used (blind area within 50 km due to the Paris urban centre, scarce raingauge network), a daily time step was used. A statistical method based on a multivariate objective analysis was shown to be appropriate for merging gauge and radar information and testing their consistency (5).

The results obtained during these short experimental periods required confirmation by longer experiments more specifically tailored to hydrological needs. A second generation of experiments was therefore designed.

For the urban application, Trappes remained the most appropriate radar site. The "La Seine Saint-Denis" and "Le Val de Marne" authorities decided to fund its insertion in the ARAMIS network. The CERGRENE and the regional Technical Services were thus able to continue integrating radar data to the control of sewage systems on an operational basis (see (6),(7)(8)).

On a rural scale, the areas most exposed to severe hydrological conditions were not properly covered by ARAMIS. The main reason was that the network had been designed for mesoscale meteorological observation rather than for hydrologic monitoring. Another reason was, and still is, that these most exposed areas are mainly in southern of France where radar use is complicated by mountain topography. To tackle the problem, a three year experiment was funded by the Ministry of the Environment and the Climatological CEC Program in order to explore radar capabilities for flash flood forecasting. In particular, a pilot radar system was installed in the Cevennes Region, near Nimes. The IMG, the LCPC and the Laboratoire de Météorologie Physique de Clermont-Ferrand worked together in this experiment, called Cevennes 86-88 (see (9)).

Based on the results of these second generation experiments, this paper reviews three stages which were distinguished in the hydrologic use of radar data :

i) the pre-processing stage, i.e. the correction of various radar assessment errors without external information.

ii) the calibrating stage involving the merging radar and raingauge information

iii) the short-term forecasting stage, predicting the rainfield behavior a few time steps ahead.

In spite of their relevance, other operational applications of the ARAMIS network are not presented here since they were not part of a research process (e.g. the Communauté Urbaine de Bordeaux or the Service Hydrologique de la Garonne). Nevertheless, they are additional signs of interest in weather radar on the part of the hydrologic community.

2. Pre-processing of radar data

Owing to their indirect nature, radar measurements of rainfall are affected by many sources of errors (see for instance (10)). The effects of these errors must first be reduced i) by the choice of a suitable site in terms of distance from the area of interest, absence of ground clutters and screening and ii) by the choice of an appropriate operating protocol. These preliminary conditions were dealt with differently in the two experiments.

In the "La Seine Saint-Denis" experiment the pre-processing step mainly dealt with the advection problem (discussed in (a) below) since the rainfall cell movements were significant and generally determinable.

In the Cevennes region, the mountainous context introduced additional difficulties and a more complex pre-processing procedure was necessary to reduce ground clutter and screening problems (discussed in (b) below) and to take into account the vertical variability of the reflectivity (discussed in (c) below).

a) Correction of the advection effect

Even when the radar data is acquired with an appropriate frequency (for instance, one exploration each five minutes), rainfall cells moving at a "normal" speed (for instance, 40 km/h) are not tracked with enough continuity regarding the total rainfall assessment. (11) showed that the radar assessment is improved when the totaling algorithm takes into account the advection effect. This result, of course, supposes a correct identification of the advection vector. Section 4 gives details on some advection detecting routines used in nowcasting and their efficiency.

b) Ground clutters and partial screening

Many methods can be applied to identify ground clutters. In the Cevennes experiment very similar ground clutter maps were obtained by i) clear air pictures, ii) analysis of the variability of the differential reflectivity (see (12)) and iii) numerical simulation using a digitized terrain model (see (13)). This convergence is not surprising since the relief varies significantly and the antenna diagram is well known.

The correction procedure applied relies on a multi-elevation exploration. Only two sites were used since the ground detection area no longer decreased at elevation angles greater than 3 degrees. The disturbed low elevation measurements were replaced by a high elevation measurement, when possible, or by using a simple interpolation procedure.

A ground clutter elimination will be included in the new CASTOR processor to be implemented on the ARAMIS network. The major part of the ground noise should then be filtered out (see (14)). Its ability to keep out rain reflectivity needs to be tested.

Partial (or total) screening is naturally a consequence of ground detection by the main lobe - a frequent problem in mountainous regions. To deal with this, a trade-off must be found concerning the altitude of the radar site. On one hand, the higher the radar, the less its beam is intercepted by the hills. On the other hand altitude increases i) the extent of side lobe reflections and ii) the atmospheric problems related to the altitude of the beam -partial filling, bright band, transformation between aloft and ground level rain intensities, etc. The correction procedure used simply determines the portion of the power which is blocked by the sky line and applies a proportional correction beyond the obstacles. A digitized terrain model is of great interest in the determination of the sky line and the correcting factors (see (12)).

c) Vertical profiles of reflectivity (VPR)

Taking the vertical variations of the reflectivity into account has been pointed out by (15) as very important especially when radar is set up at high altitudes. The reason is that the above mentioned problems of partial beam filling and bright band are then encountered frequently within a short range. The use of VPR obviously requires determination of the current profile, depending on the meteorological situation. A determination method based on multi-elevation exploration is proposed in (12). The robustness and the sensitivity of this method remain to be tested as well as the degree of improvement of the corrected measurements.

3. MERGING RAINGAUGE AND RADAR DATA

Raingauge data are often used as a reference for the control and calibration of radar assessments. Various methods based on multiplicative correcting factors have been proposed to rectify radar pictures (see (16) and (17)). The most simple use constant factors while more complex methods take into account the spatial and/or temporal variations of the factors. More recently, the objective analysis scheme has been used for merging ground and remote sensed information (see (18) and (19)). In fact this approach turns out to be equivalent to an additive correction varying in time and space (see (5)).

The complementarity of radar and gauge data is easy to exploit over long time steps that is to say on daily totals or storm sequence totals. For these cases, the rainfall fields can be considered as stationary random processes and any kind of correction factor works relatively well (see (20)). For short time steps, that is to say one hour or less, things are slightly different. At urban scales (3) showed that taking the spatial variability of multiplicative factors into account does not provide any improvement over a constant correction. Moreover, the assesment of constant correcting factors becomes very difficult when rainfall patterns become spotty as in convective situations.

4. SHORT-TERM FORECASTING

Hydrologists sometimes need more than a mere measurement of rainfall. For some applications (real-time control of sewer networks, flah-flood warning), reliable short-term forecasts may be of the utmost importance. The possibility of using radar data to make such forecasts 30 to 60 minutes ahead on drainage catchments measuring few tens of km² has been investigated at CERGRENE since 1985. Research has been carried out in two directions.

a) A forecasting method (named SCOUT), based on advanced pattern recognition techniques, was designed and tested using data from the Trappes radar (see (8) and (21)). This method seems to work better than the simpler cross-correlation technique described by (22) for some hydrologically important rainfall events (see (6)). It has been implemented in the sewer network real-time control system in the "La

Seine Saint-Denis" area where it is being tested under operational
conditions.

b) The reliability of radar rainfall forecasts is not constant from one
meteorological situation to another : this well-known fact can be a
serious drawback if an automatic forecasting procedure is needed.
Simulation results have been used on past events to automatically
generate rules for the a priori estimation of forecast reliability from
characteristics of rainfall areas and atmospheric vertical structure.
Encouraging results have been obtained (see (7) and (23)). This research
has also led to a new, user-oriented approach to the crucial problem of
forecat evaluation (see (24)).

While decisive progress has been made, many questions remain
concerning for example the effectiveness of radar forecasting techniques
for exceptional rainfall events. Moreover, work is going on to improve
the SCOUT method with representation techniques of expert system
knowledge and machine learning algorithms.

5. CONCLUSION

In the future, the work of hydrologists in the field of weather
radar will continue at both research and application levels.

Their research efforts need to be extended, especially to analyse
the impact of this new data on rainfall-runoff modelling. However this
will first require full confirmation of the accuracy of radar rain
intensity assessment at short time steps and the availability of other
distributed parameters.

At the application level, hydrologic research results need to be
applied to a wider range of operational situations. On the ARAMIS
network, the upcoming implementation of new processors (CASTOR) with a
special hydrologic module (CALAMAR) is very promising in this respect.

REFERENCES

(1) DUPOUYET, J.P. (1983). Regard sur une expérimentation française
     déjà ancienne, celle de Grèze dans le Bassin de la Dordogne.
     Session n° 123 du Comité technique de la Société Hydrotechnique de
     France, Juin 1983, Paris.
(2) FROMENT, G. (1979). Projet Hydromel-bilan des observations et
     résultats actuels. IXème Assemblée du Conseil Supérieur de la
     Météorologie.
(3) ANDRIEU, H. (1986). Interprétation des mesures du radar Rodin de
     Trappes pour la connaissance en temps réel des précipitations en
     Seine-Saint-Denis et Val de Marne. Thèse de Docteur-Ingénieur,
     CERGRENE/Ecole des Ponts et Chaussées, Paris, 182 pp.
(4) JACQUET, G., ANDRIEU, H. and DENOEUX, T. (1987). About radar
     rainfall measurement. Proc. of IV Int. Conf. on Urban Storm
     Drainage, Lausanne, 25-30.
(5) CREUTIN, J.D., DELRIEU, G. and LEBEL, T. (1988). Rain measurement
     by radar-raingage combination : A geostatistical approach. J. Atm.
     and Oceanic Tech. 5(1), 102-115.
(6) EINFALT, T. (1988). Recherche d'une méthode optimale de prévision
     de pluie par radar en hydrologie urbaine. Thèse de Doctorat de
     l'ENPC, CERGRENE, Noisy-le-Grand, 189 pp.
(7) DENOEUX, T. (1989). Fiabilité de la prévision de pluie par radar en
     hydrologie urbaine. PhD Thesis, Ecole Nationale des Ponts et
     Chaussées, Noisy-le-Grand, 246 pp.
(8) EINFALT, T. and DENOEUX, T. (1987a). Radar rainfall forecasting for
     real-time control of a sewer system. Proceedings Fourth Conference
     on Urban Storm Drainage, Lausanne, 47-48.
(9) ANDRIEU, H., CREUTIN, J.D., LEOUSSOFF, J. and POINTIN, Y. (1988).
     Cévennes 86-88 : a French hydrometeorological experiment to
     evaluate the weather radar capabilities for medium mountain
     hydrology. Int. Workshop on Hydrol. of Mountainous Areas, 6-11
     June, Strbske Plesso, Czechoslovakia.

(10) ZAWADSKI, I. (1984). Factors affecting the precision of radar measurements of rain. Proc. of the 22nd Conf. on Radar Meteorology, Zurich, 251-256.

(11) AGOSTINI-BLANCHET, B. (1988). Incertitudes liées à la mesure radar de la pluie. Rapport de DEA, CERGRENE/ENPC.

(12) FOURNIER, T. et ANDRIEU, H. (1988). Qualification d'un radar météorologique pour la mesure des précipitations dans la région des Cévennes : pré-traitement des images. Rapport IMG, 106 pp.

(13) ROUX, C. (1988). Simulation des zones d'échos de sol à l'aide d'un modèle numérique de terrain. Diplôme d'Etudes Approfondies d'Hydrologie, ORSTOM, Montpellier, 62 pp.

(14) RESIBOIS, M. (1985). Eliminateur d'échos fixes pour radar météorologique à détection d'amplitude. Revue Technique Thomson CSF, Vol. 17, n° 3, 567-589.

(15) JOSS, J. and WALDVOGEL, A. (1987). Précipitation measurement and hydrology - a review. Proc. of the 40th Anniversary Conference on Radar Meteorology, Boston.

(16) WILSON, J.W. and BRANDES, E.A. (1979). Radar measurement of rainfall - a summary. Am. Met. Soc. Bull., 60(9), 1048-1057.

(17) COLLIER, C.G., LARKE, P.R. and MAY, B.R. (1983). A weather radar correction procedure for real-time estimation of surface rainfall. Quart. J.R. Met. Soc., 109, 589-608.

(18) CREUTIN, J.D., LACOMBA, P. and OBLED, Ch. (1986). Spatial relationship between cloud-cover and rainfall fields : a statistical approach combining satellite and ground data. IAHS Publ. n° 160, 81-90.

(19) KRAJEWSKI, W.F. (1987). Co-kriging radar-rainfall and raingage data. J. of Geoph. Res. 92 (D8), 9571-9580.

(20) DELRIEU, G., BELLON, A. et CREUTIN, J.D. (1988). Estimation de lames d'eau spatiales à l'aide données de pluviomètres et de radar météorologique. J. of Hydrol., 98, 315-344.

(21) EINFALT, T. et DENOEUX, T. (1987b). Utilisation d'images radar en prévision de pluie. Actes du sixième congrès "Reconnaissance des formes et Intelligence Artificielle", AFCET, Antibes, ed. Dunod, 467-472.

(22) AUSTIN, G.L., BELLON, A. (1974. The use of digital weather radar records for short-term precipitation forecasting. Quart. J. R. Met. Soc., Vol. 100, 658-664.

(23) DENOEUX, T., EINFALT, T. et JACQUET, G. (1989a). Reliability of radar rainfall forecasts. Proceeding of the International Conference on Topical problems in urban drainage and in industrial plant systems, held in Strbské Pleso, Dom Techniky CSVTS, Bratislava, 73-75.

(24) DENOEUX, T., EINFALT, T. et JACQUET, G. (1989b). On the evaluation of radar rainfall forecasts. Proc. of the International Symposium on Hydrological Applications of weather radar, University of Salford.

RADAR DATA FOR HYDROLOGICAL USERS

Guy JACQUET - RHEA
1 bd Albert Einstein
CITE DESCARTES
CHAMPS SUR MARNE
77436 MARNE LA VALLEE CEDEX 2

J. CHEZE - SETIM
METEOROLOGIE NATIONALE
TRAPPES

Ten years ago, radar data from the French National
Weather Service Network (ARAMIS) started to be evaluated
and improved - in order to give areal rainfall estimates
and forecasts - by meteorological and hydrological
research centers : their main results are described in H.
ANDRIEU (et al)'s paper at the same conference.
A specific use of these estimates and forecasts is flash
flood forecasting, whether catchments are rural or urban ;
and for this specific use, this paper will consider the
past result evaluation, the information request and the
main design characteristics of a hydrological service as
derived from the survey of past results and of
information request.

I - REVIEW OF PAST RESULTS

I.1. PRELIMINARY RESULTS

Point rainfall measurements have shortcomings in forecasting flash floods, such as :

-- poor coverage of an area for a specific storm

-- weak efficiency in rainfall forecast

Supplemental information from radar use has been evaluated first in 1979 by the French National Weather Service (Froment).

Data sources were

* S band radar (OMERA, manufacturer, Type MELODI) situated in Dammartin (North of Paris)

* Existing NWS raingauges scattered around (loose-grid 1000 Km2)

Froment reported this radar was not suited for hydrological use, acknowledging a mismatch between the two data sources.

He blamed the poor experimental conditions :

- on one hand the radar had been istalled only for detection of severe meteorological conditions and numerous ground clutters spotted the 50 Km crown around the radar ;

- on the other hand, raingauge data were not collected automatically and paper registration and manual information extraction led to errors (asynchronisms...)

Positive results however existed :

* a better match was obtained where raingauge density was higher

* at the 50-100 km range from the radar, a relation ship was acknowledged between rainfall values over tens of Km2 and tens of minutes either resulting from radar pixel integration or from raingauge sims.

* close range calibration by raingauges reduced the discrepancy between close range raingauges and radar.

I.2. RAISING EXPERIENCE

New experiments were them attempted with the help of hydrologists :

- the radar in Dammartin and 180 raingauges were again used in 1982 ; DELRIEU (1986) proved the rainfall sampling was too spotty and he drew conclusions of untrustworthiness of radar rainfall measurements, by observing poor spatial cofluctuations with raingauges. As experimental conditions were simular to Froment's ones, such a result seemed attached to them.

- In 1986, ANDRIEU -- on one hand -- and DAVID - on the other -- reached more optimistic conclusions from an experiment involving the radar (in Trappes, 30 Km south west of Paris) ; it is a C band radar with a thinner beam and whose discretization system authorized 800 m x 800 m pixels.

Moreover, his installation offers much less ground clutters than the one in Dammartin.

ANDRIEU reached a good fit within some homogeneous rainfall event with a raingauge network in the range 20-50 km from the radar (4 km grid size, 2 minutes time intervals between images).

DAVID -- using 21 raingauges in the range 20--100 km and 15 minutes radar images - demonstrated that 15 minutes time sampling was responsible of radar-raingauge main mismatch and a better fit was observed by interpolating images using advection speed. Echo speeds could be used for a priori estimate of radar discrepancy.
Such results built up new interest within the hydrologist community, so that new experiments were held by hydrologist researchers (Cevennes, Lempdes, Grenoble) whose results are shown in H. ANDRIEU's paper.

1.3 - LESSONS OF OPERATIONAL USE

In the mean time, numerous developments were conducted to improve the daily operation of ARAMIS data.

- A real time processor prototype was built by the Research Center of the French National Weather Service (NWS) (DE CARVALHO 1986) : it can operate with any ARAMIS radar and with an hourly remote access to the NWS raingauge network. Such a system operated in 1986 with 15 minutes images from a S band radar in Bordeaux and with hourly raingauge interrogation.

- Results proved that ARAMIS improvement was needed to get rid of such limitations as abnormal propagation and as 15 minutes time interval between images and that a constant hourly correction of radar by a loose raingauge networck was inadapted.

- Bordeaux experiments (Marc AUBRY 1987) with a close range raingauge network led to the conclusion that 2 radars beam elevation should be used to avoid most ground clutters.

- Since 1986, use of the radar in Trappes is operationnal for qualitative flash flood forecast on the Seine Saint Denis and the Val de Marne counties ; the Seine Saint Denis county, with his own remotely connected raingauges (23 of them in 250 Km2 every 5 minutes) checks the rainfall estimates and forecasts algorithm built by the CERGRENE, a Research Center of the Ecole Nationale des Ponts et Chaussées (as exposed in connected H. ANDRIEU's paper). Algorithm improvements are continuing, following real time observed deviations, so that no long term results is yet available.

II – <u>**INFORMATION REQUEST FOR FLASH FLOOD FORECAST**</u>

II.1. **INTRODUCTION**

Flash flood occur at the outlets of catchments less than 10000 Km2, averaging 100 Km2, with numerous ones below 10 Km2.

More sensitive catchments are urbanized areas whose imperviousness and extended drainage increase the total output and the peak flow.

"Natural" catchments will respond in the same way if imperviousness is important due to geology (granite,...) or to soil conditions (spring cereals,...) and if flow speed is accelerated by slopes over 1 %.

Flash flood catchments are much more sensitive to convective storms which last less than a few hours for two main reasons :

- soil saturation is quickly, if not immediately, expected at the rain onset

- all parts of the catchments contribute at the same time to the flow at the outlet, due to the flow speed increase : this concentration effect allow many catchments to be sensitive to half-an-hour storms, due to short life meteorological cells (see fig. 2.1.)

Flow speed accelerating pipe construction increase
drastically this sensitiveness : 5 Km path will be run in
half an hour instead of being run in three hours due to
ponding...

Even larger catchments will be more sensitive to non-
extensive but heavy rainfall.

**Fig.2.1. SCHEMATIZED EFFECT OF CONCENTRATION OF FLOWS DUE
TO CANALISATION OF RIVER REACH**

(O'0 : 2 km;speed increase 1 to 2 m/s)

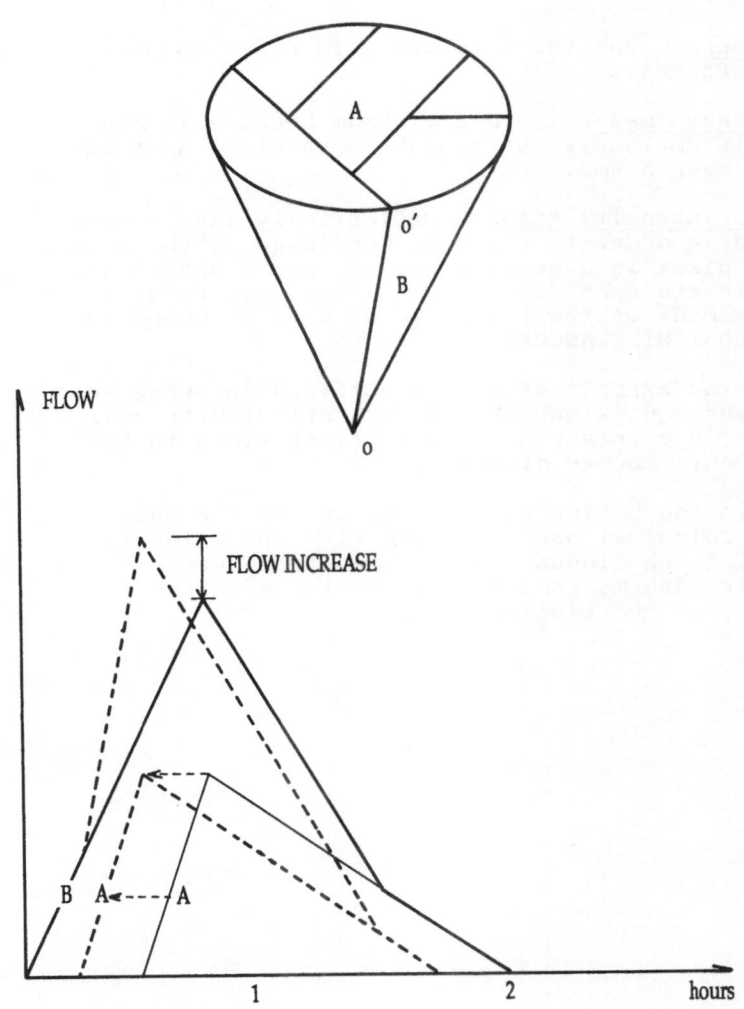

II.2. FLASH FLOOD FORECAST RADAR RAINFALL MEASUREMENTS REQUIREMENTS

Areal rainfall measurements are required to forecast these flash floods in operational river management or in real time control of urban drainage systems : so radar measurements could be welcome if they are tailored for these.

Considering that a one square kilometer area may contribute several m3/s for a yearly or decennal flood -- if it is fully impervious and drained -, a square kilometer grid is needed, and this induces also the need for

- low radar beam, over the catchment, in order to master the drift effects.

- high frequency images : 5 min minimum because 15 min frequency is obviously not enough to sample a half an hour storm over a Km2.

- narrow rain intensity steps : reflectivity levels must be selected in order to transmit the image to the users; 4 bits per pixel is a usual standard, and a better use of the 16 levels range is needed so that the rains which cause most of the flash floods will be properly sampled (AGOSTINI BLANCHET).

Fig 2.2. give an example of errors obtained in using the Marshall Palmer equivalent of a 4 dBZ reflectivity scale to discretize 30 events resulting in flash flood in the Seine Saint Denis county district.

Yet, the tailoring of the radar image is not the only need of a hydrological user involved with operational management of flash floods. There is a strong need to link the radar informations with other informations available to this hydrological user.

Fig. 2.2. ERRORS DUE TO DISCRETIZATION IN 16 RAINFALL INTENSITY LEVELS

Over a sample of 30 historical rainfalls (measured with a 12 raingauge network and selected for their flooding effects in the Seine Saint Denis department) the discretization effect has been simulated. Figure shows sharp error rise of 30 min. rainfall amounts due to "meteotel" scale compared to "optimal" scale.

"METEOTEL" SCALE : each of the 16 levels is 4dBZ equidistant after the reflectivity (Z) intensity (R) inverse transform of Marshall Palmer's law (Z : 200 R**1.6)

"OPTIMAL" SCALE : 12 of the 16 levels are in the range (2,100) MM/H.

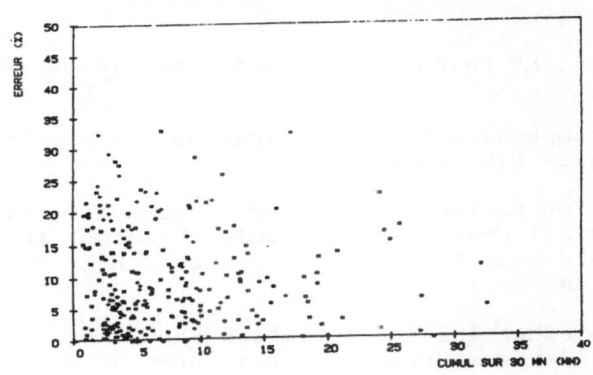

CUMUL SUR 30 MN; ECHELLE METEOTEL

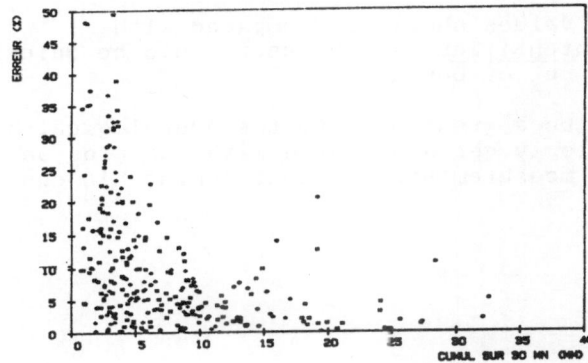

CUMUL SUR 30 MN; NOUVELLE ECHELLE

II.3 LINKING RADAR MEASUREMENTS WITH OTHER INFORMATIONS

Radar rainfall measurements are teledetection measurements which always need <u>ground proof and ground corrections</u> :

- A perfect knowledge of orography, and of masks due to short range beam obstacles, should be included within the radar image processing

- A correction processing with raingauges should be imbedded so that fluctuations of measurements due to poor radar stability should be taken care of. Yet, as there are large fluctuations between raingauge and radar due to the very different way of sampling the rains, the correction should avoid to spread these errors :

* correcting raingauges should be selected so that out of order ones, and those ones whose conjugated radar values are instable (or in a very heterogeneous area), should be eliminated.

* number of correcting raingauges selected should be minimum of 4

* time span for compansion between raingauges and radar should be adequate (15 minutes)

* if this correction factor is computed at each time step, and if the rainfall character is slowly changing, this correction should even be useful to reduce Z-R fluctuations (fig.2.3.)

<u>Topography</u> is the other information which should be linked with the radar measurement : it is needed to sum up the radar values over each initial catchment during the critical time (concentration time of this catchment).

Finally the areal values should be compared with <u>frequency areal intensities</u>, so the user would be quickly alerted by values out of bounds.

This linking will be a great help to the hydrological user which will slowly get acquainted with the pros and cons of the radar measurements, without losing his own measurements.

Fig.2.3. EVOLUTION OF CORRECTION FACTOR EVERY 15 MINUTES (from 0.8 to 0.3) DURING NIGHT DECREASE OF STRONG CONVECTIVE STORM.

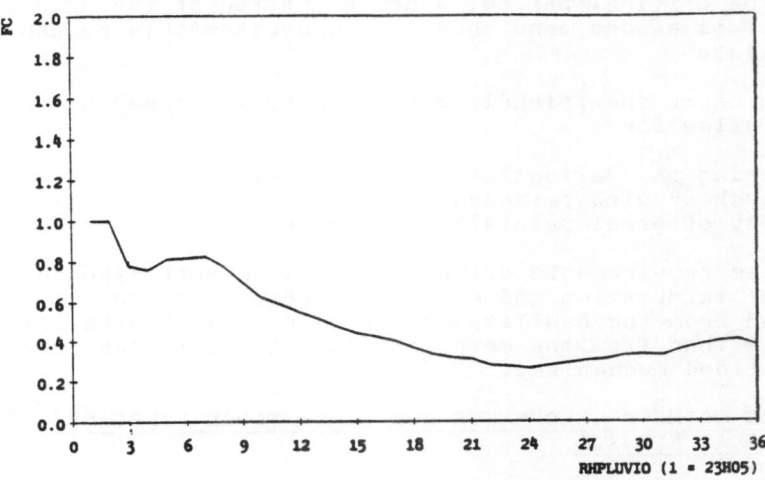

FACTEUR DE CORRECTION 21/07/82

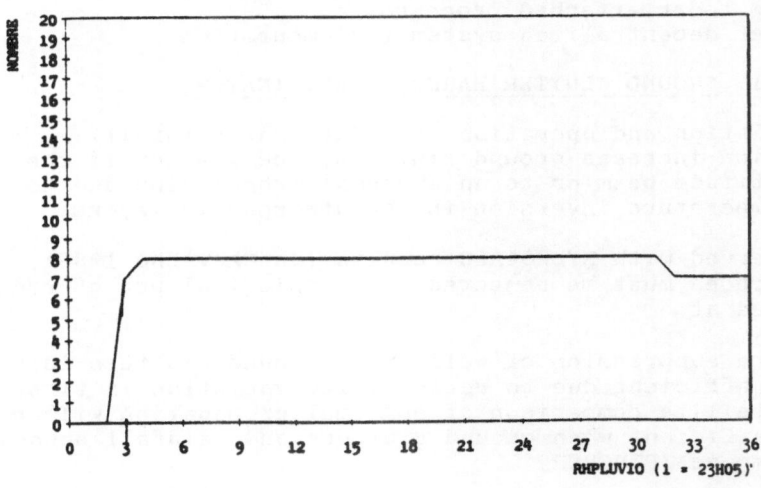

NOMBRE DE PLUVIOS VALIDES 21/07/82

II.4. USER'S INTERFACE

Hydrological user need real-time information and a friently-user display in order to be able to make up his mind in sending flood warnings or in controlling storage basins.

After the critical period, a demonstration of the actual rainfall is needed, and should be understandable by non specialists.

In both cases the friendly user display is needed and should allow for

- filtering and validating the information
- cross check with raingauges
- display of areal rainfall over catchments

All these requirements drive towards a decentralized use of radar information and more improvements can be expected from the hability of the user to deal with his own data than from the meteorologist, which is less aware of the flood mechanisms.

III - ON GOING DEVELOPMENTS FOR FLASH FLOOD FORECAST RADAR SERVICE

III.1. DESIGN CHARACTERISTICS

Design of this new radar service includes

- a ground clutter hardware eliminator (whether close range or abnormal propagation)
- a new radar attached processor
- a user decentralized system implementation

III.1.1. GROUND CLUTTER HARDWARE ELIMINATOR

Installation and operation of radars for rainfall detection increase ground clutters, due whether to the low altitude beam or to an abnormal propagation due to the temperature inversion in low atmospheric layers.

As observed with HYDROCALC results (de CARVALHO 1986), such echoes must be rejected if hydrological use of radar is aimed at.

Software suppression of well known ground clutters is not always efficient due to reflectivity variation in time, and satellite comparison of abnormal propagation will not be as efficient when ground clutters and rainfall echoes superimpose (PIRCHER)

However original signal si very different when due to a
fixed obstacle or to hydrometeors with non Doppler radar,
such as ARAMIS ones, the signal variance is much more
important with hydrometeor retro diffusion :

- with the same average intensity value, the standard
  deviation of the 300 Hz signal is much less if
  originating from the ground

- the ground clutter signal fluctuations are very slow

- the ratio of standard deviation by average value is
  constant if the signal is from hydrometeors (Raileigh
  law)

So the standard deviation estimation will allow the
estimate of the hydrometeor average reflectivity and
therefore 40 dB attenuation of ground clutters is
theoretically possible ; in practice, such ground clutter
elimination will reject echoes up to 30 dB (RESIBOIS),
but it is very efficient also for abnormal propagation
(Fig.3.1.)

Thomson has developped such a hardware device for his
radars RODIN belonging to ARAMIS and they will be adapted
on the other radar types.

III.1.2. A NEW RADAR ON SITE PROCESSOR (CALLED CASTOR)

Hydrological information request (parag. 2.2.) have been
taken into account within the specifications of this on-
site processor. The processed data will allow different
information packing for not only flash flood forecast
uses, but for detection of critical phenomena and for
meteorological analysis applied to short term forecast.

They induce specific functions

- 8 or 12 bits digitalization

- azimut integration with the use of data converted in
  rainfall intensities (R**1.6) and of best fit
  coefficient of recursive filters

- recursive filter use for polar to cartesian coordinate
  transform

- ability to correct partial mask effects

Efficiency of the ground clutter hardware eliminator is
improved by the median method of integration along the
radius and by the ability to deal with differents
scanning elevations of the radar.

Fig.3.1. EVIDENCE OF GROUND CLUTTER SUPPRESSION BY HARDWARE ELIMINATOR

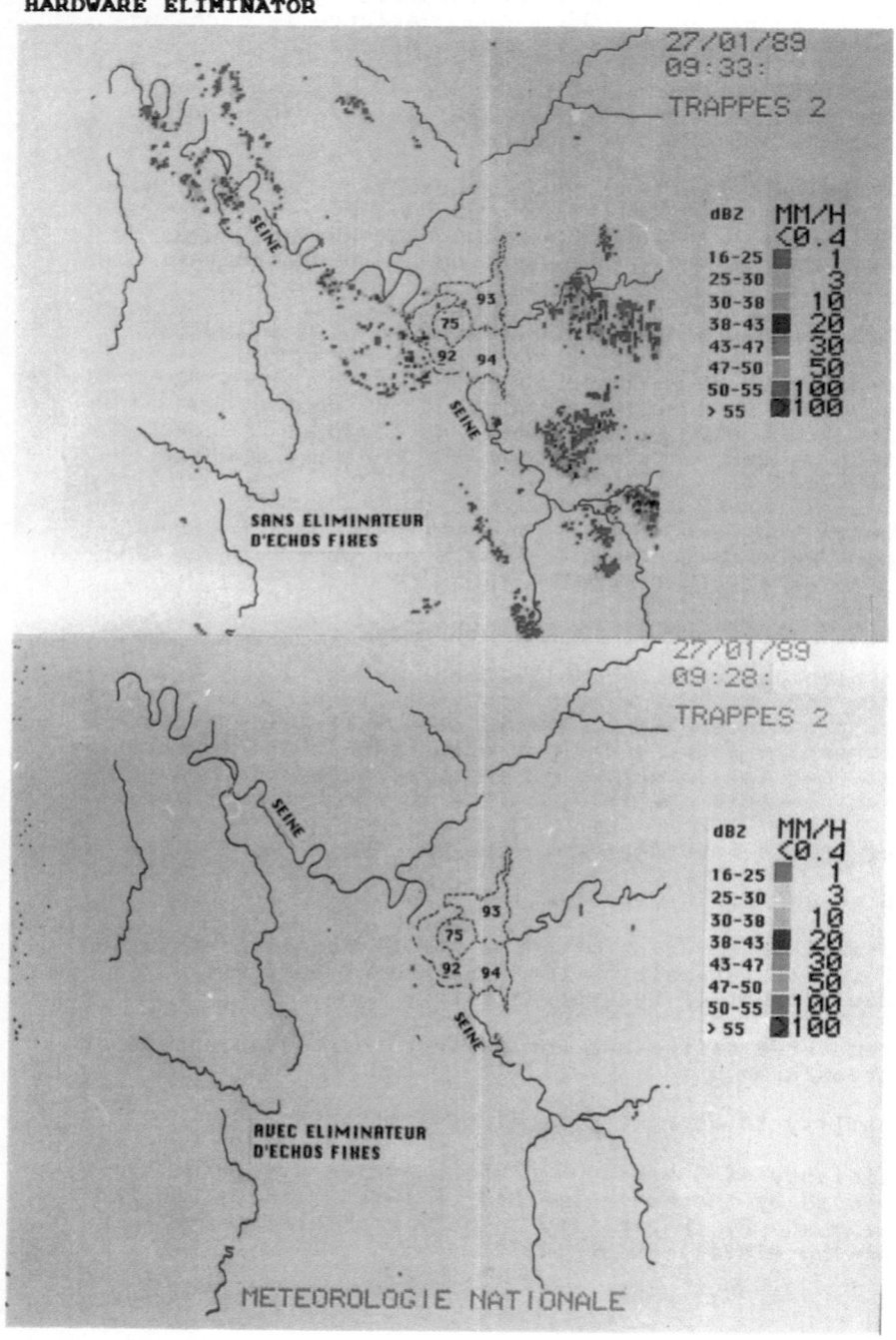

Eventually, the built in test equipment (BITE) will allow for automatic electronic radar calibration. Emision will be controlled by a miniwattmetre and reception will be controlled after sending a calibrated signal due to a hyperfrequence generator. Control results will be used by the processor to reduce a slow zero shift.

III.1.3. CALAMAR : A DECENTRALIZED SERVICE FOR HYDROLOGICAL USERS

The design of a decentralized service has taken care of most of the hydrological user's request :

- a processing "at home" allowing for selection of its own calibrating system and of slow habit of radar use

- a custom-tailored processing which includes estimates over actual catchments (far from the grid type of estimates).

- a user-friendly interface with mouse driven menus, fast speed radar display, easy sequence and zoom selection, spatial and temporal animation

The decentralized service is also a much better protection against radar abnormal measurements or against rainfall poor measurement (whatever radar or raingauge) :

- As an example, radar abnormal beam propagation will not be such a problem if the hydrological user may check this condition with his own raingauge and his own judgment by looking out of the window during a misty day.

- A far-off radar user will be aware of the size of the radar pixels and of the wideness of the radar beam and will make a better use of this information, by knowing its limits.

- A much better understanding of this rainfall sampling limit (radar or raingauge) and of the radar signal variability will be at the user's reach when he can in the same time visualize

* an unusual measurement or a discrepancy between radar and raingauge

* the quick moving shower cell which passes above the pixel with a very heterogeneous and fast changing shape.

III.2. AGREEMENT RHEA/FRENCH NATIONAL WEATHER SERVICE

Goals of RHEA, a hydrologist consulting firm specialized
in real time assistance, and the French National Weather
Service were common, and both could benefit of the other
efforts :

* Efforts have been coordinated to deliver a more secure
  radar information, both with a better radar signal
  processing (National Weather Service) and with a local
  radar check with synchronized raingauges (RHEA).

* Increased valorisation for the hydrological user
  resulted from the use of the National Weather Service
  dissemination system (METEOTEL), of the National
  Weather Service Radar Network (ARAMIS) and of the
  finely tuned information processing and user display.
  The latter resulted from RHEA hydrologist experience,
  based on years of research within various french
  research centers (CERGRENE, LCPC, IMG) as being
  explained in ANDRIEU's concommutant paper.

* hydrological research will be enhauced from the
  availability of more appropriate data due to the
  dissemination system and should allow for new advances
  in the field of flood forecast.

Main features of the agreement are :

the distinction of hydrologist radar image (CALAMAR) from
other radar products used for

- warning of dangerous phenomena as hail, fast winds for
  plane safety or agricultural protection

- large scale rainfall resulting in increase of
  groundwater ressources

- road-wetting drizzle resulting in many traffic crashes.

the planning of both developments altogether (ARAMIS &
CALAMAR) aiming at offering a full service to flash flood
forecasters.

BIBLIOGRAPHIE

AGOSTINI/BLANCHET B. (1988)
"Incertitudes liées à la mesure de la pluie",
Rapport de DEA Techniques et Gestion de l'Environnement ENPC - CERGRENE,
52 p + annexes.

BLANCHET B., NEUMANN A., ANDRIEU H., JACQUET G.,
"Improvement on rainfall measurements due to accurate synchronisation of raingauges and due to advection use in calibration - Salford, August 1989.

ANDRIEU H., JACQUET G., BACHOC A. (1985)
"Les débuts de l'utilisation du radar météorologique en hydrologie urbaine. COST project 72, Proceedings of the final seminar, ERICE, ITALY, 30 sept. - 3 oct. 1985, 96 - 106.

ANDRIEU H., (1986)
"Interprétation des mesures du radar Rodin de Trappes pour la connaissance en temps réel des précipitations en Seine Saint Denis et Val de Marne".
Thèse présentée pour l'obtention du diplôme de Docteur-Ingénieur, Ecole Nationale des ponts et Chaussées, Paris.

CARVALHO F.C., (1986)
"Hydrocalc - calculateur temps réel de la lame d'eau par radar".
Thèse présentée pour l'obtention du diplôme de Docteur Ingénieur, Université de Paris Sud.

CICCIONE M., PIRCHER V. (1984)
"Preliminary assessment of very short term forecasting of rain from single radar data".
Nowcasting II, ESA, Norrkopping, Sweden, 3-7 Sep. 1984, 241-246.

COLLIER C.G., KNOWLES J.M., (1986)
"Accuracy of rainfall estimates by radar, part III : application for short-term flood forecasting". Journal of Hydrology, 1983, 237-249.

DAVID P., MUSIEDLAK J.P., BISSONNIER P., (1987)
"Utilisation du radar Rodin en pluviométrie - résultats des mesures de 1982".
Note Technique n° 12, Direction de la Météorologie Nationale, Boulogne.

DELRIEU G., BELLON A., CREUTIN J.D., (1987)
"Estimation de lames d'eau spatiales à l'aide de données de pluviomètres et de radar météorologique. Accepté pour publication dans le Journal of Hydrology.

DELRIEU F., (1986)
"Evaluation d'un radar météorologique pour la mesure des précipitations : validation et étalonnage par technique géostatique ; application au bassin parisien. Thèse présentée à l'Université Scientifique et Médicale de Grenoble pour obtenir le titre de Docteur Ingénieur, 28 février 1986.

EINFALT T. (1988)
"Recherche d'une Méthode Optimale de Prévision de Pluie par Radar en Hydrologie Urbaine", Thèse de doctorat de l'ENPC, CERGRENE, Noisy-leGrand, 189 p.

EINFALT T., DENOEUX T. (1987 a)
"Radar Rainfall Forecasting For Real-time Control of a Sewer System".
Proceedings Fourth Conference on Urban Storm Drainage, Lausanne, sept. 1987, pp. 47-48.

EINFALT T., DENOEUX T. (1987 b)
"Utilisation d'Images Radar en Prévision de Pluie",
Actes du sixième congrès "Reconnaissance des formes et Intelligence Artificielle", AFCET, Antibes, nov. 1987, ed. Dunod, pp. 467-472.

FROMENT G., (1977)
"Rapport Hydromel". Météorologie Nationale.

GAILLARD C., PIRCHER V., PAILLISSE R. (1986)
"The French ARAMIS project : use of microprocessor based systems for real time broadcasting of weather radar and satellite pictures. Proc. 2d
Inter Conf. on Interactive Information and Processing systems for Méteorology, Oceanography and Hydrology, AMS, Miami, Fla.

GILET M., et CICCIONE M., (1983)
"Le projet ARAMIS et la prévision à courte échéance". La houille blanche, n° 3/4, 1983.

GILET M., (1985)
"Le réseau Français de radars météorologiques. Proc. of the third WMO Tech. Conf. on Instruments and Methods of Observation, TECIMO III, Ottawa, Canada, 8-12 july 1985, 175-180.

JACQUET G., ANDRIEU H., DENOEUX T., (1987)
"About Radar Rainfall Measurements"
4 th Int. Conf on Urban Draimage Lausanne 1987.

JACQUET G., EINFALT T., DENOEUX T., ANDRIEU H., PIRCHER V., DAVID P.,
"Principes d'évaluation des méthodes de mesure et de Prévision de pluie par radar en hydrologie urbaine"
Colloque Eau et Informatique - Paris 28-30 mai 1986.

MUSIEDLAK J.P., (1985)
"Un système éliminateur d'échos fixes pour les radars précipitations. Third WMO tech. conf. on Instruments and Methods of Observation, Ottawa, Canada, 8-12 july - 1985, 181-185.

PIRCHER V., (1987)
"Combined use in operational forecasting of animated imagery and NWP fields. Preprints of the EUMETSAT workshop on satellite and radar imagery interpretation, Reading, U.K., 20-24 July 1987, 365-383.

RESIBOIS M., (1985)
"Eliminateur d'échos fixes pour radar météorologique à détection d'amplitude", Revue Technique THOMSON-CSF, Vol 17 n° 3 septembre 1985

SAUVAGEOT H., 1981 "Radar Météorologie" . Eyrolles, 1981.

URBAN HYDROLOGY AND HAIL DETECTION EXPERIMENTS MADE WITH A RAINGAGE-HAILPAD NETWORK AND WITH A DUAL POLARIZATION RADAR

Y. POINTIN[1], D. HUSSON[2], J. FOURNET-FAYARD[1] and M. MESSAOUD[1]

[1] LAMP/OPGC, Observatoire de Physique du Globe de Clermont-Ferrand, 12, Avenue des Landais, 63000 Clermont-Ferrand, FRANCE

[2] GNEFA, Groupement National d'Etude des Fléaux Atmosphériques, 63170 Aubière, FRANCE

Summary

Small time (1 to 15 min) and space (0.5 to 1.5 km) measurements of the mean rainfall rate have been made within 10 km from a dual polarization radar, during several stratiform and convective precipitation events. At first, the time and space characteristic scales of the gage and of the radar data are evaluated in order to find the minimum time interval Δt and spatial resolution Δx for which the mean rainfall rate values have a large enough statistical significance. Then, the quantitative comparisons are made between the mean rainfall rate deduced from the gage data and that deduced from the radar measurements, either without or with using the differential reflectivity Z_{DR}. During the same experiments, the hailfall parameters (total hailstone number N_t, total hailfall kinetic energy E_t) have been estimated from the data of a hailpad network situated within 30 km from the radar. The values of these hailfall parameters have been approximated by functions of several radar parameters, deduced from the time evolution of the radar reflectivities (Z_H and Z_{DR}) above each hailpad.

1. INTRODUCTION

The very short time (a few minutes) estimation of the mean rainfall rate over a small catchment, of an area of a few square kilometers, can be made by a dense raingage network, provided that the gage spacing is smaller than the size of convective cells; namely less than one kilometer. In the present experiments, described in the next section, 14 1-minute raingages have been set up over a 25 km² area, giving a mean gage spacing of 1.5 km. Therefore, the first goal of this paper is to evaluate the smallest time interval Δt and spatial resolution Δx (1) for which the true mean rainfall rate R_m can be significantly estimated from the gage data. This evaluation is made in the third section.

These short time and space resolutions can also be provided by meteorological radars which lead to an estimation of the rainfall rate from the cloud reflectivity measurements. Numerous works have been devoted to the radar estimation of the rainfall rate (2, 3, 4, 5, among many others) for time intervals generally greater than 1 hour. However, for short time intervals, the cloud reflectivity Z is only weakly related to the rainfall rate, because the drop size distribution $N(D)$ shows quite a large time variability (6, 7). A dual polarization radar (8) measures two independent cloud reflectivities, namely the horizontal reflectivity Z_H, similar to the cloud reflectivity Z measured by any conventional radar, and the differential reflectivity Z_{DR}, which is defined from the ratio of the cloud reflectivity Z_H for an horizontally polarized electromagnetic wave, to that Z_V for a vertically polarized one.

$$Z_{DR} = 10 \log (Z_H/Z_V) \tag{1}$$

Theoretically, for clouds containing big oblate raindrops, the differential reflectivity Z_{DR} reaches 3 to 4 dB and is related to the mean drop diameter, while it remains below 1 dB for clouds containing only ice particles. Therefore, the differential reflectivity Z_{DR}, together with the horizontal reflectivity Z_H, can provide a good estimate of the rainfall rate (6, 8) and is very usefull in microphysical studies (6, 7) and for the ice phase detection (hail, graupel and ice crystals) in convective clouds (9, 10). However, most dual polarization measurements made so far have been compared to individual raingage measurements for very short time intervals (a few minutes). These comparisons suffer from the fact that both measurements have very different sampling volumes and may not have the same statistical significance. Therefore, the other goal of this paper is to compare, for the shortest time intervals Δt and highest spatial resolutions Δx available from the gage network, the mean rainfall rate R_G deduced from the gage data to that deduced from the radar measurements, either without (R_Z) or with (R_D) using the differential reflectivity Z_{DR}. These quantitative comparisons are made in section 4.

Similarly, the quantitative measurement, by meteorological radars, of the hailfall parameters remains a difficult problem and the best experimental results, so far, have been obtained for the estimation of the global kinetic energy of the hailfalls, by using a non-polarized radar together with ground hail spectrometers and hailpads (11). The first analyses (10) of the present experiment have revealed that mixed phase precipitations (hail and rain) are characterized by large values of the horizontal reflectivity ($Z_H > 50$ or 55 dBZ) while the corresponding differential reflectivity Z_{DR} is smaller than in rain type precipitations. The further analyses done in this paper aim at testing the improvement that the differential reflectivity Z_{DR} can bring to the estimation of hail parameters through a relationship of the form:

$$P_h = f\left[(P_{R\,i}, i = 1, \cdots, N)\right] \quad , \tag{2}$$

where P_h is the value of the hail parameter deduced from the ground hailpad data, and $P_{R\,i}, i = 1, \cdots, N$ are N parameters deduced from the radar data. These parameters are described in section 5 and the resulting relationships are given in section 6.

2. DATA DESCRIPTION

The present experiment (10) has been conducted in the vicinity of Clermont-Ferrand (France) between May 1985 and August 1988. During this period, 48 precipitation events, including some very intense convective hailstorms (10), have been fully recorded.

2.1. Raingage data

As shown in Fig. 1, the network covers a 25 km² area, lying between 5 and 10 km from the radar. In the following, the comparisons are made within the dashed line square, in order to avoid extrapolating the gage data. For each selected time interval Δt (5, 15, 30 and 60 min), and for each gage located at the point M_i, the mean rainfall rates $R(M_i, t_k)$ are computed for all time steps $t_k = (k-1)\Delta t + t_0; k = 1, \cdots, N$, as an average during the corresponding time interval Δt centered around t_k.

2.2. Dual polarization radar data

The ANATOL radar (12) is a 10 cm dual polarization radar which has a 1.8 degrees half power beam-width antenna and which rotates at about $8^\circ \cdot \mathrm{s}^{-1}$ (i.e. 1.3 rpm). In this experiment, the data have been almost continuously recorded, either during PPI scans made at a constant low elevation, or during volume scans made above the basin. In the estimation of the radar derived rainfall rate, only the data recorded below 1.5 km above the ground have been used, and most of them have been recorded below 1 km. Since the biggest raindrops take less than 3 minutes to fall from the top of the scanned volume to the ground and since the time interval Δt used in the comparison is greater than 15 minutes (section 3), no correction has been made for the horizontal advection, nor for the modification of the drop size distribution during this fall.

2.3. Hailpad data

A hailpad network, located between 5 and 30 km from the radar, provides direct informations about the existence of hailstones in convective cells. Each of the 160 hailpads gives the number and the diameter of the hailstones, from which the hailfall parameters can be evaluated.

3. TIME AND SPACE SCALE OF RAINFALL FIELDS

For each selected time interval Δt (5, 15, 30 and 60 min), the inter-correlation coefficient $\rho_r(M_i, M_j)$ has been computed as a function of the time lag τ, from the two time series $(t_k = 1, \cdots, N)$ of the mean rainfall rate values $R(M_i, t_k)$ and $R(M_j, t_k + \tau)$ of the two gages located at points M_i and M_j, respectively. Generally, the inter-correlation coefficient increases with the time interval Δt, and takes its maximum value for a null time lag, except for the 5-minute time interval. These properties result from the fact that a time of 15 minutes appears to be much less than the time needed for a precipitating cloud to cross the 5 km wide area. From the zero-lag inter-correlation coefficient $\rho_0(M_i, M_j)$ we define the neg-correlation $\omega(M_i, M_j)$ by :

$$\omega(M_i, M_j) = 1 - \rho_0(M_i, M_j) \quad . \tag{3}$$

The interest in this neg-correlation stems from the fact that, for ergodic, homogeneous and stationary fields $Z(M)$, this neg-correlation is egal to the ratio, by the field variance, of the semi-variogram $\gamma(M_i, M_j)$ (13), also called the structure function, and used in the kriging interpolation technique. Despite the fact that, in this study, the observed rainfall field can hardly be considered as ergodic, homogeneous or stationary, we shall use the neg-correlation function in place of the semi-variogram in the kriging interpolation technique (13).

The neg-correlation values for every gage pairs in the network are plotted versus the distance between the corresponding gages in Fig. 2 for the rainfall rate data relative to the month of April 1986. In this figure, the different symbols indicate the time interval Δt used in the computation of the correlogram values. Despite an appreciable scattering of the points, which decreases when the time interval Δt increases or for a less convective situation, the data points are clustered along the curve of the neg-correlation function $\omega_s(h, \Delta t)$, defined (1), for each time interval, only as a function of the distance between the two gages. Indeed, for this small network, no clear dependence upon the relative gage orientation have been found, even for a 5-minute time interval for which the influence of the advection effects should have been the largest. For each time interval, the coefficients a, b and h_p of a spherical function have been non-linearly adjusted :

$$\omega_s(h, \Delta t) = \begin{cases} a + (b - a)\left[\frac{3}{2}(h/h_p) - \frac{1}{2}(h/h_p)^3\right] & \text{if } h < h_p \\ b & \text{otherwise} \end{cases} , \tag{4}$$

where h_p is the range value beyond which the correlation takes the constant value b, and a represents a discontinuity for short distances which reflects a white noise effect. Indeed, the comparison of collocated gages shows very large differences for short time intervals (14, 15), mainly due to very small scale inhomogeneities (turbulence).

The large scattering of the points around the neg-correlation function may induce large errors in the spatial interpolation of the gage data, since each gage data is then representative of an area which is not well defined. This is specially true for the 5-minute data, which show also very small correlations (less than 0.70, corresponding to $\omega > 0.30$) for distances (0.75 km) of the order of half of the mean distance (1.5 km) between gages in this network. This implies that, for this time interval, the data of any raingage cannot be statistically extended to the area that each gage is supposed to cover. However, for time intervals greater than 15 minutes, the correlation values are above 0.8 (corresponding to $\omega < 0.20$) for distances shorter than half the network mean distance, suggesting that spatial interpolation may be statistically valid for these largest time intervals. As shown in Fig. 2, for hourly data, the correlation remains high for larger distances than for the 15-minute data, suggesting that the mean distance between gages could be of the order of 10 km. This implies also that the interpolation of hourly gage data is only statistically significant if the gage density is above 1 gage per 78 square kilometers.

4. COMPARISONS BETWEEN ESTIMATED RAINFALL RATES

For each time interval Δt, the gage data are spatially interpolated by using the kriging interpolation technique (13). The interpolated value $R_G(M, t_k)$ at any point M is a linear combination of the N measured values $R(M_i, t_k)$ at all gage points M_i :

$$R_G(M, t_k) = \sum_{i=1}^{N} \lambda_i R(M_i, t_k) \quad , \tag{5}$$

where the weighting coefficients λ_i are the elements of the N vector solution of a matrix system in which the coefficients are equal to the neg-correlation between the gage stations M_i and M_j, i.e. $\omega_s(\|\overrightarrow{M_i M_j}\|, \Delta t)$. The interpolated value $R_G(M, t_k)$ is computed at every center points of a regular grid which is included in the full line square drawn in Fig. 1, and with grid sizes $\Delta x = \Delta y$.

The radar reflectivity values $Z_H(\text{mm}^6 \cdot \text{m}^{-3})$ are converted to equivalent rainfall rates $R_Z(\text{mm} \cdot \text{h}^{-1})$ by the relationship :

$$Z_H = a R_Z^b \quad , \tag{6}$$

where the chosen coefficients are, with the used units, $a = 200$, $b = 1.6$ for stratiform conditions, and $a = 486$, $b = 1.37$ for convective conditions. Similarly, the following relationship is used for the equivalent rainfall rate R_D derived from the radar measured horizontal Z_H and differential Z_{DR} reflectivities :

$$R_D = \frac{a Z_H}{c + Z_{DR}^d} \quad , \tag{7}$$

where the coefficients are, with the used units, $a = 0.0033, c = 0.55, d = 2.33$. These coefficients have been non-linearly adjusted by using the equilibrium raindrop axial ratio relationship and 1419 drop spectra measured by 2 disdrometers every 5 minutes, during the most intense precipitation events which have appeared in all the ANATOL experiments of the last 4 years.

The equivalent rainfall rates are accumulated over all measurements made during the corresponding time interval Δt and over each grid mesh. It has been found that, for about $(0.3 \text{ km})^3$ radar sampling volume, the time series of the equivalent rainfall rate above a fixed point have a time scale of about 10 minutes. This means that an instantaneous radar measurement, made at short distance, is statistically representative of a 1-minute average (correlation above 0.9), and that a perfect time coverage of 100% of the basin can only be obtained if the antenna rotates at more than 1 rpm, as the ANATOL antenna does. From the recorded data, the radar time coverage is estimated by the relative duration of the union of every 1-minute intervals centered at each instant that the radar beam passes over a fixed given point. Due to the scanning procedures used in the data collection, and to some necessary interruptions, the time coverage for the 15-minute time intervals varies between 35 and 75%.

For a given time interval Δt, the quantitative comparisons are made by plotting either the R_Z values of the mean rainfall rate estimated from the radar data by using Eq. (6), i.e. without using the Z_{DR} values, or the R_D values estimated from the radar data by using Eq. (7), including the Z_{DR} values, versus the corresponding R_G values of the mean rainfall rate estimated from the gage data. These comparisons for the R_Z and R_D values are illustrated by Figs. 3 and 4, respectively, for the 15-minute time interval between 1500 and 1515 UTC on 3 July, 1986. In Figs. 3 and 4, the dashed line indicates the least-squares linear relationship between the gage and the radar values for all grid cells.

In this example, as shown in Fig. 3, most of the R_Z values are below the R_G values, except for rainfall rates of the order of 70 mm·h^{-1}. The correlation coefficient ρ, given among other parameters in Table I for the different time steps of different days, is 0.89 in this example, and the root mean square error σ (with respect to the least-squares relationship) is 8.6 mm·h^{-1}. This error could only be obtained if an appropriate $R - Z_H$ relationship, similar to Eq. (6), is used, but the coefficients of this relationship would be different for each time step, as shown in Table I. When using a fixed $R - Z_H$ relationship, although different for stratiform and convective conditions, the estimation error is of the order of the root mean square difference E_{GZ} between the gage value R_G and the radar derived value R_Z. This mean difference E_{GZ} is always greater than the root mean square error σ, and is above 22 mm· h^{-1} in this case.

The comparison of the raingage R_G and the radar R_Z values from the other time steps, shown in Table I, indicates that the correlation coefficient is above 0.7 when the maximum rainfall rate is large (above 5 mm·h^{-1}) and when the radar time coverage is above 60%. For small values of the maximum rainfall rate, the measurement errors due to the raingage tipping discretization (0.66 mm·h^{-1} for a 15-minute time interval) dominate the root mean square error σ. On the other hand, a small radar time coverage, particularly if the radar data are not evenly distributed over the time interval Δt, may prevent the radar to record some intense transient precipitation events which last only a few minutes, as happened on 3 July, 1986. For the analysed data set, the mean difference E_{GZ} is always above 24% of the maximum mean rainfall rate during the corresponding time step, and may reach 30% when the radar coverage is not large enough. As shown in Fig. 3, the pointwise errors are mainly below 100% but they may be as large as 500% in some cases.

The comparison of the raingage R_G values and the R_D values, deduced from the radar data by using Eq. (7), i.e. including the differential reflectivity Z_{DR} values, is illustrated in Fig. 4. In this figure, the points are closer to the main diagonal, with a larger number above this line, than in Fig. 3. This fact is confirmed by the comparison of Tables I and II which shows that in all cases, the mean difference E_{GD} between the gage value R_G and the radar derived value R_D, using Z_{DR}, is smaller than E_{GZ}, which is the mean difference given by a conventional radar. For example, the data illustrated in Figs. 3 and 4 lead to mean difference values of $E_{GD}=$ 17.0 and $E_{GZ} =$ 22.7 mm.h^{-1}, respectively. Therefore, the use of the differential reflectivity Z_{DR} values provides a 25% decrease in the mean difference between the raingage and the radar rainfall rate estimates, for a time interval $\Delta t = 15$ minutes. However, the correlation coefficient ρ is sometime slightly decreased, and the root mean square error σ is generally increased when using the Z_{DR} values. This results from the small scale microphysical inhomogeneities, not detected by the gage network or by a conventional radar, but which are revealed by the Z_{DR} values.

5. HAIL AND RADAR PARAMETERS

The most important hail parameters P_h are, for each hailpad and over the whole hailfall :

N_t : total number of hailstones per unit area (m^{-2}; in fact, the logarithmic values are used in order to approach the normal distribution),

E_t : total kinetic energy of the hailstones per unit area (J. m^{-2}; the logarithmic values are also used),

D_m : maximum recorded hailstone diameter (mm).

These hail parameters are estimated from the values of some radar parameters which are deduced from the time evolution of the radar reflectivities values recorded at low altitude (from 0.4 to 2.5 km) above each hailpad. The main radar parameters which have been tested are :

Z_{Hm} : maximum value of the horizontal reflectivity (dBZ),

$Z_{DR}(Z_{Hm})$: corresponding value, at the same time and location, of the differential reflectivity (dB),

$T_{58\ dBZ}$: time interval during which the reflectivity Z_H is above 58 dBZ (min),

All these parameters have been computed for the 25 hailpads dented by a unique well defined cell which appears either on 3 July, 1986 (1 cell), or on 14 August, 1988 (2 cells).

6. RADAR-BASED HAIL PARAMETER ESTIMATES

In order to test the improvement that the differential reflectivity Z_{DR} can bring to the estimation of the hail parameters P_h by a relationship of the form given by Eq. (2), such relationships involving 1 or 2 among the radar parameters have been obtained for the hail parameters, at first without using the Z_{DR} values (Table III), and then by using these values (Table IV). In Table IV, only the best two parameters are retained in the stepwise regression, since the number of data points (25) is too small to define a 3-parameter relationship.

The comparison between the 2-parameter relationships given in these 2 tables shows that the introduction of the $Z_{DR}(Z_{Hm})$ parameter increases, from $\rho = 0.798$ to 0.851 (from $\rho = 0.701$ to 0.866, respectively), the correlation between the radar estimated values of $log(E_t)$ (D_m respectively) and the hailpad measured values. On the opposite, no improvement is obtained for the radar estimation of the total number N_t of hailstones, by the use of the $Z_{DR}(Z_{Hm})$ parameter. The measured values of the kinetic energy E_t are plotted versus the radar estimated values in Fig. 5, when the 2-parameter relationship given in Table III is used, and in Fig. 6 for the corresponding relationship of Table IV. The comparison between Figs. 5 and 6 shows that, for the large measured values of the kinetic energy ($log(E_t) > 0.5, E_t > 3$ J.m^{-2}), the scattering of the points is smaller when the $Z_{DR}(Z_{Hm})$ parameter is used (Fig. 6), than when only non-polarized data are used (Fig. 5). For smaller measured values, the scattering remains important in both figures. Similarly, the radar estimated value of the maximum recorded diameter D_m is slighly better when one of the 2 parameters is the $Z_{DR}(Z_{Hm})$ parameter. However, for this D_m parameter, the correlation increase, brougth by the second radar parameter, is not statistically significant. These results suggest that the use of the differential reflectivity Z_{DR} values does improve the radar estimated values of the hail parameters E_t and D_m.

7. SUMMARY AND CONCLUSIONS

Measurements of the rainfall rate have been made by a network of 14 1-minute raingages, located at less than 10 km from a dual polarization radar, during several stratiform and convective precipitation events.

Analyses have been made of the inter-correlation function of the time series of the mean rainfall rates measured by any two gages during consecutive time steps t_k of size Δt. These analyses show that a high enough correlation ($\rho_0(M_i, M_j)$ greater than 0.8) can only be obtained for a distance between the gages located at M_i and M_j of the order of half the network mean distance, if the time interval Δt is greater than 15 minutes. This implies that, for time intervals smaller than 15 minutes, the mean rainfall rate measurements made by any raingage is not statistically representative of a mean value over the area that the gage is supposed to cover (1).

The comparisons between the mean rainfall rate values show that the correlation between the R_G values, interpolated by the kriging method from the gage data, and the R_Z values, deduced from the radar data without using the Z_{DR} values, is above 0.7 when the mean rainfall rate values are large, and when the radar time coverage is above 50%. The mean difference between the corresponding R_G and R_Z values is of the order of 25% of the maximum rainfall rate values. When using the Z_{DR} values, the corresponding mean difference decreases to 20%, implying that the use of the Z_{DR} values improve the estimation of the mean rainfall field for small time intervals. Indeed, the differential reflectivity Z_{DR} gives valuable information on small scale microphysical heterogeneities, but the raingage network appears not quite dense enough for a complete comparison at very short time intervals. Experiments are now planned in order to obtain accurate rainfall rate measurements at the ground which could be statistically representative for these very short time intervals.

Similarly, comparisons have been made between the hail intensity parameters, deduced from hailpad data, and their estimated values deduced from the dual polarization radar data. These comparisons show that the use of the differential reflectivity Z_{DR} values can significantly improve the radar estimation of these hail parameters. Preliminary analyses, based on a small number of hailpad data, indicates that the $Z_{DR}(Z_{Hm})$ value (value that Z_{DR} takes when the Z_H reflectivity reaches its maximum value Z_{Hm}) may already be a good radar parameter for the estimation of the total kinetic energy E_t of the hailfall, and of the maximum recorded hailstone diameter D_m. New radar parameters, which takes into account the time variation of the reflectivities Z_H and Z_{DR} during the whole hailfall, might be usefull for the estimation of time integrated hail intensities such as the total kinetic energy E_t of the hailfall.

Acknowlegments : The experiments have been carried out under the French Ministère de la Recherche et de la Technologie (M.R.T.) Contract 84F 0497, and with the partial support of the "Atmosphère Météorologique" program committee of the Centre National de la Recherche Scientifique (C.N.R.S.), and of the supporting institutions of the Groupement National d'Etude des Fléaux Atmosphériques (G.N.É.F.A.). The authors are indebted to all campaign participants, to R. Pejoux and F. Besserve for their help in the data treatments, and to O. Guillot and J. Squarise for editing this manuscript. Stimulating discussions with Drs. D. Creutin, H. Andrieu, J.F. Mezeix and D. Ramond, and with Prof. C. Obled are gratefully acknowledged.

REFERENCES

(1) EDDY, A., 1976 : Optimal raingage densities and accumulation times: a decision-making procedure. J. Appl. Meteor., 15, 962-971.

(2) WILSON, J.W., and E.A. BRANDES, 1979 : Radar measurement of rainfall - A summary. Bull. Amer. Meteor. Soc., 60, 1048-1058.

(3) DOVIAK, R.J., 1983 : A survey of radar rain measurement techniques. J. Climate Appl. Meteor., 22, 832-849.

(4) AUSTIN, P.M., 1987 : Relation between measured radar reflectivity and surface rainfall. Mon. Wea. Rev., 115, 1053-1070.

(5) CREUTIN, J.D., G. DELRIEU and T. LEBEL, 1988 : Rain measurement by raingage-radar combination : a geostatistical approach. J. Atmos. Oceanic Technol., 5, 102-115.

(6) SELIGA, T.A., K. AYDIN and H. DIRESKENELI, 1986 : Disdrometer measurements during an intense rainfall event in Central Illinois : Implications for differential reflectivity radar observations. J. Climate Appl. Meteor., 25, 835-846.

(7) ILLINGWORTH, A.J., J.W.F. GODDARD and S.M. CHERRY, 1987 : Polarization radar studies of precipitation development in convective storms. Quart. J. Roy. Meteor. Soc., 113, 469-489.

(8) SELIGA, T.A., and V.N. BRINGI, 1976 : Potential use of radar differential reflectivity measurements at orthogonal polarizations for measuring precipitation. J. Appl. Meteor., 15, 69-76.

(9) AYDIN, K., T.A. SELIGA and V. BALAJI, 1986 : Remote sensing of hail with a dual linear polarization radar. J. Climate. Appl. Meteor., 25, 1475-1484.

(10) HUSSON, D., Y. POINTIN and D. RAMOND, 1989 : Discrimination between hail and rain precipitation types from dual polarization radar, raingage and hailpad data. J. Theor. Appl. Climat., in press.

(11) WALDVOGEL, A., and W. SCHMID, 1983 : Single wavelength radar measurements of hailfall kinetic energy. Proc. 21st Conf. on Radar Meteorology. Edmonton. Amer. Meteor. Soc., 425-428.

(12) POINTIN, Y., D. RAMOND and J. FOURNET-FAYARD, 1988 : Radar differential reflectivity Z_{DR} : A real-case evaluation of errors induced by antenna characteristics. J. Atmos. and Oceanic Technol., 5, 416-423.

(13) CREUTIN, J.D., and C. OBLED, 1982 : Objective analysis and mapping techniques for rainfall fields : an objective comparison. Water Resour. Res., 18, 413-431.

(14) WOODLEY, W.L., A.R. OLSEN, A. HERNDON and V. WIGGERT, 1975 : Comparison of gage and radar methods of convective rain measurement. J. Appl. Meteor., 14, 909-928.

(15) HUFF, F.A., 1970 : Sampling errors in measurement of mean precipitation. J. Appl. Meteor., 9, 35-44.

Interval $\Delta t = 15$ min M/D/Y : Time steps	$R_{G\ max}$ $mm \cdot h^{-1}$	a	b $mm \cdot h^{-1}$	ρ	σ $mm \cdot h^{-1}$	E_{GZ} $mm \cdot h^{-1}$	R.T.C. %
6/15/86 : 1830 − 1845	39.1	0.50	2.46	0.83	2.4	8.4	48
6/15/86 : 1845 − 1900	9.3	−0.20	1.80	−0.57	0.2	4.1	10
7/03/86 : 1400 − 1415	8.5	0.36	5.25	0.28	2.6	3.4	16
7/03/86 : 1415 − 1430	15.9	0.50	2.86	0.29	2.9	5.1	55
7/03/86 : 1430 − 1445	20.4	0.33	2.62	0.71	1.5	8.0	66
7/03/86 : 1445 − 1500	59.5	0.03	33.65	0.04	9.2	14.0	33
7/03/86 : 1500 − 1515	80.0	1.31	−36.40	0.89	8.6	22.7	70
7/03/86 : 1515 − 1530	28.2	0.61	−1.39	0.94	1.7	6.9	40
7/03/86 : 1530 − 1545	2.9	0.26	−1.97	0.52	0.3	1.2	62
7/03/86 : 1545 − 1600	1.8	0.35	0.29	0.42	0.3	0.4	74

TABLE I : Results of the comparison between the mean rainfall rate values deduced from the gage data R_G and that deduced from the radar data R_Z, i.e. without using the differential reflectivity Z_{DR}. For the different time steps t_k, values of the maximum mean rainfall rate $R_{G\ max}$, of the least-squares coefficients $R_Z = aR_G + b$, of the correlation coefficient ρ, of the root mean square error σ, of the mean difference $E_{GZ} = R.M.S.(R_G - R_Z)$, and of the estimated Radar Time Coverage R.T.C.

Interval $\Delta t = 15$ min M/D/Y : Time steps	$R_{G\ max}$ $mm \cdot h^{-1}$	a	b $mm \cdot h^{-1}$	ρ	σ $mm \cdot h^{-1}$	E_{GD} $mm \cdot h^{-1}$	R.T.C. %
6/15/86 : 1830 − 1845	39.1	0.78	5.46	0.88	3.0	3.5	48
6/15/86 : 1845 − 1900	9.3	−0.22	1.86	−0.41	0.3	4.2	10
7/03/86 : 1400 − 1415	8.5	−0.01	11.64	0.00	6.9	9.4	16
7/03/86 : 1415 − 1430	15.9	0.53	5.23	0.25	3.5	3.9	55
7/03/86 : 1430 − 1445	20.4	0.34	3.82	0.71	1.6	6.7	66
7/03/86 : 1445 − 1500	59.5	0.22	33.90	0.32	6.5	10.6	33
7/03/86 : 1500 − 1515	80.0	1.41	−34.11	0.88	9.9	17.0	70
7/03/86 : 1515 − 1530	28.2	0.48	−0.39	0.93	1.5	7.3	40
7/03/86 : 1530 − 1545	2.9	0.33	0.20	0.61	0.3	0.9	62
7/03/86 : 1545 − 1600	1.8	0.37	0.46	0.44	0.3	0.4	74

TABLE II : Results of the comparison between the mean rainfall rate values deduced from the gage data R_G and that deduced from the radar data R_D, using the differential reflectivity Z_{DR}. For the different time steps t_k, values of the maximum mean rainfall rate $R_{G\ max}$, of the least-squares coefficients $R_D = aR_G + b$, of the correlation coefficient ρ, of the root mean square error σ, of the mean difference $E_{GD} = R.M.S.(R_G - R_D)$, and of the estimated Radar Time Coverage R.T.C.

$log(N_t) =$	-7.58	$+ 0.166\ Z_{Hm}$		$\rho = 0.718$
$log(E_t) =$	-15.9	$+ 0.276\ Z_{Hm}$		$\rho = 0.772$
$D_m =$	-80.8	$+ 1.55\ Z_{Hm}$		$\rho = 0.682$
$log(N_t) =$	-4.19	$+ 0.108\ Z_{Hm}$	$+ 0.054\ T_{58\ dBZ}$	$\rho = 0.754$
$log(E_t) =$	-11.3	$+ 0.197\ Z_{Hm}$	$+ 0.074\ T_{58\ dBZ}$	$\rho = 0.798$
$D_m =$	-57.1	$+ 1.15\ Z_{Hm}$	$+ 0.383\ T_{58\ dBZ}$	$\rho = 0.701$

<u>TABLE III</u> : Estimation relationships, and the corresponding correlation coefficient ρ, of the hail parameters N_t, E_t, and D_m as functions of 1, or of 2 of the radar parameters selected among Z_{Hm}, and $T_{58\ dBZ}$, without using the differential reflectivity Z_{DR} values.

$log(N_t) =$	-7.58	$+ 0.166\ Z_{Hm}$		$\rho = 0.718$
$log(E_t) =$	1.85	$- 0.547\ Z_{DR}(Z_{Hm})$		$\rho = 0.795$
$D_m =$	19.7	$- 3.75\ Z_{DR}(Z_{Hm})$		$\rho = 0.856$
$log(N_t) =$	-4.19	$+ 0.108\ Z_{Hm}$	$+ 0.054\ T_{58\ dBZ}$	$\rho = 0.754$
$log(E_t) =$	-7.74	$- 0.344\ Z_{DR}(Z_{Hm})$	$+ 0.151\ Z_{Hm}$	$\rho = 0.851$
$D_m =$	18.2	$- 3.33\ Z_{DR}(Z_{Hm})$	$+ 0.253\ T_{58\ dBZ}$	$\rho = 0.866$

<u>TABLE IV</u> : Estimation relationships, and the corresponding correlation coefficient ρ, of the hail parameters N_t, E_t, and D_m as functions of 1, or of 2 of the radar parameters selected among $Z_{Hm}, Z_{DR}(Z_{Hm})$ and $T_{58\ dBZ}$, i.e. including the differential reflectivity Z_{DR} values.

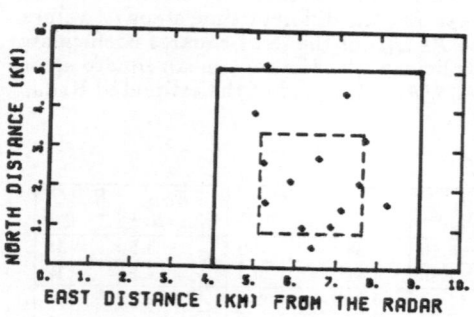

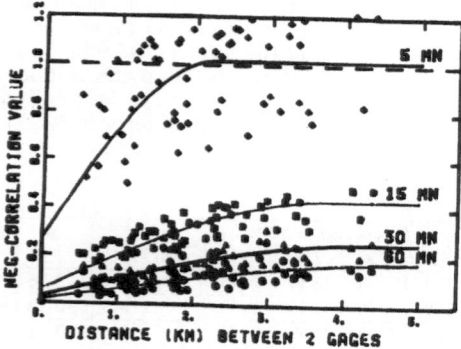

<u>FIG. 1.</u> Location of the 14 1-minute raingages with respect to the radar position. The quantitative comparisons are made within the dashed line square, in order to avoid extrapolating the gage data.

<u>FIG. 2.</u> Neg-correlation $\omega(M_i, M_j) = 1 - \rho_0(M_i, M_j)$ coefficients between the mean rainfall rate R_G time series recorded during April, 1986, by all gage pairs (M_i, M_j) of the network, plotted versus the distance between the 2 gages of the pair. The different symbols refer to different time intervals Δt ($\bullet = 5$, $\blacksquare = 15$, $\blacktriangle = 30$, $\bullet = 60$ min) over which the mean rainfall rates are averaged. The solid lines represent the adjusted neg-correlation functions $\omega_s(h, \Delta t)$ drawn versus the gage distance h, for the different Δt.

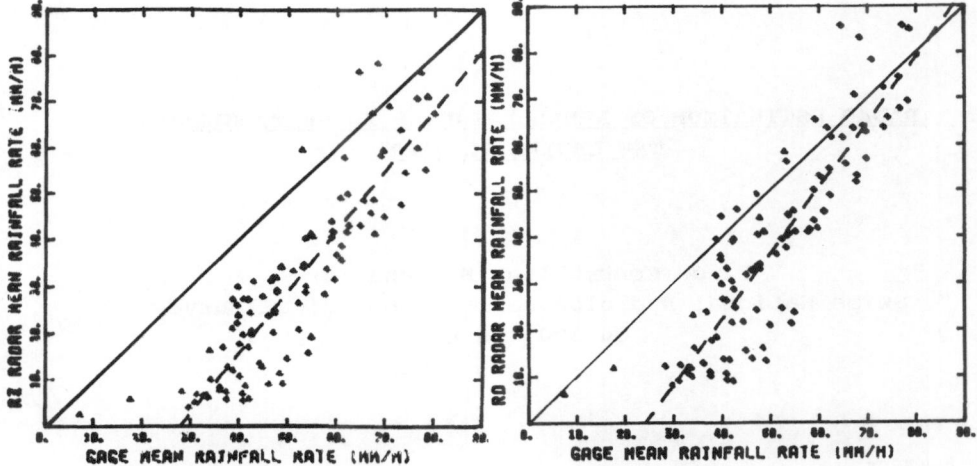

FIG. 3. Comparison between the mean values, over the 15-minute time interval lasting from 1500 to 1515 UTC on 3 July, 1986, for the different grid meshes of sizes $\Delta x = \Delta y = 0.25$ km, of the mean rainfall rate R_Z, deduced from the radar data, plotted versus the corresponding mean rainfall rate R_G, deduced from the gage data.

FIG. 4. Idem Fig. 3 for the comparison between the mean rainfall rate R_D, deduced from the dual polarization radar data, and the corresponding mean rainfall rate R_G, deduced from the gage data.

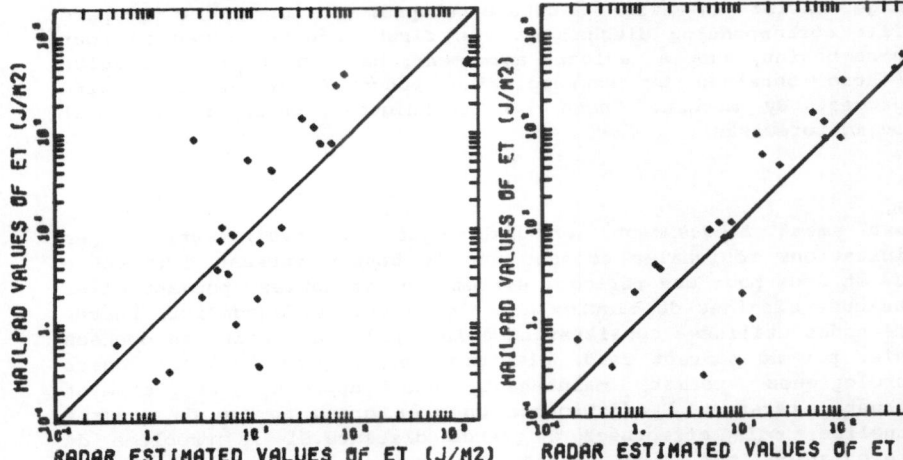

FIG. 5. Value, for each of the 25 hailpads, of the measured total kinetic energy E_t versus the radar estimated value deduced from the 2-parameter relationship of Table III: $\log(E_t) = f[Z_{Hm}, T_{58dBZ}]$

FIG. 6. Idem Fig. 5 for the 2-parameter relationship of Table IV: $\log(E_t) = f[Z_{DR}(Z_{Hm}), Z_{Hm}]$ i.e. including the differential reflectivity values.

RADAR ESTIMATION OF AREAL RAINFALL IN SWITZERLAND - THE GENIRA PROJECT

F. de Montmollin, B. Schädler
Swiss National Hydrological and Geological Survey
CH-3003 Berne

Summary

In hydrology, an estimate of areal rainfall over drainage basin has to be frequently given for very varying periods, any time pace from about ten minutes to a few years. Up to now, every method has been carried out by interpolation based on data obtained through often scarce raingauges. Since the advent of meteorological radar, a completely different technique can be used, which in fact means calculating very numerous punctual measurements at a great distance. Although the problem of interpolation can thus be avoided, others nevertheless remain to be solved. These two methods are now being applied to measure areal rainfall amounts within about fifteen drainage basins in Swiss mountain slopes, in order to compare specific corresponding discharges. The first results, based on four of these basins, show a mediocre agreement, between the radar results and those obtained by extrapolating punctual measurements with raingauges. By merging these two techniques, there is hope of improving forecasts

Résumé

On est amené fréquemment, en hydrologie, à devoir estimer les précipitations régionales tombant sur le bassin versant d'un cours d'eau, et ceci pour des périodes extrêmement variables, pouvant aller de quelques dizaines de minutes à quelques années. Jusqu'ici, toutes les méthodes utilisées consistaient à interpoler à partir des données fournies par de souvent rares pluviomètres. L'apparition des radars météorologiques permet maintenant une approche complétement différente, revenant à additionner un très grand nombre de mesures ponctuelles, mais effectuées à grande distance.Si le problème de l'interpolation est ainsi évité, de nombreux autres se posent néanmoins. Les précipitations régionales d'une quinzaine de bassins versants hydrologiques suisses sont actuellement déterminées suivant ces deux voies pour être comparées aux débits spécifiques correspondants. Les premiers résultats, obtenus à partir de 4 de ces bassins, montrent une concordance médiocre entre les résultats du radar et ceux obtenus en extrapolant les mesures ponctuelles des pluviomètres. On peut néanmoins espérer améliorer la qualité des prévisons hydrologiques en combinant les deux techniques.

1. INTRODUCTION

Besides supplying rough data information on water levels and discharges about swiss water courses, the "Hydrology Division" (HD) of the "Swiss National Hydrological and Geological Survey" (SNHGS), also produces quite elaborate analysis and research work, some of which require a thorough knowledge of the precipitation field over catchment basins. This is true also for the operational hydrological forecasts for the Rhine which are issued daily, with a lead time of three days. For this purpose the information is obtained by telemetry with the raingauges of a special automatic network, belonging to the Swiss Meteorological Institute (SMI) and called "ANETZ". To asses the areal average precipitation over a basin, some method of interpolation is needed, followed by an integration. Several interpolating techniques are in use from the simplest (such as Thyssen's polygons (1), correlation with the altitude (2) to the most complex (Universal Krigage or Matheron's "Theory of Regionalised Variables" (3) to (5)), but in fact none of these techniques can really make up for the extremely low sampling ratio between the few square decimetre surface, measured by ever too scarce raingauges and some hundred square kilometres measured by the drainage basins.

The use of weather radar leads the way to a completely different approach which enables the hydrologist to obtain an instant overall coverage of the precipitation field. A computer can cope with such an image, for instance, by gathering precipitation quantities in a given area. In order to further such possibilities, the "Hydrology Division" (HD) got in touch with research workers responsible for the introduction of weather radar network in Switzerland, notably J.Joss and G.Kappenberger of the Swiss Meteorological Institute (SMI), working at Osservatorio Ticinese di Locarno-Monti and A.Waldvogel of the Swiss Federal Institute of Technology in Zurich. A joint project called "GENIRA" was the result of a close collaboration between the HD and these different workers. ("GENIRA" stands for the german words "Gebietsniederschlaege mit Radar", radar areal average precipitation evaluation).

The "GENIRA" project uses fifteen test basins for which the SMI works out the average areal rainfall with whatever desirable pace from ten minutes up to a year. From its automatic ground-based network ("ANETZ") the SMI also provides recorded corresponding values in about fifteen selected stations. At present for four of these basins a comparison between radar areal rainfall data measurements and those given by the "Rhine Short-range Runoff Forecasing Model" (12) has been carried out. The results provided by both techniques are to be compared with the measure of runoff volumes at the basin outlet, converted to millimeters depth on the basis of the catchment area.

2. THE THREE TERMS OF THE COMPARISON

Likely errors made in quantitative radar measurements of rainfall have been examined by J. Joss and A. Waldvogel (6) who have also carefully checked up on future correcting techniques. At the conference on radar meteorology which took place in march 1989 in Florida (7), these workers submitted a paper discussing among others various correction methods relying on statistical analysis of vertical radar reflexion profiles. On glancing through these texts, you may have the impression that an unreliable measurement technique is being compared with a perfect method represented by the raingauge measurement. In actual fact the truth seems slightly different if we are to believe in what J.C. Rodda says (8): "Mesurements of precipitations are perhaps the supreme exemple of data accepted at their face value. For these measurements which are employed in so many of the activities of the hydrologists, meteorolgists and almost everyone concerned with the use of water, have invariably been assumed to be accurate. This assumption is not valid, however, for there is no means of assessing the accuracy of the gauges comprising the bulk of most national raingauge networks - the so-called 'standard' gauges". More recently, B. Sevruk (9) checked up on different techniques used for correcting measurements by raingauges: It can indeed happen that raingauges register 30 percent less than the actual rainfall amounts. But so far, it is only a matter of measurements with just one raingauge. The fine framework of the rainfall field seems to show quite an irregular aspect, as it has been proved by statistics carried out on very dense networks of raingauges, the work done by Jacquet (10) is an example. The performance of pin-pointed raingauge mesurements is to be therefore wondered about when comparing with radar measurements for which one point is a pixel of 2 x 2 km.

By some means or other, an interpolation has to be operated to estimate rainfall amounts as no information can be obtained otherwise in the vast expanse between the raingauges. A special attention should be paid to topographical relief, as it plays an important part. But on what other basis should the missing measurements be worked out ? The only obvious solution for the moment to this problem is by radar estimations !

As for the discharge measuring, it can be considered as relatively accurate, although very rare examples of errors may occur. The natural process of converting precipitations into runoff remains and will probably remain yet for a long time quite unforeseeable, because of the reaction of a basin varying constantly from one rainy period to another, as recalled by V.N. Gupta and O.T. Mesa (11).

As doubt subsists about the reliability of the measurements of the three terms compared here , it is not so surprising that no satisfactory agreement has been reached yet.

3. FIRST RESULTS

By integrating numerically composite radar imagery, SMI experts working in Locarno-Monti, managed to calculate areal rainfall with a combination of radar echoes transmitted from La Dôle, summit of the Jura, north of Geneva and from the Albis, a hill near Zurich. The areal rainfall, interpolated or extrapolated from raingauges of the SMI's automatic telemetry network "ANETZ" was carried out owing to a technique adjusted, for a Rhine forecasting model, by Lang, Jensen and Grebner (12), Swiss Federal Institute of Technology in Zurich, Department of Geography, Hydrology Section. The model in which this technique is embedded is now operated by the Hydrology Division of the Swiss Hydrological and Geological Survey for forecasts on the Rhine discharge. The method especially involves interpolation of rainfall amounts from raingauges network for the centres of gravity of about a hundred km^2 elementary basins. Raingauges are not all located within the 4 chosen basins as shown here below, which tends to greater inaccurate estimates. For the hydrological year 1987-88 (Oct.-Sept.), the following values have been established :

Drainage basins:	Raingauges: Basin:	Neighbg.:		Depths of runoff (in mm) P - Q = E				
Thur-Halden	2	3	ANETZ:	1618	-	1191	=	427
			RADAR:	1481	-	1191	=	290
Thur-Andelfingen	3	3	ANETZ:	1407	-	951	=	456
			RADAR:	1393	-	951	=	442
Langeten-Lotzwil	0	4	ANETZ:	1120	-	696	=	424
			RADAR:	2198	-	696	=	1502
Kleine Emme-Littau	2	3	ANETZ:	1394	-	989	=	405
			RADAR:	2342	-	989	=	1353

where P = precipitation, Q = discharge and E = evaporation

Some characteristics of the basins mentioned above appear in the following table:

Basins	Average height (m)	Surface (km2)	Distance (km) Dôle:	Albis:	Visible height (m) Dôle:	Albis:
Thur-Halden	910	1120	259	54	-	1500
Thur-Andelfingen	770	1752	252	45	-	1100
Langeten-Lotzwil	713	116	151	57	800	900
Kleine Emme-Littau	1050	484	164	49	1600	1700

Values of the term "evaporation" calculated on the basis of radar precipitation for Lotzwil and Littau are very plainly over-estimated and differ much from usual values for Switzerland which are between 250 and 700 m per year, as pointed out by Schaedler (13). The radar seems to detect an important background for these two basins, even during dry period (figure 2). It is also possible that the position of the raingauges could be the cause for certain under-estimation of precipitations, fortuitously combined with a roughly equivalent under-estimation in the runoff, owing to ground-water underflow. Comparison of monthly available amounts is shown by figure 1 and a comparison of daily values, in figure 2, during a flood, in June 1988.

4. CONCLUSIONS

It is too early to draw a definitive conclusion from this experiment. The corrections proposed by Joss and Waldvogel (6,7) will certainly improve to a large extent the radar surveys. A good estimation of areal precipitations should be expected from a joint radar and raingauge method (bivariate radar/raingauge analyses has been reported by Hall & Barclay (14)). Even with simplified corrections the radar should already improve the methods for solving numerous hydrological problems.

ACKNOWLEDGMENTS

We are particularly grateful to A. Pittini of the Osservatorio Ticinese di Locarno-Monti for preparing radar data and ANETZ, also to R. Bigler for calculating the total discharges and for preparing the diagrams.

REFERENCES

(1) REMENIERAS,G.,   1965:   L'hydrologie   de   l'ingénieur (Méthode   rapide   de   Thyssen,   pp.   155-157.) Eyrolles,Paris, 1965

(2) MONTMOLLIN, F. DE, OLIVIER, R. & ZWAHLEN F., 1979: Utilisation d'une grille d'altitudes digitalisées pour la cartographie d'éléments du bilan hydrique. Journal of Hydrology, 44 (1979), pp. 191-209

(3) MATHERON, G., 1972: Théorie des variables régionalisées. In: Laffitte P., Traité d'informatique géologique, Masson, Paris, pp. 306-378

(4)   DELFINER,P. & DELHOMME, J.P. 1975: Optimum interpolation
      by Kriging. In: Davis J.C. & McCullagh, M.J., Display
      and Analysis of Spatial Data. NATO Advanced Study
      Institute, John Wiley & Sons.

(5)   MONTMOLLIN, F. DE, OLIVIER, R., SIMARD, R. & ZWAHLEN, F.
      1980: Evaluation of a precipitation map using a
      smoothed elevation-precipitation relationship and
      optimal estimates (Kriging), Nordic Hydrology 11, 1980,
      pp. 113-120.

(6)   JOSS,J. & WALDFOGEL, A. 1988: Preciptation measurements
      and hydrology, a review. To appear in the Battan
      Memorial and 40th Anniversary Radar Meteorology Volume,
      Editor : David Atalas.

(7)   JOSS,J. & WALDFOGEL, A. 1989: Areal preciptation
      measurements and vertical reflectivity profile
      corrections. Paper for the 24th Conference on Radar
      Meteorology, March 27-31 1989 in Tallahassee, Fla.

(8)   RODDA, J.C., 1971: The precipitation measurement paradox
      - the instrument accuracy problem. Reports on WMO/IHD
      Projects No 16, WMO-No 316, Geneva, 42 pp.

(9)   SEVRUK, B., 1982: Methods of correction for systematic
      error in point precipitation measurement for
      operational use. Operational Hydrology Report No. 21,
      WMO-No.589, Geneva 91 pp.

(10) JACQUET, J., 1960: Répartition spatiale des
      précipitations à l'échelle fine et précision des
      mesures pluviométriques. Publication No 53 de
      l'Association internationale d'hydrologie scientifique
      (A.I.H.S.), Commission d'Erosion Continentale, pp.
      317-342.

(11) GUPTA,K. & MESA,O.J.1988: Runoff generation and
      hydrologic response via channel network geomorphology -
      recent progress. J. of Hydrology, 102(1988) pp. 3-28

(12) LANG H., JENSEN, H. & GREBNER, D. 1987: Short-range runoff
      forecasting for the River Rhine at Rheinfelden:
      experiences and present problems. Hydrological Sciences
      Journal 32,3, 9/1987, pp. 385-397

(13) SCHÄDLER, B. 1985: Der Wasserhaushalt der Schweiz.
      Mitteilung Nr 6 , Service hydrologique et géologique
      national, 3003 Bern

(14) HALL, A.J. & BARKLAY, P.A. 1980: Design of operational
      areal rainfall networks using ground-based and radar
      data. "Hydrological forecasting (Proceedings of the
      Oxford Symposium, April 1980) : IAHS Publ. no 129,
      pp.   51-56.

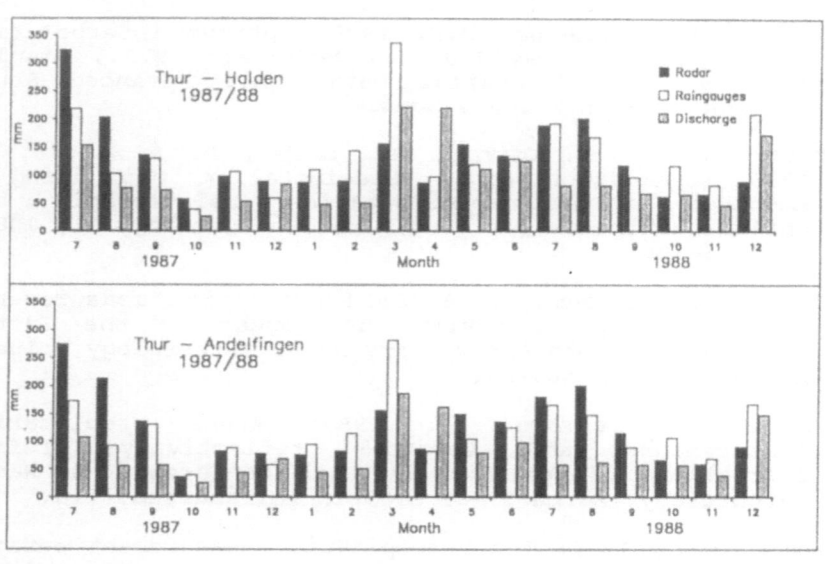

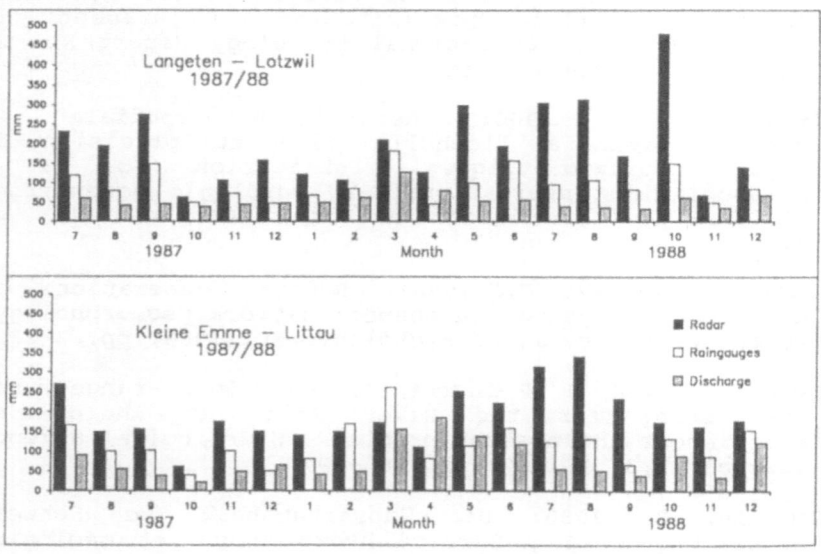

Fig. 1   Comparison of monthly areal precipitations according to radar and raingauges methods

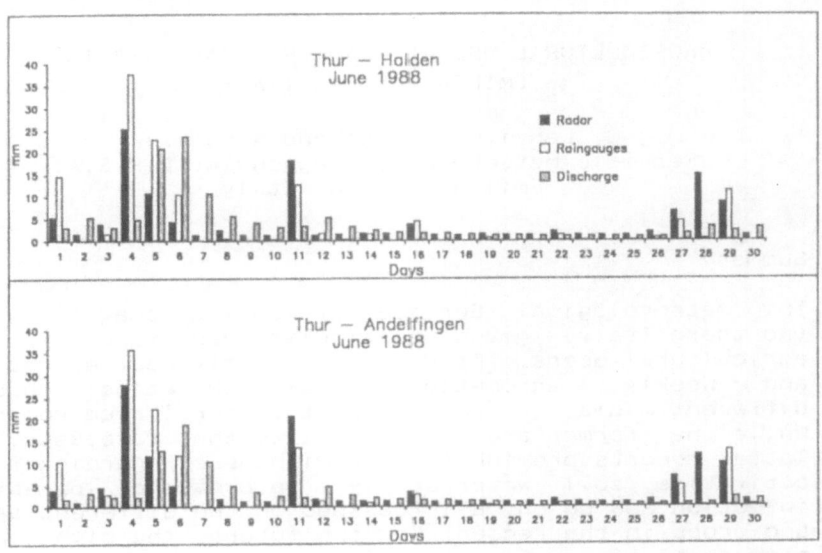

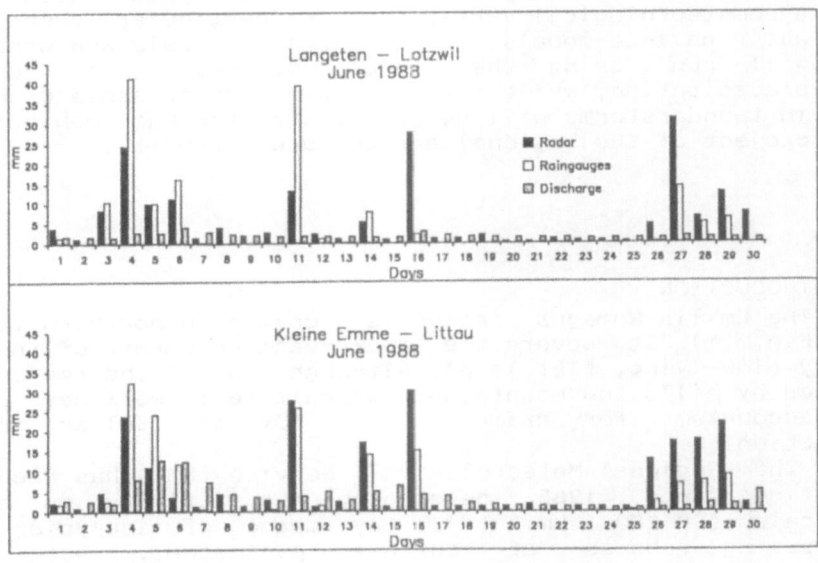

Fig. 2 Comparison of daily areal precipitations according to radar and raingauges methods

AGRICULTURAL USE OF WEATHER RADAR DATA IN EMILIA-ROMAGNA ITALY

G. Lenzi, S. Nanni and A. Salsi
Servizio Meteorologico Regionale, E.R.S.A.
Emilia-Romagna, Italy

Summary

The Meteorological Service of Emilia-Romagna region (northern Italy) gives particular assistance to the agricultural users. It boadcasts daily weather reports and weekly agrometeorological bulletins through different media in order to meet farmers' requirements. While the former are traditional weather forecasts, the latter reports provide farmers with news regarding field activities, soil water supply, outbreaks of parasitic infection and so on, all focussed on the different areas and crops in the region. For the future, the high space-time resolution of weather radar data (a C band doppler radar is being installed at S. Pietro Capofiume, near Bologna) will make it possible to improve the assistance to agriculture by enabling some input data for agrometeorological variables. In particular crop-soil water balance models, epidemiological models and others, will run using the radar estimation of areal precipitation, while information on the presence of hail in thunderstorms will be helpful for the hail monitoring project of the regional agriculture authority.

1. INTRODUCTION

The Emilia-Romagna region is located in northern Italy (see Fig. 1). It covers the south-eastern corner of the Po Valley (low-lying, flat land), although most of the region is covered by hills and mountains. Agriculture is well developed and accounts, for example, for 40% of Italian fruit production.

The Regional Meteorological Service (SMR) has been in operation since 1985, developing its activity on beta mesoscale (20-200 km, (1)) by means of analysis and forecasting schemes of surface parameters, such as temperature (2) and precipitation (3). These parameters can affect the crops in several ways: the low temperatures during critical periods (particularly in the spring season) may result in frost, while high temperatures, combined with some other conditions, may produce dangerous diseases, such as grape downy mildew and apple scab.

With regard to precipitation, the problem of its insufficiency in the soil may be overcome by irrigation, which may be optimized using crop soil water balances, while the occurrence of too much rain on certain days may cause serious damage to, or even destroy, hay crops. Too much rain may also delay sowing and harvesting. Another serious problem related to precipitation is that of hailstorms. This has been discussed with reference to the whole of the Po Valley (4) and, more recently, with reference to Emilia-Romagna in particular (5, 6).

The availability of duration and intensity precipitation data at high space resolution (1 km²) through the new radar will be very helpful to more accurately define some of these agrometeorological problems.

2. CHARACTERISTICS OF THE NEW RADAR

The radar being installed at the Meteorological Center of S.Pietro Capofiume (near Bologna) is a GPM 500 C constructed by COTIM, a consortium of Italian companies, who also delivered the software according to COST-72 guidelines. The main features of this radar (7), which is a C band Doppler radar with double polarization, are as follows:
- radar beam 0.9 degree
- antenna gain > 46 dB, secondary lobes < - 30 dB, levels of cross polarization < -27 dB peak
- transmitted power 500 kw, pulse length .5 or 1.5 or 3 microsec., with corresponding PRF of 1200, 600, 300 Hz.

A complete volume scan can be made by the radar every 5 min. The radar signal processor (8) gives in real time and simultaneously for 1024 range gates the following parameters: Z, Zdr, V, Vstd .The radar data processor can provide in real time CAPPI and RHI of the above listed parameters, on a cartesian grid up to 1024x1024 points resolution.

The computer system provides moreover the meteorological products according to COST-72 guidelines, within 1 minute from the end of the volume scan

With regard to the precipitation measurements, the radar data will be compared with the amounts received from 11 automatic weather stations of SMR network and 28 raingauges belonging to other regional institutions (see Fig. 2).

3. FUTURE APPLICATIONS OF RADAR PRODUCTS IN AGROMETEOROLOGY

The main aim of SMR is to aid farmers in taking those decisions depending on meteorological situations. For this purpose SMR broadcasts regional weather information and forecasts to several users and to Local Agrometeorological Sections (SAL), which specialize them with local agronomic information in order to take into account the different characteristics of the areas under examination.

The use of the new radar might improve regional agrometeorological assistance in different fields:
  a) crop soil water balance
  b) hail measurements
  c) epidemiological models
  d) hay curing
  e) soil tilling

f) spring sowing
g) nitrogen leaching

a) Crop soil water balance.
The increased availability of water for crop irrigation has completely changed the way of farming in our region. During the growing season the farmers have to face the following problems: when they shall irrigate, and how much water they have to use for irrigation. There are many different approaches in measuring the amount of available water in the soil, many of them are based on a crop soil water balance calculation.

Simple models have been developed by the SMR (9), and other institutions and all of them require the input of meteorological parameters, such as rain quantity (R) and intensity, or processed parameters, such as evapotranspiration (ETP). Using radar outputs and an analysis for ETP calculation, it will be possible to set up three different procedures:
1) Daily calculation of the hydrological deficit, R-ETP, for the whole of our region with a space resolution equal to that of the radar outputs.
2) Theoretical crop soil water budget estimation for some regional crops, (for example, sugar-beet and wheat), obtained from both the kind and use of soil and phenological data, which are available at a good space resolution.
3) Crop soil water balance for direct assistance to farmers, through the SAL or other institutions: those programs based on main outputs of SMR will run on resident Personal Computer and will consider all the different characteristics of the assisted farm and farmers activity (irrigation, etc.)

b) Hail measurements
Probably the Po Valley points out one of the worst hail problem in the world, with an average loss in agriculture reaching $7106 per square mile (4).
A hailpad network, managed by the Departement of "Food and Agriculture" Productive Activity of Emilia-Romagna region, is in operation since 1983 in the eastern flat part of the region (6). This network includes 426 hailpads with a chess-shaped mesh area of 16 km^2, covering 60% of the regional agriculture cultivated area, which is characterized by high Italian fruit production (60% for pear and kiwi). The main purpose of the public administration is to get an objective valuation of the priority of interventions in order to support farms damaged by hailfalls. After six year of activity, during the months from April to September 1023 hail reports were recorded relative to 129 hail day events. It is notable that 30% of these events were so localized to be recorded only by a single hailpad. During the same events the associated damages were sometimes not negligible, depending on the phenological phase of the cultivars considered.

For the future, the radar high spatial resolution together with the hailpad data will be a powerful tool to obtain a more accurate estimation of the crop loss (9).

c) Epidemiological models

Integrated pest management requires the use of models for the forecasting of parasitic infection. Most of these models use rain amount as input variable. Diffusion of daily radar outputs to those institutions taking care of agricultural assistance in this field will optimize the models forecasting capability with a high space resolution.

d) Hay curing

A model for the monitoring of hay curing through the use of meteorological parameters has been developed at SMR (11). This model calculates the moisture content of cut forage in order to collect hay when it reaches a precise humidity level, improving its conservability and quality. The effect of rain on cut forage is a function of rain quantity and forage humidity.

The time required to dry up the rainfall absorbed by forage can be computed by the model. Radar rain data will ensure direct assistance to farmers requiring it from SAL.

e) Soil tilling

The compaction of wet soil caused by heavy machinery is to be avoided as far as possible. The knowledge of rain amount and soil water balance gives some insight about the soil condition and consequently about the possibility of performing or delaying certain field activities.

f) Spring sowing

The choice of the most suitable time for spring sowing (maize, soy bean, tomato etc.) depends on soil temperature and soil water content. The latter can be estimated using soil water balance models as stated in a), above.

g) Nitrogen leaching

Nitrogen fertilization is a very important practice with economic and environmental feedback. The dynamics of nitrogen in the soil might be simulated by an appropriate model(12). For this purpose it is necessary to know the precipitation amount and intensity which cause nitrogen leaching. Such a model is at present being tested by SMR jointly with other research institutions with special care to winter cereals cultivations. The information related to the estimated nitrogen content available at the beginning of spring will improve assistance regarding soil fertilization (e.g. "how much"), thus reducing costs and avoiding negative environmental effects (such as water eutrophication).

CONCLUSIONS

A C-band Doppler radar, being installed at S. Pietro Capofiume (near Bologna) from Regional Meteorological Service of Emilia-Romagna, will supply a lot of information to agricultural users.

Some agrometeorological models developed at SMR, such as crop soil water balance and hay curing, as well as other models in progress will improve the present resolution by means of the radar areal precipitation estimation

The radar will be also very useful in the hail detection problem, providing a new support in the definition of the risk areas for the regional cultivations.

REFERENCES

(1)..ATKINSON, B.W. (1981). Meso-scale Atmospheric Circulations. Academic Press

(2) CACCIAMANI C., PACCAGNELLA T., NANNI S. and TIBALDI S.(1989) Objective Mesoscale Analysis of Daily Extreme Temperatures in the Po Valley of Northern Italy. In press on TELLUS A.

(3) CACCIAMANI C., NANNI S. and PACCAGNELLA T. (1988) Rainfall Forecast over Emilia Romagna Region (Italy) using a dynamic statistical method. IV International Conference on Meteorology and Road Safety. Florence, Italy. 8-10 Nov. 1988.

(4) MORGAN G.M.Jr. (1973) A General Description of The Hail Problem in the Po Valley of Northern Italy. Journ. of Appl.Metor, 12, 338-352 .

(5) CACCIAMANI C. and SIMONINI G. (1988) The Monitoring of Thunderstorm Activity in the Italian Region Emilia Romagna: Climatological Aspects and Operative Forecasting Systems. IV Inetrnational Conference on Meteorology and Road Safety. Florence. Italy. 8-10 Nov. 1988.

(6) VENTO D. and MALOSSINI A. (1989) Hailpad data of South East Po Valley Network. V WMO Sci. Conf. on Weather Modif. Beijing-China. 8-12 May 1989.

(7) LENZI G., VEZZANI G.F. and ROSSETTINI A. (1988) Weather Radar applications to Road Safety. IV International Confererence on Meteorology and Road Safety. Florence, Italy 8-10 Nov. 1988.

(8) BRUNKOW D. and LEE R. (1986) Chill Data System. Preprints of 23rd Conf. on Radar Meteor., Snowmass, Colorado. Amer. Meteor. Soc., 354-356.

(9) MARLETTO V. and ZINONI F. (1988) Applicazione operativa dei bilanci idrici semplificati. 1: Organizzazione, risultati e prospettive di un metodo a risoluzione territoriale. 2: Organizzazione, risultati e prospettive di un metodo a risoluzione aziendale. Bollettino Geofisico. Atti del Convegno Nazionale di Meteorologia per l'Agricoltura. Perugia, Italy 26-28 May 1988.

(10) WOJTIW L. and EWING C.G. (1983) The Use of Radar to Estimate Crop Damage for Hailstorms in Alberta, Canada.Preprints of 21rd Conf. on Radar Meteor, Edmonton, Alta Amer. Meteor. Soc., 435-441.

(11) DYER J.A. and BROWN D.M. (1977). A Climatic Simulator for Field-Drying Hay. Agricult. Meteor., 18, 37-48.

(12) GROOT J.J.R. (1987) Simulation of Nitrogen Balance in a System of winter wheat and soil. Simulation Report CABO-TT nr 13.

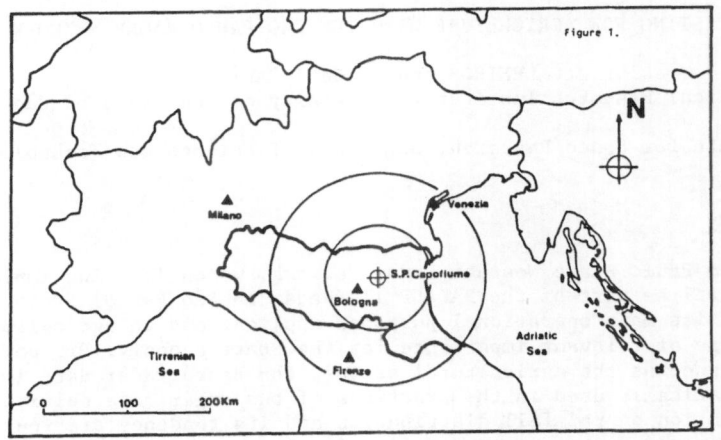

Fig. 1) The administrative border of Emilia-Romagna region in northern Italy. The cross-circle inside the region represents the location of the radar GPM 500 C being installed at S.Pietro Capofiume. The two circle represent the 64 km and 128 km range gates respectively

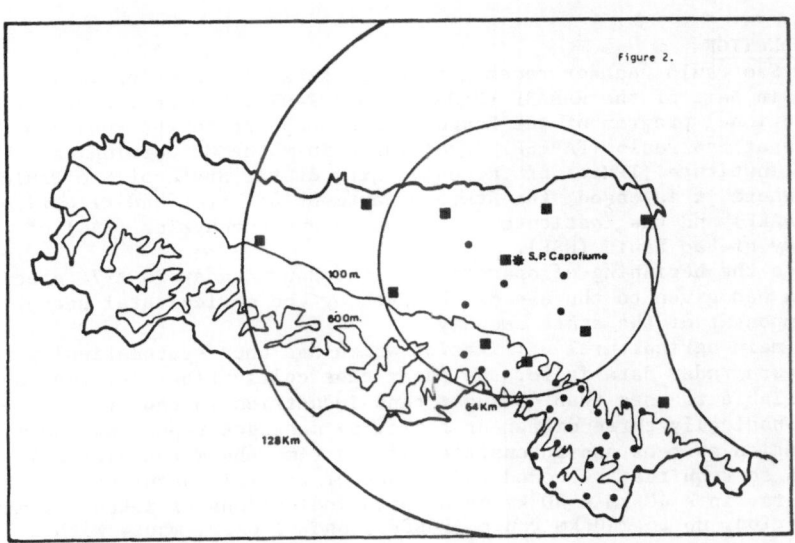

Fig. 2) The main orographic characteristics of Emilia-Romagna region. The asterisk represents the radar location, the squares represent the automatic stations, the full circles represent the raingauges.

NOWCASTING FOR AGRICULTURE WITH THE SÃO PAULO RADAR NETWORK

R.V.CALHEIROS and M. de A. ANTONIO
Meteorological Research Institute, University of the State of São Paulo
and
Institute for Space Research, Secretary of Science and Technology

Summary

The São Paulo State Weather Radar Network, which is being implemented in Brazil as part of the RADASP II (Radar in São Paulo) Project has, within its main operational purposes applications to agriculture, which is of relevant importance for the State economy. Presently, in what concerns the agricultural sector, the Bauru radar data is systematically used in the practices of the sugar cane cultivation. Information on rainfall distribution and its tendency are routinely disseminated for sugar cane refineries in support of such fundamental operations as burning, cutting and transporting of the sugar cane, the spraying of crops with chemical products and the planning of personnel transportation for work in the field. An identification was made of the applications of network data for different crops in the State and the corresponding economic value was derived from estimates of the potential benefits of meteorological information to the state agriculture. Also considered were procedures to identify the need for data, from farmers and agroindustries.

1. INTRODUCTION

The São Paulo weather radar network program (1) is being implemented as the main part of the RADASP (Radar in São Paulo) Project, a research and operational program of the Foundation of Support to the Research in the State of São Paulo (FAPESP) involving mainly the Meteorological Research Institute (IPMet) of the University of the State of São Paulo (UNESP) where it is based, the State Department of Water and Electrical Energy (DAEE) and the Institute of Astronomy and Geophysics (IAG) of the University of São Paulo (USP).

Since the beginning of operation of the Bauru radar in 1974 special attention was given to the use of the data by the agricultural sector, a major component of the state economy.

The main agricultural utilization which has been systematically made of the Bauru radar data is for the sugar cane cultivation. Information is made available to more than 40 sugar cane industries in the area of the State in basically three different forms: a) messages reporting the main precipitation systems and intensities at any time the meteorological situation so requires; b) coded radar maps of rainfall accurrence every three hours, in a 40 km x 40 km grid, with indications of intensity maxima in the period, up to 240 km range (RAREP), and c) CAPPI maps, with a 158 km range, every ten minutes, in a grid of 4 km x 4 km. The first two products are routinely disseminated in an active mode by a computer controlled telex system and the third is sitting on a VAX 11/780 where it can be accessed through a telephone - auto answer modem system.

Along the years, specific applications to different practices of the sugar cane culture were developed such as burning, cutting and transporting, spraying of cultivation and planning of personnel transportation for field work, which are presented in this paper.

The importance of radar for that agroindustry is exemplified by the 1983 anomalous autumn precipitation when intervals between rainbands sweeping the State were crucial in the determination of continuation of the operations.

Having in sight the availability of the network data, an identification was made of the cultivations more susceptible to the use of the data, and in which practices they could be benefitted. For that sake the State was tentatively divided into 7 regions.

An indication of the value of the network for the agriculture of the State was derived from an estimate of the potential benefits of meteorological information in general, made recently in the context of a proposal for a meteorological system for the State (2).

Procedures to evaluate the needs for the data from the farmers and agroindustries are also considered.

2. RADAR USE IN THE SUGAR CANE CULTURE

Radar information has different application in several practices of the sugar cane from the raising to the industrialization. Experience gained along the years of use of the Bauru radar information by the sugar-alcohol industry in the central area of the State of São Paulo allowed the detailing of that use in many aspects. A summary of that detailing is presented in the sequence for refineries in the upper range of size.

a) application of herbicides

Absence of rainfall occurrence information may lead to a loss of about 0,5 liter per hectare. This corresponds to 1000 liter in a 2000 hectare cultivation.

b) Conventional soil preparation

Considering an average daily potential of 200 hectare for the use of the grid, a lack of information of rainfall occurrence in 2 days for the period from December to March could impair the work executed on 400 hectare, with a loss corresponding to about 400 hours of operation of heavy machines.

c) Planting

The operation highly sensitive to rainfall information in this practice is the rutting. In a planting of 6000 hectare, absence of that information may lead to the need of rerutting of 400 hectare, 200 hectare of which with reapplication of fertilizers.

d) Fire on cane

Being the first phase of the harvest itself,from it derives other operations which are of capital importance for the feeding of cane to the industry.

The success of the harvest depends on it, the rainfall information being very important due to the fact that the soil behaviour is highly variable under rainfall, conditioning the possibilities of transportation. Under light rain burning must proceed as programmed; if rain is heavy burning has to be changed to sandy soil where restoration of traffic conditions is much faster. If fire is applied to the cane in the wrong type of soil and it rains, a delay will occur in the delivery of that cane to the industry forcing, also, the cutting for "cane on the straw", burdening it. Once cane is burnt, if it is not taken in time a loss of its sucrose degree will occur in the order of 10 kg of sugar per ton.

If the equivalent of two days of milling is lost, at that rate in 30000 tons of cane the loss will be 300 tons of sugar. It must be considered also the delay in the harvest.

e) Cutting
Depends on the burning to be executed;if it occurs on non burnt cane a higher work load will be required. Knowledge of rainfall occurrence will avoid costs of transportation of personnel to the field.
f) Agricultural aviation
The application of inhibitors is highly sensitive to the occurrence of rainfall since it requires a few hours for its absorption by the plant. In one hour a 60 hectare area is covered and if it rains in less than 4 hours the product and the flight are lost.
g) Application of costal herbicide
An average loss of 50 liters of this herbicide can occur in the yearly operation in an area of 1350 hectare, as a consequence of lack of rainfall information.
h) Seed bed
If irrigation is used during a continuous period of 6 months in a 30 hectare seed bed and, due to rainfall information the irrigation is not applied in the equivalent period of one week, a substantial saving will be obtained.
i) Hoeing of weeds
In the period from December to March there is a big number of workers in the cleaning of the cane breaks. If, in a period of 100 working days a gain of 10% in the correct alocation of personnel to clayey and sandy soil could be obtained as a function of the knowledge of rainfall occurrence, this would result in a saving of about 1500 men/day working in clayey soil who would be diverted to the sandy soil. In addition one working day of the whole personnel could be saved by the improvement of the knowledge on precipitation occurrence.

The radar information is disseminated to the cane industry through an agreement with the central cooperative of refineries (Copersucar), within the context of a program of meteorological vigilance (3).

The radar products are sent presently to about 40 refineries in the Copersucar system, a number which is going to increase to a total of 90 in the Cooperative.

Depending upon the meteorological situation messages are sent at any moment which contain information on precipitation distribution to 400 km range and intensities in six levels to 231 km; this information is updated every 30 minutes. Dissemination is made through a microcomputer controlled telex system.

Routinely, at every three hours, coded radar maps of occurrence of precipitation including intensity maxima in a 40 km x 40 km grid are effected and disseminated through the above mentioned telex system. Quantitative and qualitative ranges are as indicated in the description of the messages.

At about every ten minutes a 4.1 km AMSL CAPPI composed of a 4 km x 4 km grid is available at the Bauru radar central computer, a VAX 11/780, for direct coded access by the refineries through an auto answer modem connected to a dedicated telephone line. Figure 1 indicates the position of the refineries presently receiving at least one of the products. One exemple of the importance of the information to the sugar--alcohol industry is the anomalous rainfall in the State of São Paulo in May, 1983. In that opportunity, operations for supplying cane for processing could be maintained by refinaries in the intervals between rainband occurrences as informed by the radar.

3. VALUE OF NETWORK DATA

Availability of the radar network observations will increase considerably the potential benefits to the agriculture in the State. A preliminary evaluation of the use that the information could have for important cultivations was executed with that in mind. The State was divided into 7 regions, as shown in Figure 2 . The significant cultivations which could benefit more from radar information are listed in the sequence, for each region.

Region 1: Northwest
   citrus, cotton, winter bean (irrigated), cane, peanut, rubber tree.

Region 2: NE
   irrigated cereals (bean, wheat, rice), coffee, citrus, cotton, cane, parched bean, potato.

Region 3: Paranapanema Valley (region of Assis)
   wheat, parched winter cereals (oat, rye, etc) planted without irrigation, cane.

Region 4: Central-south (from the Bauru region to Capão Bonito and
                          Itararé)
   parched bean, cane, citrus, parched wheat without irrigation.

Region 5: Piracicaba, Limeira, Campinas
   cane, cotton, parched bean, water bean, vegetables, temperate fruit-growing (e.g. grapes), citrus.

Region 6: Paraíba Valley
   water bean, parched bean, irrigated rice.

Region 7: Coastal and Valley of the Ribeira
   banana, "flooded" rice, tea, cocoa bean.

The importance of the use of radar information in the cultivations is listed below.

citrus: application of defensives, irrigation in the planting out of the rainy season.

cotton: harvest and application of defensives

irrigated cereals (winter bean, wheat and rice): irrigation

peanuts: harvest and disease control (application of defensives)

coffee: application of defensives and hail

parched coffee: harvest

potato: disease control (application of defensives) and irrigation for the planting in the periods of February and March, and May and June.

parched wheat and winter cereals: planting and harvest

vegetables: irrigation, application of defensives and hail

temperate fruit-growing: application of defensives and hail

parched rice: harvest

rubber tree: hail and irrigation for seed bed.

Regarding the economic value of the meteorological information to the cultivations an estimate was effected in a proposal for a meteorological system for the State of São Paulo (4).

Table 3.1 gives data on the agricultural and bovine and swinish production in the State of São Paulo.

These estimates refer to the total meteorological information including that to be supplied by the network; however from the considerations presented by Calheiros (1) and the experience with the sugar cane cultivation the benefit-to-cost ratio of the operation of the network is highly favorable.

Table 3.1 - Agricultural and bovine and swinish production -
State of São Paulo

PRODUCTS	PLANTED AREA 1000 hectare	PRODUCTION 1000 Ton	VALUE OF THE PRODUCTION 1000 US$	% BENEF.	VALUE BENEF. 1000 US$
COTTON	325.0	540	68,571.4	3.5	2,400.0
PEANUT	130.0	154	8,470.0	3.0	254.1
RICE	299.0	540	25,848.0	10.0	2,584.8
POTATO	28.0	530	60,626.0	5.0	3,031.3
COFFEE	803.0	687	584,064.0	5.0	27,403.2
CANE	2,043.0	130,420	801,162.0	5.0	40,058.1
BEAN	447.0	300	32,264.0	10.0	3,226.4
ORANGE	762.0	9,792	497,822.0	5.0	24,891.1
CORN	1,465.0	3,921	125,034.3	3.5	4,376.2
SOYBEAN	459.0	978	72,604.0	5.0	3,630.2
BANANA	54.8	1,128	29,123.3	3.0	873.7
ONION	16.7	283	11,384.0	5.0	569.2
TOMATO	17.5	733	59,888.0	10.0	5,988.8
WHEAT	175.0	289	45,140.0	10.0	4,510.4
BOVINE AND SWINISH MEAT	-	477	301,750.0	3.0	9,052.5
TOTALS (US$ 1,000,000)			2,723.75		132.85

Obs: Data from Sept.87, corrected for Sept.88, furnished by the Agronomic
Institute of Campinas, State Secretary of the Agriculture

4. FINAL COMMENTS

Procedures for identifying agricultural users' needs and estimating
benefits, specifically in relation to the network, are being structured.

It should be taken into account in the procedures that the farmers
are less likely to be able to make full use of direct radar information
what does not happen with the big agroindustries, which agricultural
practices are highly mechanized most of the times and which have, in
general, specialized engineers and technicians. Both for farmers and
agroindustries it is necessary to understand their operations and assist
them in identifying the uses of the information and its values to them.
This had been done, to a certain extent, in the case of sugar cane.

The agricultural application of the network is much tied to a
integrated Center of Agrometeorological Information (CIIAGRO) of the
State Secretary of the Agriculture (4) which will be operated jointly by
the IPMet and the IAC and will be sitted nearby the IPMet. The Center will
provide directly agricultural counseling based on meteorological
information, what is expected to overcome most of the difficulties in the

use of the information mainly by the farmers.

Regarding the products to be generated by the network one main step will be the implementation of nowcasting techniques for the whole area of coverage of the network, of the kind of the SHARP procedure (5) presently under test with the Bauru radar system.

Another important aspect is the development of rainfall accumulation maps using radar which can improve considerably the water balance for the State, to meet the needs of many cultivations.

5. ACKNOWLEDGEMENTS

The Federal organizations FINEP (Financing Agency for Studies and Projects)and CNPq (National Research Council) contribute to the RADASP Project. Thanks are due to Carlos Alberto de Agostinho Antonio for elaborating the figures and to Marlene Sueli Moya Munhoz for typing.

REFERENCES

(1) CALHEIROS, R.V. (1989). The São Paulo Weather Radar Network Program. Elsewhere in these proceedings.
(2) IPMet (1989). Sistema Paulista de Meteorologia. Proposed to the Government of the State of São Paulo.
(3) BENETI, C.A.A. and ANTONIO, M. de A. (1987). Vigilância Meteorológica e Previsão de Precipitação de Curto Prazo: Um Esquema Operacional. Proceedings, V CONGREMET and II Congresso Interamericano de Meteorologia (Buenos Aires), CAM, 4 pages.
(4) IPMet (1987). Centro Integrado de Informação Agrometeorológica – CIIAGRO. Presented to the Government of the State of São Paulo.
(5) BELLON, A., LOVELOY, S. and AUSTIN, G.L. (1980). Combining satellite and radar data for the short-range forecasting of precipitation. Mon. Weather Rev., Vol. 108, 1554-1566.

Figure 1 - Sugar-alcohol refineries of the Copersucar presently receiving radar information on a routine basis.

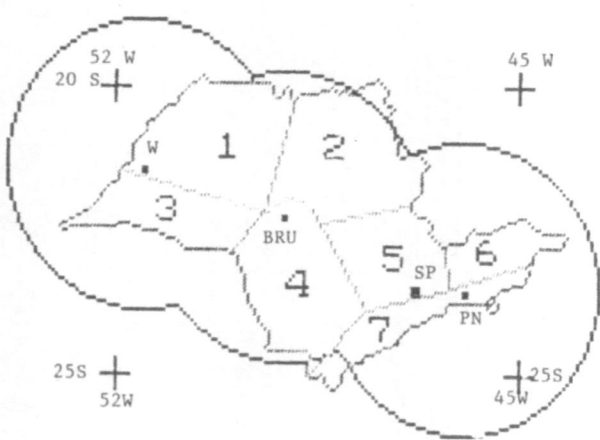

Figure 2 - Regions of the State of São Paulo for the evaluation of use of radar data to cultivations.
BRU = Bauru, SP = São Paulo, PN = Ponte Nova

THE DISTRIBUTION OF WEATHER RADAR IMAGES TO AGRICULTURAL END USERS

By Søren Overgård and Erik Wienberg
The Danish Meteorological Institute

Summary

In the summer of 1988 the Danish Meteorological Institute (DMI) entered into an agreement with two other organizations to set up a test for the distribution of weather radar images to agricultural end users. The co-partners were Landbrugets EDB-center (LEC) (the Danish Agricultural EDP-Centre) and Landskontoret for Planteavl (LK-P) (the Danish Agriculture Advisory Centre, Department of Plant Production).

A small number of agricultural end users were selected on the basis of locality and function to participate in the project and were at the same time required to keep an sort of diary of their actual use of this type of information. The purpose of this procedure was to try to establish how, on the one hand, the users' interpretation of these images and, on the other hand, the actual weather and precipitation developments corresponded to one another.

At the end of the test period a meeting was called at DMI where the end users and participant organizations discussed the pros and cons of the system and what could be done to improve the usefulness of the system.

The present paper offers a broad description of the constituent parts of the test. Some of the major ideas for the improvement of this sort of system in itself and in context with other types of data will also be presented.

1. INTRODUCTION

As already mentioned in the summary three organizational bodies were involved in the project which was aimed at a fourth body, the end users.

The Danish Agriculture Advisory Centre, Department of Plant Production (LK-P) is a research organization which serves national agriculture in an advisory capacity. Over the years LK-P has gained a thorough understanding of the needs and wants of the agriculture and have established a fine network of contacts which were invaluable to the project. Apart from aiding in the foundation of the project LK-P also worked out the basic questionnaire which was used at the end for evaluation purposes. They worked out a summary of the findings from these questionnaires.

The Danish Meteorological Institute (DMI) makes intensive use of weather radar images in its daily operations. For a couple of winter seasons DMI has had an operational warning system for the road authorities of the Danish counties against snowfall, slippery roads and so forth. Various elements constitute this service. Most importantly, the meteorologist works out a short-term forecast for the county, but some synoptic data and weather radar images are also made available to the road authorities. At intervals of half an hour a weather radar image is sent to the road authorities where the user can make some of his decisions based upon a series of the last 5 images. These images have been reduced to three levels of information, and only data for areas of detected precipitation is sent. Maps are supplied locally to reduce transmission time and cost. This solution is cheap in that it only requires that the user have a cheap IBM-compatible PC and the necessary software.

Figure 1 gives an example of a typical end user weather radar image.

The image is shown on a low-cost CGA-monitor, and the three levels of precipitation are given in the intervals 10 to 20 dBz, 20 to 40 dBz and above 40 dBz. These intervals correspond to light (in the figure LET), medium (in the figure MIDDEL), and heavy precipitation (in the figure KRAFTIG). These intervals have been chosen on the basis of a project performed in 1987 comparing raingauge measurements and weather radar measurements (1).

This concept has been well received by the users, and it was felt that this product - possibly with slight modifications - would fill some of the needs for very localized weather information which can only at great cost and with great difficulty be given by a trained meteorologist.

However, we had to find a way to distribute the weather radar images to a broad base of end users. At DMI we are fully equipped to deal with fourteen counties, but not with, say, several thousands of end users. The demands made by such a number for communications equipment, staffing and necessary accounting systems are quite heavy and not within the capacity of DMI.

PTN services (e.g. Videotex) were taken into consideration as a possible channel for distribution. These kinds of services have some of the necessary elements (accounting and a potentially large user base). However, the presentation systems have some short-comings. Prestel would be too coarse and CEPT too demanding in workload for a system with quick changes between the individual images. Furthermore, it would require some initial investment on the part of the end users which could be avoided if our system could append itself to an already existing network.

The Danish Agricultural EDP-Centre (LEC) was able to meet all of these requirements and also had a tradition of several years for catering to very broad user groups. LEC is an organization which is jointly owned by some major agricultural organizations and is very possibly the largest privately owned computing centre in all of Denmark. With its 700 employees and an annual turnover of 300 million DKr. LEC has for more than 20 years serviced the national agriculture with e.g. systems for the management of fields and livestock. 75 per cent of its annual turnover derives directly from the agricultural sector. Its computing and communications equipment would suffice for the task - and LEC, of course, also had an interest in the diversification of its line of products. What is more, LEC is currently in the process of exchanging existing 'dumb' terminals with IBM PS/2s so in time most of its natural users would be open to our concept.

In May 1988 LEC made the following survey (see table I) of the potential user base which could be accessed through its network.

Column one gives the description of the user segment, column two the total number hereof, column three the total number of terminals of the user group and column four the number of PC' (e.g.s. Thus these figures give the degree of accessibility - from left to right - from actual to principal.

Table I

User group	Total users	Total no of terminals	No of PCs hereof
Primary agriculturists	40,000	370	310
Consultants	125	600	150
Forestries,	500	75	50

nurseries

Companies dealing with feedingstuffs, seeds and fertilizers	175	1,000	75
Others	7,500		

A system of weather radar images was thought to be of interest to mainly the first four categories of users for whom it would make the rational organization of labour and machinery and other resources (i.e. toxic sprays) easier.

What we had to do to make use of LEC's network was to establish a data link between DMI and LEC which significantly reduced the initial cost. For a description of the total data flow, see figure 2 (2).

The radar which is placed at Copenhagen Airport delivers a volumetric scan that is packed in the PMERAWIS format (L.M. Ericsson Radar). This volume is then transmitted to DMI, and from it the pseudo CAPPI layer is extracted (e.g. the scan with elevation angle zero) and reduced to three levels of pixel values. The resulting image is then packed into a character-oriented format which is based upon a simple run-length code. Finally, this image is transmitted to LEC from where the end user can subsequently request it.

The end users were the fourth, but not least interesting participant. They were selected jointly on the basis of the following criteria:

a) locality, i.e. given the Danish weather radar situation (i.e. one radar at Copenhagen Airport) they had to be situated within an area of reasonable coverage by the radar and to be representative of the entire geographical area from the periphery to the centre.

b) function, i.e. we wanted an equal representation of primary and secondary agriculturists among our end users.

c) commitment and interest, i.e. they had to agree to keeping a sort of diary of their use and experience of the system for the duration of the test period. At test termination they had to send LK-P this information and, if at all possible, to attend a 'brain-storming' meeting.

A suitably small group of twelve in all was the final result. Five members were representatives of the secondary sector (e.g. consultants, machine pools), the remaining seven were primary farmers.

At this point it should be noted that the above-mentioned criteria by no means ensured a representation of typical Danish agriculture. Our users controlled larger areas of land, had heavier investments in modern technology. Consequently, they adequately represented tomorrow's agriculture (the development trends per se), and they were known to be leading members of the agrarian community.

Figure 3 shows where the active end users of the test were located geographically.

In this figure the positions of the end users and that of the radar, DMI, and LEC are marked. Also the height of the beam as a function of distance from the radar has been illustrated using concentric circles. This should give a coarse measure for the quality of the images for the various regions.

2. THE TEST PHASE

Having established the frameworks of the test we now had to deal with the more technical aspects, namely the establishment of the data link between DMI and LEC. We decided upon a cheap asynchronous 1200-baud PTN link to help keep outlays low in the test phase. Over this link a new and up-to-date weather radar image was transmitted every ten minutes. The end users would then request the latest image from LEC.

Many considerations that must by nature and necessarily go into the establishment of a full-scale operation were at this stage taken lightly. The system was not by any scale mature. Due to considerations of time and cost we had to leave what must be an integral part of any such system, fault-handling and recovery, to the care of a very small number of people involved.

At LEC various statistical material was extracted from the actual use of the system. However small the number of participants for statistical purposes, it may still be of interest to see some of this material.

Figure 4 shows the distribution of use over the weeks of the test. From week 33 to 38 there is a marked rise except for week 36 which was characterized by abnormally good weather (hardly any precipitation at all). A noticeable decline takes effect in week 39 and from then on there is a low but stable use for the rest of the test period. From reports we had this coincides with the completion of the harvest activities.

Figure 5 shows the number of images that were requested by the individual end users as a total for the entire test period. Three participants have taken only a marginal interest (the questionnaire was not returned by three participants!). The high score is held by three other users who between them represent both the primary and the secondary sector. The larger group in the middle have shown a modest interest and use. This again might be attributable to the fact that harvest was completed quickly and for all we know it might represent the differences between various crops. The fluctuations are somewhat more dramatic in the primary sector.

Figures 6A and 6B are interesting from a technical standpoint because they show the peak hours of the day, i.e. the distribution of use over the hours of the day. As can be seen the patterns vary modestly between the primary sector (6A) and the secondary one (6B). In all probability this has to do with their different job situations (field work vs. office work).

3. THE EVALUATION

From the returned questionnaires the following points could be made.

The images were easy to interpret except for weather situations of showers and thunderstorms.

An indication of the direction of movement and its velocity was needed.

As many weather situations in Denmark have a tendency to originate in the West the need for weather radar coverage of all of Jutland (and the representation hereof in the images) was an urgent requirement.
A series of images should cover from 1 to 8 hours and consist of from 3 to 10 images.

The system had been especially useful in unstable weather situations. The normal forecasts do not give information in sufficient detail for the necessary decisions of the farmers in this type of weather.

The end users had drawn great benefits from the system in the following areas: The planning of work, harvests, the gathering in of straws, beet digging, spraying, in short most every kind of field work.

In many cases the plan of work had been altered on account of the reception of the weather radar images. In about 80 per cent of the cases this alteration had been for the better, but there was a minority of cases where the opposite had been the result (3).

The users pointed to the need for additional meteorological information. The velocity and direction of the
area of precipitation, the amount of precipitation to be
expected, the temperature and the velocity of the wind,
the relative humidity of the air, and information on the
degree of cloudiness of his area.

The users wanted to have their own UTM coordinates
visible in the image itself (for better orientation).

Until the test the participants had made use of available information from television and radio, newspapers
and some telephonic services which give a regional
forecast (but for too large areas) every six hours.

Therefore there was a marked interest in the continuation of the project. One participant, however, made his
interest conditional upon the development of the system
(the integration of other types of data etc.). One
consultant who took part had talked of the system at
numerous meetings of farmers - and the interest had been
overwhelming.

Our end users stated that they had only been able to
see an hour and a half to two hours into the future and
what they truly needed was knowledge of the developments
of the coming twelve hours.

The above-mentioned problems can only be solved to a
degree. The area of coverage is a political thing in
Denmark in that it requires funding of new weather
radars. The twelve-hour prognosis period cannot be
obtained from weather radar images by themselves. Many
other types of data already exist and can be made available to the farmers. Some other types (e.g. the indication of direction of movement in the images) can be
brought about possibly through some kind of mathematical
model and possibly through the intervention of a trained
meteorologist. However, one way or another additional
funding and cost is implied - and the agriculturists,
of course, wanted to minimize their own cost. We shall
shortly return to these problems.

4. THE MEETING

After a short evaluation of the above-mentioned material
we called a meeting at DMI in which both users and
organizations took part.

Mostly this meeting stressed the points made in the
papers, but some new items were brought into focus.

For one thing, it was evident that DMI had gained the
benevolence of the user corps by promoting this initiative. One user claimed that the initiative was revolu-

tionary! Although we know that this type of information has its shortcomings, it must be understood that its availability by end users is better than no information at all. The users are out there with decisions to be made and they need all the help they can get even if it may at times be far from perfect. As long as it is only misleading in a minority of cases it may be preferable to no information at all.

For another, once such a system has been put at the public's disposal there is a very definite chance that additional effort will be invested in the building of models etc.

The users were impressed with the accuracy and resolution of information. Some of them had verified the information by calling people at different sites to check it - and had never found it to be false! Indeed it had been far more accurate than the regional telephonic services.

They also pointed to some of the other outdoor activities where it could be brought to good use (construction etc.).

On a more technical level they wanted a redefinition of the way they made their requests. Instead of requesting one (the latest) image they wanted to be able to define a packet request. The user would indicate the number of images, the intervals between these and have it sent in one packet. This will be implemented in the summer of 1989.

Other types of existing forecasts will be integrated into the service. For instance, the user can have a five-day forecast and a six-hour regional forecast included with his request. This, too, will be implemented in the summer of 1989.

If cost were not the issue the involvement of a trained meteorologist for commenting on the individual images could be a viable solution much along the same lines as DMI's winter warning system for the counties.

A member of the staff at LK-P gave some thought-provoking figures. Danish agriculture spends some 2 billion DKr. On spraying. Some sprays take effect only with precipitation, others only with no precipitation so many sprays are totally wasted if administered at the wrong time. Through improved forecasts, he ventured a guess, this amount could be reduced by 5 per cent, i.e. 100 million DKr. The annual expenditure on the improved forecasts might be about 2.5 million DKr. Given a group of users who each paid 1000 DKr. some 2-3000 users would be required to make it a profitable investment. Not to mention what could be saved by public services dealing with pollution much of which stems from the use of fertilizers and sprays by national agriculture.

5. THE IMMEDIATE FUTURE

As this is written we are in the middle of preparing for our first open season where every interested party can subscribe to the service.

Some of the improvements above have been integrated into the system - but only the more basic ones. We have duplicated the data link between DMI and LEC so that we have one link partly as a possibility for backup in fault situations, partly for test and development purposes. At the involved computer sites (LEC and DMI) much thought has gone into the handling of fault and recovery situations in order to make the system more or less immune from these, at least when seen from the user's perspective.

During the entire process we have been surprised at the amount of interest we met with from many sides. A short press-release was published in the majority of Danish newspapers and without further effort on our part it led to a surprising amount of interviews and articles in a great number of these. One could say that the idea has performed its own marketing to a high degree.

With the coming of the first open season additional effort will be put into marketing, but it remains to be seen what the real interest is. Hopefully, we will be able to give some figures for the first open season at the time of the COST meeting. The involved organizations and the original end users have agreed to follow up the developments in much the same way as was the cases in the test period.

Reference

(1) Madsen, H., Overgaard, S., Vejen, F., Sammenligning af data fra vejrradar og automatiske nedbørsmålere, The Danish Meteorological Institute, December 1987, in danish.

(2) Overgaard, S and Wienberg,E. A Technical Destription of the Dristributation of Weather Radar Images, The Danish Meteorological Institute, Technical Report, in press.

(3) Andersen, B, Agriculture Advisory Centre, Department of Plant Production, personal communication.

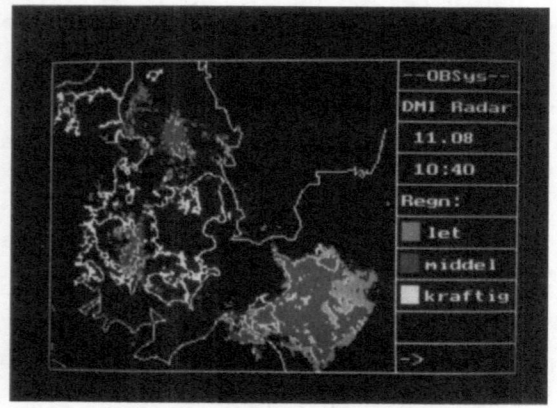

Figure 1 : Display at the end user

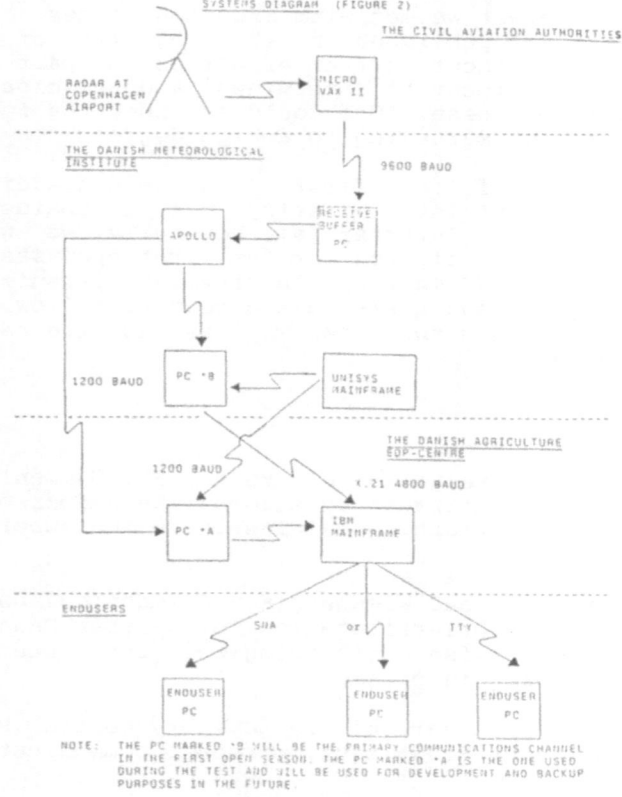

Figure 2 : System diagram

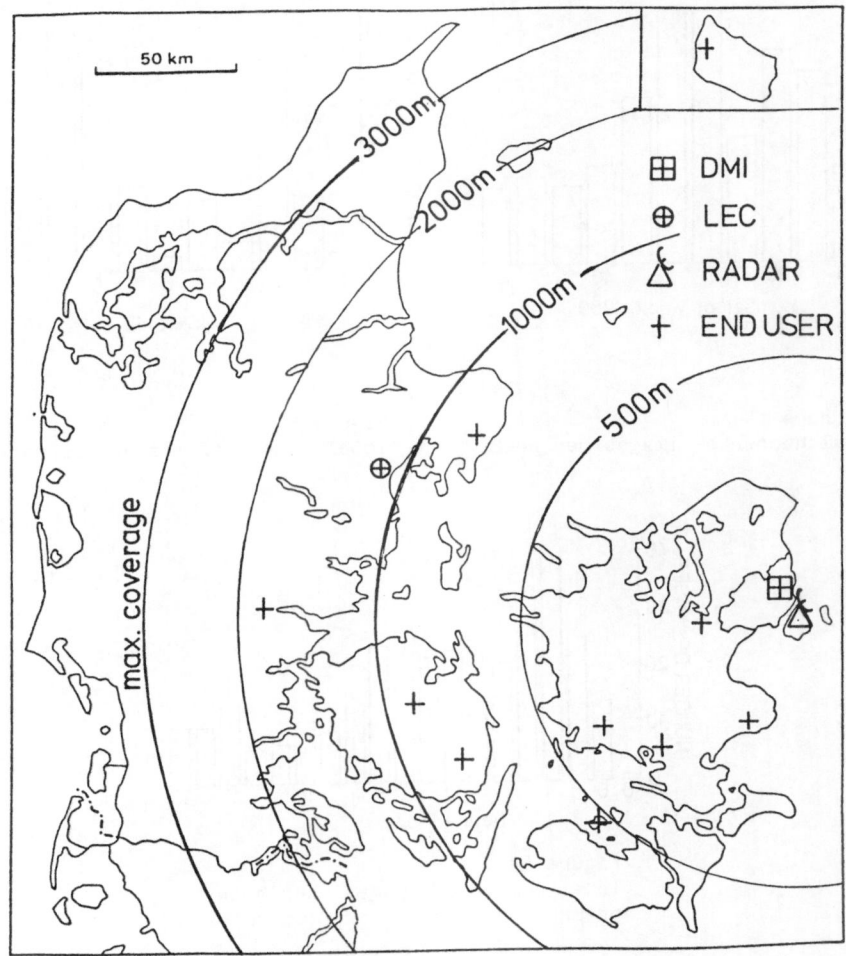

Figure 3 : Location of end users and hight of radar bean above ground. A curvature of 4/3 of each radius and a exponential refractive index-model is used for calculation.

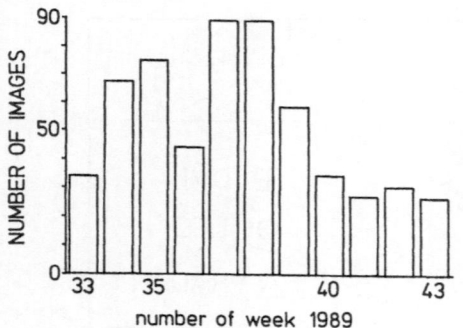

Figure 4 :
Total number of images per week

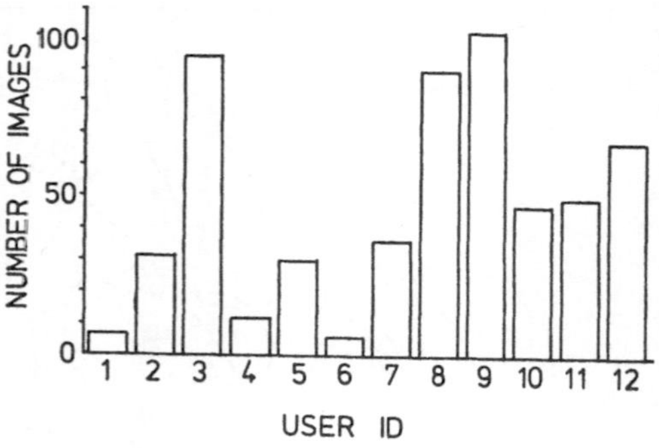

Figure 5 :
Total number of images per user

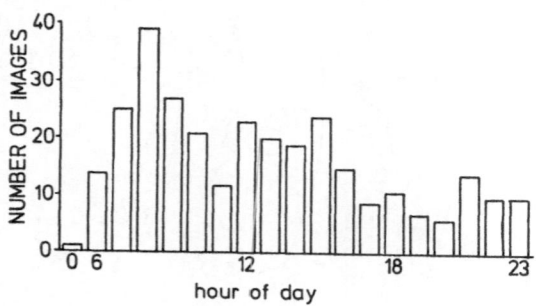

Figure 6A :
Total number of images per hour,
primary users

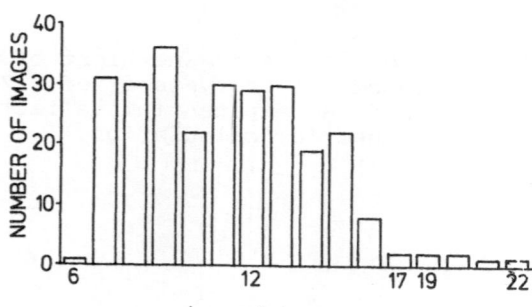

Figure 6B :
Total number of images per hour,
secondary users

THE USE OF WEATHER RADAR DATA IN ROAD WEATHER SERVICES: PRESENT AND FUTURE NEEDS

E. NYSTEN and R.H. KING
Finnish Meteorological Institute
Helsinki, Finland

Summary

Road weather services vary in nature across the length and breadth of Europe, but operationally they are mainly concerned with three hazards: the effects of snow and ice in winter, the effects of heavy rain in the warmer months of the year, and the occurrence of fog and bad visibility around the year, with the emphasis, however, on the colder season. For the minimization of the effects of these hazards, weather radar can supply direct information for the first two cases, and can also sometimes provide indirect information of forecasting value in the case of fogs. In many countries, precipitation information has been extracted from weather radar data for more than a decade, and the limitations of radar data in terms of quantitative accuracy have been widely reported. The usefulness of radar information rests to a great extent on its timeliness, based on near-real-time acquisition, and on its three-dimensional coverage of a large volume of the atmosphere. These advantages can be put to good use in an operational road weather service. This paper reviews the problems of providing accurate and timely information on snowfall to road-clearance authorities in winter, and projects some services that could be offered to road users in the warmer season, based upon weather radar data, either alone or in conjunction with other meteorological data sources. The part to be played by an operational European radar network is discussed.

1 INTRODUCTION

It is surely a commonplace to every road user, and especially to anyone who drives for a living, that weather affects his safety on the highway in a variety of ways, including winter problems of low braking effectiveness and poor out-of-vehicle visibility due to ice, snow and slush, difficulties of visibility in fog and drizzle, leading occasionally to spectacular motorway chain-crashes, and summertime dangers such as aquaplaning after heavy rainfall. In some parts of Europe, road maintenance also has to cope with such natural phenomena as flash floods, avalanches and snowdrifts. In countries such as Finland which rely upon the use of studded tyres for improved winter-time traction and braking, the effect of winter weather in the climatological sense carries over to other times of the year in the form of unevenly-worn road surfaces, which themselves represent a hazard by lowering the safety factor when driving on wet roads in cross-winds. Over most of Europe severe road hazards directly due to weather are, happily, for most of the year fairly infrequent, but nevertheless, when critical weather conditions do occur, they may be the potential cause of destructive accidents involving life and limb, as well as incurring large costs in material and other resources.

Recognition in Europe of the part played by the weather in affecting road safety is evidenced by the amount of effort being put into understanding and reducing its effects, either in such direct EC specialist projects such as COST 309 (Road Meteorology and Maintenance conditions, the successor of COST 30), or as a significant factor in the work of such groups as the OECD Road Transport Research Program. Initiatives such as the EC project DRIVE and the EUREKA project PROMETHEUS represent overall systems approaches to questions of road transport safety, including the weather problem, from the infrastructure and vehicle viewpoints. The possibility of giving timely warnings to motorists through the use of modern technology is already significant in reducing accidents, and further dramatic improvements in this field may be expected as these aforementioned projects come to fruition. Meanwhile, much can still be done by weather services to help in improving the efficiency of road maintenance, especially in its work in the winter season, when considerable sums are spent in many countries throughout Europe, mainly on the laying of salt and on snow clearing.

Day-to-day decisions which have to be made by the regional road maintenance authorities usually rely on several sources of information. Road weather stations provide point measurements from which estimates can be made (e.g. using thermal mapping) of road surface conditions over wider areas, and the probable changes of these conditions over a short period ahead. Many countries provide, through their meteorological offices, weather forecast services for their road authorities (e.g. in Britain, the Meteorological Office's "Open Road" package). Such services may also include actual weather radar data in some form or another, while in other cases the weather radar data are used by forecasters to monitor and update their road weather forecasts. The usefulness of conventional weather radar in the context of road weather stems from its following characteristics:

- extensive areal coverage (up to 300 km radius) from a single point
- rapid updating of data in time (between 10 and 30 minutes)
- fine horizontal resolution (1 - 5 km)
- the correlation observed between weather radar echoes and surface rainfall, enabling estimation of the rainfall intensity

Despite such seemingly attractive properties, there is clearly less application of weather radar data to the problems of road weather than one might expect, considering that weather radars have been widely used in routine forecasting for well over twenty years, and that radar data in digital form has been available in many European countries for ten years or more. The purpose of this paper is to review the factors involved in the use of conventional weather radars to provide precipitation information for road weather services, with particular reference to the winter season. This paper is based mainly on the experience gained by the writers through association with the COST projects 30, 309 and 73, and with weather radars in Finland over a period of 20 years.

2 WEATHER RADAR OPERATION: AN OVERVIEW

The operation of many, if not most weather radars today is the responsibility of the national agency providing weather services. Weather radar data is used either alone, or in company with other weather data, to provide the basis of services to a wide range of users, e.g. primarily for analysis and forecasting in the national weather agency itself (general use), and then secondarily for special services to e.g. hydrological applications, building and civil engineering, aviation (briefing and air traffic), land and sea transport, air quality monitoring (pollution and radioactivity), sports and other outside functions, etc. The requirements of such a broad range of users are somewhat conflicting, especially as regards the repetition rate and freshness for data collection. Most users seem to be fairly satisfied with a horizontal resolution of 2 - 5 km. Regarding frequency of acquisition of new data sets, some users (e.g. air traffic control) demand update rates of a few (5) minutes or so, with fast computing to give a fresh, rapidly-updated image. Other users (such as hydrologists) may not require such rapid updating or such rapid computing, but wish for high rainfall estimation accuracy for runoff calculations. The requirements of aviation, as well as those of general meteorology emphasize the importance of three-dimensional data collection, whereas the hydrologist is interested solely in the precipitation reaching the ground. As a result of these various demands, radars are operated a little differently in each of the European countries, depending on the main emphasis on radar use in that country. Not only the operational use of the radar, but also the actual technical characteristics of the radars are a compromise, too, especially regarding wavelength and antenna beamwidth used.

Questionnaires within COST 73 indicate that most countries employ a 10 to 15-minute update time, with a computing time of around 5 minutes. Almost all countries make observations at more than one elevation angle, and the majority make genuine three-dimensional data collections. Many countries also now regularly contribute their radar data to form the composite COST 73 image, which is constructed in computers of the UK Meteorological Office, and redistributed from there to many parts of Europe. Many countries also have distribution channels available through which pictorial data can be sent to specialized users having e.g. microcomputer terminals for the display of such data. Thus it seems that for the majority of countries the basic ingredients for providing the road maintenance authorities with a service employing weather radar data already exist, or are available at small cost.

3 PROBLEMS IN THE USE OF WEATHER RADAR DATA

In view of the apparently promising situation outlined above, it is felt useful here to review the known problems which stand in the way of the direct use of weather radar data in road weather services. These difficulties (which may be compared to the list of favourable characteristics earlier mentioned) may be enumerated as follows:
- interpretation of radar intensity field as observed
- decrease of echo intensity/rainfall intensity correlation as a function of range
- decrease of simple forecasting method usefulness as a function of time
- hydrometer phase changes as a function of height

It should also be remembered that all national weather services are under pressure, to a greater or less degree, to rationalize their use of human resources, so that there is a continual emphasis on the use of automatic computerized methods wherever feasible. In many cases it has been found that something that is relatively easy for a human to accomplish by eye and hand is extremely difficult (i.e. consuming of human resources!) to design to perform reliably by computer. Nevertheless, the following discussion aims at a system which normally functions unattended, i.e. as far as possible without subjective intervention.

3.1 Interpretation Of Radar Echo Intensity Field

As is well known, the weather radar echo intensities give an indirect measure of the number and size of the water droplets in the measurement volume, when such droplets are present. Unfortunately, echoes also arise from other sources, such as birds, aircraft, refractive index boundaries in clear air, and, worst of all, from ground sources. In certain conditions the latter can appear at anomalous ranges and elevation angles, and with an intensity which would normally indicate moderate or strong rainfall. Much work has been done to find methods of eliminating such echoes, using both signal-handling and software-based techniques. The most promising indicate that use can be made of satellite IR-data (see, e.g. (1)), although this has still to be demonstrated in winter conditions with the deep inversions typical of Northern Europe. Probably improvements can be expected from a decision scheme using several other meteorological data sources, such as satellite IR, forecast cloudiness from models and synoptic cloud observations. Point anomalies attributable to e.g. aircraft are rather easily removed by certain filters.

Another potential source of error in the interpretation of weather radar data, especially that taken direct from a PPI scan, whose height increases with range, is the effect of the melting layer (when this exists). As a result of the change of shape, fall-speed and liquid-water coating accompanying the melting of solid-phase precipitation as it falls through a zero degree isotherm, a so-called "bright band", i.e. shallow layer of enhanced echo intensity, is formed. On a PPI image, this layer appears more or less as an enhanced average intensity between two ranges. Because of the normally-occurring spatial variability of the echo intensity and the variability of the phenomenon itself, it is rarely observed as an unbroken ring, except at high elevation angles in conditions of uniform rainfall. Due to its nature, the phenomenon is not seen in normal conditions of snowfall.

Perhaps the most severe difficulty in the interpretation of a radar echo intensity field, and one which causes radar meteorologists to be wary of distributing uninterpreted data to non-specialists, is that of how to interpret areas having no echoes. If there are no echoes observed at a particular range, azimuth and elevation, then it is clear that the radar echo strength from that observation volume does not exceed the threshold intensity for that range. In this case, the following (non-exclusive) possibilities are, in general, open:

- the cloud or precipitation particles (if they exist) in the observation volume are too small and/or too few to be observed by the radar at that range, but may however constitute or produce at extreme ranges, even slight to moderate rain. The simple effect of range attenuation ($1/R^2$) is compounded by that due to gases and also to cloud and precipitation at closer ranges, as well as by the effect of imperfect beam-filling.
- although there are no appreciable meteorological targets at that height, there may be such below the radar beam. Due to the curvature of the earth's surface, at increasing ranges greater and greater depths of atmosphere remain below even the lowest radar beam and are thus unobservable.
- there may be targets above the beam which are not observed.

3.2 Range-dependence Of Echo Intensity/rainfall Intensity Correlation

A tremendous amount of research has been made into the correlation between the precipitation intensity measured at ground level and the intensity of the radar echo associated with this precipitation, referred to hereafter as the Z/R relationship (see (2) for a good summary of the work and its results up to the beginning of this decade). There are a number of factors responsible for the variation in this correlation, and a full analysis cannot be attempted here. It will perhaps be sufficient to draw attention to those which particularly influence the estimation of intensity of snowfall, which is especially important from the standpoint of road maintenance.

Many of the experiments to determine the Z/R relationship(s) were made at short ranges and in rather ideal, uniform conditions. The drop spectrum of the rainfall below the melting layer was assumed (and in some cases, shown) to be rather constant with height, at least in steady frontal rain, and so stable values of the Z/R relationship were often found, and related to the type of precipitation (steady rain, drizzle, showers). In operational conditions one must attempt to make an estimate of precipitation intensity at ranges beyond the optimum. Even using the lowest possible radar beam elevation, the radar measurement is made at an ever increasing height above radar datum level, and so the apparent vertical variation of the Z/R relationship becomes of paramount importance. In this may also be included the special case of the "bright band", which, if interpreted in terms of rainfall intensity, will produce grossly excessive estimates. Above the bright band, where the hydrometeors are solid, there is usually a very considerable vertical gradient of echo intensity, as shown e.g. in the measurements of Joss and Waldvogel (3). This reflectivity gradient is physically a consequence of the hydrometeor growth process in cloud. The profile of reflectivity naturally depends also on the type of processes; and is different for stratiform cloud and for cumuliform clouds in different stages of growth. Because of radar beamfilling considerations, the gradient itself when actually measured also appears to be range-dependent.

Despite the weakening of the correlation between rainfall and radar-measured echo intensity with range due the variance introduced by the use of a second correlation, that between the radar echo intensity at a higher and that a lower level, Joss and Waldvogel (3) believe that the use of a correction based on a real-time radar estimation of the vertical reflectivity gradient nearer to the radar can result in an improvement in the rainfall intensity estimate. Fortunately, road authorities are not so insistent on absolute accuracy in precipitation estimates, but even in this application, range corrections for the estimation of snowfall intensity must be considered as very desirable.

3.3 Usefulness Of Simple Forecasting Methods

In order to use radar data effectively in a road weather service, the use of past weather images e.g. of estimated rainfall intensity is insufficient. Decisions on salting and snow removal are best made several hours ahead of the need, and thus ideally the road authorities need forecast information for the roads in their area of responsibility. Simple forecasting methods, based on estimates of the velocity of movement of radar echoes, have been in use for many years (see e.g. Bellon and Austin (4), or Collier (5) for a review of the methods). Using single radars, the usefulness of forecasts becomes questionable for lead times of more than about 90 minutes, which is not sufficient for road weather applications, taking into account not only the desired lead time of 1 - 2 hours, possibly for the whole area of coverage of the radar, but also the time delay between the data collection and the provision of the forecast (15 - 30 min.). In winter conditions, due to the shallow nature of many precipitation systems and their lower reflectivity (especially at higher levels, as already noted in the previous section), appreciable areas of echo may not be observable until within a range of, say, 100 km from the radar, severely restricting the lead times possible for areas upwind of the radar. When the precipitation area covers the area around the radar, it is commonly observed that the PPI (and CAPPI) images are roughly circular echoes centred on the radar, from which reliable motion vectors cannot be extracted until a rear edge to the echo field starts to appear. In northern climates, where appreciable snowfall can be obtained from relatively low (3 - 4km deep) cloud systems, whose echo may only extend some 80 - 100km from the radar, the limitations of single radar observations for providing estimates of echo velocity are clear.

In a report on trials of an echo-centroid tracking forecast method using a network of four radars in the UK, however, Browning et al (6) show that the period of usefulness (lead time) of such forecasts can even be as much as 6 hours in certain weather situations. Bellon and Austin (4) make the important point that the methods should be applied to echoes in a CAPPI (constant- altitude) section (3 km in their case) in order to avoid range-dependent effects present in PPI sections. The biassing effects of image edges and permanent echoes must also be eliminated. Non-echo areas may possibly be modified by the addition, for forecasting purposes, of echoes advected out of earlier images. This could be especially valuable in winter situations of shallow precipitation systems. Over larger areas covered by a network of radars, the echo motion vector must be determined for smaller sub-areas, by analogy with the similar determination of cloud motions in satellite applications, to allow for the natural variation of the motion vector over the scene.

3.4 Hydrometeor Phase Changes In The Vertical

As discussed earlier, the change of form and phase of the hydrometeors as they fall causes difficulty through the occurrence of the "bright band", and by a variation in the Z/R relationship. There is another factor, which is especially trying in periods when the surface temperature is at, or a few degrees above, zero. Then, except for areas very close indeed to the radar, the bright band is not in evidence, and the precipitation is assumed to be snow. However, the actual hydrometeors reaching the ground may vary between rain, probably giving no road maintenance problems (except possibly washing salt away, or increasing the slipperiness of icy roads), through sleet to wet snow. In countries with considerable orography the effect is, of course, magnified. A separate problem is what happens to the precipitation when it arrives at the ground, i.e. will falling snow lay, or melt, or form slush, will rain drain away or form slippery ice? The latter questions are perhaps best solved though use of the road weather station data and experience, whereas the first problem is basically a meteorological one. It may be possible to overcome the difficulties by combining topographical data, sounding data and surface station data to produce fine-scale surface temperature maps to be combined with the radar data to determine a precipitation map which includes the probable phase of the hydrometeors at road level. In steady precipitation the zero-degree isotherm tends to descend, a factor which must also be taken into account here.

4 FORMATTING OF DATA FOR ROAD MAINTENANCE AUTHORITIES

Many countries, including Finland, already provide a special weather service for local road maintenance directors (so-called road masters), and these services have probably in all cases been developed in cooperation with the authorities concerned. Here a few comments are offered on how the data provided by weather radars could possibly be presented to advantage in these weather services.

It is suggested that the data should be provided mainly in "pre-digested" form, i.e. rather in the form of special diagrams or tables giving specific information on precipitation expected, rather than just the radar echo image, from which the conclusions should be drawn on a "do-it-yourself" basis by the road-masters themselves. The arguments for this are as follows:

- in view of what has been presented, the interpretation of any radar echo image, be it PPI or CAPPI or some other, requires quite a degree of specialization, which cannot reasonably be required of all the personnel involved in road maintenance
- the number of factors involved in making a forecast (albeit a machine one) are such that a computer can, and should, be employed to take them all into account objectively, and present the results in an easily-usable form
- because the main task for many road maintenance teams is concentrated on certain highways, a detailed areal distribution of all the information is unnecessary. Also its distribution (and possibly display) takes up more time, and its possible storage locally on disk takes more space.
- decision-making at the local level involves many more factors than purely weather-related ones, and so there is an advantage in presenting the weather data in as completely-processed and ready form as possible.

Even so, there is a case to be made for a limited amount of data in relatively unprocessed form to allow for a quick check on the overall reasonableness of the machine forecast results. For radar data this could take the form of a relatively low-resolution image (of, say, videotex-quality) showing the latest radar-network composite product covering the maintenance area and its surroundings. Diagrams and tables should include such forecast information as: time of start of precipitation, duration of precipitation at various intensity levels, end of precipitation, total amount of precipitation, phase-change times. Additionally, error limits and the actual situation should be displayed to enable an assessment of the forecast's quality. Tables can display the information for fixed points of strategic interest, while graphs of time-series can show nicely the course of the precipitation at these points together with the forecasts. Results may also be cast into map form, showing e.g. isochrones of snow arrival and cessation on a stylized road network. Colour systems provide considerably more versatile displays.

Up to this point very little mention has been made of the use of weather radar data for other than winter road weather services. During the summer period the most common hazards to the road user stem from heavy showers or continuous moderate or heavy rain, giving rise to local accumulations of water which may either flood the road itself, or at least produce conditions conducive to aquaplaning, or may even lead to flash floods and destruction of structures. Such precipitation occurrences, directly affecting the safety of road users, are the result of cloud formations whose echoes can be detected to a very high degree of probability by weather radar. A warning scheme, to be successful, must cut the time between radar observation and reception of the warning by the road user to an absolute minimum. Forecasts of severe cumuliform events are much less successful than those due to layer clouds, so that the basis of warnings would seem to be actual weather, or very short period (15 - 30 min) forecasts. The road weather communication services which are being developed promise to be ideal for this type of warning, direct to the moving vehicle, and not relying on e.g. the driver having his radio on and tuned to the correct frequency. Roadside electronic speed placards need perhaps to be supplemented with textual possibilities to allow for a fuller announcement of the situation. The weather services would need a direct computer link into any road warning system to permit the maximum speed of response to radar-observed hazards. In order to minimize the false-alarm rate, it is probable that radar data would have to be used in conjunction with other meteorological data sources, such as satellites and automatic weather stations.

5 CONCLUSIONS

This paper has attempted to show that although there are numerous difficulties in the way of using weather radar data to support road weather services, both to road maintenance authorities and direct to road users, there is already sufficient meteorological as well as technical know-how available to permit the use of this data to supply or improve those services. One of the key factors in providing short-range precipitation forecasts is the use of radar data not from single radars, but from radars in an operational network. Experience in the use of such a network is already being gained within COST 73, which also intends in the near future to enlarge the area covered by the network, and also (of prime importance for road weather forecasts) to increase the updating frequency and shorten the production time of the composite network image. Additionally, proposals for new image products, including a forecast image, are under consideration. The provision of these new products would undoubtedly benefit road transport, especially in the supra-national era that is now opening in Europe.

REFERENCES

(1) FIORE, J.V. and R.K. FARNSWORTH, 1986. Quality Control of Radar-Rainfall Data with VISSR Satellite Data. 23rd. Conference on Radar Meteorology and Conference on Cloud Physics, Snowmass, Colorado. American Meteorological Society.

(2) WILSON, J.W., 1979. Radar Measurement of Rainfall - a Summary. Bull. Amer. Met. Soc. Vol.60, No.9, pp.1048 - 1058.

(3) JOSS, J. and A. WALDVOGEL, 1989. Precipitation Estimates and Vertical Reflectivity Profile Corrections. 24th. Conference on Radar Meteorology, Tallahassee, Fla. American Meteorological Society.

(4) BELLON, A. and G.L. AUSTIN, 1979. The Evaluation of Two Years of Real-Time Operation of a Short-term Precipitation Forecasting Procedure (SHARP). J. Appl. Meteor., Vol.17, pp. 1778 - 1787.

(5) COLLIER, C.G., 1978. Objective Forecasting using Radar Data: a Review. Research Report No.9. Meteorological Office Radar Research

Laboratory.

(6) BROWNING, K.A., C.G. COLLIER, P.R. LARKE, P. MENMUIR, G.A. MONK and R.G. OWENS, 1980. On the Forecasting of Frontal Rain using a Weather Radar Network. Research Report No.22. Meteorological Office Radar Research Laboratory.

SOME CONCLUDING REMARKS

C.G. COLLIER
Chairman COST 73
Meteorological Office, Bracknell, UK

The COST-73 International Seminar on Weather Radar Networking covered a wide range of topics. In the many papers presented the following points have been identified as representative of key questions to be answered or ways forward for research and development work. The list was presented by the Chairman of COST-73 C G Collier (UK) and represents a personal view of the material discussed.

(a) **Measurement of precipitation**

- Frequency and polarization matter
- The vertical profile of reflectivity **VERY** important: further work needed.
- Occultation/screening are important; siting is critical.
- There is considerable interest in dual frequency techniques.
- Anomalous propagation remains a problem. Several techniques offer partial solutions, and it is likely that they will all have to be used operationally.

(b) **Dual polarisation**

- ZDR at C-band does not give a measurement of the unambiguous drop size distribution, and therefore may not be of great value for measurement. However this is not the case for S-band, although at this frequency propagation effects may be difficult to cope with.
- Dual polarization techniques need further work to establish reliable operational algorithms.
- Limited range (-40 km?) for operational use.

(c) **Doppler radar**

- Doppler radar is ready for operational deployment.
- Doppler radar offers good clutter removal, but the technology remains immature for operational mesoscale wind analysis.
- Dual PRF unfolding may introduce errors into derived radial velocity fields.

(d) **Electronically scanning antennae**

- Does this represent a technology for a future generation of radar systems ?

(e) <u>Networks</u>

- Rapid developments will occur in next few years.
- Dense radar networks are needed.
- Care must be taken in blending data.
- A large variety of products will be available, <u>but</u> how do forecasters use them ?

(f) <u>Future of COST-73</u>

- Transfer to a fully operational international radar network involving WMO (GTS) and/or CEC.
- There is much interest in weather radar data exchanges between Western and Eastern Europe.
- International projects on (say) severe weather forecasting and international river management may utilize European-wide radar network data. Such projects will be linked to national projects.

(g) <u>Users</u>

- Users <u>**must**</u> be listened to !
- It is often difficult for radar people to understand what hydrologists want - there must be dialogue between the two communities.
- Sometimes, users fail to understand the problems of measurement.
- There is money to be made in exploiting radar data (e.g. Danish farmers).

M.E. ALMEIDA TEIXERA
Commission of the European
Communities - DGXII E-2
Rue de la loi 200
B-1049 Brussels

G. AURIAUX
Direction Eau & Assainissement
99, Avenue du Général de Gaulle
F-93110 Rosny-sous-Bois

P. BACON
Plessey Radar Limited
Oakcroft Road
UK-Chessington KT9 1QZ

S. BARBOSA
Instituto Nacional de Meteorologia
E Geofisica
Rue C do Aeroporto
P-1700 Lisboa

H.J. BATZER
Bundesantalt für Flugsicherung
Opernplatz 14-Postfach 10 04 44
D-6000 Frankfurt am Main

H. BAUER
Deutscher Wetterdienst
Frankfurterstrasse 135
D-6050 Offenbach/Main

R. BENZINGER
Unisys Corporation
Marcus Avenue
USA-11020 Great Neck-New York

L. BERGEAS
Air Force Staff
Ängsvägen 3
S-19630 Kungsängen

B. BERINGUER
SETIM Météorologie Nationale
7, rue Teisserenc-de-Bort
F-78195 Trappes Cedex

J. BIOUCAS DIAS
Complexo I do INIC
Instituto Superior Tecnico
Av. Rovisco Pais
P-1000 Lisboa

H.P. BIRON
Quebec Weather Centre
Environment Canada-St Laurent
Quebec
CDN-Canada H4M 2N8

J. BOOT
K.N.M.I.
Postbus 201
NL-3730 AE De Bilt

P.F. BORROWS
National Rivers Authority
Thames Region
Kings Meadow House
Kings Meadow Road
UK-Reading Berks

R. BROWN
Met 0 24 C
Meteorological Office
London Road, Bracknell
UK-Berkshire RG12 2SZ

J.L. BROWNSCOMBE
Meteorological Office
Met 0 24 D Room R 325
London Road, Bracknell
UK-Berkshire RG12 2SZ

J.O. BRUNSBERG
Meteorological & Hydrological
Inst.
Folkborgsvägen 1
S-60176 Norrköping

M. BÄCKSTRÖM
Ericsson Radar Systems AB
P.O.Box 1001
S-43184 Mölndal

L. CALCERON1
DATAMAT
Via Simone Martini 126
I-00143 Roma

R.V. CALHEIROS
Meteorological Research Inst.
UTE/UNESP & INPE/MD
CP 473
BR-17033 Bauru

G. CANNIZZARO
Telespazio
Via A. Bergamini
I-00159 Roma

L. CASARSA
I.T.A.V.
Piazza degli Archivi 37
I-00100 Roma - EUR

I.J. CAYLOR
U.M.I.S.T.
P.O.Box 88, Sackville Street
UK-Manchester M60 1QD

M. CHAPUIS
Commission of the European
Communities - DGXII G-1
Cooperation COST
Rue de la loi 200
B-1049 Brussels

J.C. CHEN
General Sciences Corporation
6100 Chevy Chase Drive
USA-Laurel 20707 MD

J.L. CHEZE
SETIM Météorologie Nationale
7, rue Teisserenc-de-Bort
B.P. 202
F-78195 Trappes Cedex

G. CLIFT
M.R. Technical Services
1, Villiers Mead - Wokingham
UK-Berkshire RG11 2UB

I.D. CLUCKIE
Department of Civil Engineering
University of Salford
UK-Salford M5 4WT

C.G. COLLIER
Chairman COST 73
AD Met O(NS), Met O 24
Meteorological Office
London Road, Bracknell
UK-Berkshire RG12 2SZ

B.J. CONWAY
Met O 24 D
Meteorological Office
London Road, Bracknell
UK-Berkshire RG12 2SZ

C. CORETTI
Telespazio
Via A. Bergamini
I-00159 Roma

E. CROSBY
Software Sciences Ltd.
Farnborough
UK-Hampshire GU14 7NB

J. CUNHA SANGUINO
Complexo I do INIC
Instituto Superior Tecnico
Av. Rovisco Pais
P-1000 Lisboa

L. DAHLBERG
Ericsson Radar Systems AB
P.O.Box 1001
S-43184 Mölndal

P. DE ANGELIS
DATAMAT Ingegneria dei Sistemi SPA
Via Simone Martini 126
I-00142 Roma

F.J. DE MIGUEL
ISEL s.a.
c/Orense 6-2°
E-28020 Madrid

F. DE MONTMOLLIN
Hydrologie & Géologie Nationale
Hallwylstrasse 4
CH-3003 Berne

R. DE RICHTER
KMI / IRM
Ringlaan 3
B-1180 Brussels

C. DE RIDDER
Ministère des Communications
RLW/RVA
Météo - 7th Floor
Brussels Airport
B-1930 Zaventem

G. DE SADELEER
IRM
Avenue Circulaire 3
B-1180 Brussels

F.P. DE TROCH
State University Ghent
Lab. of Hydrology
Coupure Links 653
B-9000 Ghent

D. DE VOS
Ministère Communications
RLW/RVA
Météo-7th Floor
Brussels Airport
B-1930 Zaventem

H.E. DEISENHOFER
Bayer.Landesamt
Wasserwirtschaft
Lazarettstrasse 67
D-8000 München

Y. DEWORM
Dienst Hydrologie
Oswald Ponettestraat 14A
B-9600 Ronse

M. DIVJAK
Hydrometeorological Institute SRS
Vojkova 1B
YU-Ljubljana

F. DOMBAI
HPR - Central Inst. Weather
Forecast. Budapest XVIII
Tatabanya ter 15-18,P.O.Box 32
H-1675 Budapest

K. DUVAL
Commission of the European
Communities
Secrétariat COST 73
DG XII G 1
200 Rue de la Loi
B-1049 Brussels

L. EWALD
AB TELEPLAN
P.O.Box 1310
S-17125 Solna

H. FAAS
Eumetsat
Am Elfengrund 45
D-6100 Darmstadt

C.A. FAIR
Meteorological Office
Beaufort Park - Easthampstead
UK-Wokingham RG11 3DN

P.M. FASELLA
Directeur Général DGXII-
Commission of the
European Communities
Rue de la Loi 200
B-1049 Brussels

A. FIUMARA
Telespazio
Via Bergamini 50
I-00159 Roma

K. GERDIN
S.M.H.I.
S-60176 Norrköping

D. GODDARD
Met O 24 D
Meteorological Office
London Road, Bracknell
UK-Berkshire RG12 2SZ

J.H. GOLDEN
National Weather Service NOAA
Dept. of Commerce
8060 13th Street, Rm 715
USA-Silver Spring, MD 20910

L. GUSTAVSSON
Ericsson Radar System AB
P.O.Box 1001
S-43184 Mölndal

M. HAGEN
D.L.R.
Institut für Physik der
Atmosphäre
D-8031 Oberpfaffenhofen

R. HARRIS
Software Science
Meudon Avenue
UK-Farnborough Hants GU14 7NB

J.J. HEEK
MULTIHOUSE TSI B.V.
Techn.Scientific and Ind.Systems
Schakelstraat 16
NL-1014 AW Amsterdam

R. HEYLEN
KMI/IRM
Ringlaan 3
B-1180 Brussels

J. HEYNDERICKX
Ministère des Travaux Publics
Service Etudes Hydrologiques
Deinsesteenweg 20
B-9810 Ghent

E. HOFSTEE
Netherlands Meteorological Inst.
Aeronautical Dept.
P.O.Box 7625
NL-1118 ZJ Schiphol

A.R. HOLT
University of Essex
Dept. Mathematics
UK-Colchester CO4 3SQ

R. HOOGEWIJS
Secr. D'Etat-Pol. Scient. Belg.
5, avenue Galilée
B-1030 Brussels

E. HUDSON
Unisys Corporation
Marcus Avenue
USA-11020 Great Neck-New York

D. HUSSON
G.N.E.F.A.
24, Avenue des Landais
F-63171 Aubieres

A.J. ILLINGWORTH
U.M.I.S.T.
P.O.Box 88, Sackville Street
UK-Manchester M60 1QD

G. JACQUET
RHEA
1, bld Albert Einstein
F-77420 Champs-sur-Marne

C.S. JENSEN
Civil Aviation Administration
Luftfartshutet - P.O.Box 744
DK-2450 Copenhagen SV

P. JOE
Atmosphere Environment Service
4905 Dufferin Street
CDN-M3H 5T4 Toronto

B. JOHANSSON
AB TELEPLAN
P.O.Box 1310
S-17125 Solna

J.E. JOHNSEN
Norske Meteorologiske Instituttet
Postboks 320-Blindern
Niels Henrik Abelsvei 40
N-0314 Oslo 3

L. JONES
Plessey Radar Limited
Newport Road
UK-Cowes PO31 8PF

J. JOSS
Swiss Meteorological Institute
Osservatorio Ticinese
Sezione Fisica delle nubi
CH-6605 Locarno-Monti

J. JUEGA
Inst. Nacional Meteorologia
Paseo de las Moreras S/N
E-28071 Madrid 3

KAVOURAS
Kavouras Inc.
Federal Aviation Building 6301
34th Ave 50
USA-Minneapolis, MN 55450

R.H. KING
Finnish Meteorological Institute
c/o FMI, PL 503
SF-00101 Helsinki

J. KIRBY
Meteorological Office
Beaufort Park
Easthampstead
UK-Wokingham RG11 3DN

A. KLOSE
Commission of the
European Communities
Coop. COST-EFTA (DGXII-G1)
Rue de la Loi 200
B-1049 Brussels

J.M. KNOWLES
National Rivers Authority
Northwest Region
Buttermarket Street,P.O.Box 12
UK-Warringtonn WA1 2QG

F.S. KOLOBE
Ministry of Lesotho
Highlands Water and Energy
Affairs
c/o United Nations
CH-1211 Geneva

G.A. MC DONALD
Meteorological Office
Casement Aerodrome-Baldonnel
IRL-CO Dublin

A.G. MAENHOUT
IRM/KIM
Avenue Circulaire 3
B-1180 Brussels

H. MALCORPS
Directeur de l'Institut Royal
Météorologique de Belgique
Avenue Circulaire 3
B-1180 Brussels

M. MALKOMES
Gematronik GmbH
Raiffeisenstrasse 10
D-4040 Neuss 21

J.L. MARIDET
SETIM Météorologie Nationale
7, rue Teisserenc-de-Bort
F-78195 Trappes cedex

G. MILILLO
Meteorological Service
Piazzale degli Archivi 34
I-00144 Roma

F. MARTIN
Inst. Nacional Meteorologia
Apartado 285
P de las Moreras S/N
E-28071 Madrid 3

M. MOLLER
Civil Aviation Administration
Ellebjergvej 50 - P.O.Box 744
DK-2450 Copenhagen

C. MARTINEZ
Inst. Nacional Meteorologia
Paseo de las Moreras S/N
E-28071 Madrid 3

M. MONAI
C.S.I.M. - Regione Veneto
Via Euganea 19
I-35037 Teolo (PD)

A. MASPRONE
President of the Economic
and Social Committee of the
European Communities
2 Rue Ravenstein
B-1000 Brussels

S. NANNI
ERSA-SMR
Servizio Meteorologico
V.S. Felice 25
I-40122 Bologna

T. NEVADO
ISEL S.A.
c/Orense 6-2°
E-28020 Madrid

S. MATTINGLY
Software Sciences
Meudon Avenue
UK-Farnborough,Hampshire GU14 7NB

D.H. NEWSOME
CNS Scientific & Engineering
Service, Tresillian House
20, Eldon Road, Reading
UK-Berkshire RG1 4DL

R. MC GUINNESS
University of Essex
Dept. Mathematics
UK-Colchester CO4 3SQ

C. NIEWOEHNER
GEMATRONIK GmbH
Raiffeisenstrasse 10
D-4040 Neus 21

J. MERCHAN
Inst. Nacional Meteorologia
Paseo de las Moreras S/N
E-28071 Madrid 3

S. NILSSON
Meteorological & Hydrological
Inst.
Folkborgsvägen 1
S-60176 Norrköping

MME F. MEULENBERGHS
IRM/KIM
Avenue Circulaire 3
B-1180 Brussels

J. NUNOS LEITAO
Complexo I do INIC
Instituto Superior Tecnico
Av. Rovisco Pais
P-1000 Lisboa

E. NYSTEN
Finnish Meteorological Institute
P.O.Box 503
SF-00101 Helsinki

G.O.P. OBASI
World Meteorological Org.
Postal Box 5
CH-1211 Geneva 20

H. OTTOY
KMI/IRM
Ringlaan 3
B-1180 Brussels

S. OVERGAARD
Danish Meteorological Institute
Lyngbyvej 100
DK-2100 Copenhagen

K. PANOSCH
B.A.Z.-Flugwetterdienst
Andr. Hoferstr. 39
1300 Flughafen Wien
A-3400 Klosterneuburg

N. PAPAMANOLIS
E.L.G.A.
Greek National Agricultural
Insurance Institute
GR-54635 Thessaloniki

A. PENEDA
L.C.T.A.R.
4, rue Nieuport
F-78140 Velizy

D. PODHORSKY
Slovensky Hydrometeorologicky
Ustav
Jenseniova 17
CS-83315 Bratislava

Y. POINTIN
LAMP/O.P.G.C.
12, Avenue des Landais
F-63000 Clermont-Ferrand

S. PROIETTI
Telespazio
Via Alberto Bergamini 50
I-00159 Roma

J. RAKOVEC
Iskra Delta Computers
Stegne 15
YU-31620 Netlj

W.L. RANDEU
Technische. Universität Graz
Inst. of Applied Systems Techn.
Inffeldgasse 12
A-8010 Graz

M. RESIBOIS
THOMSON - CSF
Route du Conquet
F-29200 Brest

J. RIEDL
Deutscher Wetterdienst
Meteorol. Observatorium
Albin Schwaiger-Weg 10
D-8126 Hohenpeissenberg

J.L. RIVERO ANTONIO
SATESA
Rosario Pino 14-16 2ePl
E-28020 Madrid

D. RÖSGEN
Commission of the
European Communities
- SCIC B - Centre
de Conférences A. Borschette
36, rue Froissart, Bur 6/24
B-1040 Brussels

M.P. ROSA DIAS
Instituto National de Meteorologia
E Geofisica
Rue C do Aeroporto
P-1700 Lisboa

A. SALSI
Servizio Meteorologico Regionale
Via.S. Felice 25
I-40122 Bologna

T.W. SCHLATTER
NOAA Forecast Systems Lab
NOAA/ERL/FSL R/E/FS2
325 Broadway
USA-Boulder Colorado 80303

U. SCHLUP
Bundesamt für Strassenbau
CH-3003 Bern

M. SCHÖNBÄCHLER
S.M.A.
Krähbülstrasse 58
CH-8044 Zürich

F.A. SENADA SILVA
Complexo I do INIC
Instituto Superior Tecnico
Av. Rovisco Pais
P-1000 Lisboa

N. SHIRAKAWA
FRICS
Kojimachi Hiraoka Bldg.
1-3 Kojimachi Chiyoda-KU
J-102 Tokyo

J. SHULSTAD
Mitre Corporation
Robert Stolzstrasse 20
D-6232 Bad Soden/TS

G. SLATER
Kavouras Inc.
FAA Building 6301
34th Avenue 50
USA-Minneapolis, MN 55450

R. SORANI
Vice-president COST 73
National Meteorological Service
Piazzale Degli Archivi 34
I-00144 Roma

G. SUTTON
Met O 24 D Room R 325
Meteorological Office
London Road, Bracknell
UK-Berkshire RG12 2SZ

T. STENSTRÖM
Defense Materiel Administration
Klubbacken 39
S-12656 Hägersten

J. SVENSSON
Meteorological & Hydrological
Inst.
S.M.H.I.
S-60176 Norrköping

H. TENT
Directeur Général Adjoint
Commission of the
European Communities - DG XII
rue de la Loi 200
B-1049 Brussels

P.A. TROCH
State University Ghent
Lab. of Hydrology
Coupure Links 653
B-9000 Ghent

L.R. TROVATI
U.N.E.S.P.
Faculdade Engenharia Ilha
Solteira
BR-15378 Sao Paulo

T.C. TSEHLO
Ministry of Lesotho
Highlands Water & Energy
Affairs c/o United Nations
CH-1211 Geneva

G. VALENTINI
Commission of the European
Communities - DGXII G
rue de la Loi 200
B-1049 Brussels

S. VAN DEN ASSEM
Agricultural University
Nieuwe Kanaal 11
NL-6709 PA Wageningen

D. VAN ERDEGHEM
State University Ghent
Lab. of Hydrology
Coupure Links 653
B-9000 Ghent

J.J. VAN GORP
K.N.M.I.
P.O.Box 201
NL-3730 AE De Bilt

A. VAN GYSEGEM
Koninklijk Meteorologisch
Instituut
Ringlaan 3
B-1180 Brussels

M. VAN LOEY
KMI/IRM
Ringlaan 3
B-1180 Brussels

M.E. VAN ZELLER MACEDO
Natural Resources
Av. Gago Coutinho 30
P-1000 Lisboa

G.F. VEZZANI
S.M.A.
Via Del Ferrone
Casella Postale 200
I-50100 Firenze

L. VICAK
Slovensky Hydrometeorologicky
Ustav
Jenseniova 17
CS-83315 Bratislava

H. WESSELS
K.N.M.I.
P.O.Box 201
NL-3730 AE De Bilt

H. WICKSELL
Defense Materiel Administration
Spelvägen 14
S-14200 Trängsund

E.S. WIENBERG
Danish Meteorological Institute
Hoejgaardstoften 396
DK-2630 Taastrup

L. WILLESTOFTE
Civil Aviation Administration
Ellebjergvej 50 - P.O.Box 744
DK-2450 Copenhagen

B.J. WRIGHT
Met O 11
Meteorological Office
London Road, Bracknell
UK-Berkshire RG12 2SZ

MR T. YOSHIDA
Japan Radio Co., Ltd.
Mitaka Plant
1-1 Simorenjaku 5 Chome
J-181 Mitaka-Shi, Tokyo

INDEX OF AUTHORS